Invasive Alien Plants in Urban Ecosystems

Rishikesh Singh · Rahul Bhadouria ·
Sachchidanand Tripathi · Ravinder Kumar Kohli ·
Hema Singh
Editors

Invasive Alien Plants in Urban Ecosystems

Emerging Concerns and Socio-Ecological Dimensions

Springer

Editors
Rishikesh Singh
Amity School of Earth and Environmental Sciences
Amity University Punjab
Mohali, Chandigarh, India

Sachchidanand Tripathi
Department of Botany
Deen Dayal Upadhyaya College
New Delhi, Delhi, India

Hema Singh
Department of Botany
Banaras Hindu University
Varanasi, Uttar Pradesh, India

Rahul Bhadouria
Department of Environmental Studies
Delhi College of Arts and Commerce
New Delhi, Delhi, India

Ravinder Kumar Kohli
Department of Botany
Panjab University
Chandigarh, India

ISBN 978-3-032-00996-8 ISBN 978-3-032-00997-5 (eBook)
https://doi.org/10.1007/978-3-032-00997-5

This Springer imprint is published by the registered company Springer Nature Switzerland AG
The registered company address is: Gewerbestrasse 11, 6330 Cham, Switzerland

Preface

Urbanization is one of the leading anthropogenic phenomena that the world is witnessing today. Substantial land use changes and massive movement of people from rural to urban areas are leading to further expansion of cities. These land use changes are continuously hampering the species composition and services of the urban-natural interfaces, and thus, interfering with the ecology of the natural ecosystems globally. Urban ecosystems are not only dominated by concrete buildings but also comprise a variety of land uses such as urban forests, managed and unmanaged grasslands, water bodies, agricultural lands, industrialized areas, wastelands, roads and railway tracks, etc. Even under the realm of climate change, urban ecosystems hold crucial significance. Slightly higher air temperature in the urban ecosystems due to urban heat island effects substantiates these ecosystems as a natural laboratory for studying the possible impacts of climate change on future ecosystems. Overall, urban ecosystems may play a vital role in understanding the key emerging issues related to local, regional, and global challenges. Therefore, urban ecology is gaining wider attention from the scientific community nowadays.

Plants are one of the vital components of an ecosystem and are required for the well-being of living organisms. Urban ecosystems are characterized by a variety of plant species comprising native as well as alien species. However, due to massive land use change, the original vegetation of a region may completely or substantially vanish during the process of urbanization. Moreover, urban residents and land planners generally prefer a variety of exotic (fancy and fast-growing alien species) avenue trees and shrubs for enriching their green spaces (e.g., water bodies, home gardens, parks, forests, and roadsides), compared to the native vegetation. The exotic alien plants present in the urban ecosystems provide several ecosystem services, such as aesthetic appeal, recreational activities, and ecological services and benefits to the residents. However, some of these alien species have the potential to become naturalized, escape from their original place, and spread to the open spaces and wastelands. Urban settlements are characterized by smaller fragments/patches of vegetation and provide empty niches to invasive species. These smaller fragments/patches and empty niches in urban areas harbor a higher density of alien plant species as compared to the larger fragments, and act as a hotspot for the dissemination of invasive propagules to the natural areas. In addition, massive urban mobility and migration of people further mediate the spread of these alien species in different urban settings. Once an alien plant species succeeds in establishing a reproductive population, it may spread to nearby areas, thereby becoming invasive. Therefore, urban ecosystems act as a breeding ground for the alien species to get established, naturalized and become invasive in the long-term. The spread of invasive alien species in urban areas, followed by their infestation and spread in the surrounding rural and natural vegetation, is, thus, a serious issue. Nowadays, researchers are working on the assessment and understanding of the role of invasive alien species in urban and peri-urban areas in terms of their impacts on the ecology, economy, and social well-being. Studies reported that urban ecosystems are facing severe direct and indirect challenges due to the spread of invasive species in terms of declining aesthetic values, chocking waterways, reduction in native biodiversity of birds, insects and animals, alternation in feeding and nesting habits of birds, arthropods and pollinators communities, which further affect the food chain of the urban systems, etc. The eradication and management of invasive alien species are further leading to the channelization of a substantial amount of funds and posing massive economic pressure on a global scale.

Overall, the current scenario of rapid urbanization and anthropogenic disturbances across the globe is contributing to the expansion of invasive alien plant species. However, urban ecosystems are comparatively under-explored in terms of a constantly developing understanding of the alien species invasions. This book provides a collection of contemporary information on research focused on different facets of challenges, impacts, and precautionary measures to be taken to address invasive alien plant species in urban systems. In this book, topics related to alien plant species occurrence in urban ecosystems, process of becoming alien to invasive, their ecosystem services and disservices, their comparative assessment to native species, challenges to the urban systems under changing climatic scenarios, the impact of alien plants on different components of urban areas, aesthetic values, livelihoods and health of urban dwellers, management approaches and policies related to alien plants in urban systems have been highlighted. The book will be helpful in understanding the invasion process in urban centers, as human-affected landscapes in urban areas are expected to support a larger chunk of alien species, and are instrumental in the dissemination of invasive propagules to nearby natural landscapes. It is predicted that climate change-associated disturbances or extreme events may play a supportive role in colonization and spread of invasive alien plants in urban ecosystems.

The book has a wider approach by targeting scientists, environmental researchers, and policy makers working and augmenting state-of-the-art knowledge in the areas of urban ecology and plant invasions. The book comprises 15 chapters contributed by 60 authors hailing from eight countries, viz., India, Pakistan, Türkiye, China, Kenya, Canada, the United States, and Algeria. Based on the content, the chapters are further distributed into four parts, viz., Part I: Invasive Alien Plants and Urban Ecosystem: Emerging Concerns (Chaps. 1–4); Part II: Major Serious Invasive Alien Plants in Urban Areas (Chaps. 5–8); Part III: Socio-ecological Impacts of Invasive Alien Plant Species in Urban Ecosystems (Chaps. 9–11); and Part IV: Threat Assessment and

Management of Invasive Alien Plants in Urban Ecosystems (Chaps. 12–15). An insight into the content of different chapters is given as follows:

The opening chapter (Chap. 1) of the book, entitled "Introduction of the Invasive Alien Plant Species in Urban Ecosystems: Comprehending the Process", contributed by Rahul Bhadouria et al., from India, highlights the processes that drive the introduction, establishment, and spread of invasive alien plant species within urban ecosystems. The chapter also provides insight into the characteristics of urban ecosystems, levels of disturbance, multiple vectors of species movement, pathways of introduction of invasive species in urban areas, challenges posed by these species in urban landscapes, management perspectives, and policy frameworks for urban planners. The second chapter (Chap. 2) entitled "From Alien to Invasive Alien Plant Species: An Array of Events" by Sidra Batool et al. from China and Pakistan explores the ecological theories that underpin the ecology of plant species, from being exotic/alien to becoming invasive, and analyzes the adaptive strategies of invasive alien plant species. The chapter comprehends the anthropogenic and environmental factors responsible for the success of these plants in establishing and dominating the novel ecosystems by taking examples of case studies. The chapter emphasizes the need for early detection, risk assessment, public education, adaptive management, collaborative efforts, and mitigation strategies for these species. Chapter 3, entitled "Invasive Alien Plant Species in Urban Ecosystems" by Manasvini Singh Sambyal et al. from India, focuses on how urbanization has contributed to the spread of alien invasive species. Authors further discussed the benefits and impacts of invasive species on the ecosystem, biodiversity, economy, and human health. Chapter 4, entitled "Emerging Invasive Alien Plants in Urban Settlements: Establishment and Impacts" by D. Inci et al. from the United States and Türkiye, highlights the concerns related to invasive alien species in urban settlements, their adaptation strategies, and adverse impacts on human beings under the changing climate scenarios. The chapter also emphasized the early detection, rigorous control efforts, public awareness, and unique and case-specific management approaches for mitigating the impacts of invasive alien species.

Chapter 5 in Part II of the book, entitled "Plant Invasion in the Urban Landscape: An Ecological Synthesis" by Manoj Kumar Jhariya et al., from India, describes how changes in land use patterns are mediating ecological invasion in the urban landscape. The authors identified peri-urban areas as highly susceptible to ecological invasion and emphasized giving priority to these areas while formulating policies to combat ecological invasion. In Chap. 6, entitled "Functional Traits Reveal the Success of Alien Plant Species Along the Invasion Continuum", Priya Kaushik et al., from India, explore different socio-economic and ecological dimensions with respect to the dynamics of biological invasions within urban ecosystems. The chapter also examines the ways in which humans modify environments and facilitate plant invasion in light of climate change scenarios. The chapter has provided a model for an integrated understanding of invasive species in forest ecosystems for developing appropriate efforts for their conservation and management. In Chap. 7, entitled "Ecosystem Services and Disservices Linked to Invasive Alien Plants", Raees Khan et al. from Pakistan, Algeria, and China highlight the dual roles, viz., positive (i.e., ecosystem services, such as habitat protection, soil stabilization and erosion control, nutrient cycling, pollination, etc.) and negative (i.e., ecosystem disservices, such as invasive effect on native vegetation, biodiversity loss, resource competition, economic loss, etc.) played by invasive alien plant species in urban ecosystems, with the help of case studies. In Chap. 8, entitled "Impacts of Invasive Plant Species on Fishes, Birds, and Animals", Shalini Ojha et al., from India, emphasized the impacts of invasive alien plants on fish, birds, and animal species present in urban ecosystems. Appropriate management approaches and scenarios for invasive alien species that combine mechanical, biological, and chemical control measures with restoration, good governance, and community engagement have also been highlighted.

Chapter 9 in Part III, entitled "Economic Impacts of Invasive Alien Plant Species on Urban Ecosystems" by Prakash Rajak et al. from India, presents an in-depth exploration of the economic implications, such as city budgets, recreational opportunities, urban aesthetics, and public services affected by invasive alien plant species in urban ecosystems. The chapter also offers insights into efficient, long-term control solutions using real-world examples and management techniques. The outcomes of the chapter may assist environmental managers, city planners, and legislators in providing workable solutions for maintaining the resilience and health of urban ecosystems. Chapter 10, entitled "Invasive Alien Weeds in the Urban Environment: A Case Study of Chandigarh" by Shalini Ojha et al. from India, highlights the complex ecological, socio-economic, and health-related impacts posed by invasive alien species in the city of Chandigarh, India. The chapter documents the patterns of invasion, ecological traits, and dispersal mechanisms of selected invasive alien plant species, and further assesses the profound implications for urban biodiversity, public health, agriculture, and infrastructure. The key outcomes of the chapter emphasized the necessity of integrated weed management, restoration ecology, interdisciplinary research, and policy implementation to mitigate threats of invasive plant species and preserve the ecological integrity of urban landscapes. In Chap. 11, entitled "Some Approaches of Indigenous Peoples and Local Communities to Invasive Alien Species with Remarks on Urban Communities", Wycliffe Wanzala and Sheila I. Minyoso from Kenya have outlined the role of Indigenous Peoples and Local Communities (IPLC) imparting ethnoknowledge about the environment and ecological systems, including local biodiversity, for their future generations. These IPLCs can easily detect the introduction of alien species via the trade and transport system, and can help in the mitigation and/or management of these species at an early stage in the region.

The Chap. 12 in Part IV, entitled "Mapping and threat assessment of invasive alien plants in urban ecosystem" by Pragya Samaridhi et al. from India, outlined the role of geospatial techniques such as Geographic Information Systems (GIS) and Remote Sensing, and modern computational tools such as Species Distribution Modeling (SDM) for assessment of the distribution and threat analysis of invasive alien plants in different novel ecosystems. The chapter also emphasized the System for Conservation Planning (SCP) software tools, such as *Marxan* and *Prioritizr* (the R packages) prioritizing the conservation activities and control of invasive species at spatial scales.

Inclusion of these tools and techniques will help in identifying the existing dispersion and projecting the future expansion, which are the critical components for the successful management and mitigation of invasive alien plant species. Chapter 13, entitled "Incorporating Phenological Traits for Mapping of Invasive Plant Species in Urban Ecosystems" by Kariya Ishita Bhaveshkumar and Laxmi Kant Sharma from India, has highlighted how the incorporation of phenological traits of plants in Remote Sensing helps in precise mapping, developing indicators and characterization of ecosystem functions of the invasive species in urban areas. The penultimate chapter (Chap. 14) of the book, entitled "Management Approaches and Policy Regulation for Invasive Alien Plants in Urban Ecosystems" by Omar Aziz et al. from Pakistan and China, discusses about the developing technology, management strategies, and regulatory issues to control invasive alien plant species in urban ecosystems. The authors emphasized proactive steps, early identification, integrated approaches such as biological, chemical, mechanical, and physical control adapted to species and habitat for the effective management of invasive alien species in urban ecosystems. In addition, public education initiatives, inclusive decision-making procedures, and context-specific tactics are necessary for the successful management of these species. The ultimate chapter (Chap. 15) of the book, entitled "Management Approaches for the Restoration of Invaded Urban Plant Communities" by Leeladarshini Sujeeun from Canada, provides an outlook on the active management strategies for the restoration of the invaded urban ecosystems. The modern ecological principles, such as planting strong native competitors, increasing functional diversity of native plants, suppressing the invasive seed bank, and adding soil amendments such as biochar, can be used to enhance the recovery of native plant communities in invaded urban ecosystems and increase their resilience against re-invasion by the exotic species.

Overall, the deliberations of the book will be suitable for students (undergraduate and postgraduates) of urban planning and architecture, urban agriculture, urban forestry, urban ecology, and environmental sciences; teachers, researchers, climate change scientists working in different academic and research institutions for enriching their knowledge and understanding on the interaction between urbanization and spread of invasive alien plant species. It will also be applicable to town planning and environmental management agencies, government agencies and policy makers for sustainable urban planning, green space development, as well as mitigation and restoration of the invaded urban ecosystems.

Strength of the Book

- The book provides an excellent introduction and basic knowledge of the principles and the science of plant invasions in urban ecosystems. The content of the book offers a mechanistic understanding of the invasion process of the alien plant species in urban ecosystems.
- The book describes the environmental sustainability and future impacts of invasive alien plants in urban setups, and highlights the challenges and opportunities related

to the socio-economic feasibility of management and the ecological threats of invasive alien plants in the urban environment.

- The chapters in the book have been contributed by authors from various parts of the world, and will be helpful in policymaking and for research related to environmental science and urban environment management.
- Undergraduate and postgraduate students, university research scholars, teachers, policy makers, ecologists and early career researchers, especially working in urban ecology and urban environmental management, will be benefitted by going through the enriched content of the book.

Mohali, India — Rishikesh Singh
Delhi, India — Rahul Bhadouria
Delhi, India — Sachchidanand Tripathi
Chandigarh, India — Ravinder Kumar Kohli
Varanasi, India — Hema Singh

Contents

Editors and Contributors

About the Editors

Dr. Rishikesh Singh is an assistant professor at Amity School of Earth and Environmental Sciences, Amity University Punjab (AUP), Mohali, India. Dr. Singh completed his Ph.D. from the Institute of Environment & Sustainable Development (IESD), Banaras Hindu University (BHU), Varanasi, and a Post-doctoral fellowship (NPDF) from the Department of Botany, Panjab University, Chandigarh. His research interests include various dimensions of ecology and environmental science, viz., vegetation analysis, tree seedling ecology, invasion ecology, urban ecology, soil carbon dynamics, land use change and management, biochar technology, etc. He has worked on different ecosystems, including dry deciduous forests and savannas, riparian landscapes, and urban and agroecosystems of the tropical region. His major contribution to science is in "Biochar and Agro-ecology". Dr. Singh received the Young Scientist Award from the National Environmental Science Academy (NESA), New Delhi, in 2022. He is credited with publishing over 90 Scopus-indexed publications in various international journals and books published by Elsevier, Springer Nature, Wiley and CRC Press, etc. He is also serving as an editorial board member of several reputed international journals.

Dr. Rahul Bhadouria is an assistant professor at the Department of Environmental Studies, Delhi College of Arts and Commerce (University of Delhi), India. He obtained his doctorate in Environmental Science from the Department of Botany, Banaras Hindu University, India. His areas of research are plant ecology, soil ecology, ecophysiology, and urban ecology. He has published more than 25 papers, 30 book chapters, and 15 edited books in internationally reputed journals/publishers. His current research activities are related to management of soil C dynamics to mitigate climate change, a perspective on tree seedling survival and growth attributes in tropical dry forests under the realm of climate change, plant community assembly, functional diversity, and soil attributes along the forest-savanna-grassland continuum in India, recovery of degraded mountains in the central Himalayas and urban ecology. He is serving as an editorial board member of several reputed international journals.

Dr. Sachchidanand Tripathi is an associate professor at the Department of Botany, Deen Dayal Upadhyaya College, University of Delhi, Delhi, India. He obtained a doctoral degree from the Department of Botany, Banaras Hindu University, India. His research interests include plant ecology, soil ecology, ecophysiology, and urban ecology. He has published over 50 publications (including research publications, books and book chapters, conference proceedings) in reputed international journals and publishers. He is serving as an editorial board member of several reputed international journals.

Prof. Ravinder Kumar Kohli serves as Senior Emeritus Ecologist of ESA, USA, at the Department of Botany, Punjab University, Chandigarh, India. He has been a JC Bose National Fellow of the Government of India (for two terms of 5 years each)—completed on Dec 2, 2021. He is ranked among the top 2% of scientists worldwide as per Stanford University, USA, for the 6th consecutive year. He is a member of the Executive Council of the International Union of Biological Sciences (France) (2023–26). He had served as Vice-Chancellors of (a) Amity University Punjab, Mohali, India (Private University) from August

22, 2020 till May 2, 2025, (b) Central University of Punjab, Bathinda, India from Sept 5, 2014 till Aug 22, 2020 (Govt. of India), and (c) DAV University, Jalandhar (Private University) from April 13, 2013 till Sept 4, 2014. Prof. Kohli is a fellow of all the Major Science Academies (FRS-C, FRS-B, FNA, FASc, FNASc, FNAAS). He had been a chair of the IUFRO's Working Party 8.02.02 (Ecology of Alien Invasives for 10 years and a member IUFRO International Task Force on Forests and Biological Invasions (2015–2020). His current research interests include: managing noxious weeds using environment-friendly techniques, impact of invasive alien plants on ecology, how plants interact in the environment, urban forestry and weed ecology, and how mobile phones and their towers affect ecology and environment. He has guided 54 Ph.D. and 29 M.Phil. theses in Ecology, Environment & Forestry and published more than 365 research publications in journals of international repute. He has published more than 14 edited books with various national and international publishers.

Prof. Hema Singh currently serving as a professor in the Department of Botany at Banaras Hindu University in Varanasi, India, has made significant contributions to the field of agroecosystems ecology. Her research primarily focuses on microbial management and soil fertility restoration through various tillage practices and combinations of fertilizers and plant residues. Additionally, she has conducted studies on the ecophysiology of crops and their dominant weeds. Her current research interest also includes the ecology of invasive plants under different climatic scenarios. Professor Singh's research findings have been published in esteemed journals such as the Journal of Applied Ecology, Soil Biology & Biochemistry, Applied Soil Ecology, Biology & Fertility of Soils, Soil & Tillage Research, Journal of Sustainable Agriculture, Current Science, Environmental Monitoring & Assessment, Plant Biosystems, International Journal of Pest Management, Rice Science, and others.

Contributors

Sheikh Zain Ul Abidin Institute of Biological Sciences, Gomal University D I Khan, Dera Ismail Khan, Pakistan

Preeti Agarwal Department of Botany, Gargi College, University of Delhi, New Delhi, India

Necmi Aksoy Department of Forest Engineering, Faculty of Forestry, Düzce University, Düzce, Türkiye

Mehmet Arslan Department of Biosystem Engineering, Faculty of Agriculture, Erciyes University, Kayseri, Türkiye

Omar Aziz Soil & Water Testing Laboratory, Lahore, Pakistan

Arnab Banerjee Department of Environmental Science, Sant Gahira Guru Vishwavidyalaya, Ambikapur, Chhattisgarh, India

Daizy R. Batish Department of Botany, Panjab University, Chandigarh, India

Sidra Batool China University of Geosciences, Wuhan, Hubei, China

Safia Belhadj Faculty of Nature and Life Sciences, University of Djelfa, Djelfa, Algeria

Rahul Bhadouria Delhi College of Arts and Commerce, University of Delhi, Delhi, India

Kariya Ishita Bhaveshkumar Remote Sensing and GIS Lab, Department of Environmental Sciences, Central University of Rajasthan, Kishangarh, Rajasthan, India

Shazia Erum National Herbarium of Pakistan (Stewart Collection), Plant Genetic Resources Institute, PARC-National Agricultural Research Centre, Islamabad, Pakistan

Sara Fatma Ecosystem Analysis Laboratory, Department of Botany, Banaras Hindu University, Varanasi, India

Yuvna Puri Goswami Department of Botany, Gargi College, University of Delhi, New Delhi, India

Deniz Inci Department of Plant Sciences, University of California, Davis, CA, USA

Manoj Kumar Jhariya Department of Farm Forestry, Sant Gahira Guru Vishwavidyalaya, Ambikapur, Chhattisgarh, India

Shalinder Kaur Department of Botany, Panjab University, Chandigarh, India

Simarpreet Kaur Department of Botany, Panjab University, Chandigarh, India

Priya Kaushik Ecology Laboratory, Department of Botany, Dr. Harisingh Gour Vishwavidyalaya, Sagar, Madhya Pradesh, India

Sushma Kerketta Department of Farm Forestry, Sant Gahira Guru Vishwavidyalaya, Ambikapur, Chhattisgarh, India

Amjad Khan National Herbarium of Pakistan (Stewart Collection), Plant Genetic Resources Institute, PARC-National Agricultural Research Centre, Islamabad, Pakistan;
Department of Plant Sciences, Quaid-I-Azam University—Islamabad, Islamabad, Pakistan

Irum Khan Department of Biosciences, COMSATS University Islamabad, Islamabad, Pakistan

Mohammed Latif Khan Ecology Laboratory, Department of Botany, Dr. Harisingh Gour Vishwavidyalaya, Sagar, Madhya Pradesh, India

Raees Khan CAS Key Laboratory for Plant Diversity and Biogeography of East Asia, Kunming Institute of Botany, Chinese Academy of Sciences, Kunming, China;
Germplasm Bank of Wild Species & Yunnan Key Laboratory of Crop Wild Relatives Omics, Kunming Institute of Botany, Chinese Academy of Sciences, Kunming, China;
National Herbarium of Pakistan (Stewart Collection), Plant Genetic Resources Institute, PARC-National Agricultural Research Centre, Islamabad, Pakistan

Pramod Kumar Khare Ecology Laboratory, Department of Botany, Dr. Harisingh Gour Vishwavidyalaya, Sagar, Madhya Pradesh, India

Bhimappa Honnappa Kittur Faculty of Technical Forestry, Indian Institute of Forest Management, Bhopal, Madhya Pradesh, India

Ravinder Kumar Kohli Department of Botany, Panjab University, Chandigarh, India

Rahul Kumar Rajdhani College, University of Delhi, New Delhi, India

Rajat Kumar Amity School of Earth and Environmental Sciences, Amity University Punjab, Mohali, Punjab, India

Jie Liu CAS Key Laboratory for Plant Diversity and Biogeography of East Asia, Kunming Institute of Botany, Chinese Academy of Sciences, Kunming, China;
Germplasm Bank of Wild Species & Yunnan Key Laboratory of Crop Wild Relatives Omics, Kunming Institute of Botany, Chinese Academy of Sciences, Kunming, China

Sheila I. Minyoso Matungu Rural Herbal and Research Centre, Mumias, Kenya;
The Scholarship Network Centre, Education and Research Division, Nairobi, Kenya

Reema Mishra Department of Botany, Gargi College, University of Delhi, New Delhi, India

Shambhu Nath Mishra ICFRE—Institute of Forest Productivity, Ranchi, Indian Council of Forestry Research and Education—Lalgutwa, Ranchi, Jharkhand, India

Anindya Shankar Nandi Ecosystem Analysis Laboratory, Department of Botany, Banaras Hindu University, Varanasi, India

Shalini Ojha Department of Botany, Panjab University, Chandigarh, India

Amrit Pal Department of Botany, Panjab University, Chandigarh, India

Pranab Kumar Pati Ecology Laboratory, Department of Botany, Dr. Harisingh Gour Vishwavidyalaya, Sagar, Madhya Pradesh, India

Abhishek Raj Pt. Deendayal Upadhyay College of Horticulture and Forestry Dr. Rajendra Prasad Central Agriculture University, Samastipur, Bihar, India

Prakash Rajak Ecosystem Analysis Laboratory, Department of Botany, Banaras Hindu University, Varanasi, India

Pooja Rani Department of Botany, Panjab University, Chandigarh, India

Mahrine Rashid National Herbarium of Pakistan (Stewart Collection), Plant Genetic Resources Institute, PARC-National Agricultural Research Centre, Islamabad, Pakistan

Muhammad Riaz Guangdong Engineering and Technology Center for Environmental Pollution Prevention and Control in Agricultural Producing Areas, College of Resources and Environment, Zhongkai University of Agriculture and Engineering, Guangzhou, China

Pragya Samaridhi ICFRE—Institute of Forest Productivity, Ranchi, Indian Council of Forestry Research and Education—Lalgutwa, Ranchi, Jharkhand, India

Manasvini Singh Sambyal Department of Botany, Gargi College, University of Delhi, New Delhi, India

Syed Mukaram Shah Centre of Plant Biodiversity, University of Peshawar, Peshawar, Pakistan

Laxmi Kant Sharma Remote Sensing and GIS Lab, Department of Environmental Sciences, Central University of Rajasthan, Kishangarh, Rajasthan, India

Bhupender Singh Department of Botany, Panjab University, Chandigarh, India

Harminder Pal Singh Department of Environment Studies, Panjab University, Chandigarh, India

Hema Singh Ecosystem Analysis Laboratory, Department of Botany, Banaras Hindu University, Varanasi, India

Khushwinder Singh Department of Botany, Panjab University, Chandigarh, India

Rishikesh Singh Amity School of Earth & Environmental Sciences (ASEES), Amity University Punjab, Mohali, India

Ritturaj Singh Ecosystem Analysis Laboratory, Department of Botany, Banaras Hindu University, Varanasi, India

Soner Soylu Department of Plant Protection, Faculty of Agriculture, Hatay Mustafa Kemal University, Hatay, Türkiye

Leeladarshini Sujeeun University of Toronto, Institute of Forestry and Conservation, Toronto, ON, Canada

Sachchidanand Tripathi Deen Dayal Upadhyay College, University of Delhi, Delhi, India

Asad Ullah Centre of Plant Biodiversity, University of Peshawar, Peshawar, Pakistan

Ahmet Uludağ Department of Plant Protection, Faculty of Agriculture, Çanakkale Onsekiz Mart University, Çanakkale, Türkiye

İlhan Üremiş Department of Plant Protection, Faculty of Agriculture, Hatay Mustafa Kemal University, Hatay, Türkiye

Wycliffe Wanzala Department of Biological Sciences, School of Pure and Applied Sciences, Maasai Mara University, Narok, Kenya;
Matungu Rural Herbal and Research Centre, Mumias, Kenya

Shaopeng Zhao Guangdong Engineering and Technology Center for Environmental Pollution Prevention and Control in Agricultural Producing Areas, College of Resources and Environment, Zhongkai University of Agriculture and Engineering, Guangzhou, China

Part I
Invasive Alien Plants and Urban Ecosystem: Emerging Concerns

Chapter 1
Introduction of the Invasive Alien Plant Species in Urban Ecosystems: Comprehending the Process

Rahul Bhadouria, Sachchidanand Tripathi, Ravinder Kumar Kohli, Hema Singh, and Rishikesh Singh

Abstract Urban ecosystems are increasingly becoming hotspots for the introduction and proliferation of invasive alien plant species (IAPS), posing significant threats to the biodiversity, ecosystem services, and human well-being. This chapter delves into the multifaceted processes that drive the introduction, establishment, and spread of IAPS within urban ecosystems. By exploring the unique characteristics of urban environments—such as high levels of disturbance, diverse land uses, and the presence of multiple vectors for species movement—the chapter offers a comprehensive understanding of how these factors contribute to the vulnerability of cities to biological invasions. We examine key pathways of introduction, including horticultural trade, landscaping practices, and accidental transport, and analyze the ecological, economic, and social impacts of IAPS in urban areas. Furthermore, the chapter discusses management strategies and policy frameworks aimed at mitigating the risks associated with IAPS in cities. This chapter provides critical insights for urban

R. Bhadouria (✉)
Delhi College of Arts and Commerce, University of Delhi, Delhi, India
e-mail: rahulbhadouriya2@gmail.com

S. Tripathi
Deen Dayal Upadhyay College, University of Delhi, Delhi, India
e-mail: snt191@gmail.com

R. K. Kohli
Department of Botany, Panjab University, Chandigarh, India
e-mail: rkkohli45@yahoo.com

H. Singh
Ecosystems Analysis Laboratory, Department of Botany, Banaras Hindu University, Varanasi, India
e-mail: hema.bhu@gmail.com

R. Singh
Amity School of Earth & Environmental Sciences (ASEES), Amity University Punjab, Mohali, India
e-mail: rishikesh.iesd@gmail.com

R. Singh et al. (eds.), *Invasive Alien Plants in Urban Ecosystems*,
https://doi.org/10.1007/978-3-032-00997-5_1

planners, ecologists, and policymakers to effectively address the challenges posed by invasive species in the urban landscapes.

Keywords Biological invasions · Ecological resilience · IAPS · Urban biodiversity · Urban ecosystems · Urban planning

1 Introduction

Non-native plants that are brought into a new habitat with the potential to expand quickly and have a substantial negative influence on the environment, the economy, and society are known as invasive alien plant species (IAPS). IAPS are typically defined by their ability to outcompete native species, thrive in a wide range of environmental conditions, and reproduce rapidly (Cristina-Maria et al., 2023; O'Briain et al., 2023). IAPS are able to take over and change the local ecosystems because, in contrast to many native species, they frequently lack natural predators or diseases in their new settings. Urban systems, characterized by high human activity, frequent disturbances and diverse land uses, provide ideal conditions for the introduction and establishment of IAPS (Afonso et al., 2021; Potgieter et al., 2017). The dynamics of urban environments—such as altered microclimates, fragmented habitats, and constant movement of people and goods—facilitate the spread of these species, often at the expense of native biodiversity.

The presence of IAPS in urban areas can have profound consequences on different ecological and socio-economic processes (Branco et al., 2023; Potgieter et al., 2019). Ecologically, these species can lead to reduced biodiversity, altered ecosystem functions, and the homogenization of urban flora (Potgieter et al., 2017; Rai & Singh, 2020). For example, when IAPS are introduced, native flora may be displaced, soil composition may change, water cycles may be altered, and ecosystem services that are essential to urban life—like air purification, climate regulation, and recreation—may be disrupted (Langemeyer & Gómez-Baggethun, 2018). Socio-economically, the impact of IAPS is felt in increased maintenance costs for green spaces, damage to infrastructure, reduced property values, and potential health risks to humans (Potgieter et al., 2017; Seboko et al., 2024). Given the complex interactions between IAPS and urban ecosystems, it is crucial to develop a comprehensive understanding of the processes that drive their introduction, spread, and impact (Potgieter et al., 2018, 2019; Rai & Singh, 2020). Effective management strategies must be informed by a deep knowledge of urban ecology, invasion biology, and the socio-economic context of cities. With this background, and raising a need for the exploration of IAPS in urban ecosystems, this chapter provides a detailed insight into the characteristics of urban ecosystems, which act as a productive ground for the biological invasion, followed by a brief account on different disturbance events which facilitate the biological invasion in urban settings, as well as processes and pathways of introduction of invasive species in urban environments. In addition, the last few sections of the Chapter emphasizes on the presence and impacts of native and non-native species in urban ecosystems, followed by management and control strategies as well as policy and governance-related aspects on the judicious utilization and management of IAPS in the urban ecosystems. By going through in details and addressing the challenges posed by IAPS, urban planners, ecologists, and policymakers can work together to protect and enhance the resilience of urban environments.

2 Characteristics of Urban Ecosystems with Respect to Plant Invasion

Urbanization has an immense effect on natural landscapes and ecosystems, making it one of the major drivers of environmental change in the contemporary world. As cities expand to accommodate growing human populations, the transformation of habitats are inevitable (Grimm & Schindler, 2018). This process of urbanization brings about a series of changes that create both opportunities and challenges for the species that inhabit these areas, particularly in the context of IAPS (Early et al., 2016; McKinney, 2002, 2006). Some of the key processes (see: Fig. 1) comprehending the spread of invasive species in urban ecosystems are outlined as following:

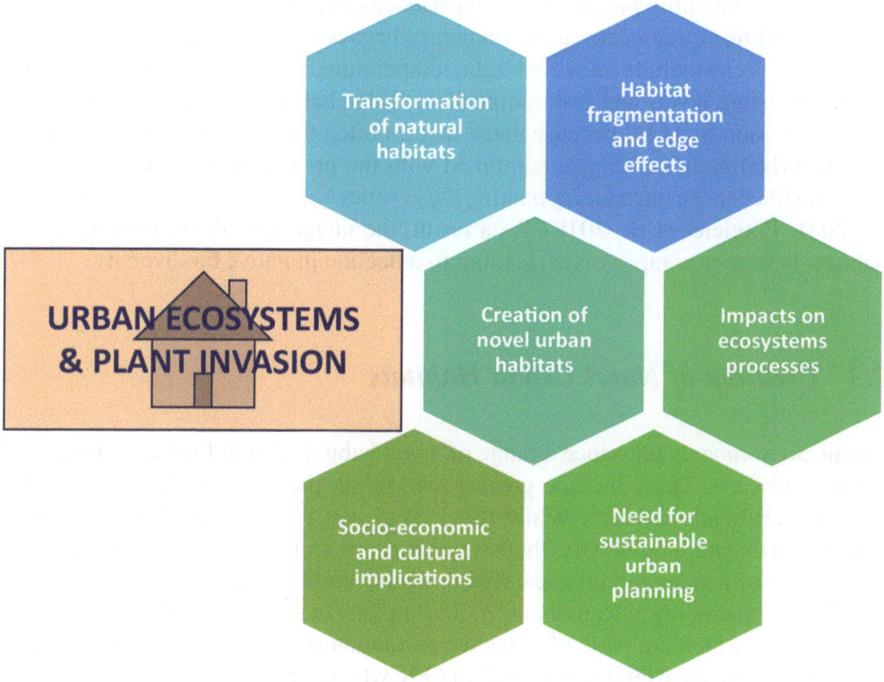

Fig. 1 Some of the key characteristics of urban ecosystems that mediate the introduction and spread of invasive alien plants

2.1 Transformation of Natural Habitats

Natural habitats including forests, waterbodies, and grasslands, are often converted into constructed settings that are dominated by residential neighbourhoods, industrial zones, and infrastructure as a result of urbanization (Bai et al., 2019; Wang et al., 2022; Zipperer et al., 2020). This transformation drastically reduces the extent and quality of natural habitats, leading to habitat fragmentation and loss. The resulting urban landscape is often a mosaic of remnant natural patches, managed green spaces (like parks and gardens), and highly modified areas. The reduction of natural habitats not only displaces native species but also alters the ecological conditions that define these environments. Changes in soil composition, hydrology, and microclimate are common, creating new niches that are often more suitable for non-native species, including IAPS, than for the original native flora (Kashe et al., 2024; Okoye, 2016; Urgenson, 2011).

2.2 Habitat Fragmentation and Edge Effects

As natural habitats are broken up into smaller, isolated fragments due to urban development, the remaining patches often experience edge effects—changes in environmental conditions that occur at the boundary between natural and developed areas. These effects can include increased light, temperature, and wind exposure, as well as altered moisture levels and soil composition. Edge habitats are particularly vulnerable to invasion by IAPS because these species often thrive in disturbed conditions. The disturbed nature of edges, combined with the proximity to urban areas where IAPS are likely to be introduced, making these zones hotspots for invasion (Afonso et. al., 2021; Potgieter et al., 2018). As a result, the integrity of the remaining natural habitats is further compromised, leading to a decline in native biodiversity.

2.3 Creation of Novel Urban Habitats

Urban environments introduce a range of novel habitats that did not exist in the pre-urban landscape. These include green roofs, urban parks, roadside verges, vacant lots, and artificial wetlands. While these areas can provide valuable green space and support some biodiversity, they are also prone to invasion by IAPS due to the frequent disturbances and unique environmental conditions they present (Afonso et. al., 2021; Appalasamy & Ramdhani, 2020; Potgieter et al., 2018). Urban heat islands, for example, which are parts of cities where human activity causes temperatures to rise, can provide ideal circumstances for IAPS, which are adapted to warmer climates (Chen et al., 2023; Mtengwana et al., 2021). Similarly, the altered water regimes in cities, characterized by rapid runoff and pollution, can enable the spread of IAPS adapted to disturbed or degraded aquatic habitats (Appalasamy & Ramdhani, 2020; Potgieter et al., 2018).

2.4 Impact on Ecosystem Processes

The transformation of habitats through urbanization not only changes the physical landscape but also disrupts ecosystem processes such as nutrient cycling, water filtration, and carbon sequestration (Svirejeva-Hopkins et al., 2004; van Delden et al., 2016; Wang et al., 2017). The development and growth of IAPS is often helped by changes made to these processes. For instance, urban soils are frequently compacted, contaminated, and subject to increased nutrient inputs from pollution, creating conditions that can be exploited by invasive species (Afreen et al., 2023; Ladd, 2016; Lone et al., 2019). Moreover, the reduction in native plant diversity and the alteration of species interactions in urban areas can weaken ecosystem resilience (Fineschi & Loreto, 2020; Miles et al., 2019), making it more difficult for these systems to resist invasion and recover from disturbances. This can lead to a feedback loop where the presence of IAPS further degrades the ecosystem, making it even more susceptible to future invasions.

2.5 Socio-Economic and Cultural Implications

The transformation of habitats through urbanization also has socio-economic and cultural implications (Arnaiz-Schmitz et al., 2018; Long et al., 2014; Tu et al., 2018). Urban green spaces are vital for human well-being, providing recreational opportunities, aesthetic value, and ecosystem services that enhance the quality of life (Davids et al., 2021; Jabbar et al., 2022; Kothencz et al., 2017; Pinto et al., 2022). However, the invasion of these spaces by IAPS can diminish their ecological and social value, leading to increased management costs and reduced public enjoyment (Potgieter et al., 2019; Rai & Singh, 2020). Further, the cultural preferences for certain non-native ornamental plants in urban landscaping can unintentionally contribute to the spread of IAPS (Potgieter et al., 2017, 2019, 2022). As these plants escape cultivation and establish in the wild, they can become invasive, further transforming urban habitats and posing challenges for biodiversity conservation.

2.6 The Need for Sustainable Urban Planning

Addressing the challenges posed by habitat transformation in the urban areas requires a holistic approach to urban planning and management (Agudelo-Vera et al., 2011; Mersal, 2016; Naess, 2001; Yigitcanlar & Teriman, 2015). Integrating biodiversity conservation into urban development can help mitigate the impacts of habitat loss and fragmentation, while promoting the creation of resilient green spaces that support both native species and human communities (Balfors et al., 2016; Di Giulio et al., 2009; Gordon et al., 2009). Strategies such as the use of native plants in landscaping, the restoration of degraded habitats, and the design of green infrastructure that mimics natural processes can all contribute to reducing the vulnerability of urban systems to invasion by IAPS (Mostert, 2016; Rai, 2022; Rai & Singh, 2020).

Overall, urbanization and the resulting transformation of habitats have profound implications for the introduction and spread of IAPS. In order to effectively manage IAPS and safeguard urban biodiversity and, ultimately, make sure that cities continue to be lively, sustainable, and resilient in the face of continuous environmental change, it is imperative that we have a thorough understanding of these changes. The next section will provide a brief understanding on different disturbance regimes mediated by these characteristics of urban ecosystems which facilitate the IAPS invasion in these areas.

3 Disturbance Regimes in Urban Areas

Disturbance regimes—defined as the patterns and frequencies of events that disrupt ecosystems—play a crucial role in shaping urban environments. In urban areas, disturbances are often more frequent and varied than in natural landscapes, significantly influencing ecological dynamics, particularly the introduction and spread of IAPS. Understanding these disturbance regimes (see: Fig. 2) is essential for managing urban ecosystems and mitigating the impacts of IAPS.

3.1 Types of Disturbances in Urban Areas

Urban areas experience a wide range of disturbances, both natural and anthropogenic, that can profoundly alter local ecosystems. Some of the most common disturbances include:

3.1.1 Construction and Development

The creation and expansion of infrastructure such as buildings, roads, and utilities, are among the most significant disturbances in the urban areas. Disturbed environments are more vulnerable to invasion by IAPS because of these activities, which frequently entail the loss of vegetation, compaction of the soil, and changes in land use (Dhakal et al., 2024; Foxcroft et al., 2017).

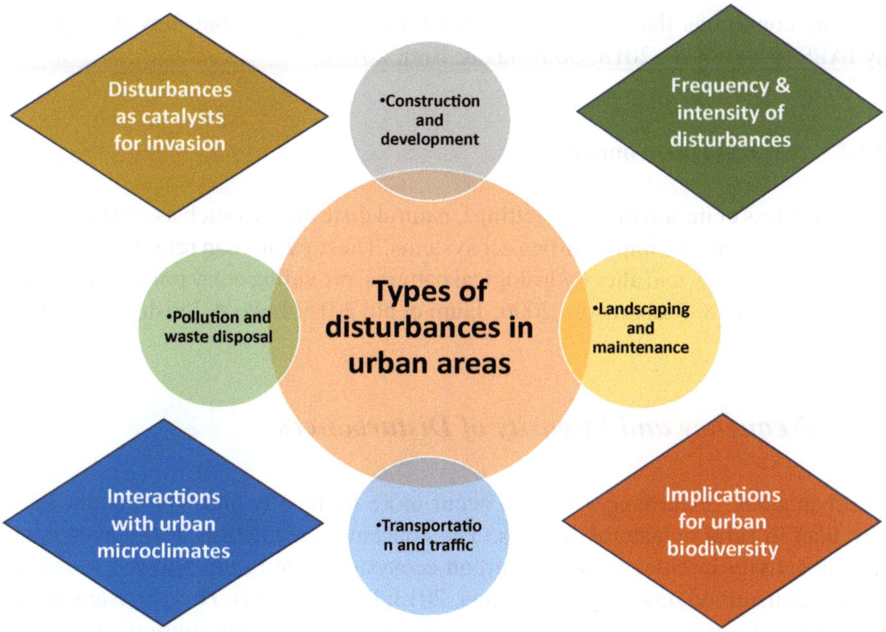

Fig. 2 An overview on different types of disturbance regimes in urban areas which facilitate the spread of invasive alien species

3.1.2 Landscaping and Maintenance

Routine urban management activities, such as mowing, pruning, and the application of herbicides and fertilizers, can create temporary disturbances that favour invasive species. These activities often disrupt native plant communities and provide opportunities for IAPS to establish and spread (Mayfield et al., 2021).

3.1.3 Transportation and Traffic

The movement of vehicles, goods, and people within urban areas can introduce disturbances through pollution, road construction, and the accidental transport of plant materials (Kowarik & Lippe, 2008; Lemke et al., 2021). Roadsides and railways often serve as corridors for the spread of IAPS due to frequent disturbance and the movement of propagules.

3.1.4 Pollution and Waste Disposal

Urban areas are hotspots for various types of pollution from industrial activities, traffic, and waste disposal. These pollutants can alter soil and water chemistry, creating conditions that are unfavourable for native species but may be exploited by IAPS (Plaza et al., 2018; Sołtysiak & Brej, 2014).

3.1.5 Natural Disturbances

Although less common in urban settings, natural disturbances such as storms, floods, and droughts can also impact urban ecosystems. These events can remove vegetation, create bare ground, and alter hydrological patterns, providing entry points for invasive species (Alston & Richardson, 2006; Jauni et al., 2015; Lake & Leishman, 2004).

3.2 Frequency and Intensity of Disturbances

In urban areas, disturbances tend to occur more frequently and with greater intensity than in rural or natural settings. The constant construction, maintenance, and human activity in cities mean that urban ecosystems are rarely allowed to reach a stable, undisturbed state (Adler & Tanner, 2013; Rebele, 1994). This persistent disturbance creates a dynamic environment where invasive species can continuously exploit newly available niches. The high frequency and intensity of urban disturbances often exceed the adaptive capacities of native species, which are typically evolved to withstand natural, periodic disturbances (Gaertner et al., 2017). In contrast, many IAPS are well-suited to thrive in these environments due to their rapid growth rates, high reproductive output, and the ability to colonize disturbed areas quickly.

3.3 Disturbances as Catalysts for Invasion

Disturbance regimes in urban areas are key catalysts for the introduction and spread of IAPS (Larios & Suding, 2013). Disturbances create conditions that reduce competition from native species, such as bare soil, increased light availability, and altered nutrient cycles (Sukopp & Starfinger, 1999; Ward et al., 2020). These conditions are ideal for invasive species, which can rapidly colonize and establish themselves in the absence of strong competition. Moreover, urban disturbances often lead to fragmented habitats, which can isolate native species populations and reduce their resilience to invasion (de Barros Ruas et al., 2022). Fragmentation creates more edges and interfaces between developed and natural areas, where invasive species are more likely to gain a foothold.

3.4 Interaction with Urban Microclimates

Urban disturbances, such as the urban heat island effect, which increases temperatures in cities relative to surrounding rural regions, frequently combine with microclimatic changes brought on by urbanization. These increased temperatures can extend the growing season and create conditions more favourable to IAPS, particularly those adapted to warmer climates (Mavimbela et al., 2018; Shrestha & Shrestha, 2019). Additionally, changes in hydrology due to urban development, such as altered water flow patterns and increased runoff, can create new moisture regimes that benefit invasive species over native ones.

3.5 Implications for Urban Biodiversity

The frequent and varied disturbance regimes in urban areas pose significant challenges for maintaining native biodiversity. Native species, which often rely on stable conditions or specific disturbance regimes, may be outcompeted by IAPS that are better adapted to the altered conditions (Potgieter et al., 2018, 2019). The resulting shift in species composition can lead to reduced biodiversity, the loss of ecosystem services, and the homogenization of urban flora.

3.6 Managing Disturbance Regimes to Control IAPS

Effective management of disturbance regimes is crucial for controlling IAPS in urban areas. Strategies for managing disturbance regimes to control IAPS include (see Fig. 3):

- **Minimizing Unnecessary Disturbances**: Reducing the frequency and intensity of disturbances where possible, such as through careful planning of construction activities and the use of sustainable landscaping practices, can limit opportunities for IAPS to establish (Afonso et. al., 2021; Potgieter, 2019).
- **Restoration and Rehabilitation**: Following disturbances, active restoration efforts that prioritize the re-establishment of native species can help resist invasion (Gaertner et al., 2012). This may involve replanting native vegetation, improving soil conditions, and controlling invasive species during the recovery phase.
- **Creating Buffer Zones**: Designing urban green spaces with buffer zones that reduce edge effects and provide more stable conditions can help protect native species from the impacts of adjacent disturbances (Bretzel et al., 2016).
- **Monitoring and Rapid Response**: Establishing monitoring programs to detect early signs of invasion in disturbed areas allows for rapid response and management before IAPS gets established and spreads (Potgieter et al., 2018).

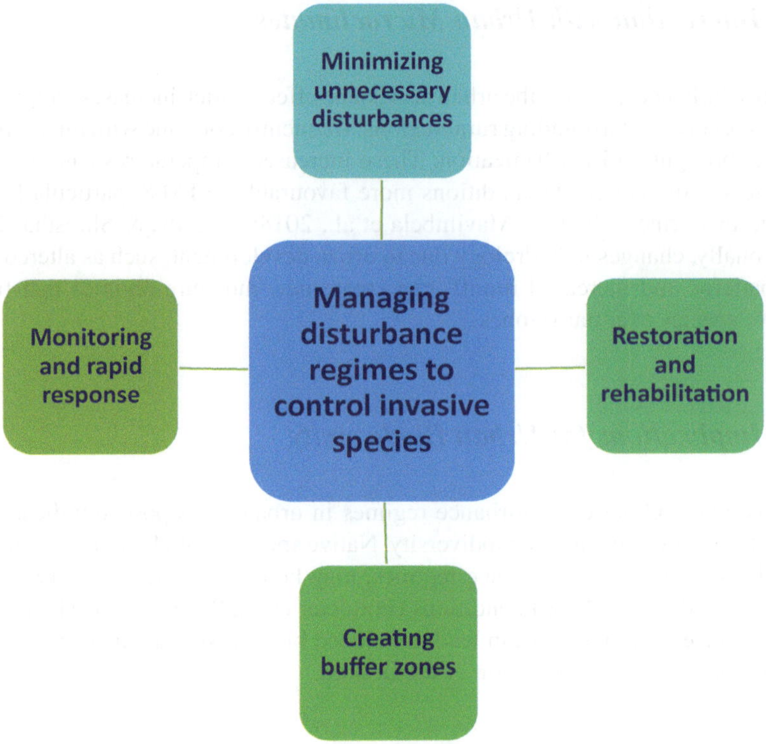

Fig. 3 Some of the key approaches for managing disturbance regimes to control the spread of invasive alien species in urban ecosystems

Overall, disturbance regimes are a defining characteristic of urban ecosystems, significantly influencing the dynamics of IAPS. By understanding and managing these disturbances, urban planners and ecologists can help mitigate the risks posed by IAPS and promote healthier, more resilient urban environments. After discussing the characteristics and disturbance regimes of the urban ecosystems, understanding the pathways of introduction of IAPS in these ecosystems are also a crucial step for their judicious management which has been elaborated in the coming section.

4 Pathways of Introduction of the IAPS in Urban Ecosystems

The introduction of IAPS into urban ecosystems is facilitated by a variety of pathways, both intentional and unintentional. Understanding these pathways is crucial for developing effective prevention and management strategies to mitigate the risks associated with IAPS. The main pathways of introduction include horticultural trade,

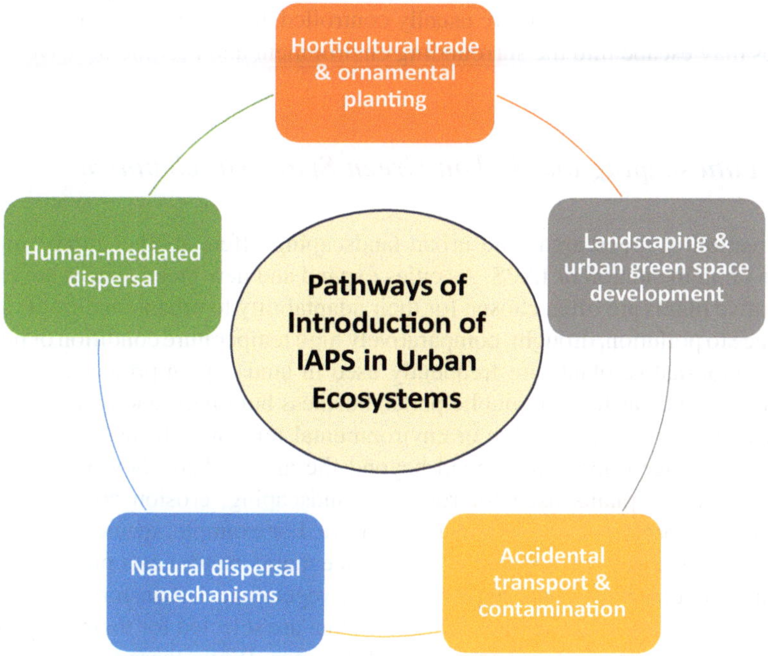

Fig. 4 An overview on the important pathways by which invasive alien species are spreading in the urban ecosystems

landscaping practices, accidental transport, and natural dispersal mechanisms (see: Fig. 4) which are described as:

4.1 Horticultural Trade and Ornamental Planting

The horticultural trade is one of the main ways that brings IAPS into the cities (Afonso et al., 2021; Potgieter et al., 2019; Seboko et al., 2024). The global demand for ornamental plants, exotic flowers, and unique garden species has led to the widespread movement of plants across borders and regions. Many non-native plants are introduced intentionally for their aesthetic value, ease of cultivation, or unique characteristics that make them popular in urban landscaping (Wichlan, 2022; Yang et al., 2024). However, some of these species can escape cultivation and establish themselves in surrounding natural or semi-natural habitats, where they may become invasive. Further, commercial nurseries and garden centers often stock a wide variety of non-native plants, some of which have the potential to become invasive if they escape into the wild. The lack of stringent regulations and awareness about the invasiveness of certain species can contribute to the unintentional spread of IAPS.

Although these introductions are usually controlled, there is always a risk that some species may escape into the surrounding environment and become invasive.

4.2 Landscaping and Urban Green Space Development

The creation of green areas and urban landscaping offers another important route for the implementation of IAPS. As cities expand and new green spaces are created, non-native plants are often chosen for their adaptability to urban conditions, such as tolerance to pollution, drought, comparatively high temperature condition or poor soil quality. Non-native plants are frequently used in landscaping projects, both public and private, such as those in public parks, business buildings, and residential areas, because of their aesthetic value or environmental tolerance. If these plants are not carefully managed, they can spread beyond the areas where they were originally planted. Further, plants used for roadside landscaping, erosion control, or green infrastructure projects can also become invasive. For example, species that are introduced for their ability to stabilize soil or reduce maintenance costs might proliferate in nearby natural areas, outcompeting native vegetation. Urban forestry initiatives sometimes introduce non-native tree species that are selected for their rapid growth, shade provision, or resistance to pests and diseases. While these trees can provide immediate benefits, they may also outcompete native trees and contribute to the spread of invasive species.

4.3 Accidental Transport and Contamination

Accidental transport is another major pathway for the introduction of IAPS into urban environments. This often occurs when plant propagules (seeds, roots, or fragments) are unintentionally transported via vehicles, goods, or human activity. Roads, railways, and waterways act as corridors for the movement of people and goods, inadvertently facilitating the spread of IAPS. Vehicles, machinery, and cargo can carry seeds or plant fragments over long distances, depositing them in new locations where they can establish and spread.

4.4 Natural Dispersal Mechanisms

Once introduced into an urban area, IAPS can spread further through natural dispersal mechanisms. These mechanisms are often enhanced by the unique conditions found in urban environments, such as altered wind patterns, water flows, and the presence of wildlife.

4.5 Human-Mediated Dispersal

The origin and dissemination of IAPS are mostly caused by human activities. Apart from the direct pathways mentioned above, there are additional human-mediated factors that contribute to the dispersal of invasive species in urban areas.

Overall, understanding the pathways of IAPS introduction is crucial for identifying and controlling these species. Urban planners, ecologists, and policymakers can develop more effective strategies to prevent the introduction and spread of IAPS by clearly identifying the pathways of introduction. This includes implementing stricter regulations on the import and sale of non-native species, promoting the use of native plants in landscaping, and raising public awareness about the risks of invasive species. By addressing these pathways, cities can better protect their biodiversity, ecosystem services, and the well-being of their residents.

5 Native Versus Non-Native Species in Urban Areas

Urban biodiversity encompassing native species that evolved in the local environment and non-native species that have been introduced, either intentionally or accidentally, from other regions. The interaction between native and non-native species in urban areas is complex and has significant implications for ecosystem health, resilience, and the quality of life for urban residents. Therefore, understanding the role and the interactive interplays of native and non-native species in the urban ecosystems is of particular importance for their control and management, as well as deriving utmost ecosystems services from these species in the urban settings. A brief insight has been given on the role and impacts of different native and non-native species and their interactions in the urban ecosystems mediating different ecological processes.

5.1 Native Species in Urban Areas

Native species are those that are found natively in a particular place and have developed there throughout time in response to the unique environmental circumstances and biological relationships that exist there (Almas & Conway, 2016). In urban environments, native species contribute to the local identity and biodiversity, playing critical roles in ecosystem functioning and providing essential services, such as pollination, water filtration, and climate regulation (Elmqvist et al., 2012; Langemeyer & Gómez-Baggethun, 2018). However, the urban landscape presents numerous challenges for native species. Habitat loss, fragmentation, pollution, and competition from non-native species can all threaten their survival (Alizadeh & Hitchmough, 2019; Chalker-Scott, 2015; Zeunert, 2014). Native species are often adapted to specific habitats and conditions that may no longer be available in urban settings, making them more vulnerable to extinction or displacement. Despite these challenges, native species can still thrive in cities, particularly in well-managed green spaces, such as parks, gardens, and nature reserves, where efforts are made to preserve or restore natural habitats. The presence of native species in urban areas is essential for maintaining ecological balance and ensuring the resilience of urban ecosystems.

5.2 Non-Native Species in Urban Areas

Species that have been brought to an area outside of their normal range are referred to as non-native, alien, or exotic species. In urban environments, non-native species can be introduced through various pathways as described earlier, including horticulture, landscaping, agriculture, and accidental transport via vehicles, goods, or people (Gaertner et al., 2017; Padayachee et al., 2017). Non-native species can play diverse roles in urban ecosystems. Some are harmless or even beneficial, providing aesthetic value, food resources, or contributing to ecosystem services. However, when non-native species become invasive—meaning they spread rapidly and outcompete native species—they can disrupt local ecosystems and lead to significant ecological and socio-economic problems.

5.3 Interactions Between Native and Non-Native Species

The interaction between native and non-native species in urban areas is a dynamic and often competitive process (Fig. 5). Key factors influencing these interactions include (see: Fig. 5a):

5.3.1 Competition for Resources

Native species tend to be in competition with non-native species for resources including light, water, nutrients, and space, particularly when the latter become invasive (Knauf et al., 2021; Minden & Gorschlüter, 2016; Rodriguez, 2006). In many cases, non-native species have competitive advantages, such as faster growth rates, broader ecological tolerances, or a lack of natural predators, allowing them to dominate and displace native species.

5.3.2 Alteration of Ecosystem Processes

Key ecological processes can be altered by non-native species in a way that disadvantages native species (de Carvalho et al., 2022; Pilliod et al., 2012). For example, invasive plants may alter soil chemistry, water availability, or fire regimes, creating

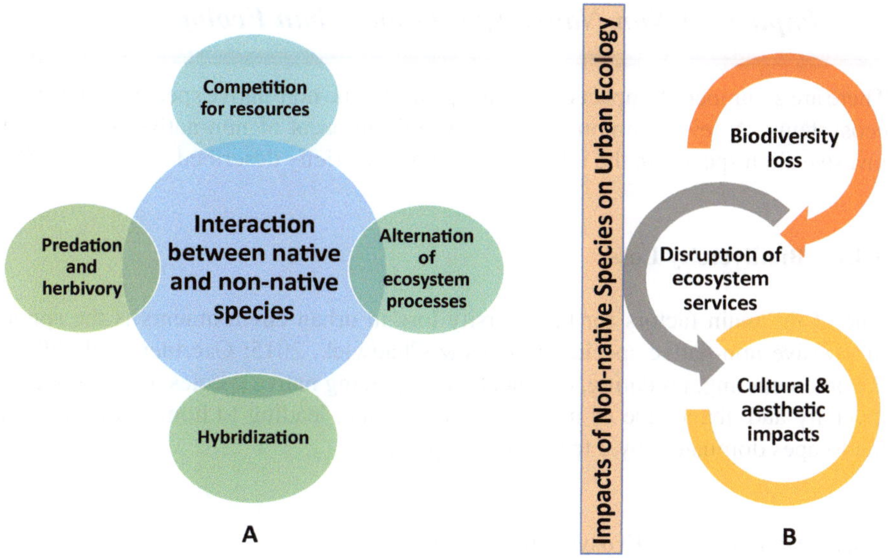

Fig. 5 Native versus non-native species in urban ecosystems. **a** Interaction between native and non-native species, and **b** major impacts of non-native species on urban ecology

conditions that are less favourable to the native species and more conducive to their further invasion.

5.3.3 Hybridization

Hybridization can occur sometimes when closely related native species breed with non-native species (DeVos et al., 2023; Ruhsam et al., 2023; Schierenbeck, 2021). This can result in the loss of genetic diversity and the unique characteristics of native species, potentially leading to their decline.

5.3.4 Predation and Herbivory

Through herbivory or predation, non-native species can also have an effect on native species (Engelkes et al., 2012; Theodorou, 2022). For example, non-native herbivores might preferentially feed on native plants, reducing their populations and allowing non-native plants to proliferate.

5.4 *Impacts of Non-Native Species on Urban Ecology*

There are a number of impacts which are posed by the non-native species in the urban ecosystems. A few common and most visible impacts of non-native, particularly invasive alien species in the urban ecosystems are briefly described as (see: Fig. 5b):

5.4.1 Biodiversity Loss

One of the main factors for biodiversity loss in urban environments is the spread of invasive non-native species (Francis & Chadwick, 2015; Gaertner et al., 2017). By outcompeting, predating, or otherwise displacing native species, they reduce the diversity and abundance of native flora and fauna, leading to homogenized urban landscapes dominated by a few resilient species.

5.4.2 Disruption of Ecosystem Services

The availability of ecosystem services that are essential to human well-being in cities can be disrupted by the loss of native species and the domination of non-native species (Castro-Díez et al., 2019; Davids et al., 2021; Kueffer & Kull, 2017). For example, the loss of native pollinators due to competition from non-native species can reduce pollination services, affecting food production and the health of urban green spaces.

5.4.3 Cultural and Aesthetic Impacts

Native species are often integral to the cultural and natural heritage of a region. Their decline can lead to a loss of local identity and aesthetic value in urban landscapes (Bryant, 2006; Müller et al., 2013; Southon et al., 2017). Conversely, the proliferation of non-native species can alter the character of urban green spaces, sometimes in ways that are less desirable or less compatible with local traditions.

5.5 *Balancing Native and Non-Native Species in Urban Planning*

Urban biodiversity conservation critically depends on maintaining the proper ratio of native to non-native species in metropolitan environments. Key strategies for balancing the native and non-native species while urban planning include (see: Fig. 6):

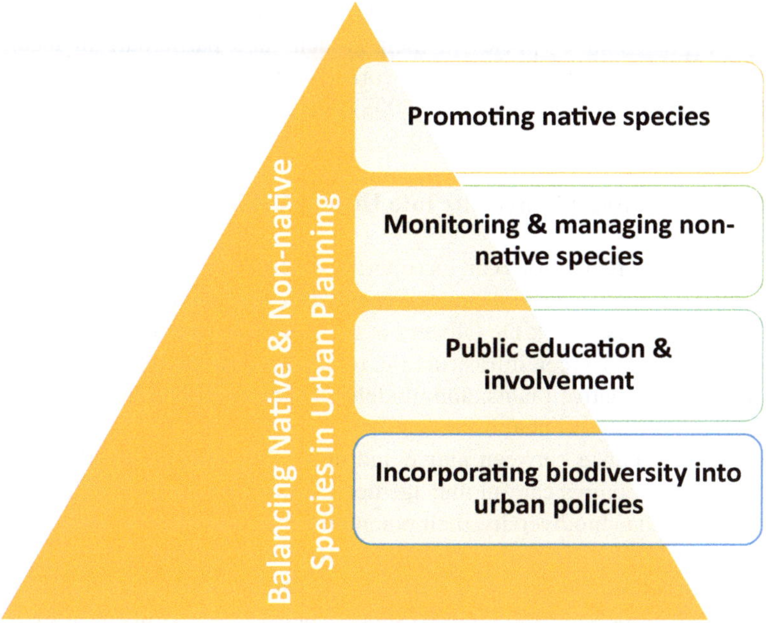

Fig. 6 Strategies which help in balancing the native and non-native species ratio in the urban planning for preventing and regulating the spread of invasive alien species in urban areas

5.5.1 Promoting Native Species

Urban planners and landscape designers can prioritize the use of native plants in public and private green spaces, which supports local biodiversity and helps maintain ecological functions (Aronson et al., 2017a, 2017b). Initiatives to restore native habitats and connect fragmented green spaces can enhance the survival and reproduction of native species.

5.5.2 Monitoring and Managing Non-Native Species

Regular monitoring of urban ecosystems is essential to detect and manage non-native species before they become invasive (Gaertner et al., 2017; Kowarik, 2011). This involves early detection, rapid response strategies, and ongoing management to control the spread of invasive species.

5.5.3 Public Education and Involvement

Educating the public about the importance of native species and the risks associated with non-native invasions can foster community support for conservation efforts

(Bremner & Park, 2007; Gozlan et al., 2013; Verbrugge et al., 2013). Encouraging residents to plant native species in their gardens and participate in local biodiversity initiatives can help create a network of urban habitats that support native wildlife (Markovchick-Nicholls et al., 2008).

5.5.4 Incorporating Biodiversity into Urban Policies

Urban policies and regulations can incorporate biodiversity conservation goals, such as protecting native species and controlling invasive species, in their core functional components (Beatley, 2000; De Oliveira et al., 2011; Gurran et al., 2015). This can include incentives for green infrastructure, regulations on the sale and use of potentially invasive non-native plants, and guidelines for sustainable urban development that prioritizes ecological health.

Overall, the balance between native and non-native species in urban areas is a delicate one that requires careful management. While non-native species can sometimes enhance urban biodiversity, their potential to become invasive and disrupt local ecosystems makes it essential to prioritize native species and manage non-native species carefully. By understanding the interactions between native and non-native species and implementing strategies to support native biodiversity, cities can create healthier, more resilient urban ecosystems that benefit both nature and people. Based on these discussions, it is imperative to understand the introduction process of IAPS in urban ecosystems to develop appropriate control and management approaches for the sustainable urban planning and development. A brief insight on this aspect has been given in the next section.

6 Understanding the Introduction Process of IAPS for Their Prevention, Early Detection and Management in Urban Ecosystems

The introduction of IAPS into urban ecosystems is a critical issue that demands careful consideration and action. It is crucial to comprehend the mechanisms behind the introduction of these species for a number of reasons, all of which have significant effects on public health, urban ecology, and economic stability. A few key measures related to mitigation of IAPS introduction in urban ecosystems are as following (see: Fig. 7):

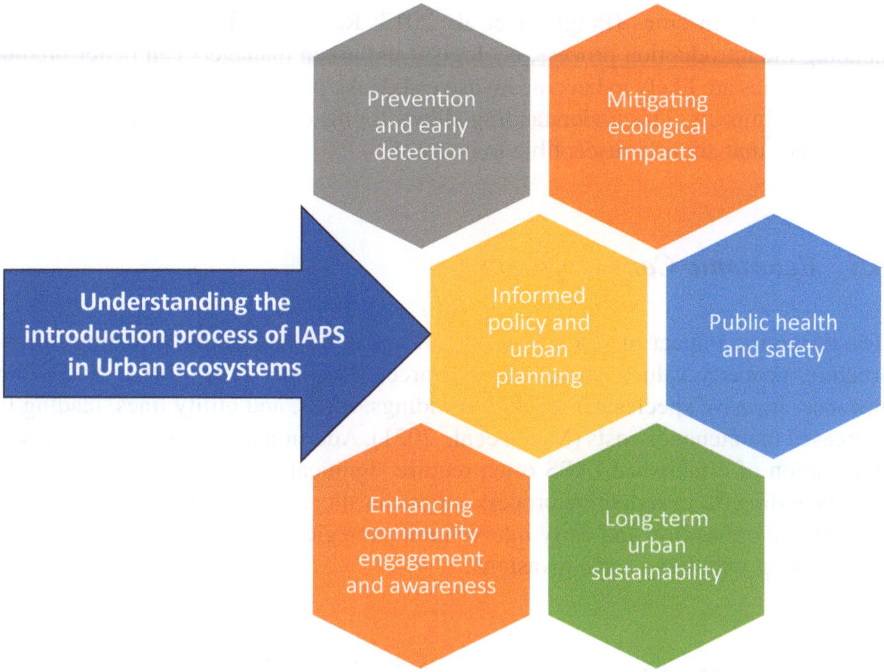

Fig. 7 Understanding the introduction processes of invasive alien plant species in urban ecosystems so that their prevention, detection and management can be done at an early stage

6.1 Prevention and Early Detection

Understanding how IAPS are introduced into urban environments is the first step in developing effective prevention strategies. By identifying the primary pathways through which these species enter cities—such as horticultural trade, landscaping practices, and accidental transport—urban planners and policymakers can implement targeted measures to minimize the risk of new invasions (Jelaska, 2021; Mudavanhu, 2018). Early detection is key to preventing the establishment of IAPS, and a thorough understanding of introduction pathways enables the development of monitoring systems that can identify and respond to new incursions quickly.

6.2 Mitigating Ecological Impacts

IAPS can drastically alter urban ecosystems by outcompeting native species, disrupting food webs, and altering ecosystem processes such as nutrient cycling and water flow (Potgieter, 2019; Rai & Singh, 2020). These alterations may result in a decline in ecosystem services and a loss of biodiversity, both of which are critical to those living in cities (Potgieter et al., 2017; Rai & Singh, 2020). By comprehending the introduction process, ecologists and urban managers can better predict which species are likely to become invasive and take proactive steps to mitigate their ecological impacts. This understanding allows for the design of more resilient urban landscapes that are less susceptible to invasion.

6.3 Economic Considerations

The economic impact of IAPS in urban areas can be substantial, affecting infrastructure, property values, and public resources (Potgieter et al., 2018, 2022). For instance, invasive species can damage buildings, roads, and utility lines, leading to increased maintenance costs (Marsh et al., 2021). Additionally, the management and eradication of established IAPS often require significant financial investments. By understanding the introduction process, cities can allocate resources more efficiently, prioritizing prevention and early intervention strategies that are more cost-effective than dealing with large-scale invasions.

6.4 Public Health and Safety

Some IAPS pose direct threats to public health, either through the production of allergens, toxins, or by creating habitats conducive to pests that spread diseases (Bengyella, 2022; Rai & Singh, 2020). For example, certain invasive plants can exacerbate respiratory issues or increase fire hazards. Understanding the introduction process allows for better risk assessments and the implementation of public health measures to protect urban populations from these threats.

6.5 Informed Policy and Urban Planning

Policymakers and urban planners play a crucial role in managing the risks associated with IAPS. A deep understanding of the introduction process informs the creation of policies and regulations that control the movement of potentially invasive species, promote the use of native plants in urban landscaping, and enforce biosecurity measures (Banerjee & Sankaran, 2023; Banerjee et al., 2021). Additionally, urban planning that incorporates green infrastructure designed to resist invasion can create more sustainable and resilient cities.

6.6 *Enhancing Community Engagement and Awareness*

Effective IAPS management requires both community engagement and public awareness. By understanding the introduction process, educational programs can be tailored to inform citizens about the risks associated with IAPS and how they can contribute to prevention efforts (Cordeiro et al., 2020; Seboko et al., 2024). Engaged communities are more likely to support and participate in initiatives aimed at controlling invasive species, such as reporting sightings, participating in removal efforts, and choosing native plants for their gardens.

6.7 *Long-Term Urban Sustainability*

Finally, the introduction of IAPS poses a significant challenge to the long-term sustainability of urban environments. Cities that fail to address the introduction and spread of invasive species may face ongoing ecological degradation, reduced quality of life for residents, and escalating management costs (Afonso et. al., 2021; Rai, 2022; Rai & Singh, 2020). By prioritizing an understanding of the introduction process, cities can develop more effective, sustainable approaches to urban biodiversity conservation and enhance their resilience to future ecological challenges.

Overall, understanding the introduction process of IAPS in urban systems is crucial for preventing and managing invasions, protecting urban biodiversity, safeguarding public health, and ensuring the economic and environmental sustainability of cities (Potgieter et al., 2018; Rai & Singh, 2020; Seboko, 2022). This knowledge forms the foundation for strategic action and informed decision-making in the face of one of the most significant ecological challenges of our time.

7 Management and Control Strategies for IAPS Present in Urban Ecosystems

Here are some of the management and control strategies for IAPS present in the urban environments (see: Fig. 8):

7.1 *Early Detection and Rapid Response*

1. **Monitoring**: Regularly monitor urban areas for signs of IAPS establishment, spread, and impact (Appalasamy et al., 2020).
2. **Surveillance**: Implement surveillance programs to detect IAPS early, using techniques such as remote sensing, aerial photography, and ground surveys (Dvořák et al., 2015).

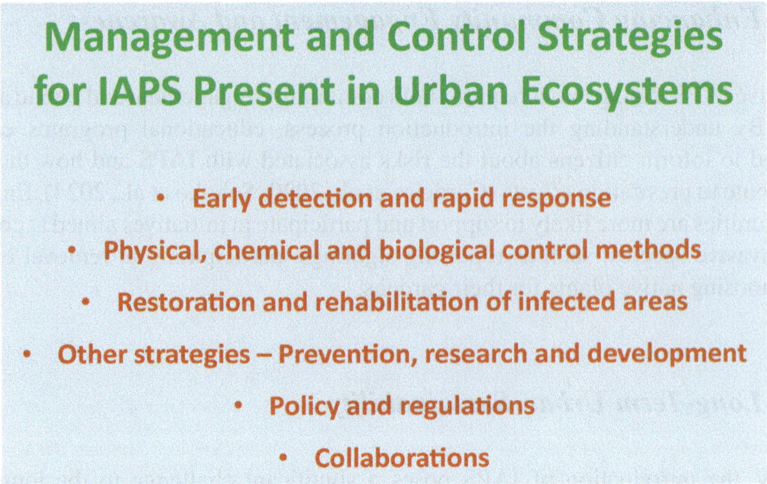

Fig. 8 An overview of different management and control strategies for the judicious management of invasive alien plant species present in urban ecosystems

3. **Rapid response**: Develop rapid response protocols to quickly contain and control IAPS infestations (Potgieter et al., 2018).
4. **Public awareness**: Educate the public about the importance of early detection and rapid response to prevent IAPS spread (Novoa et al., 2017; Potgieter et al., 2019).

7.2 Physical, Chemical, and Biological Control Methods

1. **Physical control**: Use physical barriers, such as mulching, mowing, or hand-pulling, to remove IAPS (Francis and Chadwick, 2015).
2. **Chemical control**: Apply herbicides to control IAPS growth, but use with caution and follow safety protocols (van Wilgen et al., 2001).
3. **Biological control**: Introduce natural predators or competitors of IAPS to control their populations (Culliney, 2005).
4. **Integrated pest management (IPM)**: Use a combination of physical, chemical, and biological control methods to manage IAPS. Biopesticides or related products having ability to destroy or inhibit the growth of invasive weeds can also be used as a potential technique for their regulation (Vreysen et al., 2007).

7.3 Restoration and Rehabilitation of Infected Areas

1. **Ecosystem restoration**: Restore native ecosystems by reintroducing native plant species, removing invasive species, and rehabilitating soil and habitat (Mostert et al., 2018).

2. **Soil remediation**: Remediate soil contaminated by IAPS through processes such as tilling, composting, or bioremediation (Rai and Kim, 2020).
3. **Habitat restoration**: Restore habitats for native animals and plants affected by IAPS (Rai and Kim, 2020).
4. **Community engagement**: Engage local communities in restoration efforts to promote ownership and long-term maintenance (Potgieter et al., 2019).

7.4 Other Strategies

1. **Prevention**: Prevent IAPS introduction through measures such as inspection of plants at ports of entry, quarantine of plants, and education on responsible plant trade practices (Potgieter et al., 2018).
2. **Research and development**: Research and create fresh approaches, technology, and management techniques on a constant basis to deal with IAPS (Potgieter et al., 2018).
3. **Policy and regulation**: Develop policies and regulations to prevent IAPS introduction, spread, and impact (Banerjee, 2021).
4. **Collaboration**: Collaborate with stakeholders, including government agencies, researchers, land managers, and local communities, to develop effective management and control strategies.

8 Policy and Governance: Role of Urban Planners and Policymakers for the Management of IAPS

1. **National and international laws**: Countries can implement national and international laws to regulate the introduction and spread of IAPS (Sokolova et al., 2020).
2. **Environmental regulations**: Environmental regulations can be implemented to control the cultivation, sale, and transportation of IAPS (Sokolova et al., 2020).
3. **Biosecurity measures**: Biosecurity measures can be implemented to prevent the introduction of IAPS into new areas (Black and Bartlett, 2020).
4. **Penalties for non-compliance**: Penalties can be imposed for non-compliance with regulations, such as fines or imprisonment (Cronin et al., 2017).
5. **Coordination between agencies**: Coordination between agencies responsible for environmental protection, agriculture, and urban planning can help ensure effective regulation (Potgieter et al., 2018).

9 Conclusions and Future Prospects

The integration of IAPS into urban ecosystems is a complicated matter that necessitates a thorough comprehension of the relevant ecological, social, and economic aspects. This chapter has highlighted the significant impacts of IAPS on urban ecosystems, including the displacement of native species, degradation of ecosystem services, and economic losses. Creating successful management plans requires an understanding of the mechanisms underlying the introduction and establishment of IAPS in urban environments. By addressing the pathways of introduction and the factors that facilitate invasion, urban areas can better prevent and mitigate the impacts of IAPS on biodiversity and ecosystem services. To effectively manage IAPS in urban systems, it is essential to adopt a holistic approach that integrates multiple strategies, including prevention, early detection, rapid response, and control. Urban planners, policymakers, and managers must work together to develop and implement effective management strategies that take into account the unique characteristics of urban ecosystems. The role of technology, community engagement, and education in IAPS management cannot be overstated. Innovative technologies, such as drones and sensor systems, can enhance monitoring and detection capabilities, while community engagement and education can promote public awareness and participation in IAPS management. Finally, research and collaboration are crucial for addressing the knowledge gaps and emerging trends in IAPS management. By working together to understand the complexities of IAPS in urban systems, we can develop more effective solutions for protecting urban ecosystems and promoting sustainable urban development.

Acknowledgements All the authors are thankful to their college / university authorities for providing necessary infrastructures and support for writing this chapter. Dr. Rishikesh Singh is thankful to the National Mission on Himalayan Studies (NMHS), Ministry of Environment, Forest and Climate Change (MoEFCC), New Delhi, for the Financial Assistance in the form of a Research Project (NMHS2024-25/SC-XII/MG95-200/SL06).

References

Adler, F. R., & Tanner, C. J. (2013). *Urban ecosystems: Ecological principles for the built environment*. Cambridge University Press.

Afonso, L., Esler, K.J., Gaertner, M., & Sjirk, G. (2021). Urban plant invasions: how to classify, prioritize and manage invasive alien plants; Cape Town as a case study. (Doctoral dissertation, Stellenbosch University).

Afreen, T., Kumari, S., Bhadouria, R., Devi, R. S., Singh, S., & Tripathi, S. (2023). Plant invasion and soil processes: A mechanistic understanding. In S. Tripathi, R. Bhadouria, P. Srivastava, R. Singh, & D. R. Batish (Eds.), *Plant invasions and global climate change*. Springer. https://doi.org/10.1007/978-981-99-5910-5_10

Agudelo-Vera, C. M., Mels, A. R., Keesman, K. J., & Rijnaarts, H. H. (2011). Resource management as a key factor for sustainable urban planning. *Journal of Environmental Management, 92*(10), 2295–2303.

Alizadeh, B., & Hitchmough, J. (2019). A review of urban landscape adaptation to the challenge of climate change. *International Journal of Climate Change Strategies and Management, 11*(2), 178–194.

Almas, A. D., & Conway, T. M. (2016). The role of native species in urban forest planning and practice: A case study of Carolinian Canada. *Urban Forestry & Urban Greening, 17*, 54–62.

Alston, K. P., & Richardson, D. M. (2006). The roles of habitat features, disturbance, and distance from putative source populations in structuring alien plant invasions at the urban/wildland interface on the Cape Peninsula, South Africa. *Biological Conservation, 132*(2), 183–198.

Appalasamy, M., & Ramdhani, S. (2020). Aliens in the city: Towards identifying non-indigenous floristic hotspots within an urban matrix. *Flora, 269*, Article 151631.

Arnaiz-Schmitz, C., Schmitz, M. F., Herrero-Jáuregui, C., Gutiérrez-Angonese, J. F. D. C., Pineda, F. D., & Montes, C. (2018). Identifying socio-ecological networks in rural-urban gradients: Diagnosis of a changing cultural landscape. *Science of the Total Environment, 612*, 625–635.

Aronson, M. F., Piana, M. R., MacIvor, J. S., & Pregitzer, C. C. (2017b). Management of plant diversity in urban green spaces. In *Urban biodiversity* (pp. 101–120). Routledge.

Aronson, M. F., Lepczyk, C. A., Evans, K. L., Goddard, M. A., Lerman, S. B., MacIvor, J. S., Nilon, C. H., & Vargo, T. (2017a). Biodiversity in the city: Key challenges for urban green space management. *Frontiers in Ecology and the Environment, 15*(4), 189–196.

Bai, L., Xiu, C., Feng, X., & Liu, D. (2019). Influence of urbanization on regional habitat quality: A case study of Changchun City. *Habitat International, 93*, Article 102042.

Balfors, B., Azcárate, J., Mörtberg, U., Karlson, M., & Gordon, S. O. (2016). Impacts of urban development on biodiversity and ecosystem services. In *Handbook on biodiversity and ecosystem services in impact assessment* (pp. 167–194). Edward Elgar Publishing.

Banerjee, A. K., Khuroo, A. A., Dehnen-Schmutz, K., Pant, V., Patwardhan, C., Bhowmick, A. R., & Mukherjee, A. (2021). An integrated policy framework and plan of action to prevent and control plant invasions in India. *Environmental Science & Policy, 124*, 64–72.

Banerjee, A. K., & Sankaran, K. V. (2023) An action plan to prevent and manage alien plant invasions in India. In *Plant invasions and global climate change* (pp. 397–418). Springer Nature Singapore.

Beatley, T. (2000). Preserving biodiversity: Challenges for planners. *Journal of the American Planning Association, 66*(1), 5–20.

Bengyella, L. (2022). Impact of invasive alien plant species and their management implications to endemic diseases, food and health security in Africa. *Annals of Multidisciplinary Research, Innovation and Technology (AMRIT), 1*(1), 1–7.

Black, R., & Bartlett, D. M. (2020). Biosecurity frameworks for cross-border movement of invasive alien species. *Environmental Science & Policy*, 105, 113-119.

Branco, D., Goncalves, O. F., & Badia, S. B. I. (2023). A systematic review of international affective picture system (IAPS) around the world. *Sensors, 23*(8), 3866.

Bremner, A., & Park, K. (2007). Public attitudes to the management of invasive non-native species in Scotland. *Biological Conservation, 139*(3–4), 306–314.

Bretzel, F., Vannucchi, F., Romano, D., Malorgio, F., Benvenuti, S., & Pezzarossa, B. (2016). Wildflowers: From conserving biodiversity to urban greening—A review. *Urban Forestry & Urban Greening, 20*, 428–436.

Bryant, M. M. (2006). Urban landscape conservation and the role of ecological greenways at local and metropolitan scales. *Landscape and Urban Planning, 76*(1–4), 23–44.

Castro-Díez, P., Vaz, A. S., Silva, J. S., Van Loo, M., Alonso, Á., Aponte, C., Bayón, Á., Bellingham, P. J., Chiuffo, M. C., DiManno, N., & Julian, K. (2019). Global effects of non-native tree species on multiple ecosystem services. *Biological Reviews, 94*(4), 1477–1501.

Chalker-Scott, L. (2015). Nonnative, noninvasive woody species can enhance urban landscape biodiversity. *Arboriculture and Urban Forestry, 41*(4), 173–186.

Chen, J., Zhang, Y., Liu, W., Wang, C., Ma, F., & Xu, H. (2023). Distribution patterns and determinants of invasive alien plants in China. *Plants, 12*(12), 2341.

Cordeiro, B., Marchante, H., Castro, P., & Marchante, E. (2020). Does public awareness about invasive plants pays off? An analysis of knowledge and perceptions of environmentally aware citizens in Portugal. *Biological Invasions, 22*(7), 2267–2281.

Cristina-Maria, A., Proorocu, M., Mălinaş, A., & Mălinaş, C. (2023). Unraveling plant invasion: Insights into major pathways and mechanisms of invasive species spread. *ProEnvironment Promediu, 16*(55).

Cronin, K., Kaplan, H., Gaertner, M., Irlich, U. M., & Timm Hoffman, M. (2017). Aliens in the nursery: assessing the attitudes of nursery managers to invasive species regulations. *Biological Invasions, 19*(3), 925–937.

Culliney, T. W. (2005). Benefits of classical biological control for managing invasive plants. *Critical Reviews in Plant Sciences, 24*(2), 131–150.

Davids, R., Rouget, M., Burger, M., Mahood, K., Ditlhale, N., & Slotow, R. (2021). Civic ecology uplifts low-income communities, improves ecosystem services and well-being, and strengthens social cohesion. *Sustainability, 13*(3), 1300.

de Barros Ruas, R., Costa, L. M. S., & Bered, F. (2022). Urbanization driving changes in plant species and communities—A global view. *Global Ecology and Conservation, 38*, Article e02243.

de Carvalho, C. A., Raposo, M., Pinto-Gomes, C., & Matos, R. (2022). Native or exotic: A bibliographical review of the debate on ecological science methodologies: Valuable lessons for urban green space design. *Land, 11*(8), 1201.

De Oliveira, J. P., Balaban, O., Doll, C. N., Moreno-Peñaranda, R., Gasparatos, A., Iossifova, D., & Suwa, A. (2011). Cities and biodiversity: Perspectives and governance challenges for implementing the convention on biological diversity (CBD) at the city level. *Biological Conservation, 144*(5), 1302–1313.

DeVos, T. B., Bock, D. G., & Kolbe, J. J. (2023). Rapid introgression of non-native alleles following hybridization between a native Anolis lizard species and a cryptic invader across an urban landscape. *Molecular Ecology, 32*(11), 2930–2944.

Dhakal, S., Shrestha, B. B., Sharma, K. P., Paudel, S., & Siwakoti, M. (2024). Grasslands are more vulnerable to plant invasions than forests in south-central Nepal. *Environmental Challenges, 15*, Article 100929.

Di Giulio, M., Holderegger, R., & Tobias, S. (2009). Effects of habitat and landscape fragmentation on humans and biodiversity in densely populated landscapes. *Journal of Environmental Management, 90*(10), 2959–2968.

Dvořák, P., Müllerová, J., Bartaloš, T., & Brůna, J. (2015). Unmanned aerial vehicles for alien plant species detection and monitoring. *The International Archives of the Photogrammetry, Remote Sensing and Spatial Information Sciences, 40*, 83–90.

Early, R., Bradley, B. A., Dukes, J. S., Lawler, J. J., Olden, J. D., Blumenthal, D. M., Gonzalez, P., Grosholz, E. D., Ibañez, I., Miller, L. P., & Sorte, C. J. (2016). Global threats from invasive alien species in the twenty-first century and national response capacities. *Nature Communications, 7*(1), 12485.

Elmqvist, T., Maltby, E., Barker, T., Mortimer, M., Perrings, C., Aronson, J., De Groot, R., Fitter, A., Mace, G., Norberg, J., & Pinto, I. S. (2012). Biodiversity, ecosystems and ecosystem services. In *The economics of ecosystems and biodiversity: Ecological and economic foundations* (pp. 41–111). Routledge.

Engelkes, T., Wouters, B., Bezemer, T. M., Harvey, J. A., & van der Putten, W. H. (2012). Contrasting patterns of herbivore and predator pressure on invasive and native plants. *Basic and Applied Ecology, 13*(8), 725–734.

Fineschi, S., & Loreto, F. (2020). A survey of multiple interactions between plants and the urban environment. *Frontiers in Forests and Global Change, 3*, 30.

Foxcroft, L. C., Pyšek, P., Richardson, D. M., Genovesi, P., & MacFadyen, S. (2017). Plant invasion science in protected areas: Progress and priorities. *Biological Invasions, 19*(5), 1353–1378.

Francis, R. A., & Chadwick, M. A. (2015). Urban invasions: Non-native and invasive species in cities. *Geography, 100*(3), 144–151.

Gaertner, M., Holmes, P. M., & Richardson, D. M. (2012). Biological invasions, resilience and restoration. In *Restoration ecology: The new frontier*, pp. 265–280.

Gaertner, M., Wilson, J. R., Cadotte, M. W., MacIvor, J. S., Zenni, R. D., & Richardson, D. M. (2017). Non-native species in urban environments: Patterns, processes, impacts and challenges. *Biological Invasions, 19*, 3461–3469.

Gordon, A., Simondson, D., White, M., Moilanen, A., & Bekessy, S. A. (2009). Integrating conservation planning and landuse planning in urban landscapes. *Landscape and Urban Planning, 91*(4), 183–194.

Gozlan, R. E., Burnard, D., Andreou, D., & Britton, J. R. (2013). Understanding the threats posed by non-native species: Public vs. conservation managers. *PloS one, 8*(1), p.e53200.

Grimm, N.B., & Schindler, S. (2018). Nature of Cities and Nature in Cities: Prospects for Conservation and Design of Urban Nature in Human Habitat. Rethinking Environmentalism: Linking Justice, Sustainability, and Diversity, Sharachchandra Lele, Eduardo S. Brondizio, John Byrne, Georgina M. Mace, Joan Martinez-Alier

Gurran, N., Gilbert, C., & Phibbs, P. (2015). Sustainable development control? Zoning and land use regulations for urban form, biodiversity conservation and green design in Australia. *Journal of Environmental Planning and Management, 58*(11), 1877–1902.

Jabbar, M., Yusoff, M. M., & Shafie, A. (2022). Assessing the role of urban green spaces for human well-being: A systematic review. *GeoJournal*, 1–19.

Jauni, M., Gripenberg, S., & Ramula, S. (2015). Non-native plant species benefit from disturbance: A meta-analysis. *Oikos, 124*(2), 122–129.

Jelaska, S. D. ed. (2021). Neobiota 2020. In *11th International conference on biological invasions: The human role in biological invasions-a case of Dr Jekyll and Mr Hyde? Vodice, Croatia*, 15–18 September 2020.

Kashe, K., Rukavo, B., Tsheboeng, G., Motsage, K., & Mmusi, M. (2024). The distribution of woody invasive alien species in wards in relation to land use within Maun Village, northern Botswana. *Ecological Frontiers, 44*(5), 1018–1026.

Knauf, A. E., Litton, C. M., Cole, R. J., Sparks, J. P., Giardina, C. P., Gerow, K. G., & Quiñones-Santiago, M. (2021). Nutrient-use strategy and not competition determines native and invasive species response to changes in soil nutrient availability. *Restoration Ecology, 29*(5), Article e13374.

Kothencz, G., Kolcsár, R., Cabrera-Barona, P., & Szilassi, P. (2017). Urban green space perception and its contribution to well-being. *International Journal of Environmental Research and Public Health, 14*(7), 766.

Kowarik, I. (2011). Novel urban ecosystems, biodiversity, and conservation. *Environmental Pollution, 159*(8–9), 1974–1983.

Kowarik, I., & der Lippe, M. V. (2008). Pathways in plant invasions. In W. Nentwig, (Eds.) *Biological invasions. Ecological studies* (vol. 193). Springer. https://doi.org/10.1007/978-3-540-36920-2_3

Kueffer, C., & Kull, C. A. (2017). Non-native species and the aesthetics of nature. In *Impact of biological invasions on ecosystem services* (pp.311–324).

Ladd, B. (2016). Nitrogen pollution and the meltdown of urban ecosystems. *Land, 5*(3), 23.

Lake, J. C., & Leishman, M. R. (2004). Invasion success of exotic plants in natural ecosystems: The role of disturbance, plant attributes and freedom from herbivores. *Biological Conservation, 117*(2), 215–226.

Langemeyer, J., & Gómez-Baggethun, E. (2018). Urban biodiversity and ecosystem services. Urban biodiversity. From research to practice, pp.36–53.

Larios, L., & Suding, K. N. (20130. Restoration within protected areas: When and how to intervene to manage plant invasions?. Plant Invasions in Protected Areas: Patterns, Problems and Challenges, 599–618.

Lemke, A., Buchholz, S., Kowarik, I., Starfinger, U., & von der Lippe, M. (2021). Interaction of traffic intensity and habitat features shape invasion dynamics of an invasive alien species (Ambrosia artemisiifolia) in a regional road network. *NeoBiota, 64*, 155–175.

Lone, P. A., Dar, J. A., Subashree, K., Raha, D., Pandey, P. K., Ray, T., Khare, P. K., & Khan, M. L. (2019). Impact of plant invasion on physical, chemical and biological aspects of ecosystems: A review. *Tropical Plant Research, 6*(3), 528–544.

Long, H., Liu, Y., Hou, X., Li, T., & Li, Y. (2014). Effects of land use transitions due to rapid urbanization on ecosystem services: Implications for urban planning in the new developing area of China. *Habitat International, 44*, 536–544.

Markovchick-Nicholls, L. I. S. A., Regan, H. M., Deutschman, D. H., Widyanata, A., Martin, B., Noreke, L., & Ann Hunt, T. I. M. O. T. H. Y. (2008). Relationships between human disturbance and wildlife land use in urban habitat fragments. *Conservation Biology, 22*(1), 99–109.

Marsh, A. S., Hayes, D. C., Klein, P. N., Zimmerman, N., Dalsimer, A., Burkett, D. A., Huebner, C. D., Rabaglia, R., Meyerson, L. A., Harper-Lore, B. L., & Davidson, J. L. (2021). Sectoral impacts of invasive species in the United States and approaches to management. In *Invasive species in forests and rangelands of the United States: A comprehensive science synthesis for the United States forest sector* (pp. 203–229).

Mavimbela, L. Z., Sieben, E. J., & Procheş, Ş. (2018). Invasive alien plant species, fragmentation and scale effects on urban forest community composition in Durban, South Africa. *New Zealand Journal of Forestry Science, 48*, 1–14.

Mayfield, A. E., Seybold, S. J., Haag, W. R., Johnson, M. T., Kerns, B. K., Kilgo, J. C., Larkin, D. J., Lucardi, R. D., Moltzan, B. D., Pearson, D. E., & Rothlisberger, J. D. (2021). Impacts of invasive species in terrestrial and aquatic systems in the United States. In *Invasive species in forests and rangelands of the United States: A comprehensive science synthesis for the United States forest sector* (pp 5–39).

McKinney, M. L. (2002). Urbanization, biodiversity, and conservation: The impacts of urbanization on native species are poorly studied, but educating a highly urbanized human population about these impacts can greatly improve species conservation in all ecosystems. *BioScience, 52*(10), 883–890.

McKinney, M. L. (2006). Urbanization as a major cause of biotic homogenization. *Biological Conservation, 127*(3), 247–260.

Mersal, A. (2016). Sustainable urban futures: Environmental planning for sustainable urban development. *Procedia Environmental Sciences, 34*, 49–61.

Miles, L. S., Breitbart, S. T., Wagner, H. H., & Johnson, M. T. (2019). Urbanization shapes the ecology and evolution of plant-arthropod herbivore interactions. *Frontiers in Ecology and Evolution, 7*, 310.

Minden, V., & Gorschlüter, J. (2016). Comparison of native and non-native Impatiens species across experimental light and nutrient gradients. *Plant Ecology and Evolution, 149*(1), 59–72.

Mostert, E. (2016). Identifying priority areas for active restoration after alien plant clearing in the City of Cape Town (Doctoral dissertation, Stellenbosch University).

Mostert, E., Gaertner, M., Holmes, P. M., O'Farrell, P. J., & Richardson, D. M. (2018). A multi-criterion approach for prioritizing areas in urban ecosystems for active restoration following invasive plant control. *Environmental Management, 62*(6), 1150–1167.

Mtengwana, B., Dube, T., Mudereri, B. T., & Shoko, C. (2021). Modeling the geographic spread and proliferation of invasive alien plants (IAPs) into new ecosystems using multi-source data and multiple predictive models in the Heuningnes catchment, South Africa. *Giscience & Remote Sensing, 58*(4), 483–500.

Mudavanhu, S. (2018). Sustainable land use planning in the aftermath of the clearing of alien invasive plant species: A system dynamics modelling approach (Doctoral dissertation, Stellenbosch University).

Müller, N., Ignatieva, M., Nilon, C. H., Werner, P., & Zipperer, W. C. (2013). Patterns and trends in urban biodiversity and landscape design. In *Urbanization, biodiversity and ecosystem services: Challenges and opportunities: a global assessment* (pp. 123–174).

Naess, P. (2001). Urban planning and sustainable development. *European Planning Studies, 9*(4), 503–524.

Novoa, A., Dehnen-Schmutz, K., Fried, J., & Vimercati, G. (2017). Does public awareness increase support for invasive species management? Promising evidence across taxa and landscape types. *Biological invasions, 19*(12), 3691–3705.

O'Briain, R., Corenblit, D., & Gurnell, A. (2023). Interacting effects of climate change and invasions by alien plant species on the morphodynamics of temperate rivers. *Wiley Interdisciplinary Reviews: Water, 10*(5), Article e1649.

Okoye, P. I. (2016). Grassland rehabilitation after alien invasive tree eradication: Landscape degradation and sustainability in rural Eastern Cape (Doctoral dissertation, Stellenbosch University).

Padayachee, A. L., Irlich, U. M., Faulkner, K. T., Gaertner, M., Procheş, Ş, Wilson, J. R., & Rouget, M. (2017). How do invasive species travel to and through urban environments? *Biological Invasions, 19*, 3557–3570.

Pilliod, D. S., Griffiths, R. A., & Kuzmin, S. L. (2012). Ecological impacts of non-native species. *Terminology, 9*, 44.

Pinto, L. V., Inácio, M., Ferreira, C. S. S., Ferreira, A. D., & Pereira, P. (2022). Ecosystem services and well-being dimensions related to urban green spaces—A systematic review. *Sustainable Cities and Society, 85*, Article 104072.

Plaza, P. I., Speziale, K. L., & Lambertucci, S. A. (2018). Rubbish dumps as invasive plant epicentres. *Biological Invasions, 20*, 2277–2283.

Potgieter, L. (2019). Assessing the impacts of invasive alien plants on urban ecosystem services (Doctoral dissertation, Stellenbosch University).

Potgieter, L. J., Gaertner, M., Kueffer, C., Larson, B. M., Livingstone, S. W., O'Farrell, P. J., & Richardson, D. M. (2017). Alien plants as mediators of ecosystem services and disservices in urban systems: A global review. *Biological Invasions, 19*, 3571–3588.

Potgieter, L. J., Gaertner, M., Irlich, U. M., O'Farrell, P. J., Stafford, L., Vogt, H., & Richardson, D. M. (2018). Managing urban plant invasions: A multi-criteria prioritization approach. *Environmental Management, 62*, 1168–1185.

Potgieter, L. J., Gaertner, M., O'Farrell, P. J., & Richardson, D. M. (2019). Perceptions of impact: Invasive alien plants in the urban environment. *Journal of Environmental Management, 229*, 76–87.

Potgieter, L. J., Shrestha, N., & Cadotte, M. W. (2022). Prioritizing terrestrial invasive alien plant species for management in urban ecosystems. *Journal of Applied Ecology, 59*(3), 872–883.

Rai, P. K. (2022). Environmental degradation by invasive alien plants in the anthropocene: Challenges and prospects for sustainable restoration. *Anthropocene Science, 1*(1), 5–28.

Rai, P. K., & Kim, K. H. (2020). Invasive alien plants and environmental remediation: a new paradigm for sustainable restoration ecology. *Restoration Ecology, 28*(1), 3-7.

Rai, P. K., & Singh, J. S. (2020). Invasive alien plant species: Their impact on environment, ecosystem services and human health. *Ecological Indicators, 111*, Article 106020.

Rebele, F. (1994). Urban ecology and special features of urban ecosystems. In *Global ecology and biogeography letters* (pp. 173–187).

Rodriguez, L. F. (2006). Can invasive species facilitate native species? Evidence of how, when, and why these impacts occur. *Biological Invasions, 8*, 927–939.

Ruhsam, M., Kohn, D., Marquardt, J., Leitch, A. R., Schneider, H., Vogel, J., Barrett, S. C., Hulme, P. E., Squirrell, J., & Hollingsworth, P. M. (2023). Is hybridisation with non-native congeneric species a threat to the UK native bluebell Hyacinthoides non-scripta? *Plants, People, Planet, 5*(6), 963–975.

Schierenbeck, K. A. (2021). The genetic consequences of hybridization between California native plant species with closely related taxa. *Protecting California's Native Flora since, 1965*, 25.

Seboko, T. C. (2022). The distribution and perceptions of invasive alien plants in small towns in the Eastern Cape province, South Africa (A Master of Science Dissertation, Rhodes University).

Seboko, T., Shackleton, C. M., & Ruwanza, S. (2024). Urban residents' knowledge of and attitudes and willingness to control woody invasive alien plants in their domestic gardens in South Africa. *People and Nature, 6*, 2077–2090.

Shrestha, U. B., & Shrestha, B. B. (2019). Climate change amplifies plant invasion hotspots in Nepal. *Diversity and Distributions, 25*(10), 1599–1612.

Sołtysiak, J., & Brej, T. (2014). Invasion of *Fallopia* genus plants in urban environment. *Polish Journal of Environmental Studies, 23*(2), 449–458.

Sokolova, A. K., & Tulina, E. I. (2020). Legal aspects of invasive alien plant species regulation. *Environmental Policy and Law, 49*(4-5), 300-305.

Southon, G. E., Jorgensen, A., Dunnett, N., Hoyle, H., & Evans, K. L. (2017). Biodiverse perennial meadows have aesthetic value and increase residents' perceptions of site quality in urban greenspace. *Landscape and Urban Planning, 158*, 105–118.

Sukopp, H., & Starfinger, U. (1999). Disturbance in urban ecosystems. *Ecosystems of the World* 397–412.

Svirejeva-Hopkins, A., Schellnhuber, H. J., & Pomaz, V. L. (2004). Urbanised territories as a specific component of the Global Carbon Cycle. *Ecological Modelling, 173*(2–3), 295–312.

Theodorou, P. (2022). The effects of urbanisation on ecological interactions. *Current Opinion in Insect Science, 52*, Article 100922.

Tu, S., Long, H., Zhang, Y., Ge, D., & Qu, Y. (2018). Rural restructuring at village level under rapid urbanization in metropolitan suburbs of China and its implications for innovations in land use policy. *Habitat International, 77*, 143–152.

Urgenson, L. S. (2011). *Ecological and social aspects of riparian restoration and non-native plant invasions: Studies from the Pacific Northwest, US and the Western Cape*. University of Washington.

van Delden, L., Rowlings, D. W., Scheer, C., & Grace, P. R. (2016). Urbanisation-related land use change from forest and pasture into turf grass modifies soil nitrogen cycling and increases N_2O emissions. *Biogeosciences, 13*(21), 6095–6106.

van Wilgen, B., Richardson, D., & Higgins, S. I. (2001). Integrated control of invasive alien plants in terrestrial ecosystems. *Land Use and Water Resources Research, 1*, 1–8.

Verbrugge, L. N., Van den Born, R. J., & Lenders, H. R. (2013). Exploring public perception of non-native species from a visions of nature perspective. *Environmental Management, 52*, 1562–1573.

Vreysen, M.J.B., Robinson, A.S., Hendrichs, J., Kenmore, P. (2007). Area-Wide Integrated Pest Management (AW-IPM): Principles, Practice and Prospects. In: Vreysen, M.J.B., Robinson, A.S., Hendrichs, J. (eds) *Area-Wide Control of Insect Pests: From Research to Field Implementation*. Springer, Dordrecht. pp. 3–33.

Wang, H., Marshall, C. W., Cheng, M., Xu, H., Li, H., Yang, X., & Zheng, T. (2017). Changes in land use driven by urbanization impact nitrogen cycling and the microbial community composition in soils. *Scientific Reports, 7*(1), 44049.

Wang, W., Deng, X., Wang, Y., Peng, L., & Yu, Z. (2022). Impacts of infrastructure construction on ecosystem services in new-type urbanization area of North China Plain. *Resources, Conservation and Recycling, 185*, Article 106376.

Ward, E. B., Pregitzer, C. C., Kuebbing, S. E., & Bradford, M. A. (2020). Invasive lianas are drivers of and passengers to altered soil nutrient availability in urban forests. *Biological Invasions, 22*(3), 935–955.

Wichlan, M. (2022). The perception and use of native plants in the landscape: Understanding the practices of landscape architects in the Southeast United States. Mississippi State University.

Yang, Y., Ignatieva, M., Gaynor, A., & Chen, C. (2024). Urban biodiversity in design: insights into the debate on native versus non-native plants and bees in Western Australia. *Urban Forestry & Urban Greening, 98*, Article 128391.

Yigitcanlar, T., & Teriman, S. (2015). Rethinking sustainable urban development: towards an integrated planning and development process. *International Journal of Environmental Science and Technology, 12*, 341–352.

Zeunert, J. (2014). Challenging assumptions in urban restoration ecology. *Landscape Journal, 32*(2), 231–242.

Zipperer, W. C., Northrop, R., & Andreu, M. (2020). Urban development and environmental degradation. *Urban Development and Environmental Degradation*. https://doi.org/10.1093/acrefore/9780199389414.013.97

Chapter 2
From Alien to Invasive Alien Plant Species: An Array of Events

Sidra Batool, Raees Khan, Irum Khan, Sheikh Zain Ul Abidin, Asad Ullah, and Jie Liu

Abstract The globalization process has facilitated the introduction of non-native plant species into novel ecosystems, leading to a marked increase in invasive alien plants (IAPs) that pose a serious threat to native biodiversity and ecological systems. While earlier research has extensively investigated the challenges posed by IAPs, there is a growing interest in explaining the fundamental mechanisms that propel certain non-native plants toward invasiveness. This chapter aims to comprehend the factors that determine the success of certain non-native species in colonizing and dominating new ecosystems. It explores the ecological theories that underpin invasion ecology and analyzes the adaptive strategies of IAPs, notably the Competitive, Stress-tolerant, and Ruderal (CSR) strategies. The chapter discusses how invasive species possess specific traits that enable them to adapt and thrive in new environments, often outcompeting native species. The ecological strategies of these plants, including Grime's CSR strategies, are analysed to understand their success in various environments. Many hypotheses have been developed to explain the success of invasive plants, including the enemy release hypothesis, fluctuation resources hypothesis, novel weapons hypothesis, evolution of increased competitive ability (EICA) hypothesis, and the rapid evolution hypothesis. IAPs must overcome multiple barriers to progress along the introduction-naturalization-invasion continuum and become

S. Batool
China University of Geosciences Wuhan, Hubei, China

R. Khan (✉) · J. Liu
CAS Key Laboratory for Plant Diversity and Biogeography of East Asia, Kunming Institute of Botany, Chinese Academy of Sciences, Kunming, China
e-mail: raeeskhan@bs.qau.edu.pk

Germplasm Bank of Wild Species & Yunnan Key Laboratory of Crop Wild Relatives Omics, Kunming Institute of Botany, Chinese Academy of Sciences, Kunming, China

I. Khan
Department of Biosciences, COMSATS University Islamabad, Islamabad, Pakistan

S. Z. U. Abidin
Institute of Biological Sciences, Gomal University D I Khan, Dera Ismail Khan, Pakistan

A. Ullah
Centre of Plant Biodiversity, University of Peshawar, Peshawar, Pakistan

R. Singh et al. (eds.), *Invasive Alien Plants in Urban Ecosystems*,
https://doi.org/10.1007/978-3-032-00997-5_2

dominant in their introduced ranges. While inherent traits are important, propagule pressure and residence time are more important in determining invasion success. A significant focus is placed on propagule pressure and its role in the establishment and spread of invasive species. Case studies, such as the one from South Africa's Agulhas Plain, are used to illustrate the impact of propagule pressure on invasion patterns. The chapter also addresses the role of environmental factors, including disturbances and climate, in facilitating plant invasions. IAPs achieve dominance in introduced ranges through complex interactions between inherent traits, rapid evolution, disturbance regimes, resource availability, biotic interactions, and introduction dynamics. Generalizing determinants requires comparing native and introduced ranges within a framework that captures the context dependencies of plant invasions. Finally, the chapter presents a comprehensive overview of the implications for conservation and management. It emphasizes the need for early detection, risk assessment, public education, integrated management approaches, adaptive management, collaborative efforts, and ongoing research and monitoring to effectively manage and mitigate the impact of invasive plant species on ecosystems. This chapter provides a detailed and research-based exploration of the journey of plant species from being exotic/aliens to becoming invasive, offering insights into their ecological impacts and the strategies required for effective management and conservation.

Keywords Adaptive strategies · Case studies · Ecological theory · Generalizing determinants · Invasiveness factors · Invasive plants

1 Exotic/Alien and Invasive Alien Plant Species: Characteristics and Introduction

The complex phenomenon of the introduction and transformation of alien plant species into invasive ones—a process significantly influenced by human activities and global trade. It begins by clarifying the terminology used to describe these species, such as "invasive," and "alien," as defined by IPBES (2019). The transition from non-native to invasive species is explored, highlighting the roles of both biotic and abiotic factors, including propagule pressure, environmental disturbances, and climate change. The early stages of invasion, including establishment and spread, are examined through the lens of biological characteristics and phenotypic plasticity.

Over many centuries, humans have moved various plant species to new parts of the world. In recent years, global trade has made it happen even faster and more commonly (Brown, 2013). These plants are often called non-native or alien species when introduced to places they do not naturally belong to. As per the IPBES Report of 2019 (IPBES 2019), the term "alien" (synonyms: non-native, exotic, introduced, non-indigenous, allochthonous) species refers to species whose presence in a region is attributable to human actions, intentional or unintentional, that enable them to overcome biogeographical barriers. The terms "alien" and "exotic" are used interchangeably. By whatever means, after their introduction into a non-native area, if they establish themselves and overpower the natives, especially in terms of number and impact on the native flora and ecology, they are termed as Invasive Alien Plant (IAP) species. Likewise, if the native plant species overpower the native vegetation for whatever reason, they are called Invasive Native Plant species. These IAP species can threaten local wildlife, food supplies, and human health (Liu et al., 2025; Pyšek & Richardson, 2010). Consequently, a species can be alien without being invasive, or, in the case of a species that is native to a region, it can become invasive without being an alien species. A plant can be intentionally introduced into the environment, such as for agriculture, decorative use, forestry, erosion control, or medical uses, or unintentionally, such as through transportation, travel, trade, biotic means (wind, water, animals) or may be due to disturbance (fire, floods) (Khan et al., 2022) (Fig. 1).

While all species engage in competition for existence, invasive species possess distinct qualities or intricate properties that enable them to efficiently compete with native species. In certain instances, this advantage might stem from their ability to reproduce and grow more rapidly than indigenous species. Plants are widely recognized as a highly invasive species worldwide due to their large biomass (Pyšek et al., 2012). The transition of alien plants into invasive plants involves a series of events, each with its complexities and ecological consequences (Mandle et al., 2011).

However, not all alien species are invasive; only 10% of the alien species that possess a specific set of traits and mechanisms that enable them to thrive and proliferate are potentially invasive and cause harm to the environment and biodiversity (Richardson et al., 2000).

2 Introduction to the Transformation Process

The introduction and transformation of alien plant species into invasive species within the complex web of global ecosystems is a fascinating and complex biological phenomenon (Dickie et al., 2014). This process, often referred to as the transition from alien to invasive, is a sophisticated journey that shapes the composition and dynamics of ecosystems on a global scale. This section aims to explain the complex events underlying this ecological transformation. It is crucial to understand the factors responsible for this transition. Several biotic and abiotic factors contribute to transforming a non-native alien plant into an invasive one (Kuebbing et al., 2014).

Regarding biotic factors, it is widely accepted that invasive species exhibit distinct combinations of characteristics that facilitate their establishment, population growth, and expansion into new areas. In certain instances, these qualities also allow them to outperform native species in competition (Tripathi et al., 2023). Moreover, certain traits confer advantages for the establishment in non-native ranges, while others may become significant during the invasion stage, as the barriers and filters characterizing these phases differ (Gioria & Osborne, 2014; Rai, 2015). Furthermore, the absence of natural predators and herbivores facilitates the success of invasion (Maron & Vila,

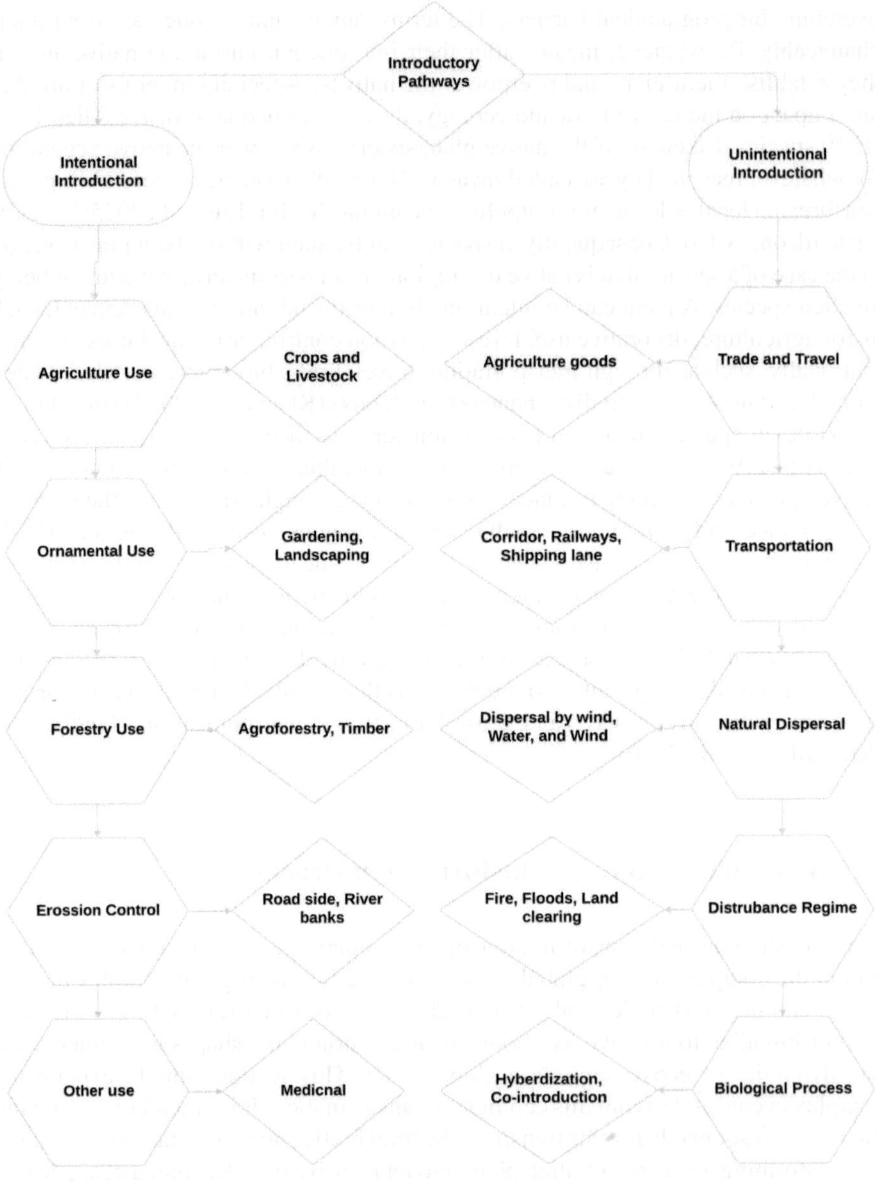

Fig. 1 Introductory pathways of invasive plants

2001). Abiotic factors also play a significant role in the transformation of aliens into invasive plants (Bajwa et al., 2016). Alien species can easily naturalize and become invasive if their fundamental ecological niche matches that of the introduced range (Mack, 1996). Thus, both biotic factors (e.g., rapid growth, phenotypic plasticity, propagule pressure, absence of natural predators) and abiotic factors (e.g., ecological niche conditions of the new region) play pivotal roles in transforming aliens into invasive plants. Several hypotheses and theories have been formulated by researchers based on these factors. Natural and anthropogenic processes mediate the transformation of aliens into invasive plants. This transition can pose a serious threat to ecosystem functioning (Fu et al., 2025; Levin et al., 2001). Therefore, comprehending the shift from alien to invasive plants is vital for conservation and management.

3 Early Stages of Invasion: Establishment and Spread

The establishment of invasive plant species depends on a complex interaction of elements, including a unique set of plant characteristics, phenotypic plasticity, ecological disturbances, and the novelty of non-native environments (Rai, 2015). During the initial phase of establishing non-native plant species, biological traits are crucial in this complex dynamic. Key biological features, such as rapid growth, resource utilization efficiency, resistance to native pests and diseases, and tolerance to environmental stressors, emerge as crucial drivers for the successful survival and establishment of alien plant species in novel habitats (Skendžić et al., 2021) (Fig. 2). These characteristics allow these plants to adapt to their non-native environment and successfully manage their ecological barriers. Consequently, this complex combination of elements, with a focus on the biological characteristics of invasive plants, highlights the underlying mechanisms at work in establishing non-native plant species, which has substantial implications for ecological research and management (Khattak et al., 2024; Kueffer et al., 2013).

3.1 Biological Characteristics and Phenotypic Plasticity

The trait-based approach states that invasion success is linked with the functional traits of alien plants (Drenovsky et al., 2012). These traits make them preadapted to non-native environments' conditions (biotic and abiotic) (Berthon, 2015; van Kleunen et al., 2011). A distinct collection of traits characterizes invasive plants, with reproductive traits essential for their successful establishment in non-native regions (Divíšek et al., 2018; Gioria et al., 2019). The reproductive traits, such as larger seeds, are the most important for early establishment (Baraloto et al., 2005; Ferreras & Galetto, 2010). At the same time, smaller-sized seeds are productive in the invasion phase.

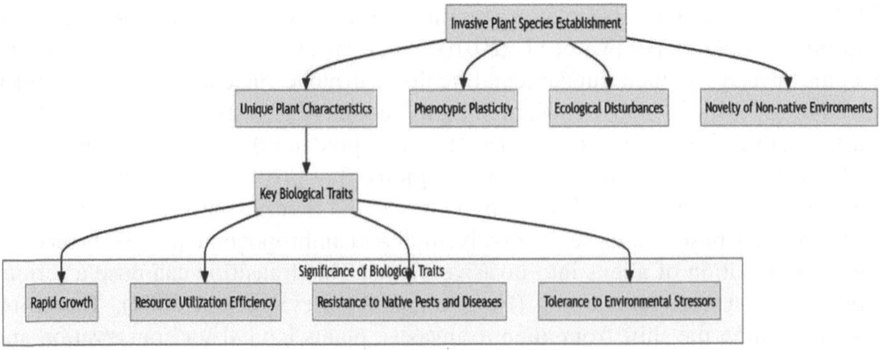

Fig. 2 Factors associated with establishment of invasive plants

Furthermore, phenotypic plasticity is essential for non-native plants to adapt to environmental changes. Phenotypic plasticity is like an organism using its genetic "tools" (traits) differently in response to different environmental "situations" to adapt and survive. Thus, by influencing how organisms adapt to their surroundings, phenotypic plasticity can explain why invasive species can thrive in a wider range of environments (Hulme, 2008; Riis et al., 2010). The early establishment depends on the phenotypic plasticity of non-native plants. Due to several characteristic features, including short durations to reproduction, high post-dispersal survival, high germination rates, human-mediated processes, and propagule pressure, invasive species frequently demonstrate fast spread rates and rapid colonization (Gioria et al., 2019; Khan et al., 2023; Skarpaas & Shea, 2007). *Acer negundo* and *Acer platanoides*, two invasive maple species, were the subjects of a study highlighting the crucial role that phenotypic flexibility plays in the early stages of establishment (Lamarque, 2013). Furthermore, polyploidy in plants tends to increase the emergence of alien species as opposed to diploid ones since it improves fitness during the establishment stage (Te Beest et al., 2012).

3.1.1 Case Study

Daisy fleabane (*Erigeron annuus*) exhibits significant phenotypic plasticity, allowing it to adapt to a broad range of environmental conditions (Levačić & Jelaska, 2022). This adaptability is evident in its successful colonization of habitats from lowlands to high-altitude regions such as the Swiss Alps. The species' ability to adjust its phenotype in response to environmental stressors is a key factor in its invasive potential, enabling it to form dense stands that can outcompete native flora and impact agricultural productivity. The pronounced plasticity of Daisy fleabane facilitates its spread and establishment in diverse environments, making it a notable example of an invasive species with a high degree of ecological adaptability (Levačić & Jelaska, 2022).

3.2 Ecological Strategies (Grime's CSR Strategies)

Plants use various ecological strategies based on their biological traits, impacting their establishment and spread (Theoharides & Dukes, 2007). The CSR ecological strategy, proposed by Grime, classifies plants into three strategies: Competitors (C): These plants excel in low-stress environments, competing for resources with traits like rapid growth, large size, and allelopathic abilities. Stress Tolerator (S): Suited for harsh conditions, they adapt with traits like small size, slow growth, and resilience to extreme factors. Ruderals (R): Flourishing in disturbed settings, they have short lifespans, high seed production, and rapid recovery (Pierce et al., 2013). CSR predicts invasive plant success: Competitors thrive in nutrient-rich areas, stress tolerators in harsh ones, and ruderals in disturbed habitats (Grime, 2006). Research on adaptive strategies in plant invasions, applying CSR and considering introduction pathways, reveals that naturalized species tend towards R-selection, while invasive species lean towards C-selection, especially deliberate introductions (Escobedo et al., 2021) (Fig. 3). Understanding these dynamics enhances our knowledge of how non-native plants establish and spread.

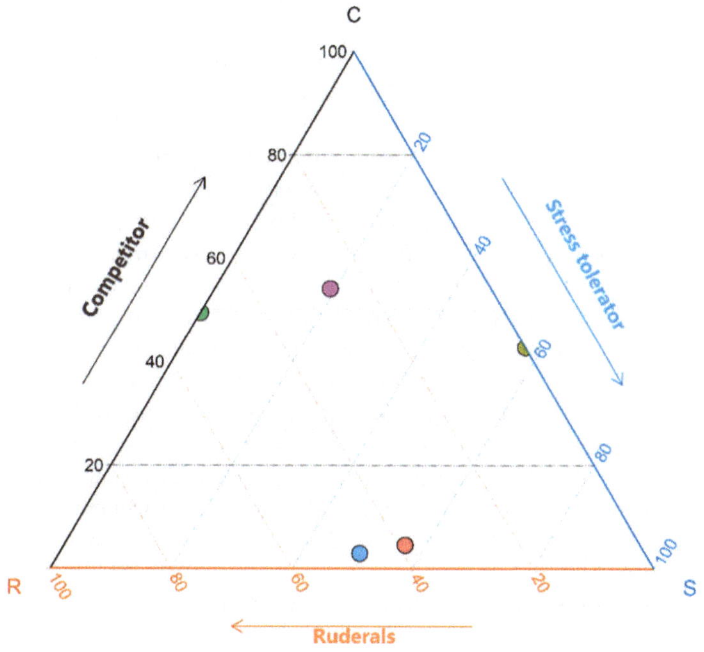

Fig. 3 Grime's CSR triangle. Adapted from Grime (1977). CSR triangle categorizing plant strategies. The triangular diagram illustrates the three fundamental plant strategies: competitors (C), stress-tolerator (S), and ruderals (R)

3.3 Propagule Pressure

Propagule pressure is intuitively a key factor in biological invasions: increased availability of propagules increases the chances of establishment, persistence, naturalization, and invasion (Dawson et al., 2009). The propagule pressure hypothesis posits that non-native species with a substantial propagule pressure are more likely to achieve successful invasion than those with a minimal propagule pressure (Dawson et al., 2009). Propagule pressure is characterized by the quantity and frequency of reproductive units entering a specific site. In studies examining what factors contribute to the successful establishment and spread of non-native species, most have traditionally focused on the traits and characteristics of these species (both biological and environmental). However, emerging empirical and statistical evidence highlights the significance of propagule pressure, which includes the size and quantity of reproductive units and the timing and spatial distribution of their arrival (Eschtruth & Battles, 2009). This concept plays a vital role in both stages of invasion.

Larger propagules increase the likelihood of successful establishment primarily by reducing the impact of random population fluctuations (Simberloff, 2009). Conversely, a higher number of propagules primarily mitigates the influence of unpredictable environmental variations (Ellis et al., 2009). Consistent and diverse arrivals of propagules, especially from various sources, can prevent or diminish the expected genetic limitation in invasions that often occur when only a few individuals initiate the process (Johnston et al., 2009). This increased genetic diversity enhances the likelihood of survival and adaptation to new environments and habitats. For certain species, recent molecular evidence suggests that ongoing propagule pressure helps invasions to spread by introducing genetic variations that are well-suited for new areas and ecological conditions (Lawson Handley et al., 2011). This phenomenon may also account for some of the delays observed between the initial establishment of a non-native species and its subsequent transformation into an invasive pest. Animals and birds are usually considered major dispersers of propagules. Propagules transported by wild boars, for instance, are likely to be deposited in disturbed habitats, such as wallowing points or trampled areas (Brescacin, 2010). These locations are often preferred sites for establishing invasive species due to the disruption of the existing vegetation and the provision of suitable conditions for non-native plant growth. The ability of invasive plants to capitalize on these disturbances and dispersal mechanisms is crucial in their successful establishment in novel environments.

3.3.1 A Propagule Pressure in Practice: Case Study

To underscore the significance of propagule pressure in biological invasions, a convincing case study was carried out on the vast 2160 km^2 Agulhas Plain in South Africa (Marini et al., 2009). This study focused on the distribution and percentage cover of alien tree species, specifically *Acacia cyclops, A. saligna*, and *Pinus pinaster*. Classifying the data based on geological, climatic, land use, and topographical factors allowed for an understanding of the distribution of these invasive species. However, the abundance, particularly the canopy cover, of these species remained a puzzle. To address this gap, a semi-mechanistic model was developed to quantify the roles of propagule pressure and environmental heterogeneity in shaping invasion patterns. The findings from this case study were remarkable. The level of propagule pressure, which is estimated by the distance from potential invasion sources, was found to be a significantly stronger predictor of canopy cover compared to any environmental element that was examined. Subsequently, an evaluation was conducted on the residuals of the model to determine the impact of environmental factors. This analysis revealed that the influence of environmental factors varied depending on the species, indicating a species-specific mediating effect. Models that integrated both propagule pressure and environmental parameters accurately predicted over 70% of the variability in canopy cover for each species. This case study provides a compelling demonstration of how the amount of propagules (reproductive structures) influences the way invasions occur.

4 The Role of Environmental Factors

The invisibility of recipient ecosystems can be shaped by various factors, including (a) The availability of resources and their fluctuations over time, the patterns of disturbance in the ecosystem, and the range of environmental conditions; (b) The diversity and species richness of native plant communities, which can affect their resistance to invasions. This resistance may be influenced by the functional or evolutionary similarities between native and incoming non-native species; (c) The presence or absence of natural predators and mutualistic relationships within the ecosystem (Fig. 4).

4.1 Disturbance

Disturbance is an essential factor in the establishment of non-native plants. Disturbance could be natural or anthropogenic (Jauni et al., 2015). Natural disturbance, i.e., fire, flood, livestock grazing, etc. Wildfire facilitates the establishment of non-native plants, providing habitat and nutrients required for ruderals. The historical use of prescribed fire to convert shrublands to grassland led to the invasion of non-native plants. In Australian savannahs, frequent and hot fires promote the establishment and spread of some aggressive grass invaders such as (*Cenchrus ciliaris*) (Marshall et al., 2012). The establishment of non-native plants on the bank of rivers is facilitated by the available resources that the floodwater deposits, thus creating fertile soil (Francis et al., 2019). Moreover, the fluctuation in resources also allows non-native plants to establish themselves. The frequent flooding events expose the soil and reduce native plant cover while providing space and resources to non-native plants, which thrive in a nutrient-poor and low-competitive environment (Hemery

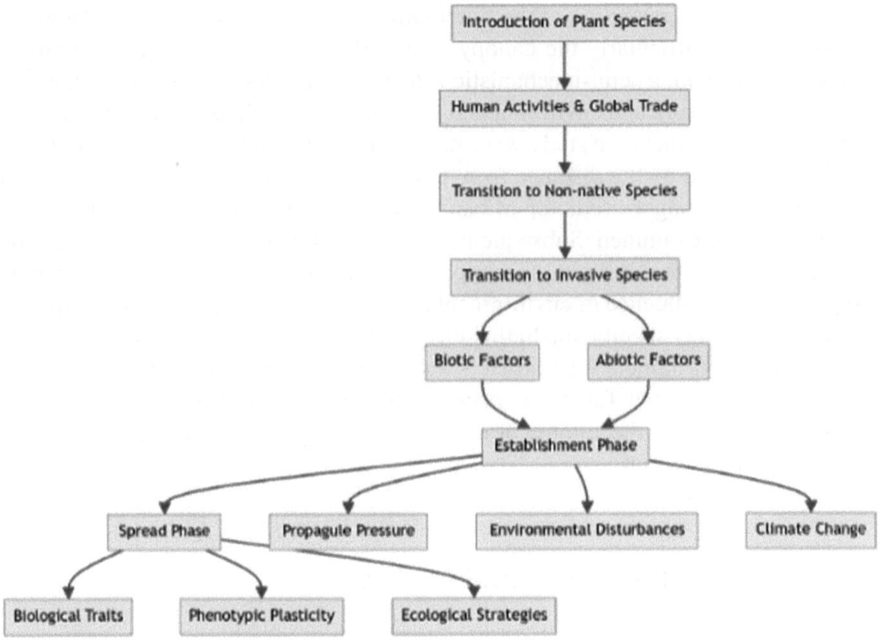

Fig. 4 Factors influencing invasive plant establishment and spread

et al., 2010). Human-mediated disturbance also promotes this process of invasion. Human use of recreational sites can disturb natural vegetation ecosystems in ways that favour establishing non-native plants. Non-native plants generally benefit from human disturbance, flourishing in disturbed habitats (Jauni et al., 2015). This includes disturbance during the construction and maintenance of tourism infrastructure such as campsites, roads, and trails.

4.2 Climate Facilitating Plant Invasion

Climate plays a pivotal role in facilitating plant invasion by shaping the availability of resources and the competitive landscape of native and invasive species (Theoharides & Dukes, 2007). Climate can affect every stage of invasion: introduction, establishment, and spread. During the establishment phase, Climate Matching plays an important role in that, those invasive species are more likely to establish if the climate of the new area is similar to their native range (Bomford et al., 2009). Moreover, human assistance also acts as a significant factor, in certain climates conducive to agriculture or transportation, human activities may inadvertently introduce invasive species (Masters & Norgrove, 2010). Apart from that invasive species often have a broad tolerance range for temperature, rainfall, and other climatic factors, aiding their establishment (Clements & Ditommaso, 2011). Mild climates with ample resources can facilitate quick establishment.

Mild or stable climates can facilitate the spread by reducing biotic resistance from native species (Mack et al., 2000). While, changing climates can open new habitats for invasion by making them more suitable for invasive species and less so for native species (Hulme, 2017). Constant warmth and abundant rainfall can offer invasive species year-round opportunities for growth and reproduction. In arid climates, stress-tolerant invasive species can outcompete native flora, especially if they are more efficient in water use (Milton & Dean, 2010). While in temperate climates, invasive species exploit the predictability of temperate climates for their growth cycles. Polar/Alpine Climate Warming trends in polar and alpine regions open new areas for invasion and constantly new invasive species are reported every year (Amin et al., 2025).

Climate significantly influences the introduction, establishment, and spread of invasive plant species. Tropical climates are highly conducive to invasion due to their constant warmth and ample resources (Kriticos et al., 2003). Arid climates can be exploited by stress-tolerant invaders, while temperate climates offer predictability that some invaders can exploit (Milton & Dean, 2010). Polar and alpine regions are becoming more susceptible due to climate change (Ernakovich et al., 2014). The interaction between climate and invasive species is complex and dynamic, shaping both the opportunities and challenges presented by different climatic zones for plant invasions.

5 Theoretical Frameworks and Models for Understanding Plant Invasion Success

Scientists have developed many hypotheses to comprehend and predict the performance of non-native plants in new environments. Different models and theories have been developed to explain how various factors interact and how success can be situation-dependent (Carpenter et al., 2009; Tromp & Sternberg, 2022). Invasive plant species must overcome multiple barriers to progress along the introduction-naturalization-invasion continuum and become dominant in their introduced ranges (Gioria et al., 2023). Trait-based approaches link invasiveness to inherent species characteristics like rapid growth, reproduction, resource acquisition, and dispersal abilities (Drenovsky et al., 2012). However, propagule pressure and residence time are more important than species characteristics in determining invasion success (Lockwood et al., 2005). Community assembly theories investigate how functional differences in resource-use-related traits between native and invasive species affect invisibility via mechanisms such as biotic resistance and empty niches (Biella, 2019). Several influential hypotheses explain key mechanisms influencing the success of an invasion. The enemy release hypothesis states that invasive species escape from their native enemies (Roy et al., 2011). The fluctuation resources hypothesis proposes that invisibility increases in response to resource fluctuations and disturbances (Davis et al., 2000). The novel weapons hypothesis attributes invasion success to the use of allopathic chemicals that are novel to native species (Pinzone et al., 2018). The evolution of increased competitive ability (EICA) hypothesis proposes that invasive species evolve increased competitive ability and decreased allocation to defence in introduced ranges (Hull-Sanders et al., 2007).

By increasing fitness, genetic diversity, and adaptation potential, rapid evolution through multiple introductions, genetic diversity, hybridization, and polyploidization frequently facilitate invasion success (Lambertini, 2019). Positive interactions with native mutualists, such as seed dispersers, can promote invasions (Traveset & Richardson, 2014; Wu et al., 2023). Invisibility is modulated by disturbance regimes, resource availability, and biotic interactions with native plant and animal species (He et al., 2019). Global changes will likely alter future invasion trajectories in complex ways through their effects on disturbance patterns, resource levels, and biotic interactions (Vilà et al., 2007). It is difficult to generalize invasiveness factors due to context dependence and introduction bias. Invasive plant species achieve dominance in introduced ranges through complex interactions between inherent traits, rapid evolution, disturbance regimes, resource availability, biotic interactions, and introduction dynamics. Generalizing determinants requires comparing native and introduced ranges within a framework that captures the context dependencies of plant invasions.

6 Implications for Conservation and Management

Numerous studies have demonstrated the complexity of managing invasive species effectively, which calls for an integrated and cogent strategy. Research by Rejmánek and Pitcairn (2002) and others provide strong evidence for the efficacy of Early Detection and Rapid Response (EDRR) tactics in controlling invasive species. By drastically lowering long-term ecological effects and management expenses, these tactics highlight how crucial it is to quickly detect and address invasive species. Regarding direct intervention techniques, Van Driesche and Hoddle (2000) reported that the use of biological control agents has proven effective in managing several invasive plant species. However, to avoid negative effects on non-target species, this strategy needs to be carefully considered. Furthermore, D'antonio and Meyerson (2002) highlight that habitat management and ecosystem restoration are critical to boosting ecosystem resilience against invasions, especially when it comes to the restoration of native plant communities that can successfully compete with invasive species. One essential element of managing invasive species is the implementation of legislative and policy measures. Regulations governing the import and export of potentially invasive species are crucial for preventing them, as noted by Perrings (2002). Because ecosystems are inherently unpredictable and management activities frequently have unknown results, these strategies must be flexible and dynamic. According to Westman (1990), this adaptive management entails constant learning and strategy modification in response to ongoing observation and research findings. The control of invasive species is complicated and necessitates cooperation between many disciplines and sectors. Lodge et al. (2006) emphasize that to successfully handle the problems caused by invasive species, coordinated efforts at the local, regional, and international levels are essential. Recent studies consistently emphasize the intricacy of controlling invasive species and the essentiality of implementing integrated methods (Siddiqui et al., 2024; van Rees et al., 2022). An example is a research study that showcases the effectiveness of using GIS and remote-sensing technology to detect and manage invasive species through early detection and rapid response (EDRR) programs, thus preventing substantial damage (Nizamani et al., 2023). Moreover, the ever-changing invasion patterns of the Black-Headed Fall emphasize the urgent requirement for swift and efficient measures to manage or eliminate it upon early detection, in order to prevent extensive ecological consequences (Shao & Pan, 2024). The Queensland Invasive Plants and Animals Strategy 2025–2030 prioritizes habitat restoration and focused surveillance, integrating climate change adaptation measures to strengthen ecosystems against the threat of invasive species. This approach involves the restoration of indigenous ecosystems to strengthen their ability to withstand challenges and prevent the intrusion of foreign species (Bhatnagar & Kumar 2024). Furthermore, the utilization of biological control agents, such as insects and pathogens, have been proven to be a successful approach in controlling invasive plant species. However, it is crucial to exercise caution in order to prevent any negative impacts on non-target species (Lake et al., 2024). The SNPLMA Round 20 Conservation Initiatives Project Addendum emphasizes the need of adaptive management tactics, which involve continuously learning and modifying strategies based on new research and observations. This method guarantees the long-term effectiveness of management procedures (BLM, 2024). This cooperation includes continuing research, which is essential to improving our knowledge of the ecology of invasive species and creating successful control plans. Simberloff (2013) argues that long-term monitoring is essential for evaluating the success of management interventions and for modifying these tactics over time to ensure they continue to work in the face of shifting ecological insights and environmental conditions (Epanchin-Niell, 2017; van Rees et al., 2022).

7 Conclusion: Synthesizing the Array of Events

This investigation of alien and invasive plant species in this chapter highlights a critical challenge of urban landscapes depicting pervasive ecological and conservation issues. The problem of these species is increasing and result of a complex interaction of ecological, geographic, and human-induced variables rather than just a biological phenomenon.

Global trade and human migration have aided in the spread of urban invasive species, which pose serious risks to native biodiversity, environmental stability, and economic well-being. Our comprehension of this process reveals the complex mechanisms that allow these plants to flourish in new regions, from the initial introduction to the eventual establishment and spread of invasive species (Fig. 4). Internationalization, economic relations, and human movement allowed them to proliferate, becoming a threat to indigenous biodiversity, destabilizing the ecosystem, and incurring economic costs. Observing this process signifies the forces of their success—from introduction to establishment and dominance—as described in models like Grime's CSR theory and traits like phenotypic plasticity. Propagule pressure, which was discovered in case studies like South Africa's Agulhas Plain, also accounts for how certain species become invasive. Combating these threats demands multifaceted strategies: early detection, threat assessment, public awareness, and adaptive management based on on-going research. In effect, conservation efforts must factor in collaboration between disciplines and nations to balance the globalization of invasive species. The transformation of alien flora into invading marauders is a testament to nature's resilience as much as humanity's inadvertent disruptions. This reality highlights the precariousness of ecological balance as much as it stresses the necessity for anticipatory, science-guided stewardship. In the future, knowledge gained through research on invasive species will guide conservation efforts in an increasingly interconnected world, channelling efforts to conserve biodiversity through the management of anthropogenic environmental change.

References

Ada, A. B., Lake, B., Bingham, B., Boise, B., Butte, C., Canyon, C., Cassia, C., Custer, E., Franklin, F., & Gemm, G. (2024). Environmental assessment rangeland grasshopper and mormon cricket suppression program.

Amin, R., Sultan, A., Khan, A., Akhtar, N., Shah, S. A., Rashid, M., Khan, R., & Erum, S. (2025). Additions to flora of Pakistan: five new records in the genera *Lepidium, Shangwua* and *Solanum* from Swat, Malakand, Dir and Chitral Districts. *Pakistan Journal of Botany, 57*, 3.

Bajwa, A. A., Chauhan, B. S., Farooq, M., Shabbir, A., & Adkins, S. W. (2016). What do we really know about alien plant invasion? A review of the invasion mechanism of one of the world's worst weeds. *Planta, 244*, 39–57.

Baraloto, C., Forget, P. M., & Goldberg, D. E. (2005). Seed mass, seedling size and neotropical tree seedling establishment. *Journal of Ecology, 93*, 1156–1166.

Berthon, K. (2015). How do native species respond to invaders? Mechanistic and trait-based perspectives. *Biological Invasions, 17*, 2199–2211.

Bhatnagar, D. G., & Kumar, S. (2024). Ecological conservation and sustainable management of marine environments. In *Nature-based solutions in achieving sustainable development goals: harmonizing nature and progress* (pp. 233–261). Cham: Springer Nature Switzerland

Biella, P. (2019). A conservation perspective on the mechanisms that influence plant-pollinator interactions.

Bomford, M., Kraus, F., Barry, S. C., & Lawrence, E. (2009). Predicting establishment success for alien reptiles and amphibians: A role for climate matching. *Biological Invasions, 11*, 713–724.

Brescacin, C. R. (2010). *The role of the feral pig (Sus scrofa) as a disturbance agent and seed disperser in central Florida's natural lands* (p. 111). University of Central Florida.

Brown, L. R. (2013). *Eco-economy: Building an economy for the earth.* Routledge.

Carpenter, S. R., Mooney, H. A., Agard, J., Capistrano, D., DeFries, R. S., Díaz, S., Dietz, T., Duraiappah, A. K., Oteng-Yeboah, A., Pereira, H. M. & Perrings, C. (2009). Science for managing ecosystem services: Beyond the Millennium Ecosystem Assessment. *Proceedings of the National Academy of Sciences, 106*(5), 1305–1312.

Clements, D., & Ditommaso, A. (2011). Climate change and weed adaptation: Can evolution of invasive plants lead to greater range expansion than forecasted? *Weed Research, 51*, 227–240.

D'Antonio, C., & Meyerson, L. A. (2002). Exotic plant species as problems and solutions in ecological restoration: A synthesis. *Restoration Ecology, 10*, 703–713.

Davis, M. A., Grime, J. P., & Thompson, K. (2000). Fluctuating resources in plant communities: A general theory of invasibility. *Journal of Ecology, 88*, 528–534.

Dawson, W., Burslem, D. F., & Hulme, P. E. (2009). Factors explaining alien plant invasion success in a tropical ecosystem differ at each stage of invasion. *Journal of Ecology, 97*, 657–665.

Dickie, I. A., Bennett, B. M., Burrows, L. E., Nuñez, M. A., Peltzer, D. A., Porté, A., Richardson, D. M., Rejmánek, M., Rundel, P. W., & van Wilgen, B. W. (2014). Conflicting values: Ecosystem services and invasive tree management. *Biological Invasions, 16*, 705–719.

Divíšek, J., Chytrý, M., Beckage, B., Gotelli, N. J., Lososová, Z., Pyšek, P., Richardson, D. M., & Molofsky, J. (2018). Similarity of introduced plant species to native ones facilitates naturalization, but differences enhance invasion success. *Nature Communications, 9*, 4631.

Drenovsky, R. E., Grewell, B. J., D'Antonio, C. M., Funk, J. L., James, J. J., Molinari, N., Parker, I. M., & Richards, C. L. (2012). A functional trait perspective on plant invasion. *Annals of Botany, 110*, 141–153.

Ellis, B. J., Figueredo, A. J., Brumbach, B. H., & Schlomer, G. L. (2009). Fundamental dimensions of environmental risk: The impact of harsh versus unpredictable environments on the evolution and development of life history strategies. *Human Nature, 20*, 204–268.

Epanchin-Niell, R. S. (2017). Economics of invasive species policy and management. *Biological Invasions, 19*, 3333–3354.

Ernakovich, J. G., Hopping, K. A., Berdanier, A. B., Simpson, R. T., Kachergis, E. J., Steltzer, H., & Wallenstein, M. D. (2014). Predicted responses of arctic and alpine ecosystems to altered seasonality under climate change. *Global Change Biology, 20*, 3256–3269.

Eschtruth, A. K., & Battles, J. J. (2009). Acceleration of exotic plant invasion in a forested ecosystem by a generalist herbivore. *Conservation Biology, 23*(2), 388–399.

Escobedo, V. M., Rios, R. S., Alcayaga-Olivares, Y., & Gianoli, E. (2021). Disturbance reinforces community assembly processes differentially across spatial scales. *Annals of Botany, 127*, 175–189.

Ferreras, A. E., & Galetto, L. (2010). From seed production to seedling establishment: Important steps in an invasive process. *Acta Oecologica, 36*, 211–218.

Francis, R. A., Chadwick, M. A., & Turbelin, A. J. (2019). An overview of non-native species invasions in urban river corridors. *River Research and Applications, 35*, 1269–1278.

Fu, L., Huang, J., Zhang, G., Han, D., Ding, L., Wei, Y., ... & Yu, H. (2025). Abrupt loss of species richness caused by ecosystem transition. *Science Bulletin, 70*(9), 1523–1532.

Gioria, M., Hulme, P. E., Richardson, D. M., & Pyšek, P. (2023). Why are invasive plants successful? *Annual Review of Plant Biology, 74*, 635–670.

Gioria, M., & Osborne, B. A. (2014). Resource competition in plant invasions: Emerging patterns and research needs. *Frontiers in Plant Science, 5*, 501.

Gioria, M., le Roux, J. J., Hirsch, H., Moravcová, L., & Pyšek, P. (2019). Characteristics of the soil seed bank of invasive and non-invasive plants in their native and alien distribution range. *Biological Invasions, 21*, 2313–2332.

Grime, J. O. (1977). Evidence for the existence of three primary strategies in plants and its relevance to ecological and evolutionary theory. *The American Naturalist, 111*, 1169–1194.

Grime, J. P. (2006). *Plant strategies, vegetation processes, and ecosystem properties*. England, John Wiley & Sons.

He, T., Lamont, B. B., & Pausas, J. G. (2019). Fire as a key driver of Earth's biodiversity. *Biological Reviews, 94*, 1983–2010.

Hemery, G., Clark, J., Aldinger, E., Claessens, H., Malvolti, M., O'Connor, E., Raftoyannis, Y., Savill, P., & Brus, R. (2010). Growing scattered broadleaved tree species in Europe in a changing climate: A review of risks and opportunities. *Forestry, 83*, 65–81.

Hull-Sanders, H. M., Clare, R., Johnson, R. H., & Meyer, G. A. (2007). Evaluation of the evolution of increased competitive ability (EICA) hypothesis: Loss of defense against generalist but not specialist herbivores. *Journal of Chemical Ecology, 33*, 781–799.

Hulme, P. E. (2008). Phenotypic plasticity and plant invasions: Is it all Jack? *Functional Ecology, 22*, 3–7.

Hulme, P. E. (2017). Climate change and biological invasions: Evidence, expectations, and response options. *Biological Reviews, 92*, 1297–1313.

IPBES, W. (2019). Intergovernmental science-policy platform on biodiversity and ecosystem services. Summary for policy makers of the global assessment report on biodiversity and ecosystem services of the intergovernmental science-policy platform on biodiversity and ecosystem services. Bonn, Germany: IPBES Secretariat.

Jauni, M., Gripenberg, S., & Ramula, S. (2015). Non-native plant species benefit from disturbance: A meta-analysis. *Oikos, 124*, 122–129.

Johnston, E. L., Piola, R. F., & Clark, G. F. (2009). The role of propagule pressure in invasion success. In: *Biological invasions in marine ecosystems: Ecological, management, and geographic perspectives*, pp. 133–151.

Khan, R., Hill, R. S., Liu, J., & Biffin, E. (2023). Diversity, distribution, systematics and conservation status of Podocarpaceae. *Plants, 12*(5), 1171.

Khan, R., Iqbal, I. M., Ullah, A., Ullah, Z., & Khan, S. M. (2022). Invasive alien species: an emerging challenge for the biodiversity. In *Biodiversity, conservation and sustainability in Asia prospects and challenges in South and Middle Asia*. Springer.

Khattak, W. A., Sun, J., Hameed, R., Zaman, F., Abbas, A., Khan, K. A., Elboughdiri, N., Akbar, R., He, F., Ullah, M. W., & Al-Andal, A. (2024). Unveiling the resistance of native weed communities: Insights for managing invasive weed species in disturbed environments. *Biological Reviews, 99*(3), 753–777.

Kriticos, D., Sutherst, R., Brown, J., Adkins, S., & Maywald, G. (2003). Climate change and biotic invasions: A case history of a tropical woody vine. *Biological Invasions, 5*, 147–165.

Kuebbing, S. E., Souza, L., & Sanders, N. J. (2014). Effects of co-occurring non-native invasive plant species on old-field succession. *Forest Ecology and Management, 324*, 196–204.

Kueffer, C., Pyšek, P., & Richardson, D. M. (2013). Integrative invasion science: Model systems, multi-site studies, focused meta-analysis and invasion syndromes. *New Phytologist, 200*(3), 615–633.

Lamarque, L. J. (2013). *Ecology and evolution of invasive maple tree species* (Doctoral dissertation. York University, Toronto, Ontario. Faculty of Graduate Studies. Graduate Program in Biology).

Lambertini, C. (2019). Why are tall-statured energy grasses of polyploid species complexes potentially invasive? A review of their genetic variation patterns and evolutionary plasticity. *Biological Invasions, 21*, 3019–3041.

Lawson Handley, L. J., Estoup, A., Evans, D., Thomas, C., Lombaert, E., Facon, B., Aebi, A., & Roy, H. (2011). Ecological genetics of invasive alien species. *BioControl, 56*, 409-428.

Levačić, D., & Jelaska, S. D. (2022). Consistent grime's CSR strategy of daisy fleabane erigeron annuus (L.) Pers. Despite its high morphological variability—a case study from Zagreb and Medvednica Mt., Croatia. *Diversity, 14*, 45.

Levin, L. A., Boesch, D. F., Covich, A., Dahm, C., Erséus, C., Ewel, K. C., Kneib, R. T., Moldenke, A., Palmer, M. A., & Snelgrove, P. (2001). The function of marine critical transition zones and the importance of sediment biodiversity. *Ecosystems, 4*, 430–451.

Liu, F. L., Mambo, W. W., Liu, J., Zhu, G. F., Khan, R., Khan, S. M., & Lu, L. (2025). Spatiotemporal range dynamics and conservation optimization for endangered medicinal plants in the Himalaya. *Global Ecology and Conservation, 57*, Article e03390.

Lockwood, J. L., Cassey, P., & Blackburn, T. (2005). The role of propagule pressure in explaining species invasions. *Trends in Ecology & Evolution, 20*, 223–228.

Lodge, D. M., Williams, S., Macisaac, H. J., Hayes, K. R., Leung, B., Reichard, S., Mack, R. N., Moyle, P. B., Smith, M., & Andow, D. A. (2006). Biological invasions: Recommendations for US policy and management. *Ecological Applications, 16*, 2035–2054.

Mack, R. N. (1996). Predicting the identity and fate of plant invaders: Emergent and emerging approaches. *Biological Conservation, 78*, 107–121.

Mack, R. N., Simberloff, D., Mark Lonsdale, W., Evans, H., Clout, M., & Bazzaz, F. A. (2000). Biotic invasions: Causes, epidemiology, global consequences, and control. *Ecological Applications, 10*, 689–710.

Mandle, L., Bufford, J. L., Schmidt, I. B., & Daehler, C. C. (2011). Woody exotic plant invasions and fire: Reciprocal impacts and consequences for native ecosystems. *Biological Invasions, 13*, 1815–1827.

Marini, L., Gaston, K., Prosser, F., & Hulme, P. (2009). Contrasting response of native and alien plant species richness to environmental energy and human impact along alpine elevation gradients. *Global Ecology and Biogeography, 18*, 652–661.

Maron, J. L., & Vila, M. (2001). When do herbivores affect plant invasion? Evidence for the natural enemies and biotic resistance hypotheses. *Oikos, 95*, 361–373.

Marshall, V., Lewis, M., & Ostendorf, B. (2012). Buffel grass (*Cenchrus ciliaris*) as an invader and threat to biodiversity in arid environments: A review. *Journal of Arid Environments, 78*, 1–12.

Masters, G., & Norgrove, L. (2010). Climate change and invasive alien species. *UK: CABI Working Paper, 1*, 30.

Milton, S. J., & Dean, W. R. J. (2010). Plant invasions in arid areas: Special problems and solutions: A South African perspective. *Biological Invasions, 12*, 3935–3948.

Nizamani, M. M., Zhang, Q., Muhae Ud Din, G., Awais, M., Qayyum, M., Farhan, M., Jabran, M., & Wang. Y. (2023). Application of GIS and remote-sensing technology in ecosystem services and biodiversity conservation, 284–321.

Pleune, J. (2024). The BLM's Public Lands Rule and. Available at SSRN 4985600

Perrings, C. (2002). Biological invasion risks and the public good: An economic perspective. *Conservation Ecology, 6*, 31.

Pierce, S., Brusa, G., Vagge, I., & Cerabolini, B. E. (2013). Allocating CSR plant functional types: The use of leaf economics and size traits to classify woody and herbaceous vascular plants. *Functional Ecology, 27*, 1002–1010.

Pinzone, P., Potts, D., Pettibone, G., & Warren, R. (2018). Do novel weapons that degrade mycorrhizal mutualisms promote species invasion? *Plant Ecology, 219*, 539–548.

Pyšek, P., & Richardson, D. M. (2010). Invasive species, environmental change and management, and health. *Annual Review of Environment and Resources, 35*, 25–55.

Pyšek, P., Jarošík, V., Hulme, P. E., Pergl, J., Hejda, M., Schaffner, U., & Vilà, M. (2012). A global assessment of invasive plant impacts on resident species, communities and ecosystems: The interaction of impact measures, invading species' traits and environment. *Global Change Biology, 18*, 1725–1737.

Rai, P. K. (2015). What makes the plant invasion possible? Paradigm of invasion mechanisms, theories and attributes. *Environmental Skeptics and Critics, 4*, 36.

Rejmánek, M., & Pitcairn, M. (2002). When is eradication of exotic pest plants a realistic goal. *Turning the tide: The Eradication of Invasive Species*, 249–253.

Richardson, D. M., Allsopp, N., D'Antonio, C. M., Milton, S. J., & Rejmánek, M. (2000). Plant invasions—the role of mutualisms. *Biological Reviews, 75*, 65–93.

Riis, T., Lambertini, C., Olesen, B., Clayton, J. S., Brix, H., & Sorrell, B. K. (2010). Invasion strategies in clonal aquatic plants: Are phenotypic differences caused by phenotypic plasticity or local adaptation? *Annals of Botany, 106*, 813–822.

Roy, H., Lawson Handley, L. J., Schönrogge, K., Poland, R., & Purse, B. (2011). Can the enemy release hypothesis explain the success of invasive alien predators and parasitoids? *Bio Control, 56*, 451–468.

Shao, F., & Pan, J. (2024). Exploring the dynamic invasion pattern of the black-headed fall webworm in China: Susceptibility to topography, vegetation, and human activities. *Insects, 15*(5), 349.

Siddiqui, J. A., Bamisile, B. S., Fan, R., Hafeez, M., Islam, W., Yang, W., Wei, M., Ran, H., Xu, Y., & Chen, X. (2024). Ant invasion in China: An in-depth analysis of the country's ongoing battle with exotic ants. *Ecological Indicators, 160*, 111811.

Simberloff, D. (2009). The role of propagule pressure in biological invasions. *Annual Review of Ecology, Evolution, and Systematics, 40*, 81–102.

Simberloff, D. (2013). *Invasive species: What everyone needs to know / Daniel Simberloff*. Oxford University Press.

Skarpaas, O., & Shea, K. (2007). Dispersal patterns, dispersal mechanisms, and invasion wave speeds for invasive thistles. *The American Naturalist, 170*, 421–430.

Skendžić, S., Zovko, M., Živković, I. P., Lešić, V., & Lemić, D. (2021). The impact of climate change on agricultural insect pests. *Insects, 12*, 440.

te Beest, M., le Roux, J. J., Richardson, D. M., Brysting, A. K., Suda, J., Kubešová, M., & Pyšek, P. (2012). The more the better? The role of polyploidy in facilitating plant invasions. *Annals of Botany, 109*, 19–45.

Theoharides, K. A., & Dukes, J. S. (2007). Plant invasion across space and time: Factors affecting nonindigenous species success during four stages of invasion. *New Phytologist, 176*, 256–273.

Traveset, A., & Richardson, D. M. (2014). Mutualistic interactions and biological invasions. *Annual Review of Ecology, Evolution, and Systematics, 45*, 89–113.

Tripathi, S., Bhadouria, R., Srivastava, P., Singh, R., & Batish, D. R. (Eds.). (2023). *Plant invasions and global climate change*. Springer Nature.

Tromp, C., & Sternberg, R. J. (2022). Dynamic creativity: A person× task× situation interaction framework. *The Journal of Creative Behavior, 56*(4), 553–565.

van Driesche, R. G., & Hoddle, M. S. (2000). Classical arthropod biological control: measuring success, step by step. In G. Gurr & S. Wratten (Eds.), *Biological control: Measures of success* (pp. 39–75). Springer.

van Kleunen, M., Schlaepfer, D. R., Glaettli, M., & Fischer, M. (2011). Preadapted for invasiveness: Do species traits or their plastic response to shading differ between invasive and non-invasive plant species in their native range? *Journal of Biogeography, 38*, 1294–1304.

van Rees, C. B., Hand, B. K., Carter, S. C., Bargeron, C., Cline, T. J., Daniel, W., & Luikart, G. (2022). A framework to integrate innovations in invasion science for proactive management. *Biological Reviews, 97*(4), 1712–1735.

Vilà, M., Corbin, J. D., Dukes, J. S., Pino, J., & Smith, S. D. (2007). Linking plant invasions to global environmental change. In *Terrestrial ecosystems in a changing world*. Springer.

Westman, W. E. (1990). Managing for biodiversity. *BioScience, 40*, 26–33.

Wu, Z. Y., Milne, R. I., Liu, J., Nathan, R., Corlett, R. T., & Li, D. Z. (2023). The establishment of plants following long-distance dispersal. *Trends in Ecology & Evolution, 38*(3), 289–300.

Chapter 3
Invasive Alien Plant Species in Urban Ecosystems

Manasvini Singh Sambyal, Yuvna Puri Goswami, Reema Mishra, and Preeti Agarwal

Abstract Invasive alien species are organisms that are non-native but dominate the native biota of an ecosystem. The damaging effects of invasive species have been documented globally. They display adverse effects on the ecosystem, and their services, biodiversity, economy and human health. The invasive species are stated to flourish more in the urban ecosystems because of several environmental disturbances and greater propagule pressure. Disturbances as a result of various anthropogenic activities like urbanization drastically alter the ecosystems. The richness of invasive species and their impact is found to increase along the urbanization gradients and they also display better physical conditions and high reproductive rates in the urban areas. Advancements in technologies has enhanced the facility for assessment and monitoring of invasive species. There is a requirement of a cohesive trans-disciplinary study for the sustainable management of invasive species. The present chapter focuses on how urbanization has contributed to the spread of alien invasive species. It also discusses the impact of invasive species on the environment, ecosystem, economy and human health. It also discusses the benefits of invasive species on the ecosystem, biodiversity and economy. It also highlights the management strategies for controlling invasive species. In addition to this, as invasive species also offer ecological and economic benefits, therefore management strategies needs to create and sustain new ecosystems that are capable of providing important ecosystem services, instead of making attempts to restore the degraded ecosystems and make it alien or invasive-free environment.

Keywords Biodiversity · Ecosystem services · Human health · Invasive species · Urbanization

All authors have contributed equally to this work.

M. S. Sambyal · Y. P. Goswami · R. Mishra · P. Agarwal (✉)
Department of Botany, Gargi College, University of Delhi, New Delhi, India
e-mail: preeti.agrawal@gargi.du.ac.in

R. Singh et al. (eds.), *Invasive Alien Plants in Urban Ecosystems*,
https://doi.org/10.1007/978-3-032-00997-5_3

Abbreviation

AIS	Aquatic Invasive Species
ES	Ecosystem Services
EDS	Ecosystem Disservices
IAPS	Invasive Alien Plant Species
IPs	Invasive Plants
IS	Invasive Species
NPS	Native Plant Species
NS	Native Species

1 Introduction

The urban ecosystem offers a unique habitat formed as a result of the change or loss of native environment or a combination of modified environmental conditions. These changes create adversities for some of the sensitive or native species (NS) while offering opportunities for alien or invasive species (IS) to invade. Urban areas are reported to be richer in IS brought as a result of human activities leading to various disturbances and thus are considered as hotspots of biological invasions (Cadotte et al., 2017).

The invasive alien plant species (IAPS) can tolerate the unique conditions of these urban areas, and thus are able to capitalize on the opportunities and become invasive. Alien species turn out to be invasive when they continue to maintain huge self-replacing inhabitants over several generations away from the place of introduction. The invasive potential of the species is associated with the traits associated with their life-history. The species displaying traits like faster growth and high fertility are expected to have high invasive potential. Many of the IS have been found to be successful in urban environments as evidenced by their high number. This is because the urban areas are under more disturbances, less competition, and newness of environmental conditions (as the habitats are severely changed because of human activities) in comparison to natural and rural areas that aid in the introduction and spread of IS (Aronson et al., 2015; Cadotte et al., 2017). It is further anticipated that extreme events (storms, floods, cyclones, droughts, fires, etc.) that are direct outcomes of changing climatic conditions also support the spread of IS (Sharma et al., 2023). Kühn et al. (2017) observed that the abundance of IS and their impact increases along the rural–urban gradient. Gaertner et al. (2016) reasoned that IS are higher in cities because of the existence of long histories of human-mediated dispersal, various disturbances and alterations in habitat that increase the chances for their establishment, reproduction and spread.

IS are reported to have injurious effects on biodiversity, ecosystem functioning and services, environment, human health and economy. Many of the IS were introduced for providing, creating, restoring or augmenting ecosystem services (ES) like trees for erosion control and timber production (Kowarik, 2011; Rai & Singh, 2020). However, some of them spread beyond their original plantation and turned into invasive and posed a negative impact on the ecosystem health and services like leading to flood and soil erosion, obstruction of canals and streams, water blockage, filtration, coastal protection and risk to fire severity (van Wilgen & Scott, 2001). Though many of IS are known to interrupt ES, there are some that are also beneficial in various ways. Therefore, if they are managed in a sustainable way, they can be used for bioremediation, air purification, biofuel, food and fodder, timber, livelihood, etc. Some authors also claim that we need to learn to "cohabitate" with these IS in cities (Gaertner et al., 2016).

The current chapter focuses on how urbanization has contributed to the introduction and spread of IAPS in urban areas. It discusses the adverse impacts as well as benefits of IAPS on the environment, ecosystem health and services and livelihood. It further highlights the management strategies and various case studies for management methods opted to control the spread of these IAPS.

2 Plant Invasion in Urban Areas: Conquering the New Area

2.1 Introduction of IAPS in Urban Areas

IAPS are introduced to urban areas in various ways that include intentional introduction or inadvertent introduction of plants. The intentional introduction of these plant species are done normally by humans to improve the visual appeal of an area (Turbelin et al., 2017). For example, *Eichhornia crassipes,* commonly known as Water Hyacinth, a free-floating aquatic plant, reported to be native to tropical South America, was brought in India during the British colonial rule as an ornamental plant. However, it turned invasive and has led to water-clogging problem in Sundarbans (Bengal) and also has negatively affected the potable drinking water (Rodríguez-Lara et al., 2022).

The accidental or inadvertent introduction of IAPS is done by many routes. The Global Shipping industry has affected the transport of the seeds of these IAPS. They attach to the ballast water discharge and shipment hulls and are then released in the port areas. An example of this is the unintentional spread of *Cirsium palustre* (Dobson, 1977). Transport linkages like airports and harbors aid in the introduction and expansion of IAPS via dispersal pathways like tourism, trade and horticulture. Many urban areas are located along the waterways that form passages for the dispersion of IS (Gaertner et al., 2016). Therefore, urban areas are cited as "immigration beachheads" for various IS (Pyšek et al., 2010).

The introduction of IAPS is also facilitated by anthropogenic activities like construction and land development in urban areas. These activities cause disturbances in the soil, making it ideal for the growth and establishment of IAPS. Another aspect of these disturbances are the inadvertent movement of the soil, which may contain seeds, rhizomes, and other vegetative parts. Their transport to similar new locations may aid in their establishment in other areas. For example, *Fallopia japonica* (Japanese knotweed) is reported to establish itself in disturbed areas like roadsides as well as riparian forests in the invaded areas of Europe and North America (Beerling et al., 1994). They are also usually present on the riverbanks, abandoned areas, and at construction sites (Smerdu et al., 2020). Most urban ecosystems have effective arrangements of seed dispersal, pollinators, and other facilitators that help in the introduction and spread of many IS (Gaertner et al., 2016). Moreover, the urban areas are characterized by high density and a lot of human activities, including the movement of commodities. Hence, immigration forms another major reason for the introduction of IS in urban areas.

2.2 Dynamics Favouring the Introduction of IAPS in Urban Areas

In urban habitats, there is an amalgamation of changed environmental conditions. They display distinctive characteristics as a result of a change in the land use pattern that significantly impact the urban biotic communities through several anthropogenic effects. The urban environment, built-up surface, and unsettling physical surroundings, ecosystem dynamics, and organism ecology alter natural landscapes, forests, and agricultural areas (Gaertner et al., 2016; Kowarik, 2011). There are primarily three factors that are regarded to support the IAPS introduction into the new areas:

(1) An urban habitat exhibits an exclusive blend of environmental conditions that usually do not occur in natural habitats. There is an alteration in soil structure, nutrient dynamics, regional climates, and hydrological regimes (in urban waterways). These conditions favour the spread of IAPS, especially those that are pre-adapted and flexible to these conditions (Gaertner et al., 2016).
(2) Urban areas are places of reduced competition, thereby offering abundant opportunities to non-native species. For example, plants are usually encircled by large-bodied competitors that permit very little light to reach the ground. However, in the urban areas, these competitors are absent and hence ground is open that allows the light to reach the soil surfaces. These conditions support the growth and abundance of weedy species which is generally excluded from the integral communities (Knapp et al., 2012).
(3) Another reason for the introduction and expansion of IAPS in the urban areas is immigration and movement of commodities among cities (by various modes like ships/trucks/aircraft etc.). These serve as means to import and introduce plants from one place to another, thus making cities "beachhead" for the biological invasions (Pyšek et al., 2010).

2.3 *Basic Mechanisms Involved in the Spread of IAPS*

Disturbances as a result of human activities have played a main role in the world-wide growth of IAPS, and also in shaping the mechanisms involved in invasion. Various hypotheses have been proposed that help in the successful invasion of IAPS in their new areas like enemy release, empty niche and novel weapon hypothesis. The IAPS become more efficacious in their new environment as they are now away from their natural enemies, like herbivores and pathogens, that are otherwise found in their natural habitats (Rai, 2015; Rai & Singh, 2020). For instance, seeds of IAPS *Impatiens glandulifera* are free from the fungal pathogens in their newly invaded area (Najberek et al., 2018). Another is allelopathy, by which invasive plants (IPs) release allelochemicals that act as a novel weapon that represses the NPS (native plant species) and paves the way for IAPS colonization in their new environment (Pinzone et al., 2018). Propagule pressure is another strategy that assists in the invasion. For example, in *Ardisia elliptica,* this has helped in its invasion in South Florida (Rai, 2015). Zheng et al., (2015a, 2015b) reported that *Ageratina adenophora* tends to distribute additional energy for resource utilization and growth to compete with the NS. In some of the IS like *Melaleuca ericifolia, Phragmites australis* and *Rumex conglomeratus*, resource competition and allelopathy act together for successful invasion (Uddin & Robinson, 2017). Also, plant–microbe or insect interactions (antagonist and mutualist) are considered essential for the invasion and spread of IAPS (Jack et al., 2017). Another is nutrient enhancement in both aquatic as well as terrestrial ecosystems that play a substantial role in the effective invasion of IAPS in their new habitats (Uddin & Robinson, 2018). For instance, an increased amount of nitrogen in the soil is known to aid the *Bromus tectorum,* commonly known as annual cheat grass to compete with the NS (Morris et al., 2016). The trade-in ornamental plant species has also assisted in the introduction of IAPS into urban areas. Many of these species remain unnoticed and unregulated and pose some serious threats to humans. A major example of this is the invasion of pines and wattles that has inadvertently increased the intensity of wildfires near the residential areas (van Wilgen & Scott, 2001).

2.4 *Deciphering the Dynamics Determining the Victory of IAPS in Urban Areas*

Urban environment leads to the creation of new ecological dynamics, arrangements and processes. It is also important to understand what leads to the assembly of communities and their changes over the period of time. There are three major factors that govern the assembly of communities: a. conditions of the local site that comprises history and local environmental surroundings; b. accessibility and species selection that will apparently inhabit the local environment and c. performance of species (Williams et al., 2009).

The local site environment includes local climatic environment, like factors and disturbances, histories associated with pollution or contamination, and resource accessibility that has an enduring effect on the forms of species that can survive and persist in the newly invaded habitat. The state of these sites also governs the forthcoming assembly factors by defining the species selection and their performance. Therefore, the local environment is assumed to be important for the successful invasion in the urban areas. It is further suggested that species in their ideal climatic situations are most expected to be affected by different interactions, particularly the inter-specific competition with the other species (Cadotte & Tucker, 2017).

Another factor is the selection of species that decides which species will be able to endure at the local site. This phase displays the relations between their trait and physical surroundings (Cadotte et al., 2017). Another factor is species performance that ensures how IAPS subsists in the local habitat, and this takes account of the consequences of the local species interactions. The comparison between the environmental conditions are not the only factor that guarantees the success of IAPS in the new areas. Another very essential factor is the reproductive success of the species relative to the other opponents and the degree of niche overlap between them (MacDougall et al., 2009). This means that if IAPS has a high niche overlap with the NS, then there will be significant competition and the species with the larger average population growth rate will outcompete the other. It is also reported that urban stress can lead to a decrease in the birth rate, growth rate and high mortality (Cadotte et al., 2017).

IAPS are well known to invade urban water bodies as well. The water bodies in urban landscapes are illustrated in the form of small ponds and lakes. These water bodies are generally dirty and have poor water quality because of high anthropogenic activities in the urban areas. The urban streams have more levels of faecal coliforms and ammonium due to contamination from the sewage (Santana Marques et al., 2020).

Prokopuk and Zub (2020) analysed the uniqueness of alien macrophytes invasions in urban aquatic ecosystems found in Kyiv city (Ukraine). They reported seven species of aquatic invasive species (AIS) that represented 11% of the aquatic flora of the city. The latest invasion representatives, i.e. *Pistia stratiotes* (water lettuce), *Egeria densa* and *Elodea nuttallii* in comparison to *Elodea canadensis* showed broader ecological valence regarding the presence of nitrogenous compounds in water. The non-native macrophyte species dispersed in the city are branded by extensive ecological amplitudes, reproduction rate, robustness to stress, and great amount of naturalization that help them to make use of the resources of the new environment that are unreachable to the NPS (Prokopuk & Zub, 2020).

Humans have purposely introduced IAPS in aquariums, ponds, and garden trades and unconsciously by introducing anglers for swimmers, and there is also a transfer through the boats and ships. This immense dispersal capacity has been kept under observations to keep the IAPS under control and prevent invasion in uninvaded areas (von der Lippe & Kowarik, 2012). Pliszko (2025) reported the first evidence of AIS *Ludwigia repens* and *Myriophyllum aquaticum* cultivated in aquaria in Poland.

The AIS present in the aquatic environments also exhibits damaging effects to the water bodies' ecology, particularly the ecological functions, and also impairs the valued ecological services. For example, *Hydrocotyle ranunculoides* L.f. is a stoloniferous perennial plant that is native to Europe and is commonly used in the aquarium. In 1983, it was found to block some urban drainage networks in Western Australia (Ruiz-Avila & Klemm, 1996). The AIS *E. nuttallii* is observed to cause eutrophication by recycling phosphorus efficiently, and thus exerts a negative effect on the water bodies (Stabenau et al., 2018). In addition to human involvement, the species are also impacted by the rising temperature in urban areas due to climate change. The rising temperature and the heat corridors favour the growth and flourishing of the heat-tolerant IAPS and exotic species. The temperature rise is not tolerated by the local plants and hence they gradually die out (Rai & Singh, 2020).

3 Influence of IAPS on Urban Area: Positive and Negative Impacts

3.1 Adverse Effect on Ecological Services and Environment

The IAPS presents a wide range of impacts on the environment and ecological services. They are documented to affect the ecosystem functioning via three mechanisms: (a) decrease in the diversity of native plants and animal species (b) variations in physical–chemical properties of soil via allelopathy and (c) improvement in environment response to changed fire regimes (Pyšek et al., 2012).

Places like cities are densely populated and already have a low density of green areas or green cover. In this situation, the presence of these IAPS is expected to reduce the local plant diversity as they competitively win over the local plant diversity (Rai & Singh, 2020). Various studies have established that urbanization leads to a direct decrease in the species abundance in extremely urbanized areas. It encourages biological invasions, where many non-native species over the years have become plentiful and commonly visible in urban areas as compared to the NS (Dyderski & Jagodziński, 2016). Recent studies have confirmed that urbanization causes biotic homogenization within the cities, indicating that the extremely urbanized spaces are more alike in their species composition in comparison to less urban areas (Lokatis et al., 2023).

IAPS brings about a change in the soil properties through the process of allelopathy (Qu et al., 2021). Allelopathy results in biotic interference due to the release of allelochemicals by the plants (Latif et al., 2017). These allelochemicals effectually suppresses the NPS, allowing their species to grow and spread in the newly invaded area. There are studies that state that IAPS deleteriously affect the nutrients of soil, its microbial communities, and subsequently affect the agricultural crops. They also affect the nutrient cycling of soil by changing its microbial population via allelopathy (Afreen et al., 2023). Studies indicate that the presence of IPs has a tendency to promote the activity of N and P metabolizing enzymes. Moreover, plant invasion substantially supports the bacterial and fungal diversity and hence has determining impact on soil processes. *Fallopia japonica*, an IP present in the United Kingdom, releases certain types of allelochemicals that considerably alter the food webs of plants that surround it (Abgrall et al., 2018). Another is *Chromolaena odorata* which releases odoratin, an allelochemical known to confer the ability to provide protection against competitors, particularly soil-borne pathogens, and thus, offers a competitive edge to IP over the NPS (Zheng et al., 2015a, 2015b).

These IAPS spread rapidly and there are recorded improvements in the ecosystems' response towards altered fire regimes. This unprecedented spread generally leads to huge socio-economic losses in the terrestrial semi-urban regions (towns or some developing regions) (Pejchar & Mooney, 2009).

Eutrophication of aquatic bodies in urban areas also contributes to the introduction of IAPS in urban areas (Pyšek et al., 2012). This invasion of the aquatic ecosystems occurs as these IAPS display traits like high biomass, high transpiration, and a deep root system. These aquatic IAPS are reported to impede the water flow, making this stagnant water a breeding ground for many diseases and also makes the water unhealthy for irrigation and drinking purposes. IAPS affects the quantity of both ground and surface water, increases the frequency of floods and decreases the water-holding capacities of soil (Pejchar & Mooney, 2009; Shackleton et al., 2019). *E. crassipes* affects the water bodies by disrupting the water flow and also reduces the oxygen content of the water (Hussner et al., 2017). Lizarralde (1993) stated that *Castor canadensis* (beavers' plant) disrupts the water quality and also increases the chances for flood risks. *Prosopis pallida*, a nitrogen-fixing IP found in arid areas of Hawaii Island uses the groundwater to such a level that it modifies the environment of the soil (Rai & Singh, 2020). Some of the IAPS use huge amounts of water that can even lead to water scarcity, and thus, bring change in socio-ecological systems (Gaertner et al., 2016; Shackleton et al., 2019).

IAPS has a significant impact on the regional soil quality as well. They affect the soil stability leading to soil erosion in that area (Pejchar & Mooney, 2009). *Acacia dealbata,* an IAPS of the Mediterranean region, diminishes the NPS by changing the soil chemistry and microbial activity (Lazzaro et al., 2014).

IAPS also disrupts biological and biochemical processes like nutrient cycling. Some of the IAPS have the ability to hold and retain nutrients to a very high capacity, thus making these nutrients indirectly unavailable for other plant species. *P. australis* is reported to invade the wetland areas and alter nutrient cycling by changing the nutrient availability in the soil. It takes up nitrogen and phosphorus thereby affecting the microbial communities in the soil (Findlay et al., 2003). IAPS also alters the decomposition rates by producing leaves and litter with different bio-chemical compositions in comparison to NPS. This affects the decomposition of organic matter and subsequently disrupts the release of nutrients into the soil. *Lantana camara* has been reported to alter the decomposition rate which affects soil fertility by disturbing the carbon and nitrogen cycle of the soil (Sharma & Raghubanshi, 2009). *F. japonica* is reported to affect soil microbial communities and the growth of NPS (Abgrall et al., 2018). IAPS also affects the soil chemistry by fixing the atmospheric nitrogen and enriching the soil with excessive nitrogen. For example, the vine, *Pueraria montana* (Kudzu), is reported to double the release of nitric oxide and also cause an increase in ozone pollution (Hickman et al., 2010). IAPS modifies the hydrological cycles by altering the water flow, retention and water availability of the area they invade. These changes affect the moisture level that is required for the growth of the NPS.

IAPS causes destruction to infrastructure like railroads, roads, ports, canal networks, harbours, disturbed areas, sites near newly constructed roads, crevices of sidewalks, cracks walls of old buildings, green parks, abandoned urban sites (Booy et al., 2017; Danin, 2000). For example, *Ailanthus altissima* (Tree of Heaven) roots can penetrate and damage the asphalt cover of the roads, leading to structural weakness (Danin, 2000). It grows in urban habitats, particularly in abandoned areas, edges of roads, alleys, and up to 100 m in altitude (Sladonja et al., 2015). It grows along the roadsides, thus posing safety hazards by blocking the views of the drivers, and the roots also break the asphalt surfaces of the roads (Burch & Zedaker, 2003). *E. crassipes* is reported to hinder transport along canals and other water bodies in Africa (Julien et al., 2006). Roots of *Triadica sebifera* (Chinese tallow) move their way through these structures (Pile et al., 2017). *I. glandulifera* (Himalayan balsam) are reported to clog the storm water drainage channel leading to unnecessary flooding in that area (Coakley & Petti, 2021).

IAPS also affects wildlife, for example in the wetland of China, *Spartina alterniflora* is reported to replace the native macrophytes—*P. australis* and *Scirpus mariqueter*, which further leads to the decline of avian populations because of restrictions in movement and feeding (Gan et al., 2009). The impact on urban wildlife causes alternation in the food webs of the invaded areas that are brought by mainly two ways. The IAPS may disrupt the already existing mutualistic relationships between the different NPS and the urban wildlife of that area. One way is, IAPS outcompetes the NS and takes over the relationships between the plant and the pollinator, the seed dispersers, and herbivores that feed on these plants (Rai & Singh, 2020). Another way of affecting the urban wildlife is by providing lower quality food (the plant itself) as compared to the NPS, which disturbs the nutritional habits of the native wildlife. For example, *Ailanthus altissima* (Tree of Heaven), an IS of Central Europe interrupts the food webs by affecting the insects, herbivores, and their associated predators in the urban ecosystems (Sladonja et al., 2015). Another is *Rhamnus cathartica* (Common buckthorn), an IS found in the Midwestern United States that alters the interactions between the plants and the insects, herbivores and their associated predators (Knight et al., 2007). Blossey and Notzold (1995) studied the ecological mechanisms used by the IPS that permit them to easily penetrate the local wetland ecosystems, particularly the urban wetlands. They hypothesized that *Lythrum salicaria* (Purple loosestrife) possesses certain traits like rapid growth, efficient resource utilization, and allelopathic properties (as discussed above) that confer them with competitive advantage over the NPS. It was documented that this plant modifies the natural ecosystem of the wetlands and overpopulates the area. The plant forms dense strands and outcompete the NPS over space, sunlight, nutrients, and water. Their profuse growth suppresses the growth of NPS, leading to a change in the flora composition of the urbanized wetland area. This further affects the habitat and food availability of the various organisms, which in turn impacts its local wildlife (Blossey & Notzold, 1995). Another is Japanese knotweed that forms dense mats and outcompete the NPS for resources. It also affects the food availability for invertebrates by eliminating the NPS which serve as the food source for some of the invertebrates. Thus, it has a negative impact on the invertebrate populations (Gerber et al., 2008). The dominance of Japanese knotweed also suppresses the other plant species leading to a decline in plant diversity. This consequently affects nutrient cycling, decomposition rates, and other ecosystem processes as well (Aguilera et al., 2010).

Many IAPS are well documented to affect the various ES like regulatory, recreational, aesthetic, and cultural services. IAPS blocks water navigation, thus adversely impacting tourism and recreation services. They also affect regulatory services like pollination, pest management, water treatment, hazard mitigation, climate change, etc. (Colautti et al., 2006; Pejchar & Mooney, 2009).

3.2 Impact of IAPS on the Economy

IAPS are reported to affect the economy in various ways. The maximum affected sectors include agriculture, aquaculture, fisheries, forestry, nature conservation, and health sectors (Pyšek & Richardson, 2010). Long-term impacts of IAPS affect a wide range of ES that underpin human well-being. It includes the provisioning services like food and fibre, regulation of the spread of human diseases and providing tourism, recreational, and visual benefits (Rai & Singh, 2020). IAPS reduces global agriculture productivity and forest diversity. The majority of the expenditures generated by invaders are the management costs involved in the extinction, control, monitoring of IAPS, and on eco-friendly education programs. In South Africa, species-specific costs of management is available for the national Working for Water program (Gaertner et al., 2016). It is reported that for 2002–2003, 57% of funds out of the total of US$48 million was disbursed on the clearance of invasive trees that were targeted because of their influence on surface water-runoff. Further, a large amount was reported to be spent on the clearance of species like *L. camara, C. odorata* and *Opuntia* spp. (Pyšek & Richardson, 2010). It was reported that in Europe, the entire cost on IS management was assessed to be a minimum of over €20 billion/year (Pyšek & Richardson, 2010). *E. crassipes* invasion in Lake Victoria has caused a decline in the production of fish, thus has imposed a big problem for the fishermen. This has further led to the loss of some important species of fish that were a major attraction for tourists (Pejchar & Mooney, 2009). The invasion of *Opuntia stricta* in the African region is reported to have an antagonistic effect on the environment as well as the economy. It has adversely impacted the livelihood of the locals via the reduction in fodder and the health of livestock (Shackleton et al., 2017).

The short-term economic impacts include IAPS that offer economic returns in some sectors, like invasive alien trees with fast growth, for use in commercial forestry or to meet the demand for garden plants and exotic products. However, studies indicate that the invasion-related costs for some of the species having commercial importance may be more in comparison to their beneficial effects. For example, *P. australis,* whose eradication from the managed wetland systems leads to the loss of capital and also resulted in increased emission of NH_4^+, causing a reduction in the denitrification process (Alldred & Baines, 2016).

Another example is of Cape Town, where IAPS, like pines (species of *Pinus*) grown in plantations and Australian Wattles (species *Acacia*) planted along the coast for the stabilization of dunes has expanded extensively. It has been further stated that 70% of funds for IS are used to control terrestrial IPs like *A. mearnsii, Acacia cyclops, Acacia saligna, Leptospermum laevigatum, Pinus halepensis, Pinus pinaster,* and *Pinus radiata.* The bulk of the fund is spent on AIS, like Water hyacinth and *Nymphaea mexicana* (Gaertner et al., 2016). Kotzé et al. (2025), based on the change in the coverage area of IS in South Africa during the period of 2008 and 2023, the results showed that attempts to control IS have not been able to manage the spread of IS at the national level other than for a few of those by biological control methods. Thus, it was suggested the utilization of limited funds should be done in priority areas, and greater importance should be given to expanding the application of biological control methods. Tabular representation of IAPS and their impact have been provided in the Table 1.

3.3 *Impact of IAPS on Human Health*

Some IAPS negatively impact human health by disturbing the natural processes involved in regulating the spread of diseases (Pyšek & Richardson, 2010). IAPS act as carriers for existing pathogens or even introduce entirely new ones, as evident with the fast rise in vector-borne illnesses (Clow, 2017). *L. camara* is a home for tse-tse fly (*Glossina* spp.), a sleeping sickness pathogen (Leak, 1999). In New Zealand, Brushtail possum aids in the transmission of bovine tuberculosis to deer and livestock, thus indirectly affecting human health via the food chain (Clout, 1999). *P. hysterophorus* aids as a vector for Malaria (Stone et al., 2018). *Ixodes scapularis* for *Borrelia burgdorferi* (Lyme disease in humans) (Clow, 2017). *P. australis* and *Typha* contribute to the establishment and reproduction of vector-borne pathogens, mainly West Nile virus (Mackay et al., 2016). *E. crassipes* aids in the spread of schistosomiasis (Stone et al., 2018). *Arundo donax* (Giant reed) is also reported to have a negative effect on health (Plaza et al., 2018). Trade of these plants can expedite the spread of vectors causing disease, thus increasing the risks for vector-borne diseases and human health (Mazza et al., 2014; Stone et al., 2018). Some of the IAPS is reported to cause allergy in humans, e.g., *Acer, Acacia, Ambrosia artemisiifolia, A. altissima, Casuarina, Eucalyptus, P. hysterophorus, Platanus, Helianthus,* and *Xanthium* (Mazza et al., 2014; Stone et al., 2018).

Table 1 List of IAPS and their impact on ecosystems and its services, biodiversity, environment, economy and human health

S. No	Species	Common name	Native place	Invasion area	Impacts	References
1	*Acacia dealbata*	Silver Wattle	Southeastern Australia in New South Wales, Victoria, Tasmania and the Australian Capital Territory	The Mediterranean, warm temperate and highland tropical landscapes	Its population emerges alongside the agricultural crops and impairs the germination as well as the growth of the main cultivated food species	Raposo et al. (2021)
2	*Ailanthus altissima*	Tree of heaven	Southeast Asia	North America and Europe	Mainly limited to urban areas, disrupts community structures, alters soil properties, affects the structure of the local vegetation and stability of the local ecosystem, nesting sites for birds and enhances the urban bird diversity. Causes damage to infrastructure and archaeological remains with its extensive roots, roots penetrate the walls and sewers and breaks the asphalt surface, and causes health issues (allergic reactions, respiratory problems and skin rashes)	Danin (2000), Gerber et al. (2008), Sladonja et al. (2015)
3	*Eichhornia crassipes*	Water hyacinth	South America	India (late 1700s)	Alters water clarity, decreases phytoplankton production, reduces the amount of dissolved oxygen, nitrogen, phosphorous, heavy metals and other contaminants, increases the incidence of infectious diseases, affects fisheries and other associated commercial activities, affects working of irrigation canals and hydroelectric programs, damages turbines in hydro-power plants, involves huge expenditure in clearing, and negatively affects navigation, recreation and tourism	Malik (2007), Villamagna and Murphy (2010), Nguyen et al. (2015), Degaga (2018)

(continued)

Table 1 (continued)

S. No	Species	Common name	Native place	Invasion area	Impacts	References
4	*Elodea nuttallii*	Western waterweed	Europe	Europe, Asia and Australia	Negatively affects water-based programs (generation of energy, conservation of nature and recreation), blocks the inflow units of hydroelectric power plants, and positively impacts the density and richness of invertebrates in the urban lakes	Zehnsdorf et al. (2015), Tan et al. (2024)
5	*Fallopia japonica*	Japanese knotweed	East Asia in Japan, China and Korea	North America and Europe	Reduces the local species diversity, slows down the nutrient cycling, and reduces the light availability to the NPs	Aguilera et al. (2010)
6	*Hedera helix*	English Ivy	Europe and Western Asia	North America	Prevents germination and re-sprouting of some of the NS via allelopathy, acts as a wildlife resource in urban and rural ecosystems, and provides pollen and nectar for insects	Jiménez-Hernández et al. (2023), Lam et al. (2024)
7	*Hydrilla verticillata*	Water thyme	Warmer Parts of Asia	United States of America (1960)	Causes financial hardships, obstructs various water uses, dislodges native aquatic plant communities and negatively affects the freshwater habitats, blocks urban drainage and irrigation canals, and increases mosquito-breeding grounds leading to higher disease risk, and raises the cost of water body maintenance	Langeland (1996)
8	*Hydrocotyle ranunculoides* L.f	Floating pennywort	Europe	Western Australia (1983)	Blocks the urban drainage network, disrupts water ecology, and affects recreation	Ruiz-Avila and Klemm (1996)
9	*Impatiens glandulifera*	Himalayan balsam	The foothills of the Himalayas in India and Pakistan	Europe, parts of North America and New Zealand	Clogs storm water drainage channels in urban areas leading to unnecessary flooding, alters the floristic composition of invaded ecosystems, negatively affects the invertebrates, as it is unpalatable to them, and leads to high soil erosion rates in the riparian areas of Switzerland and UK	Greenwood et al. (2018), Coakley and Petti (2021)

(continued)

Table 1 (continued)

S. No	Species	Common name	Native place	Invasion area	Impacts	References
10	*Lantana camara*	Big sage	Central and South America	Shivalik Hills in the North, the fragmented deciduous forests of Central India, and the Southern Western Ghats	Replaces grasses (primary fuel for forest fires) and thus changes the degree of spread of fire that occurs in the forest, causing loss and destruction of crops, and showing the conditioning capacity of air pollution in urban forests of Aizawl, Mizoram, and Northeast India	Rai and Singh (2015), Mungi et al. (2020), Singh et al. (2023)
11	*Lemna minuta*	Duckweed	Central Europe	Greece, Southern Russia, and Japan	Forms dense mats on the water surface, thus decreasing the amount of dissolved oxygen and sunlight, and severely affects the conservation of the underlying aquatic ecosystem	Ceschin et al. (2019)
12	*Lythrum salicaria*	Purple loosestrife	Europe, Temperate Asia and Northwest Africa	North America, Late 1800s	Suppresses growth of NPS, alters flora composition of urbanized wetland area, and decreases biomass of other species	Farnsworth and Ellis (2001)
13	*Nymphaea mexicana*	Water lily	Southern USA and Mexico	Australia, Europe, India, New Zealand, Spain and South Africa	Disrupts biodiversity, reduces water quality and gas exchange capacity, restricts water movement, decreases recreational value of water bodies, blocks sunlight, affects the dissolved oxygen level, leads to a decline in the fish population in urban ponds, and promotes foul smell, thus reducing water quality and increasing water treatment costs	Reid et al. (2021)

(continued)

Table 1 (continued)

S. No	Species	Common name	Native place	Invasion area	Impacts	References
14	*Opuntia stricta*	Common Prickly Pear	Americas	Africa, Australia and the Mediterranean Basin	Restricts water movement and decreases the recreational value of water bodies	Shackleton et al. (2017)
15	*Prosopis pallida*	Bayahonda	Peru, Colombia and Ecuador	Hawaii, Puerto Rico and South Africa	Reduces the density of native tree species and the cover of native herbaceous species, and causes increased mortality of native trees	Shackleton et al. (2015)
16	*Pueraria montana*	Kudzu	East Asia, primarily subtropical and temperate regions of China, Japan, and Korea	United States of America	Affects soil chemistry, doubles nitric oxide emissions and increases ozone pollution, negatively affects air quality, and modifies the nitrogen cycle	Lindgren et al. (2013)
17	*Rhamnus cathartica*	Common buckthorn	Europe, Northwest Africa and Western Asia, from the Central British Isles South to Morocco, and East to Kyrgyzstan	North America, Late 1700s	Forms dense mono-specific thickets over large areas, alters the relations between plants, insects, herbivores and their associated predators, alters the organic matter, and enhances the upper layer beneath it in the urban woodlands dominated with its growth	Heneghan et al. (2004), Knight et al. (2007)

(continued)

Table 1 (continued)

S. No	Species	Common name	Native place	Invasion area	Impacts	References
18	*Salvinia molesta*	Giant salvinia	Brazil	Tropical and Subtropical parts of the world, (early 1900s)	Forms dense mats on the surface of water and thus blocking sunlight, impedes access to water bodies and degrades the resource quality of aquatic ecosystem and urban lakes, prevents boat passage, restricts access to fishing or waterfowl hunting, and negatively impacts recreational activities in an urban lake	Coetzee and Hill (2020), Wahl et al. (2020)
19	*Trapa natans*	Water chestnuts	Europe and Asia	Massachusetts, United States of America, (late 1800s)	Causes overloading in the littoral zones of lakes, acts as a host for other IPS, impedes water quality, destroys the habitat of submerged macrophytes, reduces public use of water bodies due to clogging, and acts as a hazard to swimmers	Hummel and Kiviat (2004), Yuan et al. (2021)
20	*Triadica sebifera*	Chinese tallow	Eastern China	First introduced to South Carolina (1700s) and now found from Eastern North Carolina to Florida	Displaces native vegetation and modifies soil conditions due to the presence of high amount of tannins in the leaf litter	Cotten et al. (2012)

3.4 Beneficial Impact of IAPS: Not All Are Dangerous

There are reports that not all the IAPS pose threats to the environment (Rai & Singh, 2020). About 99% of the selected IPs are used as food crops globally, and studies also indicate their positive impact on the ecosystem, its services, livelihood, economy, etc. (Pejchar & Mooney, 2009).

These plants, under certain conditions, might add to the urban biodiversity by making up for vegetation at places where there is a loss as a result of urbanization. *Buddleja davidii,* despite being invasive, actually adds to the urban ecosystem by serving as a source of nectar for the native butterflies and other pollinators, thus positively contributes to their habitat (Dennis et al., 2006). *A. altissima* provides nesting sites for birds, and enhances the urban bird diversity by filling the vacant niches in these transformed urban environments (Gerber et al., 2008). *Eucalyptus* trees are used as roosting sites for *Falco naumannii* and as breeding sites for *Haliaeetus vocifer* (African Fish Eagle) in the northern region of South Africa (Cilliers & Siebert, 2012).

These IAPS are useful in binding the soil particles together with their deep root penetration, thus helps in reducing soil erosion. Japanese knotweed and Tree of heaven display extensive root spread that helps them in water absorption and decreasing the surface run-off of water (Swearingen et al., 2002). Thus, allowing the water to infiltrate the soil and prevent soil erosion. Another way in which the IPs positively affect the ecosystems by covering and stabilizing the slopes in the hilly regions by allowing their roots to penetrate and anchor the soil, thereby averting soil erosion. Examples are *P. montana* and *Elaeagnus angustifolia* (Bagarello, 2017). Another very important role of the IAPS is in bioremediation. The IPs bear the ability to absorb pollutants or contaminants from the soil, and thus, it has a role in phytoremediation. These plants can absorb the heavy metals and store them in their vacuoles and thus aids in soil detoxification (Rai & Singh, 2020). For example, *P. australis* can remediate benzimidazole anthelmintics from the environment (Podlipná et al., 2013) and *P. stratiotes* can accumulate silver nanoparticles (Hanks et al., 2015). *P. arundinacea* and *Chrysopogon zizanioides* help in the absorption of organic pollutants like hydrocarbons and pesticides from the soil (Patel & Patra, 2017). *P. australis* and *F. japonica* are helpful in the uptake of heavy metals present in the soil, and thus, can be used in bioremediation. *Bidens pilosa, C. odorata,* and *Praxelis clematidea* are reported to be hyper-accumulators for cadmium (Wei et al., 2018). *Prosopis juliflora* is a multi-purpose IP that can accumulate fluoride in combination with iron oxide nanoparticles (Kumari & Khan, 2018). *A. donax, E. crassipes, L. camara,* and *P. stratiotes* also impart a role in phytoremediation (Prabakaran et al., 2019; Rai & Kim, 2020).

Besides their role in phytoremediation, IPs also serve as a source for animal feed, bio-energy, bio-polymers, and in boosting the green economy (Rai & Kim, 2020). IPs also play an important role in air filtration, pollution control, microclimatic regulation, management of noise pollution, etc. (Potgieter et al., 2017). *S. alterniflora* can be used for heavy metals phytoremediation and in carbon sequestration in the bioenergy industry (Prabakaran et al., 2019). *Acer platanoides* helps in CO_2 sequestration, and thus, can be employed in climate change mitigation (Potgieter et al., 2017, 2019). IPs like *F. japonica, Heracleum mantegazzianum, I. glandulifera,* and *Solidago gigantea* produce enormous biomass with high calorific value and thus can be used for bioenergy production (Van Meerbeek et al., 2019). *E. nuttallii* efficiently recycles phosphorus that causes nutrient enrichment and consequently affects the aquatic systems. However, its biomass can be used for biogas production and compost formation (P-rich) (Stabenau et al., 2018).

Melaleuca quinquenervia plays a positive role in honey production in Florida (Serbesoff-King, 2003). *Prosopis glandulosa* is considered valuable as its wood is used and sold as fuel wood in numerous countries of Southeast Asia and Africa (Shackleton et al., 2019). Australian Acacia (in Africa and Madagascar) is also used as fuel wood, and thus, serves as an important source of livelihood for people with low-income (Shackleton et al., 2019). *P. juliflora* and *Opuntia ficus-indica* are also used as food and fodder in their invaded areas (Shackleton et al., 2019). *A. dealbata* in the Eastern Cape region of South Africa serve as a main source of livelihood (Shackleton et al., 2019). *Cenchrus ciliaris* is used by farmers for grazing purposes (Shackleton et al., 2019). In India, *L. camara* serves as a livelihood source for numerous local villagers as they are used in furniture and pulp making (Kannan et al., 2016).

Robinia pseudoacacia attracts more insect pollinators in comparison to the NP *Cytisus scoparius* (Buchholz & Kowarik, 2019). Similarly, *Parthenium hysterophorus* also attracts more bee pollinators in comparison to the NPS (Ojija et al., 2019).

These applications of IAPS associated with ES, energy production, bioremediation, and livelihood can be ensured by their sustainable management. As many of the IPs serve as valuable sources for animal feed, bio-energy and bio-polymers, therefore these can be used in the expansion of the green economy. Hence, efforts should be made to the eco-technological prospects of these IPs by exploiting their positive benefit (Rai & Kim, 2020).

A diagrammatic representation of both the positive and negative impacts of IAPS on the ecosystem and economy has been depicted in Fig. 1.

4 Management and Control of IAPS

Even though IPs offer some benefits, there is a need for management and control of IAPS. This has been an area of debate among many researchers worldwide (Potgieter et al., 2017). For example, even though IPs like *E. crassipes, L. camara, P. stratiotes* and *A. donax* are considered to be in the list of top 100 IS, they are also reported to be useful for the enhancement of vegetation through phytoremediation (Prabakaran et al., 2019; Rai & Kim, 2020). Kaushik et al. (2025) suggested that control of IAPS, like *P. juliflora,* is needed for the areas with extended hot summers and a shortage of freshwater, like tropical Asia, tropical Africa, and the Middle East, where it has

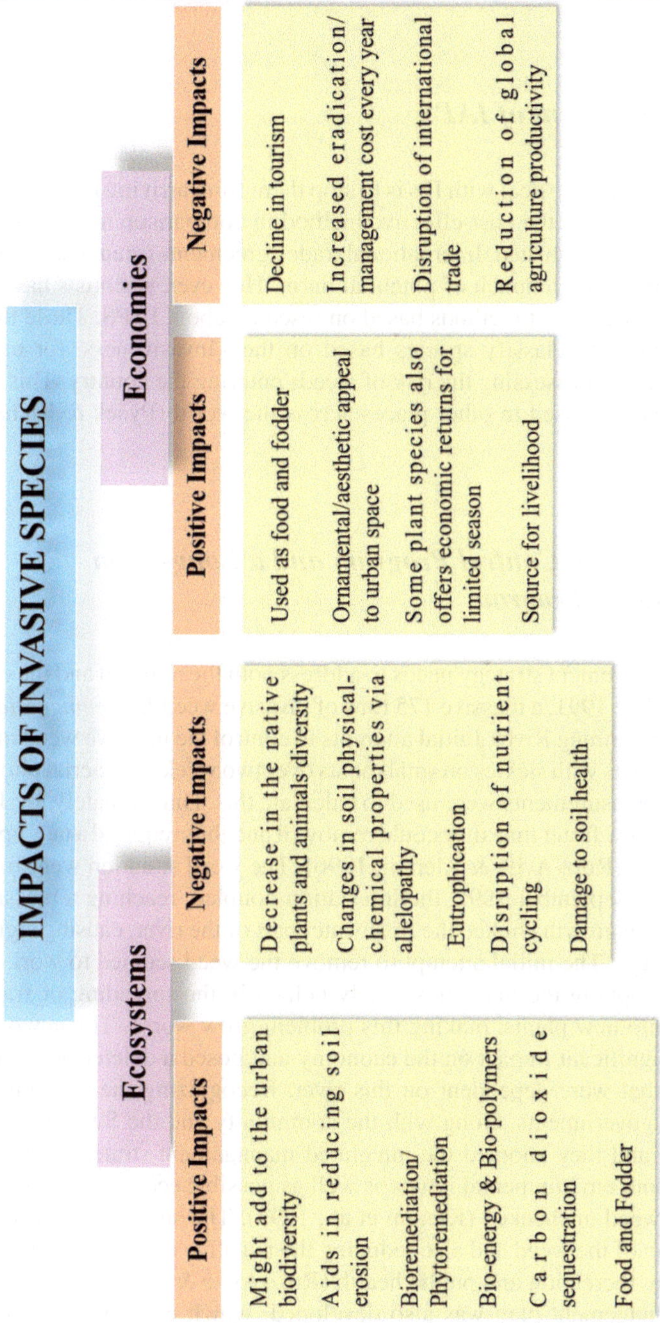

Fig. 1 Diagrammatic representation of both positive and negative impacts of IAPS on ecosystems and economies

invaded large areas as they show high transpiration that leads to lowering the soil moisture level.

4.1 Risk Assessment of IAPS

The most effective way to deal with IPs is to stop them from arriving in the first place. This is considered to be the cost-effective method in comparison to controlling the IAPS after they get established. International trade agreements often make it difficult to ban species based on a hunch of potential harm. However, scientists have developed better risk assessment methods based on research about IAPS. These methods use large databases to classify species based on their invasiveness, for example, Australia's system for assessing the risk of weeds entering the country. This system has been adapted and used in other places across the world (Pyšek & Richardson, 2010).

4.2 A Short-Term Control Program and a Long-Term Eradication Program

An effective management strategy needs to address both the removal and the cause of the invasion. In late 1991, a massive 175 tons of invasive weed *H. ranunculoides* was estimated in the Canning River. Initial attempts to control the invasive weed involved cutting floating mats with sickles on small boats over two weeks and aerial photos and weight-to-area measurements were used to calculate this. Unfortunately, by January 1992, the weed grew faster than they could remove it and thus required a new approach to solve this issue (Ruiz-Avila & Klemm, 1996). The weed situation went from bad to worse, and by September 1992, the infestation doubled, reaching a massive 420 tons. This massive growth choked the freshwater part of the river, causing significant ecological damage. The initial attempt to remove the weed seemed to work at first, but it failed. Chopping the mats most likely helped in the spreading of fragments that sprouted into new plants, making this problem grow worse. This uncontrolled growth had a significant impact on the economy and posed a deleterious impact on the industries that were dependent on this river. Recognizing the growing threat, state and local governments, along with the community and the Swan River Trust, came together, and they adopted the integrated management strategy. This helped in addressing both environmental issues as well as possible economic consequences of leaving the weed unchecked (Klemm et al., 1993). The strategy aimed for long-term eradication of the weed and also reducing the negative impact on water quality, river ecosystem, recreation and public health (Ruiz-Avila & Klemm, 1996). A two-component management plan was also developed, which comprised a short-term control program (implemented in early 1993) and a long-term eradication program.

Short-term control methods were dependent mainly on the physical way of removal systems (like those used in 1991) and the use of herbicides that helped in eliminating 2000 tons projected from the number of truckloads used in removing it. The weed was further utilized for making compost by the local council. Ecological control methods were not considered suitable for the short-term control measures and were recommended for the long-term if the eradication measures failed. After the maximum of the weed was removed, herbicide was used to avert the growth of new mats (Ruiz-Avila & Klemm, 1996). Kurose et al. (2024) studied on *Mycosphaerella polygoni-cuspidati* as a biological control to control Japanese knotweed (*Reynoutria japonica* Houtt.) and also suggested the use of exotic pathogens as mycoherbicides to control this Japanese knotweed.

The long-term control measures aimed to eliminate this weed from the river, thereby stopping their spread throughout the state. It also included physical methods of removal and application of herbicides, similar to short-term programs as well as ecological control. Ecological techniques included methods to reduce the nutrient load in the river and removal of nutrient-enriched sediments. Currently, a programme of the integrated catchment has been developed to reduce nutrient loads entering the river via the drainage network (Elbal, 2013; Stone et al., 2018). This case study served as a classic example of the role of public involvement in the control of IAS in the urban environment, as this long-term and short-term control helped in controlling the population of AIS as well as eradicating the weeds from the river.

Another classic example is a study by Gaertner et al. (2016). They investigated the challenges and intricacies of managing AIS in cities by examining the classic case of Cape Town (South Africa). In Cape Town, IAPS were introduced during the European colonization posed a major threat to its native ecosystems. IPs like species of *Pinus* (that are grown in plantations) and species of *Acacia* Australian Wattles (that are grown along the coast) for stabilization of dunes spread extensively beyond their original plantation. AIS, like *E. crassipes* were also found to block waterways and affect the water quality. The South African government-NEMBA (National Environmental Management: Biodiversity Act) imposes the responsibility for control of IS on government bodies and landowners, including Cape Town. In 2008, the Invasive Species Management Unit (budget of approximately $84,000 USD) and one semi-skilled team were established. This program included areas that were managed by several city departments (van Wilgen & Wannenburgh, 2016). It was stated that 70% of the city's budget is for IS, and the IS includes *A. cyclops, A. mearnsii,* and *A. saligna, L. laevigatum, P. halepensis, P. pinaster,* and *P. radiata,* and AIS like *E. crassipes* and *N. mexicana.* They aimed to re-establish the native biodiversity wherever possible, and also to safeguard the functioning of the ecosystem via active and passive restoration. There are several other organizations that are engaged in the control of IS, like South African National Parks, the South African National Biodiversity Institute and the Working for Water programme (Gaertner et al., 2016; Wilson et al., 2013).

The case study of Rahway River (an urban area in New Jersey) serves as another example that highlights the challenges of urban ecology (Aronson et al., 2017). In spite of massive development, the river corridor (riparian zone) offers a vivacious green space for wildlife. The most common IS found were *Artemisia vulgaris, Digitaria ischaemum, Microstegium vimineum* and *Polygonum cuspidatum* that utilized the river for seed dispersal. The research suggested that the plant communities on the riverbank and in the leaf litter are quite different. The study found a startling number of IS along the Rahway that surpassed the NPs on the riverbank, and more non-native seeds were found to germinate in the spring season. These areas have been disturbed in different ways, so the restoration of these area with NPS can help in creating conditions that are less favorable for IS to establish themselves. It was suggested that control of the IPs in the early part of the growing season (March in this case) could be more effective. Moreover, by preventing the non-native seeds from germinating and establishing roots can further help in reducing the long-term impact of these IPs (Aronson et al., 2017).

Some of the clonal perennial IS found in European riparian forests form dense stands that reach up to 3–4 m in height (for example, *Fallopi*a sp.), or 9 m for *A. donax* (Schnitzler et al., 2007). These IPs affect water movement and, consequently, the flooding regimes that are very problematic and costly to manage. Thus, planning and application of management activities for these IS should be directed at the watershed level, which in turn requires an intensive effort of all the impacted municipalities that are found along the watercourse. This will prevent the formation of invasion foci in these upland ranges that would help in eliminating the probable sources for IS propagules and their spread in the downstream region (Aronson et al., 2017).

Geospatial technologies like Remote Sensing and GIS (geographical information system), a satellites and mapping tool is also used tracking and managing the IPS. These tools can identify the types of IPS, where they're spreading, and can even document the best time of year to target them for their removal. This technology serves as an effective tool for controlling the IAPS (Khare et al., 2018, 2019). To control IPs like *H. mantegazzianum* and *F. japonica* in Europe, researchers have used GIS at specific times of the year. This timing helps to identify the plants based on their growth patterns and appearance (Mullerova et al., 2017).

Arguments in support or against the management of IAPS in urban areas rest on the ES offered by IAPS and their (EDS) ecosystem disservices (negative impacts on ES). IAPS that deliver both ES and EDS generally lead to clashes regarding their application and management. Local conditions direct the context of the multifaceted interactions that result between ecological and socio-cultural relations with invasions in urban areas. Researchers, decision-makers and managers are also focusing on another aspect, i.e. human dimensions in regard to management of IS as social dimensions facilitate response to invasions mainly in urban areas (Potgieter et al., 2019).

5 Conclusion and Future Prospects

- Biotic invasions are the major drivers of biodiversity loss and have badly affected the native biodiversity and ES. Anthropogenic disturbances are regarded as the main factors accountable for the invasion in urban areas.
- IAPS are increasing in both number and impact globally because of growing globalization. IAPS negatively impacts the environment, ecosystem health and services, biodiversity and human health.
- The knowledge regarding the mechanisms of their introduction, spread and establishment is suggested to be crucial for the management of IAPS. Therefore, current advances in risk assessment, initial detection, pathway or corridors, vector management, and fast response offer a firm basis. Prevention is suggested to be more effective in comparison to post-invasion restoration methods. Biotechnological innovations for exploiting the particular IAPS may also aid in the sustainable management of IS.
- There is a need to consider the social dimension aspect for the management of IAPS. Thus, analysis of cost–benefit is also considered crucial to safeguard for benefits associated with livelihood.
- Many researchers claim different views regarding IAPS management. Arguments in favour or against its management of urban areas depend on the IS contributions to ES and EDS. IAPS that deliver both ES and EDS usually lead to clashes regarding their usage and management.
- Every urban area represents a distinctive set of challenges, and thus, requires precise management methods. It is also recommended that management needs to focus on creating and sustaining new ecosystems capable of providing important ES, instead of making attempts to restore the degraded ecosystems and make it alien or invasive-free environment.

References

Abgrall, C., Forey, E., Mignot, L., & Chauvat, M. (2018). Invasion by *Fallopia japonica* alters soil food webs through secondary metabolites. *Soil Biology & Biochemistry, 127*, 100–109. https://doi.org/10.1016/j.soilbio.2018.09.016

Afreen, T., Kumari, S., Bhadouria, R., Devi, R. S., Singh, S., & Tripathi, S. (2023). Plant invasion and soil processes: A mechanistic understanding. In S. Tripathi, R. Bhadouria, P. Srivastava, R. Singh, & D. R. Batish (Eds.) *Plant invasions and global climate change* (pp. 227–246). Springer Nature Singapore. https://doi.org/10.1007/978-981-99-5910-5_10

Aguilera, A. G., Alpert, P., Dukes, J. S., & Harrington, R. (2010). Impacts of the invasive plant *Fallopia japonica* (Houtt.) on plant communities and ecosystem processes. *Biological Invasion, 12*, 1243–1252. https://doi.org/10.1007/s10530-009-9543-z

Alldred, M., & Baines, S. B. (2016). Effects of wetland plants on denitrification rates: A meta-analysis. *Ecological Applications, 26*, 676–685. https://doi.org/10.1890/14-1525

Aronson, M. F., Handel, S. N., La Puma, I. P., & Clemants, S. E. (2015). Urbanization promotes non-native woody species and diverse plant assemblages in the New York metropolitan region. *Urban Ecosystem, 18*, 31–45. https://doi.org/10.1007/s11252-014-0382-z

Aronson, M. F., Patel, M. V., O'Neill, K. M., & Ehrenfeld, J. G. (2017). Urban riparian systems function as corridors for both native and invasive plant species. *Biological Invasions, 19*, 3645–3657. https://doi.org/10.1007/s10530-017-1583-1

Bagarello, V. (2017). Effective practices in mitigating soil erosion from fields. *Oxford Research Encyclopaedia Environment*. https://doi.org/10.1093/acrefore/9780199389414.013.242

Beerling, D. J., Bailey, J. P., & Conolly, A. P. (1994). *Fallopia japonica* (Houtt.) Ronse Decraene. *Journal of Ecology, 82*, 959–979.

Blossey, B., & Notzold, R. (1995). Evolution of increased competitive ability in invasive nonindigenous plants: A hypothesis. *Journal of Ecology, 83*, 887–889. https://doi.org/10.2307/2261425

Booy, O, Cornwell, L, Parrott, D, Sutton-Croft, M., & Williams, F. (2017). Impact of biological invasions on infrastructure. In M. Vilà, & P. Hulme, (Eds.) *Impact of biological invasions on ecosystem services. invading nature - springer series in invasion ecology* (vol. 12, pp. 235–247). Springer. https://doi.org/10.1007/978-3-319-45121-3_15

Buchholz, S., & Kowarik, I. (2019) Urbanisation modulates plant-pollinator interactions in invasive vs. native plant species. *Science Reports, 9*, 6375. https://doi.org/10.1038/s41598-019-42884-6

Burch, P. L., & Zedaker, S. M. (2003). Removing the invasive tree Ailanthus altissima and restoring natural cover. *Arboric Journal, 29*,18–24. https://doi.org/10.48044/jauf.2003.003

Cadotte, M. W., & Tucker, C. M. (2017). Should environmental filtering be abandoned? *Trends in Ecology & Evolution, 32*, 429–437. https://doi.org/10.1016/j.tree.2017.03.004

Cadotte, M. W., Yasui, S. L. E., Livingstone, S., & MacIvor, J. S. (2017). Are urban systems beneficial, detrimental, or indifferent for biological invasion? *Biological Invasions, 19*, 3489–3503. https://doi.org/10.1007/s10530-017-1586-y

Ceschin, S., Abati, S., Traversetti, L., Spani, F., Del Grosso, F., & Scalici, M. (2019). Effects of the invasive duckweed *Lemna minuta* on aquatic animals: Evidence from an indoor experiment. *Plant Biosystem, 153*, 749–755. https://doi.org/10.1080/11263504.2018.1549605

Cilliers, S. S., & Siebert, S. J. (2012). Urban ecology in Cape Town: South African comparisons and reflections. *Ecology and Society, 17*, 33. https://doi.org/10.5751/ES-05146-170333

Clout, M. N. (1999) Biodiversity conservation and the management of invasive animals in New Zealand. In O. T. Sandlund, P. J. Schei, A. Viken (Eds.), *Invasive species and biodiversity management*. Based on papers presented at the Norway/United Nations (UN) Conference on Alien Species, 2nd Trondheim Conference on Biodiversity, Trondheim, pp. 349–361, Kluwer Academic Publishers.

Clow, K. M. (2017). The influence of abiotic and biotic factors on the invasion of *Ixodes scapularis* in Ontario, Canada. *Ticks Tick Borne Diseases, 8*, 554–563. https://doi.org/10.1016/j.ttbdis.2017.03.003

Coakley, S., & Petti, C. (2021). Impacts of the invasive *Impatiens glandulifera*: Lessons learned from one of Europe's top invasive species. *Biology, 10*, 619. https://doi.org/10.3390/biology10070619

Coetzee, J. A., & Hill, M. P. (2020). *Salvinia molesta* D. Mitch. (Salviniaceae): Impact and control. *CABI Reviews, 15*, 033. https://doi.org/10.1079/PAVSNNR202015033

Colautti, R. I., Bailey, S. A., Van Overdijk, C. D., Amundsen, K., & MacIsaac, H. J. (2006). Characterized and projected costs of nonindigenous species in Canada. *Biological Invasions, 8*, 45–59. https://doi.org/10.1007/s10530-005-0236-y

Cotten, T. B., Kwiatkowski, M. A., Saenz, D., & Collyer, M. (2012). Effects of an invasive plant, Chinese tallow (*Triadica sebifera*), on development and survival of anuran larvae. *Journal of Herpetology, 2*, 186–193. https://doi.org/10.1670/10-311

Danin, A. (2000). The inclusion of adventive plants in the second edition of Flora Palaestina. *Willdenowia, 30*, 305–314. https://doi.org/10.3372/wi.30.30209

Dennis, R. L., Shreeve, T. G., & Van Dyck, H. (2006). Habitats and resources: The need for a resource-based definition to conserve butterflies. *Biodiversity and Conservation, 15*, 1943–1966. https://doi.org/10.1007/s10531-005-4314-3

Degaga, A. H. (2018). Water hyacinth (*Eichhornia crassipes*) biology and its impacts on ecosystem, biodiversity, economy and human well-being. *Journal of Life Science Biomedicine, 8*, 94–100.

Dobson, A. T. (1977). Adventive plants. History and Science in the Cass District, Canterbury, New Zealand, 271.

Dyderski, M. K., & Jagodziński, A. M. (2016). Patterns of plant invasions at small spatial scale correspond with that at the whole country scale. *Urban Ecosystem, 19*, 983–998. https://doi.org/10.1007/s11252-015-0524-y

Elbal, P. M. A. (2013). Invasive aquatic plants and culicids: A dangerous pairing. Bulletin of the Royal Spanish Society of Natural History. *Biological Section, 107*, 5–15.

Farnsworth, E. J., & Ellis, D. R. (2001). Is purple loosestrife (*Lythrum salicaria*) an invasive threat to freshwater wetlands? Conflicting evidence from several ecological metrics. *Wetlands, 21*, 199–209. https://doi.org/10.1672/0277-5212(2001)021[0199:IPLLSA]2.0.CO;2

Findlay, S., Groffman, P., & Dye, S. (2003). Effects of *Phragmites australis* removal on marsh nutrient cycling. *Wetland Ecology and Management, 11*, 157–165. https://doi.org/10.1023/A:1024255827418

Gaertner, M., Larson, B. M. H., Irlich, U. M., Holmes, P. M., Stafford, L., van Wilgen, B. W., & Richardson, D. M. (2016). Managing invasive species in cities: A framework from Cape Town, South Africa. *Landscape and Urban Planning, 151*, 1–9. https://doi.org/10.1016/j.landurbplan.2016.03.010

Gan, X., Cai, Y., Choi, C., Ma, Z., Chen, J., & Li, B. (2009). Potential impacts of invasive *Spartina alterniflora* on spring bird communities at Chongming Dongtan, a Chinese wetland of international importance. *Estuarine, Coastal and Shelf Science, 83*, 211–218. https://doi.org/10.1016/j.ecss.2009.03.026

Gerber, E., Krebs, C., Murrell, C., Moretti, M., Rocklin, R., & Schaffner, U. (2008). Exotic invasive knotweeds (*Fallopia* spp.) negatively affect native plant and invertebrate assemblages in European riparian habitats. *Biological Conservation, 141*, 646–654. https://doi.org/10.1016/j.biocon.2007.12.009

Greenwood, P., Baumann, P., Pulley, S., & Kuhn, N. J. (2018). The invasive alien plant, *Impatiens glandulifera* (Himalayan Balsam), and increased soil erosion: Causation or association? Case studies from a river system in Switzerland and the UK. *Journal of Soils and Sediments, 18*, 3463–3477. https://doi.org/10.1007/s11368-018-2041-0

Hanks, N. A., Caruso, J. A., & Zhang, P. (2015). Assessing *Pistia stratiotes* for phytoremediation of silver nanoparticles and Ag(I) contaminated waters. *Journal of Environmental Management, 164*, 41–45. https://doi.org/10.1016/j.jenvman.2015.08.026

Heneghan, L., Rauschenberg, C., Fatemi, F., & Workman, M. (2004). European buckthorn (*Rhamnus cathartica*) and its effects on some ecosystem properties in an urban woodland. *Ecological Restoration, 22*, 275–280. https://doi.org/10.3368/er.22.4.275

Hickman, J. E., Wu, S., Mickley, L. J., & Lerdau, M. T. (2010). Kudzu (*Pueraria montana*) invasion doubles emissions of nitric oxide and increases ozone pollution. *Proceedings of the National Academy of Sciences of the United States of America, 107*, 10115–10119. https://doi.org/10.1073/pnas.0912279107

Hummel, M., & Kiviat, E. (2004). Review of world literature on water chestnut with implications for management in North America. *Journal of Aquatic Plant Management, 42*, 17–28.

Hussner, A., Stiers, I., Verhofstad, M. J., Bakker, E. S., Grutters, B. M., Haury, J., Van Valkenburg, J. L., Brundu, G., Newman, J., Clayton, J. S., & Anderson, L. W. (2017). Management and control methods of invasive alien freshwater aquatic plants: A review. *Aquatic Botany, 136*, 112–137. https://doi.org/10.1016/j.aquabot.2016.08.002

Jack, C. N., Friesen, M. L., Hintze, A., & Sheneman, L. (2017). Third-party mutualists have contrasting effects on host invasion under the enemy-release and biotic-resistance hypotheses. *Evolutionary Ecology, 31*, 829–845. https://doi.org/10.1007/s10682-017-9912-5

Jiménez-Hernández, H. A., Bonilla-Valencia, L., Martínez-Orea, Y., Zamora-Almazan, M. J., Espinosa-García, F. J., & Castillo-Argüero, S. (2023). Effects of *Hedera helix* L. removal on the understory early regeneration in an oak temperate forest in Mexico City. *Ecological Processes, 12*, 31. https://doi.org/10.1186/s13717-023-00443-y

Julien, M. H., Julien, M. H., Hill, M. P., Center, T. D., & Jianaing, D. (2006). Biological and integrated control of water hyacinth, *Eichhornia crassipes*. In *Proceedings of the second meeting of the Global Working Group for the Biological and Integrated Control of Water Hyacinth, Australian Centre for International Agricultural Research (ACIAR)*, Beijing, China, October 2000, pp. 9–12.

Kannan, R., Shackleton, C. M., Krishnan, S., & Shaanker, R. U. (2016). Can local use assist in controlling invasive alien species in tropical forests? The case of *Lantana camara* in southern India. *Forest Ecology and Management, 376*, 166–173. https://doi.org/10.1016/j.foreco.2016.06.016

Kaushik, R., Pandit, M. K., & Manish, K. (2025). Higher transpiration in plant invasive species impacts soil water. *Environmental and Sustainability Indicators, 26*, Article 100665. https://doi.org/10.1016/j.indic.2025.100665

Khare, S., Latif, H., & Ghosh, S. K. (2018). Multi-scale assessment of invasive plant species diversity using Pléiades 1A, RapidEye and Landsat-8 data. *Geocarto International, 33*, 681–698. https://doi.org/10.1080/10106049.2017.1289562

Khare, S., Latif, H., & Rossi, S. (2019). Forest beta-diversity analysis by remote sensing: How scale and sensors affect the Rao's Q index. *Ecological Indicators, 106*, Article 105520. https://doi.org/10.1016/j.ecolind.2019.105520

Klemm, V. V., Siemon, N. L., & Ruiz-Avila, R. J. (1993). *Hydrocotyle ranunculoides: A control strategy for the Canning River Regional Park*. Swan River Trust Report No. 6, Perth, Western Australia.

Knapp, S., Dinsmore, L., Fissore, C., Hobbie, S. E., Jakobsdottir, I., Kattge, J., King, J. Y., Klotz, S., McFadden, J. P., & Cavender-Bares, J. (2012). Phylogenetic and functional characteristics of household yard floras and their changes along an urbanization gradient. *Ecology, 93*, S83–S98. https://doi.org/10.1890/11-0392.1

Knight, K. S., Kurylo, J. S., Endress, A. G., Stewart, J. R., & Reich, P. B. (2007). Ecology and ecosystem impacts of common buckthorn (*Rhamnus cathartica*): A review. *Biological Invasions, 9*, 925–937. https://doi.org/10.1007/s10530-007-9091-3

Kowarik, I. (2011). Novel urban ecosystems, biodiversity, and conservation. *Environmental Pollution, 159*, 1974–1983. https://doi.org/10.1016/j.envpol.2011.02.022

Kotzé, J. D., Wannenburgh, A., & van Wilgen, B. W. (2025). Changes in the cover of selected invasive alien plant taxa between 2008 and 2023 in South Africa. *Biological Invasions, 27*, 1–7. https://doi.org/10.1007/s10530-025-03558-9

Kühn, I., Wolf, J., & Schneider, A. (2017). Is there an urban effect in alien plant invasions? *Biological Invasions, 19*, 3505–3513. https://doi.org/10.1007/s10530-017-1591-1

Kurose, D., Seier, M. K., & Evans, H. C. (2024). Exploiting exotic pathogens as mycoherbicides against invasive alien weeds: Japanese knotweed as a case study. *Pest Management Science, 80*, 87–91. https://doi.org/10.1002/ps.7510

Kumari, S., & Khan, S. (2018). Effect of Fe_3O_4 NPs application on fluoride (F) accumulation efficiency of *Prosopis juliflora*. *Ecotoxicology and Environmental Safety, 166*, 419–426. https://doi.org/10.1016/j.ecoenv.2018.09.103

Langeland, K. A. (1996). *Hydrilla verticillata* (LF) Royle (Hydrocharitaceae), "the perfect aquatic weed." *Castanea, 61*, 293–304.

Lam, M., Ryan, I., & Dicks, L. (2024). Pollinator dependence and pollination deficit in *Hedera helix*, and effects on frugivore feeding choices: Pollination and frugivory in ivy *Hedera helix*. *Journal of Pollination Ecology, 36*, 14–28.

Latif, S., Chiapusio, G., & Weston, L. A. (2017). Allelopathy and the role of allelochemicals in plant defence. In G. Becard, (Ed.), *Advances in botanical research* (pp. 9–54). Academic Press. https://doi.org/10.1016/bs.abr.2016.12.001

Lazzaro, L., Giuliani, C., Fabiani, A., Agnelli, A. E., Pastorelli, R., Lagomarsino, A., Benesperi, R., Calamassi, R., & Foggi, B. (2014). Soil and plant changing after invasion: The case of *Acacia dealbata* in a Mediterranean ecosystem. *Science of the Total Environment, 497*, 491–498. https://doi.org/10.1016/j.scitotenv.2014.08.014

Leak, S. G. A. (1999). *Tsetse biology and ecology: Their role in the epidemiology and control of trypanosmosis*. CAB International (in association with the International Livestock Research Institute, Nairobi, Kenya).

Lindgren, C. J., Castro, K. L., Coiner, H. A., Nurse, R. E., & Darbyshire, S. J. (2013). The biology of invasive alien plants in Canada. 12. *Pueraria montana* var. lobata (Willd.) Sanjappa & Predeep. *Canadian Journal of Plant Science, 93*, 71–95. https://doi.org/10.4141/cjps2012-128

Lizarralde, M. S. (1993). Current status of the introduced beaver (*Castor canadensis*) population in Tierra del Fuego, Argentina. *Ambio, 22*(6), 351–358.

Lokatis, S., Jeschke, J. M., Bernard-Verdier, M., Buchholz, S., Grossart, H. P., Havemann, F., Hölker, F., Itescu, Y., Kowarik, I., Kramer-Schadt, S., Mietchem, D., Musseau, C. L., Planillo, A., Schittko, C., Straka, T. M., & Heger, T. (2023). Hypotheses in urban ecology: Building a common knowledge base. *Biological Reviews, 98*, 1530–1547. https://doi.org/10.1111/brv.12964

MacDougall, A. S., Gilbert, B., & Levine, J. M. (2009). Plant invasions and the niche. *Journal of Ecology, 97*, 609–615. https://doi.org/10.1111/j.1365-2745.2009.01514.x

Mackay, A. J., Muturi, E. J., Ward, M. P., & Allan, B. F. (2016). Cascade of ecological consequences for West Nile virus transmission when aquatic macrophytes invade stormwater habitats. *Ecological Applications, 26*, 219–232. https://doi.org/10.1890/15-0050

Malik, A. (2007). Environmental challenge vis a vis opportunity: The case of water hyacinth. *Environment International, 33*, 122–138. https://doi.org/10.1016/j.envint.2006.08.004

Mazza, G., Tricarico, E., Genovesi, P., & Gherardi, F. (2014). Biological invaders are threats to human health: An overview. *Ethology Ecology and Evolution, 26*, 112–129. https://doi.org/10.1080/03949370.2013.863225

Morris, K. A., Stark, J. M., Bugbee, B., & Norton, J. M. (2016). The invasive annual cheatgrass releases more nitrogen than crested wheatgrass through root exudation and senescence. *Oecologia, 181*, 971–983. https://doi.org/10.1007/s00442-015-3544-7

Mullerova, J., Bruna, J., & Bartalos, T. (2017). Timing is important: unmanned aircraft vs. satellite imagery in plant invasion monitoring. *Frontiers in Plant Science, 8*, 887. https://doi.org/10.3389/fpls.2017.00887

Mungi, N. A., Qureshi, Q., & Jhala, Y. V. (2020). Expanding niche and degrading forests: Key to the successful global invasion of *Lantana camara* (sensu lato). *Global Ecol Conserv, 23*, Article e01080. https://doi.org/10.1016/j.gecco.2020.e01080

Najberek, K., Pusz, W., Solarz, W., & Olejniczak, P. (2018). The seeds of success: Release from fungal attack on seeds may influence the invasiveness of alien *Impatiens*. *Plant Ecology, 219*, 1197–1207. https://doi.org/10.1007/s11258-018-0872-9

Nguyen, T. H., Boets, P., Lock, K., Ambarita, M. N., Forio, M. A., Sasha, P., Dominguez-Granda, L. E., Hoang, T. H., Everaert, G., & Goethals, P. L. (2015). Habitat suitability of the invasive water hyacinth and its relation to water quality and macroinvertebrate diversity in a tropical reservoir. *Limnologica, 52*, 67–74. https://doi.org/10.1016/j.limno.2015.03.006

Ojija, F., Arnold, S. E., & Treydte, A. C. (2019). Impacts of alien invasive *Parthenium hysterophorus* on flower visitation by insects to co-flowering plants. *Arthropod-Plant Interact, 13*, 719–734. https://doi.org/10.1007/s11829-019-09701-3

Patel, A., & Patra, D. D. (2017). A sustainable approach to clean contaminated land using terrestrial grasses. In K. Bauddh, B. Singh, & J. Korstad (Eds.), *Phytoremediation potential of bioenergy plants* (pp. 305–331). Springer. https://doi.org/10.1007/978-981-10-3084-0_12

Pejchar, L., & Mooney, H. A. (2009). Invasive species, ecosystem services and human well-being. *Trends in Ecology & Evolution, 24*, 497–504. https://doi.org/10.1016/j.tree.2009.03.016

Pile, L. S., Wang, G. G., Stovall, J. P., Siemann, E., Wheeler, G. S., & Gabler, C. A. (2017). Mechanisms of Chinese tallow (*Triadica sebifera*) invasion and their management implications—a review. *Forest Ecology and Management, 404*, 1–3. https://doi.org/10.1016/j.foreco.2017.08.023

Pinzone, P., Potts, D., Pettibone, G., & Warren, R. (2018). Do novel weapons that degrade mycorrhizal mutualisms promote species invasion? *Plant Ecology, 219*, 539–548. https://doi.org/10.1007/s11258-018-0816-4

Plaza, P. I., Speziale, K. I., & Lambertucci, S. A. (2018). Rubbish dumps as invasive plant epicentres. *Biological Invasions, 20*, 2277–2283. https://doi.org/10.1007/s10530-018-1708-1

Pliszko, A. (2025). First record of two invasive aquatic weeds *Ludwigia repens* (Onagraceae) and *Myriophyllum aquaticum* (Haloragaceae) in Poland. *Biologia, 80*, 233–240. https://doi.org/10.1007/s11756-024-01837-9

Podlipná, R., Skálová, L., Seidlová, H., Szotáková, B., Kubíček, V., Stuchlíková, L., Jirásko, R., Vaněk, T., & Vokřál, I. (2013). Biotransformation of benzimidazole anthelmintics in reed (*Phragmites australis*) as a potential tool for their detoxification in environment. *Bioresource Technology, 144*, 216–224. https://doi.org/10.1016/j.biortech.2013.06.105

Potgieter, L. J., Gaertner, M., Kueffer, C., Larson, B. M., Livingstone, S. W., O'Farrell, P. J., & Richardson, D. M. (2017). Alien plants as mediators of ecosystem services and disservices in urban systems: A global review. *Biological Invasions, 19*, 3571–3588. https://doi.org/10.1007/s10530-017-1589-8

Potgieter, L. J., Gaertner, M., O'Farrell, P. J., & Richardson, D. M. (2019). Perceptions of impact: Invasive alien plants in the urban environment. *Journal of Environmental Management, 229*, 76–87. https://doi.org/10.1016/j.jenvman.2018.05.080

Prabakaran, K., Li, J., Anandkumar, A., Leng, Z., Zou, C. B., & Du, D. (2019). Managing environmental contamination through phytoremediation by invasive plants: A review. *Ecological Engineering, 138*, 28–37. https://doi.org/10.1016/j.ecoleng.2019.07.002

Prokopuk, M., & Zub, L. (2020). Urban ecosystems as locations of distribution of alien aquatic plants. *Folia Oecologica, 47*, 159–167. https://doi.org/10.2478/foecol-2020-0019

Pyšek, P., Jarošík, V., Hulme, P. E., Kühn, I., Wild, J., Arianoutsou, M., Bacher, S., Chiron, F., Didžiulis, V., Essl, F., & Genovesi, P. (2010). Disentangling the role of environmental and human pressures on biological invasions across Europe. *Proceedings of the National Academy of Sciences, 107*, 12157–12162. https://doi.org/10.1073/pnas.1002314107

Pyšek, P., Jarošík, V., Hulme, P. E., Pergl, J., Hejda, M., Schaffner, U., & Vilà, M. (2012). A global assessment of invasive plant impacts on resident species, communities and ecosystems: The interaction of impact measures, invading species' traits and environment. *Global Change Biology, 18*, 1725–1737. https://doi.org/10.1111/j.1365-2486.2011.02636.x

Pyšek, P., & Richardson, D. M. (2010). Invasive species, environmental change and management, and health. *Annual Review of Environment and Resources, 35*, 25–55. https://doi.org/10.1146/annurev-environ-033009-095548

Qu, T., Du, X., Peng, Y., Guo, W., Zhao, C., & Losapio, G. (2021). Invasive species allelopathy decreases plant growth and soil microbial activity. *PloS One, 16*, e0246685. https://doi.org/10.1371/journal.pone.0246685

Rai, P. K. (2015). Paradigm of plant invasion: Multifaceted review on sustainable management. *Environmental Monitoring and Assessment, 187*, 759. https://doi.org/10.1007/s10661-015-4919-1

Rai, P. K., & Kim, K. H. (2020). Invasive alien plants and environmental remediation: A new paradigm in sustainable restoration ecology. *Restoration Ecology, 28*, 3–7. https://doi.org/10.1111/rec.13058

Rai, P. K., & Singh, J. S. (2020). Invasive alien plant species: Their impact on environment, ecosystem services and human health. *Ecological Indicators, 111*, Article 106020. https://doi.org/10.1016/j.ecolind.2019.106020

Rai, P. K., & Singh, M. M. (2015). *Lantana camara* invasion in urban forests of an Indo-Burma hotspot region and its ecosustainable management implication through biomonitoring of particulate matter. *Journal of Asia-Pacific Biodiversity, 8*, 375–381. https://doi.org/10.1016/j.japb.2015.09.003

Raposo, M. A., Pinto Gomes, C. J., & Nunes, L. J. (2021). Evaluation of species invasiveness: A case study with *Acacia dealbata* link. on the slopes of Cabeça (Seia-Portugal). *Sustainability, 13*, 11233. https://doi.org/10.3390/su132011233

Reid, M. K., Naidu, P., Paterson, I. D., Mangan, R., & Coetzee, J. A. (2021). Population genetics of invasive and native *Nymphaea mexicana Zuccarini*: Taking the first steps to initiate a biological control programme in South Africa. *Aquatic Botany, 171*, Article 103372. https://doi.org/10.1016/j.aquabot.2021.103372

Rodríguez-Lara, J. W., Cervantes-Ortiz, F., Arámbula-Villa, G., Mariscal-Amaro, L. A., Aguirre-Mancilla, C. L., & Andrio-Enríquez, E. (2022). Water hyacinth (*Eichhornia crassipes*): A review. *Agronomía Mesoamericana 33*. https://doi.org/10.15517/am.v33i1.44201

Ruiz-Avila, R. J., & Klemm, V. V. (1996). Management of *Hydrocotyle ranunculoides* Lf, an aquatic invasive weed of urban waterways in Western Australia. *Hydrobiologia, 340*, 187–190. https://doi.org/10.1007/BF00012717

Santana Marques, P., Resende Manna, L., Clara Frauendorf, T., Zandonà, E., Mazzoni, R., & El-Sabaawi, R. (2020). Urbanization can increase the invasive potential of alien species. *Journal of Animal Ecology, 89*, 2345–2355. https://doi.org/10.1111/1365-2656.13293

Schnitzler, A., Hale, B. W., & Alsum, E. M. (2007). Examining native and exotic species diversity in European riparian forests. *Biologiclogical Conserversion, 138*, 146–156. https://doi.org/10.1016/j.biocon.2007.04.010

Serbesoff-King, K. (2003). Melaleuca in Florida: A literature review on the taxonomy, distribution, biology, ecology, economic importance and control measures. *Journal of Aquatic Plant and Management, 41*, 98–112.

Shackleton, R. T., Le Maitre, D. C., Van Wilgen, B. W., & Richardson, D. M. (2015). The impact of invasive alien *Prosopis* species (mesquite) on native plants in different environments in South Africa. *South+A1443 African Journal of Botany, 97*, 25–31. https://doi.org/10.1016/j.sajb.2014.12.008

Shackleton, R. T., Shackleton, C. M., & Kull, C. A. (2019). The role of invasive alien species in shaping local livelihoods and human well-being: A review. *Journal of Environmental Management, 229*, 145–157. https://doi.org/10.1016/j.jenvman.2018.05.007

Shackleton, R. T., Witt, A. B. R., Piroris, F. M., & van Wilgen, B. W. (2017). Distribution and socio-ecological impacts of the invasive alien cactus *Opuntia stricta* in eastern Africa. *Biological Invasions, 19*, 2427–2441. https://doi.org/10.1007/s10530-017-1453-x

Sharma, A., Kaur, A., Kaur, S., Kohli, R. K., & Batish, D. R. (2023). Plant invasion and climate change: A global overview. In S. Tripathi, R. Bhadouria, P. Srivastava, R. Singh, & D. R. Batish (Eds.), *Plant invasions and global climate change* (pp. 3–30). Springer Nature Singapore. https://doi.org/10.1007/978-981-99-5910-5

Sharma, G. P., & Raghubanshi, A. S. (2009). *Lantana* invasion alters soil nitrogen pools and processes in the tropical dry deciduous forest of India. *Applied Soil Ecology, 42*, 134–140. https://doi.org/10.1016/j.apsoil.2009.03.002

Singh, A., Khudsar, F. A., Nag, S., & Adhikari, B. S. (2023). Ecological and socioeconomic impacts of *Lantana camara* invasion: A case study in central India. *Indian Journal of Forestry, 46*, 187–199. https://doi.org/10.54207/bsmps1000-2024-90I720

Sladonja, B., Sušek, M., & Guillermic, J. (2015). Review on invasive tree of heaven (*Ailanthus altissima* (Mill.) Swingle) conflicting values: Assessment of its ecosystem services and potential biological threat. *Environmental Management, 56*, 1009–1034. https://doi.org/10.1007/s00267-015-0546-5

Smerdu, A., Kanjir, U., & Kokalj, Ž. (2020). Automatic detection of Japanese knotweed in urban areas from aerial and satellite data. *Management and Biological Invasions, 11*, 661–676. https://doi.org/10.3391/mbi.2020.11.4.03

Stabenau, N., Zehnsdorf, A., Rönicke, H., Wedwitschka, H., Moeller, L., Ibrahim, B., & Stinner, W. (2018). A potential phosphorous fertilizer for organic farming: Recovery of phosphorous resources in the course of bioenergy production through anaerobic digestion of aquatic macrophytes. *Energy, Sustainability and Society, 8*, 16. https://doi.org/10.1186/s13705-018-0155-2

Stone, C. M., Witt, A. B., Walsh, G. C., Foster, W. A., & Murphy, S. T. (2018). Would the control of invasive alien plants reduce malaria transmission? A review. *Parasites Vectors, 11*, 76. https://doi.org/10.1186/s13071-018-2644-8

Swearingen, J., Reshetiloff, K., Slattery, B., & Zwicker, S. (2002). *Plant invaders of mid-Atlantic natural areas* (pp. 82–85). National Park Service and US Fish and Wildlife Service.

Tan, C., Greulich, S., Medina, V., Zheng, X., Canu, P., Fritsch, A., & Wantzen, K. M. (2024). Still exotic? Comparison of macroinvertebrates associated with *Elodea nuttallii* and native macrophytes in urban artificial lakes. *Hydrobiologia, 851*, 4461–4475. https://doi.org/10.1007/s10750-024-05598-z

Turbelin, A. J., Malamud, B. D., & Francis, R. A. (2017). Mapping the global state of invasive alien species: Patterns of invasion and policy responses. *Global Ecology and Biogeography, 26*, 78–92. https://doi.org/10.1111/geb.12517

Uddin, Md., & Robinson, R. W. (2018). Can nutrient enrichment influence the invasion of *Phragmites australis*? *Science of the Total Environment, 613–614*, 1449–1459. https://doi.org/10.1016/j.scitotenv.2017.06.131

Uddin, M. N., & Robinson, R. W. (2017). Allelopathy and resource competition: The effects of *Phragmites australis* invasion in plant communities. *Botanical Studies, 58*, 29. https://doi.org/10.1186/s40529-017-0183-9

Van Meerbeek, K., Muys, B., & Hermy, M. (2019). Lignocellulosic biomass for bioenergy beyond intensive cropland and forests. *Renewable and Sustainable Energy Reviews, 102*, 139–149. https://doi.org/10.1016/j.rser.2018.12.009

van Wilgen, B. W., & Scott, D. F. (2001). Managing fires on the Cape Peninsula: Dealing with the inevitable. *Journal of Mediterranean Ecology, 2*, 197–208.

van Wilgen, B. W., & Wannenburgh, A. (2016). Co-facilitating invasive species control, water conservation and poverty relief: Achievements and challenges in South Africa's Working for Water programme. *Current Opinion in Environment Sustainability, 19*, 7–17. https://doi.org/10.1016/j.cosust.2015.08.012

Villamagna, A. M., & Murphy, B. R. (2010). Ecological and socio-economic impacts of invasive water hyacinth (*Eichhornia crassipes*): A review. *Freshwater Biology, 55*, 282–298. https://doi.org/10.1111/j.1365-2427.2009.02294.x

von der Lippe, M., & Kowarik, I. (2012). Interactions between propagule pressure and seed traits shape human-mediated seed dispersal along roads. *Perspective in Plant Ecology, Evolution and Systematics, 14*, 123–130. https://doi.org/10.1016/j.ppees.2011.09.006

Wahl, C. F., Diaz, R., & Ortiz-Zayas, J. (2020). Assessing *Salvinia molesta* impact on environmental conditions in an urban lake: Case study of Lago Las Curias, Puerto Rico. *Aquatic Invasions, 15*, 562–577. https://doi.org/10.3391/ai.2020.15.4.02

Wei, H., Huang, M., Quan, G., Zhang, J., Liu, Z., & Ma, R. (2018). Turn bane into a boon: Application of invasive plant species to remedy soil cadmium contamination. *Chemosphere, 210*, 1013–1020. https://doi.org/10.1016/j.chemosphere.2018.07.129

Williams, N. S., Schwartz, M. W., Vesk, P. A., McCarthy, M. A., Hahs, A. K., Clemants, S. E., Corlett, R. T., Duncan, R. P., Norton, B. A., Thompson, K., & McDonnell, M. J. (2009). A conceptual framework for predicting the effects of urban environments on floras. *Journal of Ecology, 97*, 4–9. https://doi.org/10.1111/j.1365-2745.2008.01460.x

Wilson, J. R., Ivey, P., Manyama, P., & Nanni, I. (2013). A new national unit for invasive species detection, assessment and eradication planning. *South African Journal of Science, 109*, 1–13. https://hdl.handle.net/10520/EJC136404

Yuan, C., Bai, X., Zhu, T., Wen, Z., Cao, T., Zhang, X., & Ni, L. (2021). Long-term effects of the harvesting of *Trapa natans* on local water quality and aquatic macrophyte community in Lake

Erhai, China. *Frontiers in Environmental Sciences, 9*, Article 706746. https://doi.org/10.3389/fenvs.2021.706746

Zehnsdorf, A., Hussner, A., Eismann, F., Rönicke, H., & Melzer, A. (2015). Management options of invasive *Elodea nuttallii* and *Elodea canadensis*. *Limnologica, 51*, 110–117. https://doi.org/10.1016/j.limno.2014.12.010

Zheng, Y., Feng, Y., Valiente-Banuet, A., Li, Y., Liao, Z., Zhang, J., & Chen, Y. (2015a). Are invasive plants more competitive than native conspecifics? Patterns vary with competitors. *Science Reports, 5*, 15622. https://doi.org/10.1038/srep15622

Zheng, Y. L., Feng, Y. L., Zhang, L. K., Callaway, R. M., Valiente-Banuet, A., Luo, D. Q., Liao, Z. Y., Lei, Y. B., Barclay, G. F., & Silva-Pereyra, C. (2015b). Integrating novel chemical weapons and evolutionarily increased competitive ability in success of a tropical invader. *New Phytologist, 205*, 1350–1359. https://doi.org/10.1111/nph.13135

Chapter 4
Emerging Invasive Alien Plants in Urban Settlements: Establishment and Impacts

Deniz Inci, İlhan Üremiş, Soner Soylu, Mehmet Arslan, Necmi Aksoy, and Ahmet Uludağ

Abstract Invasive alien plants (IAPs) are one of the greatest threats to agriculture, economy, ecosystem, food security, and animal and human health from a broad continental scale to a narrow local scale. In urban settings, human and IAPs relations are more dynamic than in rural and range lands due to the demanding needs of IAPs, such as ornamental landscapes, reforestation, or residential usage of IAPs. Rapid distribution of IAPs to a new urban ecosystem influences the local community due to the noticeable economic losses. This phenomenon requires time-sensitive tracking of new IAPs introduction and distribution in the urban systems more than the rural and/ or native range of IAPs. While some IAPs have more potential to establish themselves in a metropolitan area to invade, others may seek longer time or stay in a harmless fate. Understanding IAPs' invasiveness for establishment is a crucial component in which IAP researchers, policymakers, and landowners/managers could develop surveillance and management strategies. In this chapter, we criticize the significant

D. Inci
Department of Plant Sciences, University of California, Davis, CA, USA
e-mail: inci@ucdavis.edu

İ. Üremiş · S. Soylu
Department of Plant Protection, Faculty of Agriculture, Hatay Mustafa Kemal University, Hatay, Türkiye
e-mail: iuremis@mku.edu.tr

S. Soylu
e-mail: soylu@mku.edu.tr

M. Arslan
Department of Biosystem Engineering, Faculty of Agriculture, Erciyes University, Kayseri, Türkiye
e-mail: mehmetarslan@erciyes.edu.tr

N. Aksoy
Department of Forest Engineering, Faculty of Forestry, Düzce University, Düzce, Türkiye
e-mail: necmiaksoy@duzce.edu.tr

A. Uludağ (✉)
Department of Plant Protection, Faculty of Agriculture, Çanakkale Onsekiz Mart University, Çanakkale, Türkiye
e-mail: ahuludag@yahoo.com

R. Singh et al. (eds.), *Invasive Alien Plants in Urban Ecosystems*,
https://doi.org/10.1007/978-3-032-00997-5_4

threats posed by IAPs to urban settlements, the adaptation of IAPs to human-altered environments, and the adverse impacts of IAPs. Under the expected climate change scenarios, IAPs' introduction to non-native lands may be favored, and accelerated in certain places more than others. The importance of early detection, rigorous control efforts, and public awareness are essential strategies for mitigating the impacts of IAPs. Unique and case-specific management approaches are most likely the backbone of the near-future policies.

Keywords Abundance · Biological invasions · Community · Distribution · Urban ecosystems

1 Introduction

Invasive alien plants [hereinafter referred to as 'IAP(s)'] are non-native species, which are often abundant in urban settlements such as cities, airports, harbors, and transportation centers. National and international trade, tourism, floriculture, and agriculture highly favor the introduction and distribution of IAPs into non-native areas, which may include cities. The plant trade of horticultural and nursery production are some of the main reasons for the distribution of IAPs (Shah et al., 2025). In urban areas, IAPs encounter habitats, soils, climatic conditions, and hydrology that have been profoundly changed by human activity; therefore, they can easily promote the IAPs distribution if those species are pre-adapted to similar environmental conditions in their region of origin (Gaertner et al., 2016; Tripathi et al., 2023). The introduction of IAPs could be natural by themselves (it should be noted that there is a trace of human effect on it), unintentional by human-driven paths, and intentional for desirable purposes such as the augmentation or restoration of specific ecosystem services. While some species are displaced due to urbanization, humans deliberately or accidentally introduce a significant portion of IAPs (Seboko et al., 2024). However, besides providing the desired ecosystem services, some of the IAPs can severely introduce negative impacts on the new area and/or create novel ecosystem disservices within the urban milieu. Some of these IAPs would spread into the surrounding areas where they can alter ecosystem functions, disrupt the flows of ecosystem services, modify native biodiversity, and negatively affect local economies and/or human well-being (Potgieter et al., 2019). In this chapter, we review IAPs' introduction, establishment, and impacts in urban settlements.

In the present era, globalization and the acceleration of international trade have facilitated various human activities worldwide, making the movement of plant species more accessible. In this context, plants, mainly used in sectors such as agriculture and landscaping, are transported between countries, and different species are introduced into new regions and integrated into local ecosystems (van Valkenburg et al., 2014). However, some of the plants imported from abroad may be invasive species that threaten natural ecosystems. These plants can have undesirable and unexpected consequences, exerting significant pressure on local flora and fauna, causing considerable problems, and disrupting ecosystems by affecting the biological balance. This can lead to reduced biodiversity, degradation of ecosystem services, and reduced agricultural productivity. This section discusses the potential threats of alien invasive plants to ecosystems and strategies to address these issues.

2 Impacts of Plant Import from Abroad on Existing Flora

The trade and transport of plants around the world have become a global process, enabling the movement of different plant species to other regions and their introduction into new ecosystems (Hulme, 2009; Rai & Singh, 2020; Uludağ & Ertürk, 2012; Yazlık, 2021). However, the importation of plants from abroad can impact existing flora. The integration and controlled management of carefully selected plants can help to minimize environmental impacts (Sarı, 2019; Vilà et al., 2024; Yazlık, 2021; Yazlık et al., 2020). Collaboration between local experts, scientists, and communities is critical to sustainable plant transport. This process should be managed with careful planning, scientific assessment, and sustainable practices to ensure its success.

2.1 Potential Benefits

Conservation of local flora and ecosystems is essential for sustainable biodiversity management. However, there are some positive aspects of the newcomers. These include:

- Enrichment of biodiversity: Plants imported from abroad can enhance the genetic diversity of local flora, contributing to more resilient and diverse ecosystems.
- Agricultural and economic contributions: New plant species can be used in agriculture to adapt to different climatic and soil conditions, contributing to increased diversity and productivity.
- Innovation and aesthetic enrichment: Alien plants can add aesthetic value and diversify parks, gardens, and public spaces for landscaping and ornamental cultivation.
- Potential to support ecosystems: Some imported plants may have the potential to provide ecosystem services such as water supply, erosion control, or biological pest control, thereby contributing to their improvement.
- Positive contributions to climate change: New plant species can be used to address climate change, with more resilient or water-efficient plants helping to adapt to changing climate conditions.

2.2 Potential Issues

In the current era of globalization and rapidly increasing international trade, plant species imported from abroad are transported worldwide for various purposes, especially in agriculture. However, some of these plants have invasive characteristics that can threaten local vegetation, leading to negative impacts on biodiversity when introduced into natural ecosystems (Gaertner et al., 2009). This situation can lead to reduced biodiversity, disruption of ecosystem services, and a decline in agricultural productivity (Williams et al., 2010). Here we assess the potential problems posed by alien plants imported from abroad, and discuss measures that can be taken to address these issues (Pyšek et al., 2020; Sarı, 2019; Vilà et al., 2024; Yazlık et al., 2018).

Ecological Imbalance: Imported plants can become invasive, competing with local vegetation, and disrupting the ecosystem. For example, some invasive plants may consume more water resources, alter soil structure, or be incompatible with local climate, and plant and animal species. This can threaten the health of ecosystems by affecting other organisms within the ecosystem (Pyšek et al., 2020; Rai & Singh, 2020; Vilà et al., 2024).

Decrease in Biological Diversity: When introduced into natural ecosystems, alien plant species can compete with and outcompete native plant species, leading to a reduction in local biodiversity. The spread of these invasive plants can reduce biodiversity, contribute to the extinction of endemic/rare species, and cause imbalances in ecosystems (Simberloff et al., 2013; Vilà et al., 2024).

Genetic Pollution and Hybridization: Genetic interactions between local and alien plants can lead to genetic pollution and hybridization, threatening the genetic integrity of local species (Pyšek et al., 2020).

Impacts on Soil and Water Resources: Invasive plants may also cause soil erosion by disrupting local vegetation. They can also put pressure on water sources, affecting water quality. Invasive plants, particularly in aquatic ecosystems, can impede water circulation, degrade water quality, and disrupt aquatic ecosystems (Sarı, 2019; Vilà et al., 2024).

Fire Risk: Some invasive plants, especially in arid regions, can increase the risk of wildfires. Due to their rapid growth and flammability, these plants can dry out local vegetation, allowing wildfires to spread more quickly (Ricciardi et al., 2013).

Damage to Agricultural Production: Invasive plants can harm not only natural ecosystems, but also agricultural areas. They can carry new diseases and pests, exposing local plant populations and crops to new threats. Due to their competitiveness and rapid spread, invasive plants can make local crops less productive, invade agricultural land, and lead to economic losses in the agricultural sector (Booth et al., 2004; Yazlık, 2021; Yazlık et al., 2018).

Economic Losses: Various problems caused by invasive plants can lead to economic losses. For example, losses in the agricultural sector, the cost of restoring damaged ecosystems, and measures such as fire control can be a significant economic burden (Vilà et al., 2024; Yazlık et al., 2018).

Risks Related to Climate Change: The adaptation of invasive plants to ecosystems can increase the risks associated with climate change. These plants may adapt better to changing climate conditions, potentially reducing the competitive ability of local plant species, and making efforts to combat climate change more challenging (Yazlık, 2021).

Impacts on Human Health: Invasive plants not only affect ecosystem functions but also pose direct or indirect risks to human health. Many invasive plants may cause allergies and can indirectly host disease agents or create favorable conditions for their proliferation (Rai & Singh, 2020; Yazlık et al., 2018).

3 Mistakes in the Selection of Plants Imported from Abroad and Factors to Be Considered

With the increase in global trade and the expansion of economic activities around the world, the selection of plants imported from abroad has become a complex process that can involve various risks. Mistakes made in introducing new plant species can result in invasive species that cause damage to ecosystems and threaten biodiversity. A conscientious and careful approach should be taken to avoid these mistakes in selecting plants imported from abroad. Decisions should be based on scientific data, considering the impact on local ecosystems and communities, and the process should be managed appropriately. In this way, the selection of plants imported from abroad can be consistent with the goals of conserving biodiversity and ensuring a sustainable environment. When considering common mistakes in the selection of plants imported from abroad (Sarı, 2019; Uludağ et al., 2013; Yazlık et al., 2020):

- Failure to Conduct Compatibility Analysis and Selection of Incompatible Plants
- Inadequate Biosecurity Assessment
- Insufficient Control of Harmful Pests and Diseases for Local Crops
- Neglect of Climate and Soil Conditions
- Lack of Effective Monitoring and Evaluation
- Neglect of Economic and Agricultural Impacts
- Lack of Societal Participation, Collaboration, and Coordination

Solutions to address these shortcomings are provided below.

4 Plant Selection from Abroad: Steps Toward Ecosystem Preservation

In today's world, global trade has led to a significant increase in the utilization and the trade of plants imported from abroad. These plants are used in various fields, including landscape design, agriculture, and gardening. However, selecting the right plants require a meticulous process to ensure ecosystem preservation. Introducing new plant species from foreign countries can potentially threaten local ecosystems and biological diversity. Therefore, it is crucial to consider several factors when selecting plants for importation. The selection process should not only prioritize commercial benefits but also aim to maintain ecosystem balance and preserve biological diversity. Proper plant selection is a crucial step in sustainable environmental management. Effective communication and collaboration among local communities, experts, and authorities are essential in this process (Sarı, 2019; Uludağ et al., 2013). When selecting plants imported from abroad, it is important to consider several factors.

Assessing the Invasiveness Potential and Scientific Compatibility of Imported Plants: The invasiveness potential of plants imported from abroad determines the likelihood of these plants causing harm to natural ecosystems. Invasive plant species, characterized by rapid growth, high reproductive capacity, and a competitive advantage over other plant species, can pose a significant risk. Therefore, it is crucial to prioritize plants with low invasiveness potential to preserve local biological diversity. The initial measure to mitigate the environmental hazards of imported plants is to perform scientific compatibility assessments to evaluate their capacity to adapt to new environments. Compatibility with climate, soil, water, and other environmental factors can decrease the likelihood of plants becoming invasive (Booth et al., 2004).

Biological Security: When considering foreign plants, it is important to consider the potential for them to carry harmful insects, diseases, or other pathogens. To prevent imported plants from causing harm to natural ecosystems, biosecurity measures play a significant role. A fundamental step in ensuring biosecurity is rigorous inspection of plants during import and export processes (Farooq, 2018; Simberloff et al., 2013).

Resistance to Pests and Diseases: It is important to select plants resistant to local pests and diseases. To support the protection of local plant cover and reduce the need for pesticide use, this will aid in the healthy growth of plants (Simberloff et al., 2013).

Resilience to Climate Conditions, Ecosystem Compatibility, and Support for Biological Diversity: When introducing new plant species, it is important to carefully examine the climate conditions of the region. The resilience of plants to factors such as frost, temperature, rainfall, and other climate conditions should be evaluated to ensure ecosystem compatibility and support for biological diversity. Imported plant species must adapt to the climate, soil, water, and other environmental conditions of their new country. It is critical to ensure ecosystem compatibility to prevent the spread of invasive plants and protect local plant cover. Therefore, it is recommended to give preference to plant species that are compatible with the local ecosystem, supporting biological diversity (Pyšek et al., 2020; Vilà et al., 2024; Yazlık, 2021; Yazlık et al., 2018).

Contribution to Ecosystem Services: Selected plants should positively contribute to ecosystem services, such as preventing soil erosion, regulating the water cycle, and providing habitat. These factors are crucial for the sustainability of ecosystems (Farooq, 2018).

Public Health and Safety: When selecting plants, it is important to consider any potential risks they may pose to human health or safety. This includes considering allergen content, potential toxicity, and other health risks (Richardson & Pyšek, 2006).

Agricultural and Economic Value: When considering imported plants for production/multiplication, it is important to evaluate their agricultural and economic values. They should enhance productivity, provide economic returns, and contribute to local community development. However, it is crucial to ensure that these values do not conflict with biological diversity and ecosystem health (Pyšek et al., 2020; Yazlık, 2021).

Quarantine, Compliance with National and International Regulations: Full compliance with national and international regulations regarding plant import and export helps control invasive plant species. These regulations establish plant health, quarantine, and biological safety standards, ensuring appropriate checks. Imported plants should undergo thorough quarantine and control mechanisms to reduce the risk of carrying diseases and to keep potentially invasive plants under control. Pre-import tests to confirm the health of plants must adhere to national and international standards (Hulme, 2009).

Cultural and Aesthetic Values: It is important to consider local cultural and aesthetic values when selecting plants. Choosing plants that are appropriate to the local cultural context helps to preserve local identity in landscape design (Sarı, 2019).

Community Involvement and Social Analyses: Local communities should participate in the plant selection process. Local knowledge and experience contribute to the sustainable use of natural resources. Addressing the concerns of local communities and informing them about the ecosystem is essential for long-term success (Farooq, 2018; Yazlık, 2023).

Collaboration and Information Sharing: International cooperation and information sharing can help identify best practices in equipment transportation and reduce environmental risks worldwide. Effective communication and cooperation between countries is essential to prevent environmental risks (Farooq, 2018; Yazlık, 2023).

5 Finding and Evaluating Solutions

Invasive plants pose a significant threat to ecosystems worldwide due to their ability to disrupt the balance of natural environments (Sarı, 2019; Uludağ et al., 2013; Yazlık, 2021). Table 1 provides information on some plants with a high invasion risk worldwide and their characteristics. Given that carelessly selected plants can cause damage to local ecosystems, it is essential to establish careful planning and control mechanisms.

Here are some examples of the negative consequences that can occur if these considerations are overlooked:

5.1 *Pueraria Montana* (Lour.) Merr. (Kudzu)

Characteristics: Known for its rapid growth, Kudzu, originally from Japan, has become a major invasive species in the United States. Its uncontrolled spread can cover vast areas, particularly forested lands, leading to the displacement of local plant species.

Impact: Disrupts forest ecosystems, impedes the proliferation of native plant species and contributes to a loss of biological diversity (Tungmunnithum et al., 2020).

5.2 *Opuntia Stricta* (Haw.) Haw. (Common Prickly Pear)

Characteristics: Originally introduced from the United States to Australia, the Common Prickly Pear is characterized by its ability to spread rapidly over large expanses of land.

Impact: Poses significant challenges, covering millions of hectares and causing ecological disruptions in affected regions (Nentwig et al., 2017).

5.3 *Eichhornia Crassipes* (Mart.) Solms (Water Hyacinth)

Characteristics: Native to South America, the Water Hyacinth is recognized as one of the most harmful aquatic plants globally. It is often used as an ornamental plant due to its striking pink and violet flowers. It is now found in more than 50 countries on five continents (Tetik & Üremiş, 2015; Üremiş et al., 2014). A more careful approach to the transport and trade of such plants is essential to maintain the ecosystem's health. Adopting more effective international control and prevention measures can be crucial in addressing this issue.

Table 1 Important IAPs in the world with their characteristics (GISD, 2024; Holzmueller & Jose, 2009; Lowe et al., 2004; Nentwig et al., 2017; Rai & Singh, 2020; Uludağ et al., 2017; Vilà et al., 2024; Yazlık, 2021)

Scientific name	Family	Common name	Origin	Type
Acacia mearnsii De Wild	Fabaceae	Black wattle	Australia and Tasmania	Shrub
Acer negundo L	Sapindaceae	Box elder	North America	Tree
Acer platanoides L	Sapindaceae	Norway maple	Eastern Europe and Western Asia	Tree
Achyranthes japonica (Miq.) Nakai	Amaranthaceae	Japanese chaff flower	Korea and Japan	Herb
Ailanthus altissima (Mill.) Swing	Simaroubaceae	Tree of heaven	Northeast and Central China, and Taiwan	Tree
Albizia julibrissin Durazz	Fabaceae	Persian silk tree	Iran, Azerbaijan, China, and Korea	Tree
Amorpha fruticosa L	Fabaceae	Desert false indigo	United States and South Mexico	Shrub
Ardisia elliptica Thunb	Primulaceae	Shoe button ardisia	India, Sri Lanka, Malaysia, Indonesia, and New Guinea	Tree
Arundo donax L	Poaceae	Giant reed	Middle East	Grass
Berberis thunbergii DC	Berberidaceae	Japanese barberry	Japan and Eastern Asia	Shrub
Broussonetia papyrifera (L.) L'Her. ex Vent	Moraceae	Paper mulberry	Japan and Taiwan	Tree
Caulerpa taxifolia (M. Vahl) C. Agardh	Caulerpaceae	Killer algae	Pacific Ocean, Indian Ocean, and Caribbean Sea	Algae
Cecropia peltata L	Urticaceae	Trumpet tree	Central America	Tree
Choromolaena odorata (L.) R.M. King & H.Rob	Asteraceae	Siam weed	Americas, Mexico, and the Caribbean	Herb
Cinchona pubescens Vahl	Rubiaceae	Red cinchona	Central and South America	Tree
Convallaria majalis L	Asparagaceae	Lily of valley	Northern Europe and Asia	Herb
Eichhornia crassipes (Mart.) Solms	Pontederiaceae	Water hyacinth	South America	Aquatic
Hedera helix L	Araliaceae	English ivy	Europe and Western Asia	Climbing plant

(continued)

Table 1 (continued)

Scientific name	Family	Common name	Origin	Type
Hedychium gardnerianum Sheppard ex Ker Gawl	Zingiberaceae	Kahili ginger	Himalayas	Herb
Hemerocallis fulva (L.) L	Asphodelaceae	Orange day-lily	Asia	Herb
Heracleum mantegazzianum Sommier & Levier	Apiaceae	Giant hogweed	Western Caucasus	Herb
Hiptage benghalensis (L.) Kurz	Malpighiaceae	Hiptage	India, Southeast Asia, Taiwan, and Philippines	Shrub
Humulus japonicus Siebold & Zucc	Cannabaceae	Japanese hops	East Asia	Climbing plant
Impatiens glandulifera Royle	Balsaminaceae	Himalayan balsam	Himalayas	Herb
Imperata cylindrica (L.) P. Beauv	Poaceae	Cogon grass	Asia, Australia, Africa, and South Europe	Grass
Ipomoea triloba L	Convolvulaceae	Morning glory	Tropical Americas	Herb
Lantana camara L	Verbenaceae	Common lantana	Central and South America	Shrub
Leucaena leucocephala (Lam.) de Wit	Fabaceae	Coffee bush	Southern Mexico and Central America	Tree
Ligustrum robustum (Roxb.) Blume	Oleaceae	Ceylon privet	South and Southeast Asia	Shrub
Lonicera japonica Thunb	Caprifoliaceae	Japanese honeysuckle	East China, Korea, and Japan	Climbing vine
Lythrum salicaria L	Lythraceae	Purple loosestrife	Europe	Herb
Melaleuca quinquenervia (Cav.) S.T.Blake	Myrtaceae	Paper bark tea tree	New Caledonia, Papua New Guinea, and coastal Eastern Australia	Tree
Miconia calvescens DC	Melastomataceae	Velvet tree	Tropical America	Tree
Miconia crenata (Vahl) Michelang	Melastomataceae	Soapbush	Caribbean region	Shrub
Microstegium vimineum (Trin.) A. Camus	Poaceae	Japanese stilt grass	Asia	Herb

(continued)

Table 1 (continued)

Scientific name	Family	Common name	Origin	Type
Mikania micrantha Kunth	Asteraceae	American rope	South and Central America	Vine
Mimosa pigra L	Fabaceae	Giant sensitive tree	South America	Shrub
Myrica faya Ait	Myricaceae	Fire tree	Azores, Madeira Islands, and the Canary Islands	Shrub
Opuntia stricta (Haw.) Haw	Cactaceae	Common prickly pear	Americas	Shrub
Pinus pinaster Aiton	Pinaceae	Cluster pine	Mediterranean Basin	Tree
Prosopis glandulosa Torr	Fabaceae	Honey mesquite	United States and Northern Mexico	Shrub
Psidium cattleyanum Sabine	Myrtaceae	Cattley guava	Brazil	Shrub
Pueraria montana (Lour.) Merr	Fabaceae	Kudzu	Southeast and East Asia	Vine
Pyrus calleryana Decne	Rosaceae	Callery pear	China and Vietnam	Tree
Reynoutria japonica Houtt	Polygonaceae	Japanese knotweed	East Asia	Shrub
Robinia pseudoacacia L	Fabaceae	Black locust	North America	Tree
Rubus ellipticus Sm	Rosaceae	Golden Himalayan raspberry	South Asia	Shrub
Schinus terebinthifolius G. Raddi	Anacardiaceae	Brazilian peppertree	South America	Tree
Sphagneticola trilobata (L.) Pruski	Asteraceae	Creeping ox-eye	Mexico, Central America, and the Caribbean	Herb
Spathodea campanulata P. Beauv	Bignoniaceae	African tulip tree	West Africa	Tree
Tamarix ramosissima Ledeb	Tamaricaceae	Tamarisk	Eurasia	Shrub
Tradescantia fluminensis Vell	Commelinaceae	Small leaf spiderwort	South America	Herb
Ulex europaeus L	Fabaceae	Common gorse	Western Europe	Shrub
Wisteria sinensis (Sims) DC	Fabaceae	Chinese wisteria	China	Climbing vine

Impact: Widespread presence across five continents, disrupting aquatic ecosystems and posing a threat to water bodies (Sarı, 2019; Wagner et al., 2024).

In this context, when considering what can be done to address this significant problem:

International Collaboration: Increase international cooperation and information exchange on plant species imported from abroad.

Strict Inspection and Monitoring: Establish strict inspection and monitoring mechanisms to control the introduction of alien plant species.

Early Detection and Control: Early detection and control of invasive plant species can prevent their spread. This can be achieved through cooperation between local authorities and scientists.

Biocontrol and Integrated Pest Management: Use biological control and integrated pest management strategies to control invasive plants. Biological control methods use natural enemies to control populations of harmful plant species, providing an alternative to chemical control methods.

Education and Awareness: Making the public and the agricultural sector aware of the damage caused by invasive plants. This is essential to prevent the spread of these plants.

Legislation: Implement strict legislation to control the importation of plants from abroad and to ensure the control of invasive plants. Strong legislation can be effective in tackling this problem.

6 Genetic and Epigenetic Responses of Alien Plant Species to Urbanization

The urban green areas can suffer high air and soil pollution due to high industrial activities, disposal of city wastes, high traffic density, allocation of lands to building constructions, dense human population, high nitrogen deposition, low water availability, and depletion of natural resources (Grimm et al., 2008; Szulkin et al., 2020a, 2020b). Human activities in urban areas can generate high concentrations of carbon monoxide, carbon dioxide, dust particles, sulfur dioxide, nitrogen dioxide, higher heavy metal and NaCl make these areas unsuitable habitats for most plant species (Elmqvist et al., 2013; Schäfer et al., 2009). The profile of a city structure such as the density, width and height of buildings affect the diurnal pattern of heating and cooling (Grimmond, 2007). The urban areas have higher temperatures than the rural surroundings called the 'urban heat island' effect (Arnfield, 2003) that extends the growing period and alters the phenology of the species (Mimet et al., 2009). During urbanization, most of the existing habitats are destroyed while new habitats are created. Critical shifts in temperature, heat, and precipitation patterns may alter the behavior of species in urban areas and other potentially limiting factors such as habitat availability (Walther et al., 2009). Some of the plant species have great survival and dispersal mechanisms in those extremely stressed urban areas, and they become predominant in cities worldwide (McKinney, 2006). Some of the introduced plant species from warmer areas, such as *Ailanthus altissima*, *Robinia pseudoacacia*, and *Hordeum murinum* were better adapted to raised temperatures and drought in urban areas (Sukopp & Wurzel, 2003). To be invasive, native and non-native species have to overcome stress barriers that are caused by urbanization.

Owing to the sessile characteristics of the plants, they cannot move to a suitable habitats but have to adapt themselves to the new changed conditions to avoid extirpation. Adaptive evolution is one of the most effective strategies to survive in these stressful environments. The ability of an invasive plant species to invade an urban area can be grouped into three categories: introduction, establishment, and expansion. Some important traits related to invasiveness such as high environmental tolerance, phenotypic plasticity, reproduction rate and competitiveness are well-known functional mechanisms that are determined by the genetic makeup of invasive plants in urban as well as rural areas (Te Beest et al., 2012). Understanding of how plant species have adapted to urban environments is a main concern of urban ecologists. However, studying both urban ecosystems and detecting mechanisms involved in rapid adaptation to global climate change is still in its infancy.

Increased genetic diversity due to multiple introductions, hybridizations, and mutations stimulated to adaptation of invasive alien plant species to new environments (Ellegren & Galtier, 2016). Genetic information carried in the DNA sequence plays a major role in the adaptation potential of the plant species. Mutations, permanent alterations of a DNA sequence, are one of the major sources of genetic variations. Abiotic stress can directly affect mutation levels of specific traits by inducing stress-directed mutagenesis (Heidenreich, 2007). Stressful conditions of urban ecosystems may lead to new beneficial mutations that can facilitate the adaptability and invasiveness of a species to changing conditions. The contribution of spontaneous mutations to long-term adaptation and invasiveness depends on the number of mutated genes and their phenotypic effects (Houle, 1998). Mutant genotypes can demonstrate a dramatic advantage under multiple stress conditions.

Inter or intraspecific hybridization of introduced plant species with closely related native or other non-native species may generate novel rapid invader genotypes in a single generation in the stressful urban habitats due to increased genetic variance, suppressing deleterious recessive alleles and combining favorable genes in a genotype (Gallego-Tévar et al., 2018; Qiu et al., 2020). Increased tolerance to local temperatures (Milne & Abbott, 2000), clonal growth (Vilà & D'Antonio, 1998), and herbivore resistance (Whitney et al., 2006) are excellent examples of hybridization effects on invasiveness. However, intentional and unintentional plant introductions in urban ecosystems can threaten and endanger native plant species by genetic contamination. Putative hybridization due to genetic contamination can break ecological rules that enhance the replacement of hybrid genotypes.

Phenotypic plasticity, the ability of a genotype to produce multiple phenotypes under altering environmental conditions, facilitates the survival and invasions of plant species under biotic and abiotic stress conditions encountered in new environments (Pyšek & Richardson, 2007; Rejmánek et al., 2005). Some of the plastic responses can be inherited through epigenetic inheritance. Under variable environmental conditions, phenotypic plasticity can influence environmental tolerance and assist in successful colonization events of invasive species (Ghalambor et al., 2007). Phenotypic plasticity can be crucial in invasion when invasive species have low genetic variation.

Natural selection or genetic drift may play a significant role in the adaptive and dispersal capacity of plant species in response to sudden stresses in urban habitats. Following the introduction, selection occurs in response to temperature, salinity, drought, photoperiod, competition, and predators (Filchak et al., 2000; Linde et al., 2001). Some invasive species completely eradicate local species through selective exclusion due to increased levels of tolerance for biotic and abiotic stresses that occur in the global climate changes.

It is difficult to determine which mechanisms are likely to be effective. However, both genetic and epigenetic mechanisms play a critical role in invasion success in stressful environments, such as urban ecosystems. Natural selection, mutation, and genetic drifts are too slow to cope with the growth and survival limitations under stressed environments, especially when populations are small and genetic variation is low (Jablonka & Raz, 2009). However, epigenetic modifications are faster than genetic modifications and may be critical for fast adaptation into altered environments. During epigenetic adaptations, expressions of genes are chemically altered without altering the underlying DNA sequence (Baerwald et al., 2016; Ibrahim et al., 2021). In natural plant populations, remarkable phenotypic variations within and among populations can be seen due to epigenetic modifications. Understanding the role of epigenetics in the adaptation and dispersal mechanisms of these widespread plant species may help to conserve, restore, and expand urban ecosystems in the altered climate. Environmentally induced epimutations may be a component of the adaptive radiation of closely related species to new environments.

The urban ecology suffers from too many speculations and little experiential data on the adaptation and dispersal mechanisms of invasive plant species. Extensive research is needed to combine ecological and genetic approaches to elucidate the role of drivers of invasiveness and their mechanisms in urban environments. Recent remarkable technological advances in high-throughput sequencing, genome-wide, and epigenome-wide sequencing, metabolomics, and transcriptomics approaches can increase our understanding of how genomic and epigenomic responses influence the outcome of alien species in urbanized areas. Integrating epigenetics into urban ecosystems to elucidate how alien species become invasive, has become one of the more important research areas in the study of urban ecosystem functions.

Epigenetic modifications can cause remarkable phenotypic variance in the introduced and native populations. Epigenetics is the study of changes in organisms caused by the modification of DNA or chromatin rather than alterations or deletions of the DNA sequence. The epigenetic modifications of DNA and its chromatin environment can be accepted as an additional level of information that is carried within the genome of an organism. The well-known epigenetic modifications are DNA methylation, histone modification, microRNA and nucleosome positioning. DNA methylation, addition of a methyl group onto cytosines in the DNA, is the most well-studied epigenetic mechanisms occurring in plant populations that affect gene expression, DNA conformation, DNA stability, transposon silencing and gene imprinting by regulating the accessibility of transcription factors. Alien species can evolve more rapidly through epigenetic modifications that increase phenotypic diversity to adapt to adverse environments (Herrera & Bazaga, 2010). Epigenetically altered traits, such as drought, cold, and heat tolerance, increase the invasive potential of alien plant species. Some of the epigenetic modifications due to environmental factors are heritable and reversible (Kumari et al., 2022). The effects of environmental factors on epigenetic modifications were reported or reviewed by several authors (Ni et al., 2018; Uludağ et al., 2022). Epigenetic modifications can increase the speed of adaptation in altered urban environments; this process may be an essential mechanism for the invasive success of alien plants in urban ecosystems.

7 Conclusions and Future Perspectives

IAPs pose significant challenges to urban ecosystems and their surroundings due to their ability to adapt to human-altered environments, outcompete native species, and disrupt natural functions. Facilitated by globalization and international trade, the movement of IAPs often results in unintended ecological consequences, threatening native and/or local biodiversity, negatively impact local economies, and human well-being. Effective management of IAPs in urban areas necessitates an integrated approach, including rigorous biosecurity measures, early detection and rapid response systems, and ongoing monitoring and assessment. Collaboration among policymakers, researchers, land managers, and local communities is essential for developing and implementing sustainable management practices. Public education and awareness campaigns are crucial for mitigating the risks associated with IAPs by informing the local community about their impacts and the importance of preventing their spread. With the intensification of climate change, adaptive management strategies that are flexible and responsive to changing environmental conditions are increasingly necessary. Future policies should prioritize case-specific management approaches tailored to the unique characteristics of local ecosystems and the specific challenges posed by different IAPs. Integrating scientific research with practical management efforts is essential for mitigating the negative impacts of IAPs, promoting the resilience and sustainability of urban ecosystems, and the stewardship of IAPs prevention and control.

References

Arnfield, A. A. (2003). Two decades of urban climate research: A review of turbulence, exchanges of energy and water, and the urban heat island. *International Journal of Climatology, 23*(1), 1–26.

Baerwald, M. R., Meek, M. H., Stephens, M. R., Nagarajan, R. P., Goodbla, A. M., & Tomalty, K. M. H. (2016). Migration-related phenotypic divergence is associated with epigenetic modifications in rainbow trout. *Molecular Ecology, 25*, 1785–1800.

Booth, B. D., Murphy, S. D., & Swanton, C. J. (2004). Invasive ecology of weeds in agricultural systems. In Inderjit (Ed.), *Weed biology and management* (pp. 29–45). Springer.

Ellegren, H., & Galtier, N. (2016). Determinants of genetic diversity. *Nature Reviews Genetics, 17*, 422–433.

Elmqvist, T., Barnett, G., & Wilkinson, C. (2013). Exploring urban sustainability and resilience. In L. Pearson (Ed.), *Resilient sustainable cities*. Routledge.

Farooq, S. (2018). Experimental and ecological niche modeling approaches to predict potential distribution areas of some invasive weeds in Turkey. Ph.D. Thesis, Gaziosmanpasa University, Tokat, Turkey.

Filchak, K. E., Roethele, J. B., & Feder, J. L. (2000). Natural selection and sympatric divergence in the apple maggot *Rhagoletis pomonella*. *Nature, 407*, 739–742.

Gaertner, M., Breeyen, A. D., Hui, C., & Richardson, D. M. (2009). Impacts of alien plant invasions on species richness in Mediterranean-type ecosystems: A meta-analysis. *Progress in Physical Geography, 33*, 319–338.

Gaertner, M., Larson, B. M. H., Irlich, U. M., Holmes, P. M., Stafford, L., van Wilgen, B. W., & Richardson, D. M. (2016). Managing invasive species in cities: A framework from Cape Town, South Africa. *Landscape and Urban Planning, 151*, 1–9.

Gallego-Tévar, B., Rubio-Casal, A. E., de Cires, A., Figueroa, E., Grewell, B. J., & Castillo, J. M. (2018). Phenotypic plasticity of polyploid plant species promotes transgressive behavior in their hybrids. *AoB Plants, 10*, ply055.

Ghalambor, C. K., McKay, J. K., Carroll, S. P., & Reznick, D. N. (2007). Adaptive versus non-adaptive phenotypic plasticity and the potential for contemporary adaptation in new environments. *Functional Ecology, 21*, 394–407.

GISD. (2024). *Global invasive species database*. https://www.iucngisd.org/gisd. Accessed January 5, 2024

Grimm, N. B., Faeth, S. H., Golubiewski, N. E., Redman, C. L., Wu, J., Bai, X., & Briggs, J. M. (2008). Global change and the ecology of cities. *Science, 319*, 756–760.

Grimmond, S. (2007). Urbanization and global environmental change: Local effects of urban warming. *The Geographical Journal, 173*, 83–88.

Heidenreich, E. (2007). Adaptive mutation in Saccharomyces cerevisiae. *Critical Reviews in Biochemistry and Molecular Biology, 42*, 285–311.

Herrera, C. M., & Bazaga, P. (2010). Epigenetic differentiation and relationship to adaptive genetic divergence in discrete populations of the violet *Viola cazorlensis*. *New Phytologist, 187*, 867–876.

Holzmueller, E. J., & Jose, S. S. (2009). Invasive plant conundrum: What makes the aliens so successful? *Journal of Tropical Agriculture, 47*, 18–29.

Houle, D. (1998). How should we explain variation in the genetic variance of traits? *Genetica, 103*, 241–253.

Hulme, P. E. (2009). Trade, transport and trouble: Managing invasive species pathways in an era of globalization. *Journal of Applied Ecology, 46*, 10–18.

Ibrahim, A., Papin, C., Mohideen-Abdul, K., Le Gras, S., Stoll, I., & Bronner, C. (2021). MeCP2 is a microsatellite binding protein that protects CA repeats from nucleosome invasion. *Science, 372*, eabd5581.

Jablonka, E., & Raz, G. (2009). Transgenerational epigenetic inheritance: Prevalence, mechanisms, and implications for the study of heredity and evolution. *The Quartarly Review of Biology, 84*, 131–176.

Kumari, P., Khan, S., Wani, I. A., Gupta, R., Verma, S., Alam, P., & Alaklabi, A. (2022). Unravelling the role of epigenetic modifications in development and reproduction of angiosperms: A critical appraisal. *Frontiers in Genetics, 13*, Article 819941.

Linde, M., Diel, S., & Neuffer, B. (2001). Flowering ecotypes of *Capsella bursa-pastoris* (L.) Medik. (Brassicaceae) analysed by a cosegregation of phenotypic characters (QTL) and molecular markers. *Annals of Botany, 87*, 91–99.

Lowe, S., Browne, M., Boudjelas, S., & De Poorter, M. (2004). *100 of the World's Worst Invasive Alien Species A selection from the Global Invasive Species Database*. Published by The Invasive Species Specialist Group (ISSG) a specialist group of the Species Survival Commission (SSC) of the World Conservation Union (IUCN), p. 12.

McKinney, M. L. (2006). Urbanization as a major cause of biotic homogenization. *Biological Conservation, 127*, 247–260.

Milne, R. I., & Abbott, R. J. (2000). Origin and evolution of invasive naturalized material of *Rhododendron ponticum* L. in the British Isles. *Molecular Ecology, 9*, 541–556.

Mimet, A., Pellissier, V., Quenol, H., Aguejdad, R., Dubreuil, V., & Roze, F. (2009). Urbanisation induces early flowering: Evidence from *Platanus acerifolia* and *Prunus cerasus*. *International Journal of Biometeorology, 53*, 287–298.

Nentwig, W., Bacher, S., Kumschick, S., Pyšek, P., & Vilà, M. (2017). More than "100 worst" alien species in Europe. *Biological Invasions, 20*, 1611–1621.

Ni, P., Li, S. G., Lin, Y. P., Xiong, W., Huang, X. N., & Zhan, A. B. (2018). Methylation divergence of invasive *Ciona ascidians*: Significant population structure and local environmental influence. *Ecology and Evolution, 8*, 10272–10287.

Potgieter, L. J., Gaertner, M., O'Farrell, P. J., & Richardson, D. M. (2019). Perceptions of impact: Invasive alien plants in the urban environment. *Journal of Environmental Management, 229*, 76–87.

Pyšek, P., Hulme, P. E., Simberloff, D., Bacher, S., Blackburn, T. M., Carlton, J. T., Dawson, W., Essl, F., Foxcroft, L. C., Genovesi, P., Jeschke, J. M., Kühn, I., Liebhold, A. M., Mandrak, N. E., Meyerson, L. A., Pauchard, A., Pergl, J., Roy, H. E., Seebens, H., ... Richardson, D. M. (2020). Scientists' warning on invasive alien species. *Biological Reviews, 95*, 1511–1534.

Pyšek, P., & Richardson, D. M. (2007). Traits associated with invasiveness in alien plants: Where do we stand? In W. Nentwig (Ed.), *Biological Invasions* (pp. 97–126). Springer.

Qiu, T., Liu, Z., & Liu, B. (2020). The effects of hybridization and genome doubling in plant evolution via allopolyploidy. *Molecular Biology Reports, 47*, 5549–5558.

Rai, K. P., & Singh, J. S. (2020). Invasive alien plant species: Their impact on environment, ecosystem services and human health. *Ecological Indicators, 111*, Article 106020.

Rejmánek, M., Richardson, D. M., Higgins, S. I., Pitcairn, M., & Grotkopp, E. (2005). Ecology of invasive plants: State of the art. In H. A. Mooney, J. A. McNeely, L. Neville, P. J. Schei, & J. Waage (Eds.), *Invasive alien species: A new synthesis* (pp. 104–162). Island Press.

Ricciardi, A., Hoopes, M. F., Marchetti, M. P., & Lockwood, J. L. (2013). Progress toward understanding the ecological impacts of nonnative species. *Ecological Monographs, 83*, 263–282.

Richardson, D. M., & Pyšek, P. (2006). Plant invasions: Merging the concepts of species invasiveness and community invasibility. *Progress in Physical Geography, 30*, 409–431.

Sarı, D. (2019). Kentsel Peyzajda Kullanılan Bazı İstilacı Süs Bitkileri; Türkiye'den Örnekler. In M. Dalkılıç (Ed.), *Scientific developments* (pp. 173–192). Gece Kitaplığı.

Schäfer, A., Heywood, J. B., Jacoby, H. D., & Waitz, I. A. (2009). *Transportation in a climate constrained world* (p. 329). The MIT Press.

Seboko, T., Shackleton, C. M., & Ruwanza, S. (2024). Urban residents' knowledge of and attitudes and willingness to control woody invasive alien plants in their domestic gardens in South Africa. *People and Nature, 6*, 2077–2090.

Shah, M., Khan, S. M., Mana, F., Hussain, J., Din, Z. U., Shah, I., Arshad, M., & Ahmad, I. (2025). Role of the plant nurseries in spread of invasive alien plant species in Pakistan's subtropical region: A threat for urban greening. *Urban Forestry & Urban Greening, 105*, Article 128688.

Simberloff, D., Martin, J.-L., Genovesi, P., Maris, V., Wardle, D. A., Aronson, J., Courchamp, F., Galil, B., García-Berthou, E., Pascal, M., Pyšek, P., Sousa, R., Tabacchi, E., & Vilà, M. (2013). Impacts of biological invasions: What's what and the way forward. *Trends in Ecology and Evolution, 28*, 58–66.

Sukopp, H., & Wurzel, A. (2003). The effects of climate change on the vegetation of central European cities. *Urban Habitats, 1*, 66–86.

Szulkin, M., Garroway, C. J., Corsini, M., Kotarba, A. Z., & Dominoni, D. M. (2020a). How to quantify urbanization when testing for urban evolution? In M. Szulkin, J. Munshi-South, & A. Charmantier (Eds.), *Urban evolutionary biology* (pp. 13–35). Oxford University Press.

Szulkin, M., Munshi-South, J., & Charmantier, A. (2020b). *Urban evolutionary biology*. Oxford University Press.

Te Beest, M., Le Roux, J. J., Richardson, D. M., Brysting, A. K., Suda, J., Kubešová, M., & Pyšek, P. (2012). The more the better? The role of polyploidy in facilitating plant invasions. *Annals of Botany, 109*, 19–45.

Tetik, Ö., & Üremiş, İ. (2015). Eichhornia crassipes. In H. Önen (Ed.), *Türkiye İstilacı Bitkiler Kataloğu* (pp. 303–315). TAGEM.

Tripathi, S., Bhadouria, R., Srivastava, P., Singh, R., & Batish, D. R. (2023). *Plant invasions and global climate change* (p. 418). Springer.

Tungmunnithum, D., Intharuksa, A.,& Sasaki, Y. (2020) A promising view of Kudzu Plant, *Pueraria montana* var. *lobata* (Willd.) Sanjappa & Pradeep: Flavonoid phytochemical compounds, taxonomic data, traditional uses and potential biological activities for future cosmetic application. *Cosmetics, 7*, 12.

Uludağ, A., Aksoy, N., Yazlık, A., Arslan, Z. F., Yazmış, E., Üremiş, I., Cossu, T. A., Groom, Q., Pergl, J., Pyšek, P., & Brundu, G. (2017). Alien flora of Turkey: Checklist, taxonomic composition and ecological attributes. *NeoBiota, 35*, 61–85.

Uludağ, A., Arslan, M., Üremiş, İ, & Aksoy, N. (2022). Advances in understanding invasive characteristics in weed species. In P. Kudsk (Ed.), *Advances in integrated weed management* (pp. 121–140). Burleigh Dodds Science Publishing Limited.

Uludağ, A., & Ertürk, Y. E. (2012). İthâl ev hayvanları ve süs bitkilerinin çevreye etkileri. *Journal of History Culture and Art Research, 1*, 428–444.

Uludağ, A., Ruşen, M., Ertürk, Y. E., & Üremiş, İ. (2013). İstilacı yabancı bitkilerin Türkiye'ye girişinde ve yayılmasında süs bitkilerinin muhtemel yeri ve önleyici faaliyetler. Süs Bitkileri Kongresi, 06–09 Mayıs 2013, Yalova, Türkiye,pp. 845–851.

Üremiş, I., Uludağ, A., Arslan, Z. F., & Abacı, O. (2014). A new record for the flora of Turkey: *Eichhornia crassipes* (Mart.) Solms (Pontederiaceae). *EPPO Bulletin, 44*, 83–86.

Van Valkenburg, J. L. C. H., Brunel, S., Brundu, G., Ehret, P., Follak, S., & Uludağ, A. (2014). Is terrestrial plant import from East Asia into countries in the EPPO region a potential pathway for new emerging invasive alien plants? *EPPO Bulletin, 44*, 195–204.

Vilà, M., & D'Antonio, C. M. (1998). Hybrid vigor for clonal growth in Carpobrotus (Aizoaceae) in coastal California. *Ecological Applications, 8*, 1196–1205.

Vilà, M., Trillo, A., Castro-Díez, P., Gallardo, B., & Bacher, S. (2024). Field studies of the ecological impacts of invasive plants in Europe. *NeoBiota, 90*, 139–159.

Wagner, V., Richardson, D. M., & Pyšek, P. (2024). Plant Invasions. In S. M. Scheiner (Ed.), *Encyclopedia of biodiversity* (3rd ed., pp. 680–695). Elsevier.

Walther, G. R., Roques, A., Hulme, P. E., Sykes, M. T., Pyšek, P., Kühn, I., & Bugmann, H. (2009). Alien species in a warmer world: Risks and opportunities. *Trends in Ecology & Evolution, 24*, 686–693.

Whitney, K. D., Randell, R. A., & Rieseberg, L. H. (2006). Adaptive introgression of herbivore resistance traits in the weedy sunflower *Helianthus annuus*. *The American Naturalist, 167*, 794–807.

Williams, F., Eschen, R., Harris, A., Djeddour, D., Pratt, C., Shaw, R. S., Varia, S., Lamontagne-Godwin, J., Thomas, S. E., & Murphy, S. T. (2010). *The economic cost of invasive non-native species on Great Britain.* CABI for The Scottish Government, Department for Environment Food and Rural Affairs UK Government, and Department for Economy and Transport Welsh Assembly Government.

Yazlık, A. (2021)0 İstilacı Yabancı Bitkiler ve Etkileri. (Yabancı Ot Biliminde Güncel Konular). In H. Mennan, & F. Pala (Eds.), İksad International Publishing House, pp. 263–293.

Yazlık, A. (2023). *Current status of invasive alien species (IAS): Trends, impacts, drivers, management and challenges.* 100th Anniversary Turkish Herbology Congress, 19–21 October 2023, Şanlıurfa, Türkiye, p. 6.

Yazlık, A., Kavak, M., Aşkın, E., Külcüoğlu, N., Ersoy, Ö., Kovankaya, F., Demirtaş, E., & Aydoğdu, A. (2020). Kentsel yaşam alanında bitki çeşitliliği ve etkileri: Düzce Üniversitesi Konuralp Kampüsü örneği. *Turkish Journal of Agricultural Research, 7*, 66–77.

Yazlık, A., Pergl, J., & Pyšek, P. (2018). Impact of alien plants in Turkey assessed by the Generic Impact Scoring System. *NeoBiota, 39*, 31–51.

Part II
Major Serious Invasive Alien Plants in Urban Areas

Chapter 5
Plant Invasion in the Urban Landscape: An Ecological Synthesis

Manoj Kumar Jhariya, Arnab Banerjee, Abhishek Raj, Bhimappa Honnappa Kittur, and Sushma Kerketta

Abstract The urban landscape is more fragile, and the altered land-use system becomes more hostile and a hub of exotic species through species introduction. These species, due to their wider ecological amplitude and wider ecological adaptation become invasive by excluding the indigenous flora of the region. Urban expansion leads to the destruction of local habitats and causes the loss of valuable biodiversity in addition to ecosystem functioning and services. The rapid urbanization due to population rise enhances the infrastructure and urban development, which transformed the natural landscapes. At the present time, changes in land use patterns also promote the integration of ecological invasion in the urban landscape. It was found that humans act as a potential agent for the spreading of invasive alien plant species (IAPs). Hence, the urban and peri-urban areas become highly susceptible towards the event of ecological invasion. From management perspective, the peri-urban areas should be given prior importance along with proper species selection by policymakers and decision-making people that would help to combat the ecological invasion process. The present chapter showcase the event of ecological invasion under the context of urban setup and addresses the issues of component, ecology, problem, challenges, policy and management aspects under urban setup.

Keywords Ecology · Exotics · Invasion · Urban green space · Vegetation

---

M. K. Jhariya (✉) · S. Kerketta
Department of Farm Forestry, Sant Gahira Guru Vishwavidyalaya, Ambikapur, Chhattisgarh, India
e-mail: manu9589@gmail.com

A. Banerjee
Department of Environmental Science, Sant Gahira Guru Vishwavidyalaya, Ambikapur, Chhattisgarh, India

A. Raj
Pt. Deendayal Upadhyay College of Horticulture and Forestry, Dr. Rajendra Prasad Central Agriculture University, Samastipur, Bihar, India

B. H. Kittur
Faculty of Technical Forestry, Indian Institute of Forest Management, Bhopal, Madhya Pradesh, India

R. Singh et al. (eds.), *Invasive Alien Plants in Urban Ecosystems*,
https://doi.org/10.1007/978-3-032-00997-5_5

1 Introduction

With the gradual rise in the human population, there is a significant crisis in terms of the availability of food along with their nutritional values. According to reports, the global population will rise to 9 billion by 2050, followed by 805 million people who would be suffering from malnutrition (FAO, 2018). With the objective of an increase in productivity, the conventional system of agriculture is unable to fulfil the objective, and shows negative ecological consequences (Jhariya et al., 2024a). Gradual conversion of forest lands into agricultural lands for more productivity is a serious consequence that leads to alteration in the land use pattern of the region. Urban forestry is a new concept that eliminates the negative consequences of habitat heterogeneity of urban landscapes (Jhariya et al., 2024b). Usually, the urban landscapes host diverse forms of plant species of which some are exotics in nature. For proper management of such species, the spatial distribution and diversity is of immense importance as they often regulate the propagation of invasive alien plant species (IAPs) (Fernandez Winzer et al., 2025; Zurell et al., 2022).

From the ecology and biodiversity point of view, the urban landscape has got the least attention and is yet to be explored properly (Bhagat et al., 2024; Dickey et al., 2018). Another huge research gap in ecology and conservation approaches are lacking in urban landscape due to improper approaches of the concerned municipalities of the concerned area under urban setup. Nowadays, the cities have become rich areas of IAPs due to tourism and trade (van Wilgen et al., 2020). Various mechanism in urban areas promotes the distribution of seeds and propagules of IAPs across the urban landscape in which the peri-urban areas often act as the migratory path for such species (McLean et al., 2017; Tripathi et al., 2023). Further, alteration in the habitat structure promotes an increase in the occurrence of introduced species due to anthropogenic influence (Jhariya et al., 2022b; McLean et al., 2017). If we go for a comparison between the urban and rural areas, separate management approaches are required to combat the problem in respective areas (Banerjee et al., 2023).

On a global basis, it was recorded that the occurrence of IAPs under urban areas comprises of one-fourth of the plant population (Aronson et al., 2014). Urban landscape tends to provide specific environmental conditions for novel species occurrences within particular habitats. It was observed that human habitation acts as a potential source for the migration of alien species (Buci et al., 2025; Lososova et al., 2012). Land use changes often result in the extinction of indigenous species which is highly deleterious from the point of view of urban biodiversity (Dickey et al., 2018). Urbanization is an event that profoundly influences the trends and patterns of biodiversity under an urban setup. It leads to the ecological expansion of invasive species followed by a declining population of indigenous flora (Trentanovi et al., 2013). The urban ecosystem is the human-made ecosystem where human-made structures dominates the land surface (Pickett et al., 2011). Ecological consequences of urbanization are huge as it leads to habitat fragmentation, changes in the food chain, lesser complexity of the food web which leads to the ecological crash of indigenous species (Kowarik, 2011). Another problem is that more than half of the human global population is dependent upon the urban landscape, and it is showing a further increasing trend across the globe. Therefore, there will be more landscape transformation, habitat fragmentation, edge effect and gradual loss of indigenous vegetation with the spreading of IAPs (Pickett et al., 2011). Hence, proper care needs to be taken to combat such ecological expansion of exotic species in urban environments.

2 Biological Invasion Under Urban Setup

At present, global warming and elevated temperatures are converting urban areas into urban heat islands which is facilitating the spreading of invasive species. Further, habitat fragmentation in city areas often creates a significant level of stress or disturbances with fragmented habitats that favour the ecological invasion mechanism (Zisenis, 2015). In the central part of Europe, such invasive species are thriving well under such disturbed habitat conditions as well as higher temperature (Klotz & Kuhn, 2010). The occurrence of invasive species has become more frequent in city areas due to higher dispersal ability, the presence of a significant level of disturbance regimes followed by habitat alterations that have created opportunities for the survival of non-native species (Gaertner et al., 2016a, 2016b). Biological invasion has now become a global event gaining the attention of research organization and institutions aiming towards generating a global database named Country Compendium of the Global Register of Introduced and Invasive Species (Pagad et al., 2022). According to one report till 2020 globally, 65 countries have reflected some sort of investment for the management of ecological invasion followed by listing of invasive species as was reported from 163 countries under National Biodiversity Strategy and Action Plans (Tataridas et al., 2023). Urbanization is often associated with altered land use that has caused significant loss of biodiversity, and the prevalence of invasive species across Europe, America and Asian subcontinents (Jaureguiberry et al., 2022). Very interestingly, invasion is associated with loss of economy in terms of millions of dollars on an annual basis, and very few amounts are invested to combat the problem. This was prevalent in South America, Oceania followed by Asia (Cuthbert et al., 2022).

Ecological invasion is very much predominant in urban areas, as the human-altered land use acts as a triggering agent for the spreading of invasive species across the habitat. Due to these, areas adjacent to cities are very vulnerable towards ecological invasion. From a global perspective, a gradual rise in the human population would lead to an increase in the frequency of non-native species within the urban landscape. It was observed that exotic species are often more associated with the urban setup in comparison to the indigenous species. This is due to the fact of higher migration rate and easy transportation of propagules of exotic species under urban setup. Factors such as the heterogeneous nature of the habitat, adaptability towards disturbances, and a higher level of temperature often regulates the distribution and dispersal of exotic species (Lososova et al., 2012; Williams et al., 2015). Hence, under urban setup, there is a prevalence of diverse species under diverse habitat forms. For example, in the central part of Europe, 40% of floral contribution comes from exotic species.

Similar things were reported from the Czech Republic, representing more than 50% contribution of exotic species among 1454 total plant taxa (Lososova et al., 2012). Most interestingly, these studies reflect that most of these exotic species tend to occur in various forms of human settlement. The higher adaptability of the alien species make them competent to invade to urban ecosystem and affect the indigenous vegetation negatively (Buci et al., 2025). Similarly, landscape dynamics determine the ecological behaviours of the existing invasive plants in the urban ecosystem. Change in land use practices and conversion affects invasion diversity, richness, occurrence and ecology (Mungi et al., 2023; Vila & Ibáñez, 2011). In this context, the effects of land use changes/conversion on plant invasions are depicted in Table 1.

Table 1 Effects of land use changes/conversion on plant invasions

Invasive plants in countries	Land use conversion	Effects on invasive plants	References
A. barbata in Mediterranean region of the Australia	Agricultural to old field	Increase the persistence of the invasive plant	Standish et al. (2008)
Alliaria petiolata in temperate region of the USA	Agricultural to forest land	Affects the occurrence and abundance of the invasive plant	Burls and McClaugherty (2008)
Berberis thunbergii in temperate region of the USA	Agricultural to forest land	Increase the occurrence and abundance of the invasive plant	DeGasperis and Motzkin (2007)
Cortaderia selloana in the Mediterranean region of Catalonia	Agricultural to grassland	Increase occurrence of the invasive plant	Domenech et al. (2005)
Heracleum mantegazzianum in temperate region of Germany	Agricultural to forest land	Increased occurrence and coverage of the invasive plant	Thiele and Otte (2006)
Leucaena leucocephala in subtropical region of the Puerto Rico	Agricultural to forest land	Increase basal area and density of the invasive plant	Colon and Lugo (2006)
Opuntia maxima in Mediterranean region of the Catalonia	Agricultural to shrub land and woodland	Increased occurrence of the invasive plant	Vila et al. (2003)
Several invasive plants in temperate region of USA	Agricultural to forest land	Decreased richness of the invasive plants	Kuhman et al. (2010)

3 Ecological Invasion and Urban Landscape

Urban and peri-urban areas are the potential sites of ecological invasion. It was observed that most of the invasion has taken place through the promotion of tourism as well as trading activity of horticultural plants (van Wilgen et al., 2020). Altered land use of the natural landscape helps in the dispersal and dissemination of the invasive species propagules within urban and peri-urban areas (McLean et al., 2017). Hence, there is frequent introduction of invasive species under anthropogenic influence plants. The major consequence of ecological invasion includes homogenization of the genome and alteration in the natural gene pool at the global level (van Wilgen et al., 2020). For the spreading of propagules of the invasive species, river networks, storm-water pathways are the major transport mechanisms. They significantly facilitate the wide distribution of invasive species. According to a historical report, in South Africa, the urban areas were infested with invasive species through timber and fuel-based forest activities along with sand dunes stabilization (Richardson et al., 2003). According to Mack (2003), up to 10,000 species come under the category of ecological transit towards incorporation within the traditional cultural practices globally. In Europe, restoration of the gardens often led to the spreading and distribution of invasive species into the local traditional culture (Davoren et al., 2016). A study was conducted on plant invasion, its distribution and its impacts on ecosystem services in the urban regions of the world. Urban ecosystem services and their percentage value under different invasive plants are depicted in Fig. 1 (Potgieter et al., 2017).

4 Component of Ecological Invasion Under Urban Landscape

The major driver of plant ecological invasions can be evaluated on temporal and spatial aspects and hence requires a broader landscape. Various researchers have mentioned about the influence of landscape architecture towards ecological invasion, however, most of the analyses were mono-species specific. In the European subcontinent invasive plant group, such as archaeophytes and neophytes, have been extensively studied on the basis of their distribution pattern as well as their mode of dispersal (Chytry et al., 2008). Results reveal the ecological invasion of archaeophytes is often associated with agricultural land use (Sheppard & Brendel, 2021) whereas the distribution of neophytes is associated with urban land use. Hence, landscape structure is an important aspect in determining the invasive nature of the invasive plants. All of these are important issues and factors that govern biodiversity loss and associated ecosystem services. Moreover, the invasion is the result of anthropogenic intervention of the land use, followed by mechanisms such as changing climate. Wide levels of variation of invasive species were recorded on a spatial scale. For example, in areas such as rail and roadways, a lesser proportion of urban vegetation happens to be the hotspot of invasive species. The results often indicate that habitat quality

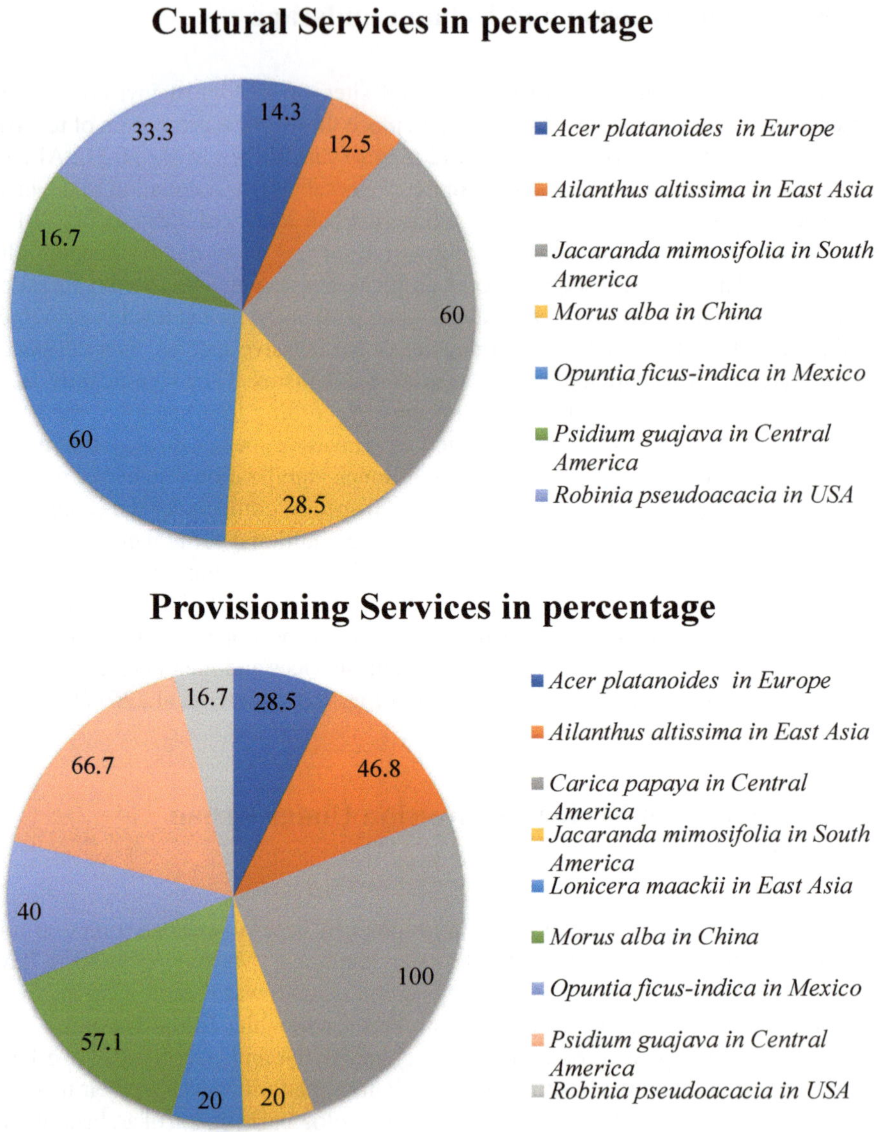

Fig. 1 Urban ecosystem services and its percentage value under different invasive plants in the world (based on Potgieter et al., 2017)

and structure are the important factors that need to be considered while going for management of alien species in an ecosystem (Stajerova et al., 2017).

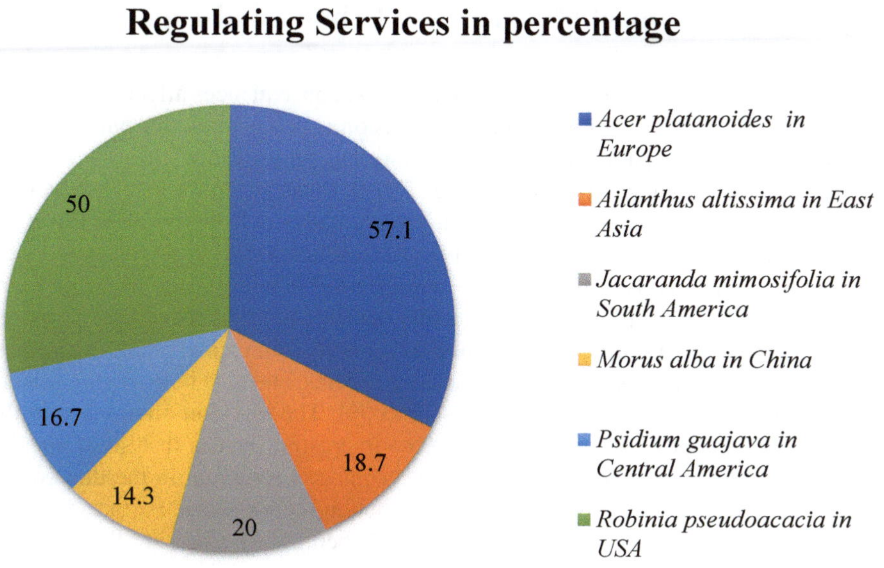

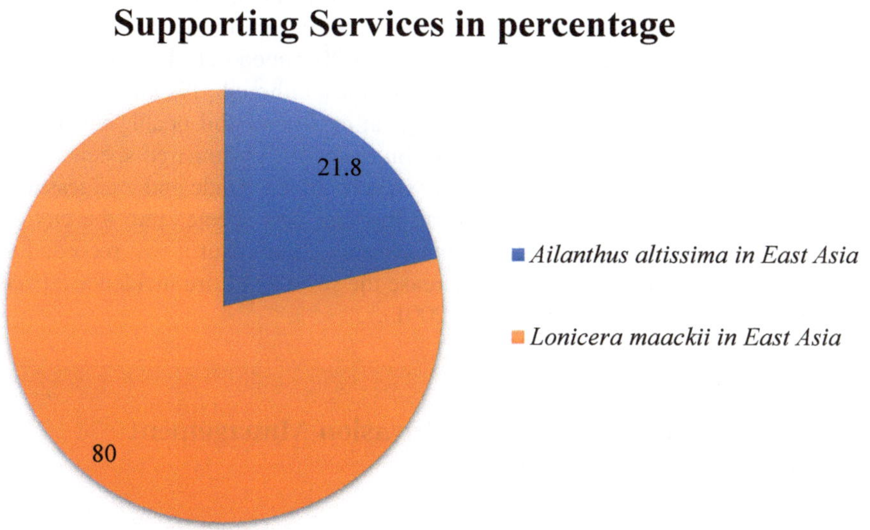

Fig. 1 (continued)

5 Ecological Impact of Invasion in Urban Landscape

Ecological invasion in cities has its own ecological consequences affecting the entire urban ecosystem. It was reported that the invasion is problematic for public's open access places such as parks, woodlots, etc. as well as for protected areas, and natural and semi-natural habitats (Trentanovi et al., 2013). Invasive plants, its consequences and management strategies are depicted in Fig. 2 (Hanley & Roberts, 2019; Raj et al., 2023a). The secondary forest's invasion by disturbance-associated species is a sign that the forest area and animal diversity are suffering in comparison to the oil palm plantations (Shafie et al., 2011). Sometimes, species used for beneficial purposes often lead to acts as a source of invasion. For example, ornamental plants in gardens, and parks can be a potential source of invasive species, and also help in a significant level of propagation (Alston & Richardson, 2006). The invasion often leads to the disruption of major ecosystem services. One interesting fact is that some species are used for net economic gain, while they become detrimental in the due course of time. Eucalyptus is a typical example of such a phenomenon. Further, invasive species under urban setup hampers human well-being by decreasing the aesthetic pleasure of the city (Kowarik, 2011).

Although most of the impacts of ecological invasion in urban areas are detrimental in nature, but they sometimes also provide positive impacts. For example, urban (exotic) trees in Florida have reported better capability of carbon sequestration in comparison to native indigenous species (Escobedo et al., 2010). Another most significant example is the Eucalyptus species, although it is known as a bane for any ecosystem as it alters the hydrological cycle and soil health but in South Africa, this species happens to be the roosting sites for endangered species Lesser Kestrel and African fish eagle (Cilliers & Siebert, 2012). Trade and business activities using pets and ornamental plants can mediate new species into the city area and tend to remain unregulated due to net economic gain. But it was recorded that some of the ornamental species may increase the severity of fire incidence adjacent to urbanized area (van Wilgen & Scott, 2001).

6 Problems and Challenges of Invasion Management Under Urban Landscape

Management of ecological invasion under an urban landscape is focussed on assessing the influence of invasive species over ecosystem services and disservices (Fernandez Winzer et al., 2025). It was observed that the conflict of interest occurs depending upon the species that are selected will impact over perceived benefits or create significant negative impacts (Vaz et al., 2017). The invasive species often create a condition regarding their use as positive aspects, along with the negative consequences they impose over a particular habitat (Novoa et al., 2018). Local conditions and existing ecological complexity often regulate urban plant invasion (Dickie et al.,

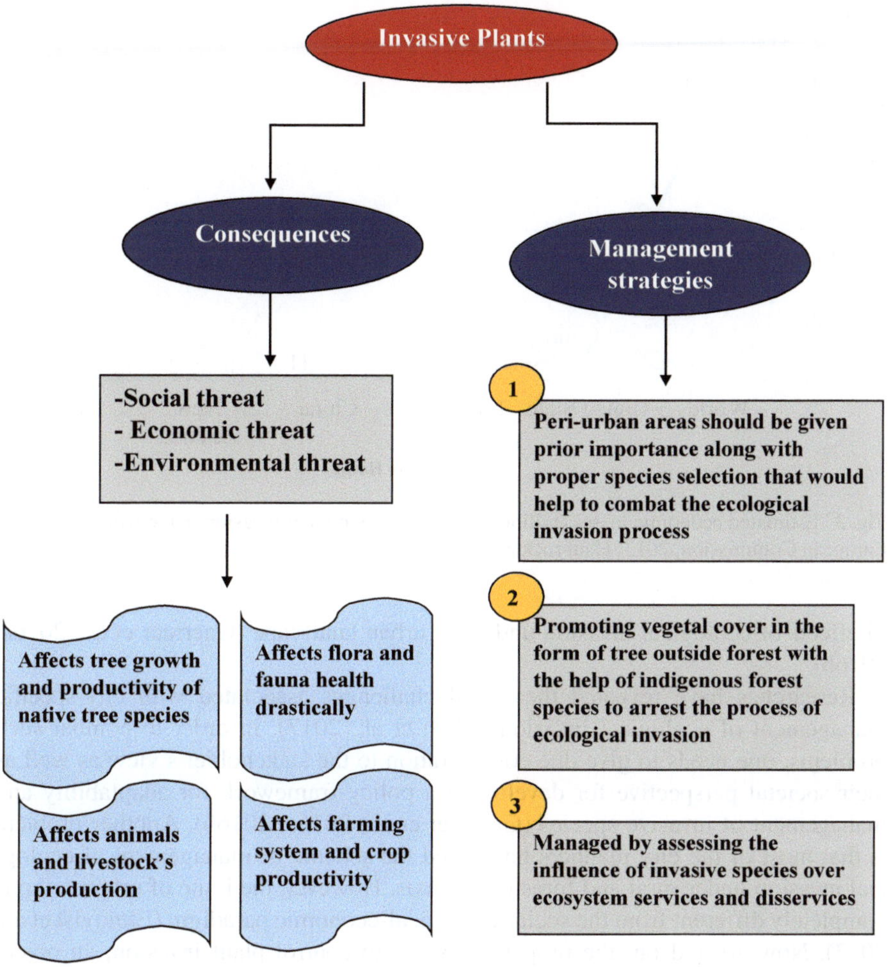

Fig. 2 Invasive plants, its consequences and management strategies (based on Hanley & Roberts, 2019; Raj et al., 2023a)

2014). As a consequence of all these, the proper management of ecological invasion becomes gloomy and proper care is not taken off. It was also observed that many of the exotic species were not regulated because of their social-political importance in a particular geographical area. This creates a huge problem for the proper management of the invasion mechanism. Estimated economic losses (euro billion/year) due to plant invasion in the world is depicted in Fig. 3 (European Commission, 2013). Management of invasive species under an urban setup often creates complications due to the cultural association with the local stakeholders and hence becomes important to eradicate or control (Warren, 2007). Such challenges need to be combated through increase in the level of awareness and education so that people can realise the

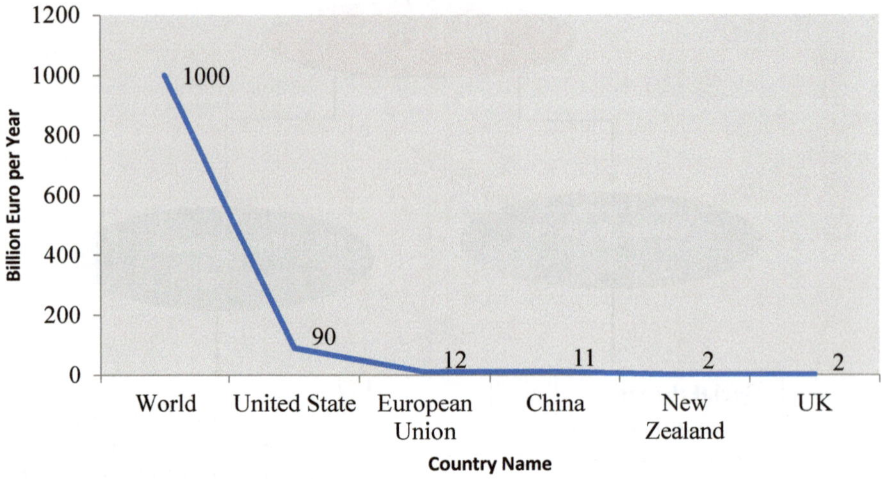

Fig. 3 Estimated economic losses (billion euro/year) due to plant invasion in the world (based on European Commission, 2013; Haubrock et al., 2021)

ill effects of ecological invasion under the urban landscape (Gaertner et al., 2016a, 2016b).

Researches have revealed the set of challenges associated with city-specific management of ecological invasion (Irlich et al., 2017). In order to combat such problems, one needs to give due consideration to the stakeholder's view as well as their societal perspective for developing a policy framework for adaptability and management of invasive species (Gaertner et al., 2016a, 2016b). Another problem is that most of the city planners have used the scheme of management of ecological invasion under rural and forestry contexts. However, the issue of urban setup is completely different from the social-ecological-economic paradigm (Elmqvist et al., 2013). Now, to find out the proper measures to control plant invasion, an objective quantification of the possible impacts is important. Knowledge regarding the database of exotic and invasive species are not only sufficient to manage the invasion process but also it requires people's participation as they would work at the grassroot level to combat the problem.

7 Management, Planning and Policy Perspective for Urban Landscape Invasion

Management or combating the ecological invasion phenomenon involves a cyclical pathway that comprises of risk evaluation, and early inventory followed by appropriate response for mitigation of such problems (Fernandez Winzer et al., 2025).

Blackburn et al. (2011) developed a framework to combat human-made biological invasion. According to this framework, the invasion process is divided into several stages. Physical, chemical and biological barriers can be employed to arrest the spreading of invasive species. Further, different methodologies such as containment, eradication, biological control, etc. can be applied. Very interestingly different approaches were observed regarding the control of invasion in urban areas across different parts of the world (van Wilgen et al., 2020). In this connection, a lack of proper funding resources for eradication and control was observed across countries in the world. Rather it is better to say that lack of awareness and interest on these issues have often led to lesser combating measures in the combating areas (van Wilgen et al., 2020).

Proper management of alien species requires the identification of specific sites and species that are vulnerable to the event of ecological invasion. Such activities would have significant socio-ecological impacts on the local stakeholders and would help for the long-term sustainability of the urban ecosystem (Banerjee et al., 2024). According to the latest report, the major aim is to restore vegetation as well as undergo plantation activities in various forms of degraded lands in order to improve the forest cover (Devi et al., 2023a; Raj et al., 2023b). Major initiatives in this regard include enhancing vegetal cover in the form of trees outside the forest with the help of the indigenous forest species to arrest the process of ecological invasion (Devi et al., 2023b). Under the scheme of the Nagar Van Project, the Indian Government has initiated the process of greening of urban space across 200 major cities of India during the upcoming five years. Such approaches have a positive turnover towards imposing prosperity and well-being of human civilization that would lead to sustainable development (Prebble et al., 2021).

Plantation in urban setup popularly known as urban forestry is expanding day by day in order to promote urban greening and develop sustainable cities. Urban forestry has got some inherent positivity in terms of the sustainable use of forest resource and optimum ecological services for the humans. Urban forestry proliferates various dimensions in terms of ecological restoration, socio-economic upliftment, building green urban infrastructure, combating climate change, developed climate resiliency and many other perspectives. However, all these approaches require spontaneous performance and participation of the local people as well as the level of awareness among them.

8 Future Roadmap

Ecological invasion is a process that hampers both the urban and other ecosystems in terms of decline of species diversity, loss of species interaction, soil quality, assets value, human health, and ecological and environmental consequences. From a future perspective, ecological invasion under an urban setup is an inevitable and irreversible phenomenon as the urban setup gradually builds up through an altered land use structure. This is a very important aspect to consider in order to combat ecological invasion under an urban setup. In this regard, proper screening and identification of species is required which is negatively influencing the altered urban food web structure. A proper ecological database on exotic species distribution in a particular habitat is also important in order to develop a suitable strategy for combating ecological invasion (Jhariya et al., 2022a). Fulfilling SDGs, urban forestry, and recreation forestry have gained significant importance in recent times. Hence, species selection and introduction should be taken care of in order to identify the risks to species of a given habitat. In the monitoring process of urban ecological invasion, early detection followed by community preparedness for arresting the spread of the species is highly important and should be prioritized in future times. Therefore, there is an urgent need to develop awareness among the people regarding the issue of urban ecological invasion followed by a significant level of R&D activities to this direction to combat the problem. Also, the identification of major causal factors of ecological invasion needs to be explored for developing sustainable management of the urban habitat conditions. People perception and awareness development is also important regarding this issue in order to involve them in the combating program and give a fruitful positive outcome.

9 Conclusions and Future Perspectives

Ecological invasion is an important event that is regulated by habitat structure, quality, landscape distribution pattern and various other intrinsic and extrinsic factors. On the other hand, the regulation and management of ecological invasion is another important issue. Such events are very severe in urban landscapes due to altered land-use patterns. Invasion often leads to loss of biodiversity, altered ecosystem structure and function and overall hampers the gene pool of an area. Proper care and management need to be formulated to address such issues in order to combat biodiversity loss as well as habitat degradation and the prosperity of human civilization. In general, management of invasion in city is a big challenge in comparison to other areas such as forests and agroecosystems. Hence, a proper framework for combating invasion under an urban setup is a scarce resource. The problem becomes more hectic when the area under consideration is of the higher value of conservation. Participation of the community stakeholders through awareness generation in relation to the invasion, and biodiversity is the major aspect that needs to be considered while planning for combating invasion. Another hard reality that needs to be understood by recognition of some widespread invasive species that are least likely to be eradicated. Under this condition, the major focus should be on minimizing the negative consequences of invasive species. Some proactive measures include an attempt to control invasive species through biological mode, restoring the original plant population, depending upon the condition, followed by replacement through species with non-invasive in nature. All these policies would help to combat the problem of ecological invasion and lead towards environmental sustainability.

References

Alston, K. P., & Richardson, D. M. (2006). The roles of habitat features, disturbance, and distance from putative source populations in structuring alien plant invasions at the urban/wildland interface on the Cape Peninsula, South Africa. *Biological Conservation, 132*(2), 183–198. https://doi.org/10.1016/j.biocon.2006.03.023

Aronson, M. F. J., La Sorte, F. A., Nilon, C. H., Katti, M., Goddard, M. A., Lepczyk, C. A., Warren, P. S., Williams, N. S. G., Cilliers, S., Clarson, B., Dobbs, C., Dolan, R., Hedblom, M., Klotz, S., Kooijmas, J. L., Kuhn, I., MacGregor-Fors, I., McDonnell, M., Mortberg, U., … Winter, M. (2014). A global analysis of the impacts of urbanization on bird and plant diversity reveals key anthropogenic drivers. *Proceedings of the Royal Society B, 281*(1780), 20133330.

Banerjee, A., Jhariya, M. K., Bargali, S. S., & Palit, D. (2023). *Ecorestoration for sustainability* (pp. 1–441). Wiley-Scrivener. ISBN: 9781119879718. https://doi.org/10.1002/9781119879954

Banerjee, A., Jhariya, M. K., Raj, A., & Mechergui, T. (2024). *Ecosystem management: Climate change and sustainability* (pp. 1–669). Wiley-Scrivener. ISBN 978-1-394-23121-8. https://doi.org/10.1002/9781394231249

Bhagat, S., Jhariya, M. K., Yadav, D. K., & Banerjee, A. (2024). Climber ecology and CO_2 mitigation potential in an urban landscape. *Energy, Ecology and Environment, 9*, 188–205. https://doi.org/10.1007/s40974-023-00297-2

Blackburn, T. M., Pysek, P., Bacher, S., Carlton, J. T., Duncan, R. P., Jarosik, V., Wilson, J. R. U., & Richardson, D. M. (2011). A proposed unified framework for biological invasions. *Trends in Ecology and Evolution, 26*(7), 333–339.

Buci, M., Jarčuška, B., Klinga, P., Ruzinska, R., Berggren, A., & Kanuch, P. (2025). Individual heterozygosity and fitness in bottlenecked populations during early colonisation. *Biological Invasions, 27*, 111. https://doi.org/10.1007/s10530-025-03571-y

Burls, K., & McClaugherty, C. (2008). Landscape position influences the distribution of garlic mustard, an invasive species. *Northeastern Naturalist, 15*, 541–556.

Chytry, M., Maskell, L. C., Pino, J., Pysek, P., Vila, M., Font, X., & Smart, M. T. (2008). Habitat invasions by alien plants: a quantitative comparison among Mediterranean, subcontinental and oceanic regions of Europe. *Journal of Applied Ecology, 45*, 448–458. https://doi.org/10.1111/j.1365-2664.2007.01398.x

Cilliers, S. S., & Siebert, S. (2012). Urban ecology in cape town: South African comparisons and reflections. *Ecology and Society, 17*(3), 33. https://doi.org/10.5751/ES-05146-170333

Colon, S. M., & Lugo, A. E. (2006). Recovery of a subtropical dry forest after abandonment of different land uses. *Biotropica, 38*, 354–364.

Cuthbert, R. N., Diagne, C., Hudgins, E. J., Turbelin, A., Ahmed, D. A., Albert, C., Bodey, T. W., Briski, E., Essl, F., Haubrock, P. J., Gozlan, R. E., Kirichenko, N., Kourantidou, M., Kramer, A. M., & Courchamp, F. (2022). Biological invasion costs reveal insufficient proactive management worldwide. *Science and Total Environment, 819*, Article 153404. https://doi.org/10.1016/j.scitotenv.2022.153404

Davoren, E., Siebert, S., & Cilliers, S. (2016). Influence of socioeconomic status on design of Batswana home gardens and associated plant diversity patterns in northern South Africa. *Landscape Ecology, 12*, 129–139.

DeGasperis, B. G., & Motzkin, G. (2007). Windows of opportunity: Historical and ecological controls on Berberis thunbergii invasions. *Ecology, 88*, 3115–3125.

Devi, A., Jhariya, M. K., Raj, A., Singh, K. P., & Singh, B. (2023a). *Acacia nilotica*: A promising species for soil sustainability. In: *Land and environmental management through forestry* (pp. 339–353). https://doi.org/10.1002/9781119910527.ch14

Devi, A., Jhariya, M. K., Yadav, D. K., & Banerjee, A. (2023b). Understory diversity and forest soil properties in different forest stands in Northern Chhattisgarh, India. *South African Journal of Botany, 162*, 171–182. https://doi.org/10.1016/j.sajb.2023.09.012

Dickey, J. W., Cuthbert, R. N., Rea, M., Laverty, C., Crane, K., South, J., Briski, E., Chang, X., Coughlan, N. E., MacIsaac, H. J., & Ricciardi, A. (2018). Assessing the relative potential ecological impacts and invasion risks of emerging and future invasive alien species. *NeoBiota, 40*, 1. https://doi.org/10.3897/neobiota.40.28519

Dickie, I. A., Bennett, B. M., Burrows, L. E., Nunez, M. A., Peltzer, D. A., Porte, A., Richardson, D. M., Rejmanek, M., Rundel, P. W., & van Wilgen, B. W. (2014). Conflicting values: Ecosystem services and invasive tree management. *Biological Invasion, 16*, 705–719.

Domenech, R., Vila, M., Pino, J., & Gesti, J. (2005). Historical landuse legacy and Cortaderia selloana invasion in the Mediterranean region. *Global Change Biology, 11*, 1054–1064

Elmqvist, T., Fragkias, M., Goodness, J., Guneralp, B., Marcotullio, P. J., McDonald, R. I., Parnell, S., Schewenius, M., Sendstad, M., Seto, K. C., & Wilkinson, C. (2013). *Urbanization, biodiversity and ecosystem services: Challenges and opportunities: A global assessment* (A part of the cities and biodiversity outlook project) (pp. 1–755). Springer, ISBN 978-94-007-7087-4, ISBN 978-94-007-7088-1 (eBook). https://doi.org/10.1007/978-94-007-7088-1

Escobedo, F., Varela, S., Zhao, M., Wagner, J. E., & Zipperer, W. (2010). Analyzing the efficacy of subtropical urban forests in offsetting carbon emissions from cities. *Environmental Science and Policy, 13*(5), 362–372. https://doi.org/10.1016/j.envsci.2010.03.009

European Commission (2013) Flash Eurobarometer 379—Attitudes towards biodiversity. https://ec.europa.eu/environment/eurobarometers_en.htm

FAO. (2018). *The future of food and agriculture: Alternative pathways to 2050* (p. 224). FAO UN.

Fernandez Winzer, L., Greve, M., le Roux, P. C., Faulkner, K. T., & Wilson, J. R. U. (2025). Using indicators to assess the status of biological invasions and their management on islands–the Prince Edward Islands, South Africa as an example. *Biological Invasions, 27*, 108. https://doi.org/10.1007/s10530-024-03463-7

Gaertner, M., Larson, B. M. H., Irlich, U. M., Holmes, P. M., Stafford, L., van Wilgen, B. W., & Richardson, D. M. (2016a). Managing invasive species in cities: A framework from Cape Town, South Africa. *Landscape and Urban Planning, 151*, 1–9. https://doi.org/10.1016/j.landurbplan.2016.03.010

Gaertner, S., Guerra, R., Rebelo, M., Dovidio, J., Hehman, E., & Deegan, M. (2016). The common ingroup identity model and the development of a functional perspective: A cross-national collaboration. In J. Vala, S. Waldzus, & M. M. Calheiros (Eds.), *The social developmental construction of violence and intergroup conflict* (pp. 105–120). Springer International Publishing AG. https://doi.org/10.1007/978-3-319-42727-0_5

Hanley, N., & Roberts, M. (2019). The economic benefits of invasive species management. *People and Nature, 1*, 124–137. https://doi.org/10.1002/pan3.31

Haubrock, P. J., Turbelin, A. J., Cuthbert, R. N., Novoa, A., Taylor, N. G., Angulo, E., Ballesteros-Mejia, L., Bodey, T. W., Capinha, C., Diagne, C., & Essl, F. (2021). Economic costs of invasive alien species across Europe. *NeoBiota, 67*, 153–190.

Irlich, U. M., Potgieter, L. P., Stafford, L., & Gaertner, M. (2017). Recommendations for municipalities to become compliant with national legislation on biological invasions. *Bothalia–african Biodiversity & Conservation, 47*(2), 1–11. https://doi.org/10.4102/abc.v47i2.2156

Jaureguiberry, P., Titeux, N., Wiemers, M., Bowler, D. E., Coscieme, L., Golden, A. S., Guerra, C. A., Jacob, U., Takahashi, Y., Settele, J., Diaz, S., Molnar, Z., & Purvis, A. (2022). The direct drivers of recent global anthropogenic biodiversity loss. *Science Advances, 8*, eabm9982. https://doi.org/10.1126/sciadv.abm9982

Jhariya, M. K., Banerjee, A., Raj, A., Meena, R. S., Khan, N., Kumar, S., & Bargali, S. S. (2022a). Species invasion and ecological risk. In *Natural resources conservation and advances for sustainability* (pp. 503–531). Elsevier. ISBN: 9780128229767. https://doi.org/10.1016/B978-0-12-822976-7.00014-4

Jhariya, M. K., Meena, R. S., Banerjee, A., Kumar, S., Raj, A. (2024a). *Agroforestry for carbon and ecosystem management* (pp. 434). Elsevier. ISBN: 9780323953931, eBook ISBN: 9780323958752. https://doi.org/10.1016/C2021-0-02916-3.

Jhariya, M. K., Meena, R. S., Banerjee, A., & Meena, S. N. (2022b). *Natural resources conservation and advances for sustainability*. Elsevier. ISBN: 9780128229767. https://doi.org/10.1016/C2019-0-03763-6

Jhariya, M. K., Raj, A., Banerjee, A., Meena, R. S., Devi, A., & Kittur, B. H. (2024b). Urban emission reduction and carbon management. In: *Agroforestry for carbon and ecosystem management* (pp. 161–171). Elsevier. ISBN: 9780323953931. https://doi.org/10.1016/B978-0-323-95393-1.00022-1

Khan, N., Jhariya, M. K., & Raj, A. (2022a). Urban greening toward sustainable development and sustainability. In M. Ozturk, S. M. Khan, V. Altay, R. Efe, D. Egamberdieva, & F. O. Khassanov (Eds.), *Biodiversity, conservation and sustainability in Asia, south and middle Asia* (pp. 345–373). Springer.

Khan, N., Jhariya, M. K., Raj, A., Banerjee, A., & Meena, R. S. (2021a). Soil carbon stock and sequestration: Implications for climate change adaptation and mitigation. In M. K. Jhariya, R. S. Meena, & A. Banerjee (Eds.), *Ecological intensification of natural resources for sustainable agriculture* (pp. 461–489). Springer.

Khan, N., Jhariya, M. K., Raj, A., Banerjee, A., & Meena, R. S. (2021b). Eco-designing for sustainability. In M. K. Jhariya, R. S. Meena, & A. Banerjee (Eds.), *Ecological intensification of natural resources for sustainable agriculture* (pp. 565–595). Springer.

Khan, N., Jhariya, M. K., Raj, A., Bargali, K., Devi, A., & Ghanshyam. (2022b). Urban forest resources: A strategy for achieving land degradation neutrality. In P. Panwar et al. (Eds.), *Land degradation neutrality: Achieving SDG 15 by forest management* (pp. 363–380). Springer. ISBN: 978-981-19-5477-1. https://doi.org/10.1007/978-981-19-5478-8_19

Klotz, S., & Kuhn, I. (2010). Urbanisation and alien invasion. In K. J. Gaston (Ed.), *Urban ecology* (pp. 120–133). Cambridge University Press.

Kowarik, I. (2011). Novel urban ecosystems, biodiversity, and conservation. *Environmental Pollution, 159*, 1974–1983. https://doi.org/10.1016/j.envpol.2011.02.022

Kuhman, T. R., Pearson, S. M., & Turner, M. G. (2010). Effects of landuse history and the contemporary landscape on non-native plant invasion at local and regional scales in the forest dominated southern Appalachians. *Landscape Ecology, 25*, 1433–1445.

Lososova, Z., Chytry, M., Tichy, L., Danihelka, J., Fajmon, K., Hajek, O., Kintrova, K., Lanıkova, D., Otypkova, Z., & Rehorek, V. (2012). Biotic homogenization of Central European urban floras depends on residence time of alien species and habitat types. *Biological Conservation, 145*(1), 179–184.

Mack, R. N. (2003). Global plant dispersal, naturalization, and invasion: pathways, modes and 557 circumstances. In G. M. Ruiz, & J. T. Carlton (Eds.), *Invasive species: Vectors and management 558 strategies*. Island Press.

McLean, P., Gallien, L., & Wilson, J. R. U. (2017). Small urban centres as launching sites for plant invasions in natural areas: Insights from South Africa. *Biological Invasions, 19*, 3541–3555.

Mungi, N. A., Qureshi, Q., & Jhala, Y. V. (2023). Distribution, drivers and restoration priorities of plant invasions in India. *Journal of Applied Ecology, 60*(11), 2400–2412.

Novoa, A., Shackleton, R., Canavan, S., Cybele, C., Davies, S. J., Dehnen-Schmutz, K., Fried, J., Gaertner, M., Geerts, S., Griffiths, C. L., Kaplan, H., Kumschick, S., Le Maitre, D. C., John Measey, G., Nunes, A. L., Richardson, D. M., Robinson, T. B., Touza, J., & Wilson, J. R. U. (2018). A framework for engaging stakeholders on the management of alien species. *Journal of Environmental Management, 205*(1), 286–297.

Pagad, S., Bisset, S., Genovesi, P., Groom, Q., Hirsch, T., Jetz, W., Ranipeta, A., Schigel, D., Sica, Y. V., & McGeoch, M. A. (2022). Country compendium of the global register of introduced and invasive species. *Sci Data, 9*, 1–13. https://doi.org/10.1038/s41597-022-01514-z

Pickett, S. T. A., Cadenasso, M. L., Grove, J. M., Boone, C. G., Groffman, P. M., Irwin, E., Kaushal, S. S., Marshall, V., McGrath, B. P., Nilon, C. H., Pouyat, R. V., Szlavecz, K., Troy, A., & Warren, P. (2011). Urban ecological systems: Scientific foundations and a decade of progress. *Journal of Environmental Management, 92*(3), 331–362.

Potgieter, L. J., Gaertner, M., Kueffer, C., Larson, B. M. H., Livingstone, S., O'Farrell, P., & Richardson, D. M. (2017). Alien plants as mediators of ecosystem services and disservices in urban systems: A global review. *Biological Invasions*. https://doi.org/10.1007/s10530-017-1589-8

Prebble, S., McLean, J., & Houston, D. (2021). Smart urban forests: An overview of more-than-human and more-than-real urban forest management in Australian cities. *Digital Geography Soc, 2*, Article 100013.

Raj, A., Jhariya, M. K., Banerjee, A., Dinesha, S., Jha, R. K., Kumar, K., & Mechergui, T. (2023a). Menace of plant invasion: A view from ecological lens. In *Plant invasions and global climate change* (pp. 57–76). Springer. https://doi.org/10.1007/978-981-99-5910-5_3

Raj, A., Jhariya, M. K., Banerjee, A., Nema, S., & Bargali, K. (2023b). *Land and environmental management through forestry* (pp. 1–372). Wiley-Scrivener, USA. ISBN: 9781119910404. https://doi.org/10.1002/9781119910527

Richardson, D. M., Cambray, J. A., & Chapman, R. A. (2003). Vectors and pathways of biological invasions in South Africa—Past, future and present. In G. Ruiz & J. Carlton (Eds.), *Invasive species: Vectors and management strategies* (pp. 292–349). Island Press.

Shafie, N. J., Mohd Sah, S. A., Abdul Latip, N. S., Azman, N. M., & Khairuddin, N. L. (2011). Diversity pattern of bats at two contrasting habitat types along Kerian River, Perak. *Malaysia. Tropical Life Science Research, 22*(20), 13–22.

Sheppard, C. S., & Brendel, M. R. (2021). Competitive ability of native and alien plants: effects of residence time and invasion status. *NeoBiota, 65*, 47–69. https://doi.org/10.3897/neobiota.65.63179

Stajerova, K., Smilauer, P., Bruna, J., & Pysek, P. (2017). Distribution of invasive plants in urban environment is strongly spatially structured. *Landscape Ecology, 32*, 681–692. https://doi.org/10.1007/s10980-016-0480-9

Standish, R. J., Cramer, V. A., & Hobbs, R. J. (2008). Land-use legacy and the persistence of invasive *Avena barbata* on abandoned farmland. *Journal of Applied Ecology, 45*, 1576–1583.

Tataridas, A., Travlos, I., & Freitas, H. (2023). Agroecology and invasive alien plants: A winner-take-all game. *Frontiers in Plant Science, 14*, 1143814. https://doi.org/10.3389/fpls.2023.1143814

Thiele, J., & Otte, A. (2006). Analysis of habitats and communities invaded by *Heracleum mantegazzianum* Somm. et Lev. (Giant Hogweed) in Germany. *Phytoenologia, 36*, 281–320.

Trentanovi, G., von der Lippe, M., Sitzia, T., Ziechmann, U., Kowarik, I., & Cierjacks, A. (2013). Biotic homogenization at the community scale: Disentangling the roles of urbanization and plant invasion. *Diversity and Distributions, 19*(7), 738–748. https://doi.org/10.1111/ddi.12028

Tripathi, S., Bhadouria, R., Srivastava, P., Singh, R., Batish, D. R. (Eds). (2023). In *Plant invasions and global climate change*. Springer Singapore. https://doi.org/10.1007/978-981-99-5910-5

van Wilgen, B., Measey, J., Richardson, D. M., Wilson, J. R., & Zengeya, T. A. (2020). Biological invasions in South Africa. In *Invading nature, springer series in invasion ecology* (INNA Vol. 14, pp. 1–975). https://doi.org/10.1007/978-3-030-32394-3

van Wilgen, B. W., & Scott, D. F. (2001). Managing fires on the Cape Peninsula: Dealing with the inevitable. *Journal of Mediterranean Ecology, 2*, 197–208.

Vaz, A. S., Kueffer, C., Kull, C. A., Richardson, D. M., Vicente, J. R., Kuhn, I., Schroter, M., Hauck, J., Bonn, A., & Honrado, J. P. (2017). Integrating ecosystem services and disservices: Insights from plant invasions. *Ecosystem Services, 23*, 94–107.

Vila, M., Burriel, J. A., Pino, J., Chamizo, J., Llach, E., Porterias, M., & Vives, M. (2003). Association between Opuntia spp. invasion and changes in land-cover in the Mediterranean region. *Global Change Biology, 9*, 1234–1239.

Vila, M., & Ibáñez, I. (2011). Plant invasions in the landscape. *Landscape Ecology, 26*, 461–472.

Warren, C. R. (2007). Perspectives on the 'alien' versus 'native' species debate: A critique of concepts, language and practice. *Progress in Human Geography, 31*(4), 427–446. https://doi.org/10.1177/0309132507079499

Williams, N. S. G., Hahs, A. K., & Vesk, P. A. (2015). Urbanisation, plant traits and the composition of urban floras. *Perspectives in Plant Ecology, Evolution and Systematics, 17*, 78–86.

Zisenis, M. (2015). Alien plant species: A real fear for urban ecosystems in Europe? *Urban Ecosystems, 18*(2), 355–370. https://doi.org/10.1007/s11252-014-0400-1

Zurell, D., Konig, C., Malchow, A. K., Kapitza, S., Bocedi, G., Travis, J., & Fandos, G. (2022). Spatially explicit models for decision-making in animal conservation and restoration. *Ecography, 2022*(4). https://doi.org/10.1111/ecog.05787

Chapter 6
Functional Traits Reveal the Success of Alien Plant Species Along the Invasion Continuum

Priya Kaushik, Pranab Kumar Pati, Mohammed Latif Khan, and Pramod Kumar Khare

Abstract Invasive plant species pose a global ecological challenge, exacerbated by our interconnected world and rapid population growth. These invasions interact with broader ecological shifts, amplifying their reach and impact. Invasive plants disrupt natural biogeographic boundaries, reducing native species' diversity and abundance, elevating extinction risks, and altering genetic traits and behaviours. They also disrupt ecosystem function, trophic relationships, and native species' interactions. However, urban ecosystems, with their unique environmental and socio-economic settings, are particularly vulnerable to such invasions. Numerous frameworks and hypotheses attempt to explain invasion success, but a thorough analysis of the hidden attributes enabling alien plants to progress through invasion stages is lacking. This chapter explores the dynamics of biological invasions within urban ecosystems, examining the ecological, socio-economic, and management dimensions. It also focuses on community-level perspectives, particularly exploring how intrinsic traits drive invasiveness and how biotic and abiotic factors influence ecosystem resistance. It examines the ways humans modify environments, at various scales, leading to plant invasions and alter future courses. Climate change's potential impact on invasive plants underscores the need for ongoing monitoring and modelling. Managing invasive species requires caution, with demand for innovative approaches to enhance control measures. Recent studies highlight the dynamic interactions between urbanization and invasive species, highlighting the human-mediated disturbances and habitat modifications. Strategies for managing invasions in urban contexts are also discussed, providing insights into maintaining biodiversity of urban areas and ecosystem services. This chapter provides a model for understanding invasive species in forest ecosystems to guide research and conservation efforts.

Keywords Anthropogenic disturbances · Climate change · Ecosystem resilience · Species interaction · Urban ecosystems

P. Kaushik (✉) · P. K. Pati · M. L. Khan · P. K. Khare
Ecology Laboratory, Department of Botany, Dr. Harisingh Gour Vishwavidyalaya, Sagar, Madhya Pradesh, India
e-mail: priyakaushiik@gmail.com

R. Singh et al. (eds.), *Invasive Alien Plants in Urban Ecosystems*,
https://doi.org/10.1007/978-3-032-00997-5_6

Highlights

This chapter presents a comprehensive analysis of biological invasions in urban ecosystems, emphasizing:

1. The unique drivers and routes of invasions in urban settings.
2. The effects of invasive species on the environment and the economy in urban areas.
3. Recent advances in management strategies tailored to urban contexts.
4. By synthesizing recent findings, this chapter underscores the importance of addressing urban invasions to protect ecosystem services and biodiversity in rapidly urbanizing landscapes.

1 Introduction

Organisms' welfare and ecosystem services are necessarily linked to biodiversity (Pati et al., 2022; Rai & Singh, 2020). Biodiversity serves as the basis for provisioning goods and services, such as medicines, and upholding ecosystem functions, contributing substantially to the global economy (Jones, 2019). Further, Biodiversity provides various strategies to mitigate global change impacts and manage natural disasters, thereby playing a pivotal role in advancing the Sustainable Development Goals.

Due to significant changes in the global environment over the past few decades, invasive alien species (IAS) have been recognized as an important risk to biodiversity (Gulzar et al., 2023). After the destruction of natural habitats, IAS are now one of the primary drivers for decreasing biodiversity, causing harmful effects on local ecosystems, environmental health, and biodiversity (Kueffer, 2017). In addition, these species pose threats to human health and result in significant economic losses. IAS is one of the main causes of biodiversity loss, according to the UN's Global Assessment Report on Biodiversity and Ecosystem Services (Jones & McDermott, 2018).

The human-driven introduction of alien species has increased significantly due to worldwide trade and tourism, resulting in an increased rate of species introductions outside their native ranges. Research suggests that the ecological and economic impacts of IAS are disproportionately substantial compared to other natural disasters, with economic costs surpassing those of major natural disasters (Epanchin-Niell et al., 2021). Global ecosystem functioning and biodiversity are being threatened by biological invasions. Recent studies have underlined the increasing prevalence of invasions in urban ecosystems, facilitated by drivers such as habitat fragmentation, climate change, and global trade (McKinney, 2002; Pyšek et al., 2020a).

Consequently, there has been a global increase in research within invasion biology to understand the traits contributing to the invasiveness of certain alien species and to monitor, manage, and predict future incursions effectively (Divíšek et al., 2018; Gulzar et al., 2023).

2 IAS and Drivers of Urban Invasions

Invasive species are non-native plants that have become established and are reproducing rapidly over various distances from their parent plants. The Global Invasive Species Programme (GISP), the Convention on Biological Diversity (CBD), and the World Conservation Union coined the term "IAS." After the successful invasion of the ecosystems invaded, these species can significantly alter mechanisms such as hydrology, nutrient cycling, fire patterns, and the abundance, survival of native species. They often employ efficient reproductive strategies, enabling them to thrive even in remote areas (Kumschik et al., 2025).

As per the report by Díaz et al., (2019), Invasive species are a hazard to about 20% of the Earth's surface, according to the United Nations Intergovernmental Platform for Biodiversity and Ecosystem Services (IPBES). The invasion of these alien species, along with anthropogenic disturbances, climate change, and biodiversity is closely interconnected (Fig. 1).

Urban ecosystems are hotspots for biological invasions due to their unique characteristics, including altered microclimates, high levels of disturbance, and diverse pathways for species introduction (Gaertner et al., 2017). In tropical cities, the impact of urban heat islands can create optimal conditions for invasive species, like *Parthenium hysterophorus*. (Sharma et al., 2019). The negative effects of invading species on human populations' health may worsen due to the rapid spread of diseases via vectors. Additionally, invasive plants may act as weeds, reducing plant diversity and agricultural productivity. Their presence is detrimental to achieving Sustainable Development Goals (SDGs) related to poverty alleviation, human well-being, and food security.

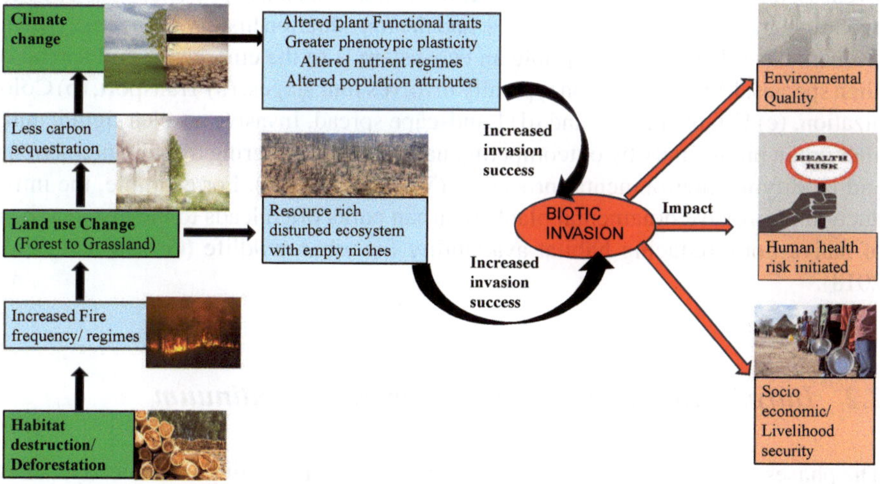

Fig. 1 A framework interlinking anthropogenic factors and global environmental changes with their impacts on socio-economy, livelihoods, and human health

Invasive species are often categorized into four groups: (i) Primary invasive plant species that require careful management in forest environments. (ii) Species that are invasive in both human-altered and natural habitats, necessitating monitoring and control efforts everywhere. (iii) Species that are invasive only in human-altered habitats, which should still be monitored. (iv) Potentially invasive species that warrant ongoing monitoring and control measures (Chauhan & Verma, 2023; Marsh et al., 2021).

2.1 Mechanism of Spread of IPS and Impact on Urban Biodiversity

Invasive plant species (IPS) possess the ability to outcompete indigenous species in their new habitats, disrupting ecological processes and potentially causing economic and ecological harm (Tripathi et al., 2023). Thus, the invasion ecology aims to study the dynamics, impacts, and invasive species management. It involves the intentional introduction of IPS by humans into areas outside of their natural range, made possible by actions that cross biogeographic boundaries. Position of a species on the naturalization-invasion continuum affects its potential to invade a new habitat (Blackburn et al., 2011; Cadotte et al., 2018; Gioria et al., 2023). Strong reproductive systems enable IPS to form self-sustaining populations and disperse even to distant places. The native vegetation of the invaded ecosystem, which has developed and flourished naturally without human interference, may be significantly impacted. The two core concepts of invasion ecology are invasiveness and invasibility. Degree to which an alien species may colonize a new environment is known as its invasiveness, and it can be affected by several attributes such as the traits of the species, evolutionary and ecological processes, and the introduction history. Conversely, invasiveness describes how susceptible an ecosystem is to the emergence and spread of alien species. IAS proliferation typically involves four stages: (a) Transport, (b) Colonization, (c) Establishment, and (d) Landscape spread. Invasive species significantly alter urban biodiversity by outcompeting native species, altering trophic interactions, and modifying environmental processes (Vilà et al., 2011). For example, the introduction of invasive ornamental plants in urban parks often leads to the displacement of native flora, reducing habitat availability for urban wildlife (Groeneveld et al., 2018).

2.2 Introduction-Naturalization-Invasion-Continuum

The phases and stages that an organism experiences upon being introduced to a new environments are referred to as the invasion continuum. For the species to advance along the continuum, it must overcome obstacles that relate to the environment and reproduction during these phases (Richardson et al., 2000). Due to this, it could be challenging for introduced species to grow into thriving populations (Simberloff, 2009). While some species establish self-sustaining communities, others only live in the new habitat for a brief period. A small proportion of these self-sufficient populations develop into invasive species that quickly expand and establish themselves over the surrounding area (Fristoe et al., 2021; Richardson & Pyšek, 2012). Few aliens progress through extended lag phases before exhibiting the fast population growth, range extension, and enhanced local dominance that many invasive plants exhibit (Osunkoya et al., 2021).

Successful invasions describe the progression of IAS along the continuum and are assessed based on intrinsic factors, such as intrinsic species attributes or functional attributes which determine invasiveness (Lonsdale, 1999; Pyšek et al., 2014). Moreover, external factors (either ecological or evolutionary) related to the recipient ecosystem functions determine invasibility. Propagule pressure is the frequency and number of introduced propagules, influences invasion probability and the colonization pressure denotes species number introduced into an ecosystem (Briski et al., 2018; Colautti et al., 2014).

2.3 *Functional Traits Behind Invasion Success*

Functional attributes are specific features of individual plants, including morpho-physiological, phenological, biochemical, structural, or behavioural traits (phenotypes) that shows their responses to the environment. Their significance varies along the path of the invasion continuum (Divíšek et al., 2018; Kaushik et al., 2022a). A great deal of effort has been given to how functional traits and their ability to adapt enable invaders succeed (Kaushik et al., 2022b; Pyšek et al., 2015; Richardson, 2013). This research dates to the discovery of ideal weed characteristics that are typically linked to human activities (Baker, 1965) (Fig. 2).

Functional trait approaches are based on the hypothesis that invasion success is based on specific traits of exotic / alien species, and also on the way they render a species better able to flourish in the abiotic and biotic environments of the introduced ecosystems (Cadotte et al., 2018; Fridly, 2013). IPS have a specific trait combination that makes it easier to establish, sustain population growth, and expand their range. This helps them outcompete native species in non-native environments (Drenovsky et al., 2012). Although some characteristics are advantageous in the establishment stage, others become essential in the invasion stage as they provide different barriers and filters (Kaushik et al., 2022b; Richardson et al., 2000).

According to a well-established conceptual framework, the naturalization of alien plant species are impacted by various environmental factors. The realized niche, i.e., the real ecological niche of plant species depends on its broad fundamental niche, which includes favourable environmental conditions required for its survival, reproduction, and interactions with other species. Environmental filtering is a crucial factor that contributes to the success of naturalization of exotic plants. To become

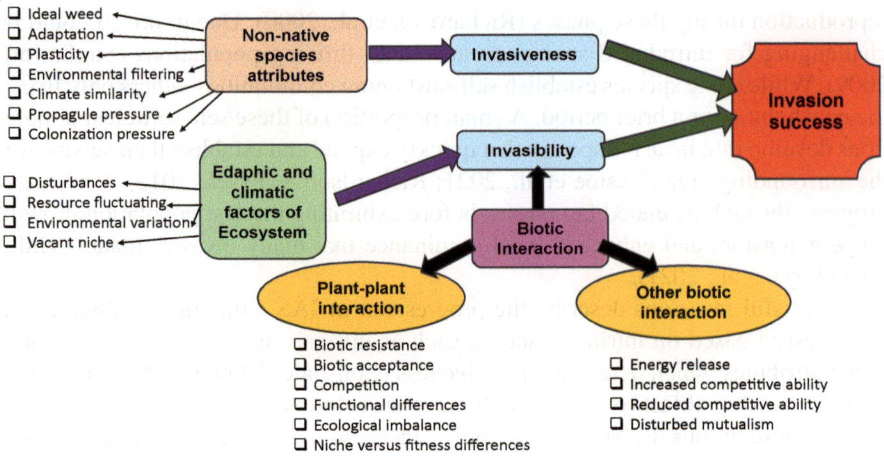

Fig. 2 Different factors (biotic and abiotic) contributing to invasion success

naturalized, a non-native species must have a fundamental ecological niche that aligns with the requirements of the new environment. This alignment depends on pre-adaptation, such as climatic matching or pre-introduction selection, and is subject to variation based on the features of the introduced ecosystem. These ideas have been explored in several studies, including those by Diez et al. (2008), Thuiller et al. (2010), Richardson and Pyšek (2012), Colautti and Lau (2016), Briski et al. (2018).

A recent study suggests how these IAS thrive in climates that are similar to their native region. Invasive plants often occupy similar niches to native species in the ecosystem, with modest niche extension across native and introduced ranges (Cadotte & Tucker, 2017). However, even when non-native species fit the ecological requirements of a specific area, they may experience biotic resistance due to competing with native species (Mitchell et al., 2006). This could be owing to new natural enemies or a lack of mutualists. Environmental filtering is one mechanism that can discriminate against non-native species, but other mechanisms, such as competition, can produce comparable patterns that are often indistinguishable (Liu et al., 2020).

2.4 *Reproductive Traits and Invasion Success*

Since the founder populations are usually small because of low propagule input or low survival rates, they are vulnerable to stochastic extinction. Overcoming obstacles to establishment is mostly dependent on reproductive characteristics (Moravcová et al., 2015; Razanajatovo et al., 2019). IPS have characteristics like high seed production and/or broad vegetative propagation and development, which are beneficial at different phases of the invasion mechanism (Jelbert et al., 2015; Richardson, 2013).

Comparative studies have underscored the significance of specific traits at the species or individual level in facilitating the establishment and naturalization of species in introduced ranges. Among these traits, factors related to regeneration mechanisms emerge as critical determinants. For instance, species with larger seed masses tend to exhibit higher rates of establishment in new environments. This is often attributed to the increased resource allocation and enhanced chances of germination and seedling survival, often associated with larger seeds. Additionally, self-compatible, and auto-fertile species possess a reproductive advantage in colonizing new areas, as they are not reliant on external pollinators for successful reproduction.

Moreover, the production of long-lasting dormant seeds represents a key adaptation that significantly promotes naturalization. Seeds capable of remaining dormant in the soil over multiple seasons provide a reservoir of viable seeds, ensuring the persistence of populations through adverse environmental conditions (Gioria et al., 2021). This resilience enhances the likelihood of successful seedling recruitment when conditions become more favourable, thereby bolstering the persistence and resilience of introduced populations.

The variation in functional strategies of native species and invasive plant species (Table 1) further illuminates the mechanisms driving successful establishment and naturalization. Native species typically exhibit functional strategies finely tuned to their native environments, while IPS may display traits enabling them to exploit novel resources or adapt to new habitat conditions (Klisz et al., 2023). Such traits may include fast growth, efficient resource utilization and high reproductive output, which contribute to the competitiveness and invasive potential of IPS. Collectively, these regeneration traits, dormancy mechanisms, and functional strategies have significant implications for the existence and resilience of introduced populations (Zirbel et al., 2017). By allowing seeds to remain viable in soil under unfavourable conditions and, enabling successful recruitment when conditions improve, these traits enhance the capability of introduced plants to persist and expand their geographical limitations over time. Consequently, understanding and managing these traits are crucial for addressing the challenges posed by invasive species to native ecosystems (Finch-Savage & Bassel, 2016; Gioria et al., 2023).

The rate of range expansion in exotic species is heavily influenced by their ability for long-distance dispersal, a crucial factor often observed in previous studies (Lavoie et al., 2016; Wyse & Hulme, 2021). However, obtaining field evidence of long-distance dispersal poses challenges and frequently yields conflicting results (Xia et al., 2020). Molecular research emerged as a pivotal tool in evaluating the worldwide, regional, and local ranges of exotic species. It sheds light on various aspects such as dispersal modes, plant lineages resulting from frequent introductions, and flexibility of reproductive traits strategies (Smith et al., 2020; Xia et al., 2020). Analyses at the genetic level play a significant role in identifying the specific genotypes of regional and global invaders that successfully colonize to a new area and flourish in new introduced ranges (Iwanycki Ahlstrand et al., 2022). Additionally, these analyses help elucidate how the invasion processes and trajectories respond to critical climatic changes along with the environmental disturbances (Dainese et al., 2017; Liu et al., 2021a, 2021b).

Table 1 Variation in functional strategies adopted by native and invasive species

Functional trait	Correlation to plant strategy/function	IAS traits
Plant canopy height**	Competitive vigour, reproductive size, light interception, fecundity, potential lifespan, and resilience	Invasive species are tall, taller than wide, more often phanerophytes, and have larger leaves (Dainese et al., 2022; Williamson & Fitter, 1996)
Specific leaf area (SLA)**	Positively associated with relative growth rate (RGR), photosynthetic rate, leaf nitrogen content (LNC), plasticity, stress tolerance, ever greenness, leaf longevity	Invasive species have been shown to possess higher SLA (Baruch & Goldstein, 1999; Burns, 2006; Durand & Goldstein, 2001; Gioria et al., 2022; Grotkopp et al., 2002; Lake & Leishman, 2004; Liu et al., 2021a, 2021b; Pattison et al., 1998)
Leaf dry matter content (LDMC)**	Negatively associated with RGR and positively associated with leaf lifespan. It also depicts leaf water relation and flammability to some extent	Invasive species generally have lower LDMC values than natives (Gioria et al., 2022; Lukács et al., 2016)
Percent Leaf carbon (LCC)**	Carbon inputs into soil, herbivory rates and biomass production	LCC of invasive herb is higher in the open meadow than in understory habitats (Dostálek et al., 2021; Sodhi et al., 2019)
Percent Leaf nitrogen (LNC)**	Photosynthetic rate positively associated with RGR	Invasives have higher N content when compared to natives (Baruch & Goldstein, 1999; Craine & Lee, 2003; Durand & Goldstein, 2001; Xie et al., 2022)
Number of leaves**	Competitive strategy of the plant and longevity of leaves	Native species that are present in the invaded communities have fewer leaves, with greater water content compared to less invaded sites (Gioria et al., 2022; Sodhi et al., 2019)
Traits of living leaves (NIRS spectrum)	Resorption of nutrients, decomposition of litter	Invasiveness is strongly related to leaf traits associated with rapid C capture (high SLA, high leaf area ratio (LAR) and fast relative growth rate (Liu et al., 2021a, 2021b; Rejmánek et al., 2005)
Seed mass*	Dispersal distance, longevity in the seed bank, establishment success, fecundity	Invasives produce more seeds per unit seed mass than natives (Mason et al., 2008). On average seed mass of an invasive species is higher in its invasive range than its native range (Daws et al., 2007; Gioria et al., 2022)

(continued)

Table 1 (continued)

Functional trait	Correlation to plant strategy/function	IAS traits
Seed shape and size*	Dispersal, establishment	Seeds with small size found to be significantly correlated with invasion success (Hamilton et al., 2005; Montesinos, 2022b)
Clonality*	Space acquisition	Aliens are more likely clonal than natives and have transient seed banks (De Kroon & Groenendael, 1997; Pyšek & Richardson, 2007; Ye et al., 2023)
Above-ground biomass (AGB)*	Competitive ability, fecundity	Fecundity has been identified as an important correlate of invasiveness in the invaded range (Baker, 1965; Burns, 2006; Mason et al., 2008; Westoby et al., 2002)
Life history*	Plant longevity, space holding ability, disturbance tolerance	Life-history traits (germination percentage and timing) are key traits that play an important role in biological invasions (Baker, 1974; Montesinos, 2022a)
Onset of flowering*	Stress avoidance, disturbance avoidance	The onset of flowering is associated with response to disturbance, establishment, and invasiveness (Rejmanek & Richardson, 1996; Xie et al., 2022)
Resprouting ability*	Disturbance tolerance	Several studies have shown that seeds, originating from introduced populations, germinated at a greater percentage, earlier and or faster (Blair & Wolfe, 2004; Erfmeier & Bruelheide, 2005; Gioria et al., 2022; Maron et al., 2007)
Stem density*	Plant longevity, aboveground carbon storage	Invasive plant formed dense, monospecific thickets with a high stem density, resulting in a closed canopy which directly influenced a reduction in light transmittance on the soil surface and low light conditions may be a crucial environmental factor preventing the seed germination and growth of pioneer and native species (Dostálek et al., 2021; Marod et al., 2012)
Stem girth*	Stability, defence, architecture, carbon gain and growth potential	Invasive species have significantly larger basal stem diameters than their non-invasive relatives (Jelbert et al., 2015; Liu et al., 2021a, 2021b)
Specific root length (SRL)***	Absorption (nutrients, water), faster root turnover, high RGR of seedlings	Invasive plants exhibit a better performance of root traits and nutrients than co-occurring common native plant species (Xia et al., 2020; Ye et al., 2023)

Source *Weiher et al., (1999); **Kleyer et al., (2008), Perez-Harguindeguy et al., (2013), ***Cornelissen et al., (2003)

In summary, while the field evidence of seed dispersal to a long-distance remains challenging to obtain, molecular research, particularly genetic analyses, offers valuable insights into the mechanisms driving the global spread of non-native species. It allows for a more precise understanding of dispersal patterns, genetic diversity, and adaptation strategies, ultimately improving our capacity to anticipate and control the detrimental effects of invasive species on many ecosystem types with ease.

2.5 Naturalization and Invasion Success

The ability to compete for resources has long been linked to the naturalization and invasiveness of species compared to their native counterparts (Gioria & Osborne, 2014). Traits associated with species performance in ecological niches such as size, fitness, biomass allocation, growth rate, and physiology are widely recognized as the reliable indicators of invasiveness (Diez et al., 2008; Lavoie et al., 2016). These characteristics are frequently linked to a species' capacity to quickly establish itself in new habitats and outcompete native vegetation. Many invasive species possess preadaptation that confers advantages in early succession stages and disturbed habitats (Catford et al., 2012; Gallien et al., 2019). Role of fast nutrient uptake, growth rate, and reproduction in describing successful invasion might be influenced by prevalence of ruderal plant species among invasive plants, particularly those colonizing nutrient-rich and highly disturbed habitats (Montesinos, 2022a). It is well known that ruderal species can exploit resource-rich environments and quickly occupy available space. Consequently, the prevalence of invasive species with these performance traits could partly stem from a bias towards species that are well-suited for human-altered landscapes and disturbed habitats.

Moreover, the intentional selection of species by humans for introduction further exacerbates this bias. Trait plasticity, which allows species to adapt to different environments and conditions, can contribute to their invasiveness and favour their introductions (Cadotte et al., 2018; Chrobock et al., 2011). Species with performance traits conducive to invasion may have been deliberately chosen for introduction due to their perceived utility or aesthetic value. This intersection between intentional introductions and performance traits describe why intentionally introduced non-native species are often more prone to becoming invasive compared to those introduced accidentally (Maurel et al., 2016). Overall, the interplay between performance traits, human selection, and environmental conditions underscore the complex dynamics driving invasion success. Understanding these factors is significant for anticipating and managing the impacts of invasive species on ecosystems.

Plant functional traits play a crucial role in determining how plants acquire resources and utilize them, and their capacity to modify the availability of resources for other plants. One such trait, nitrogen-fixing ability, is associated with invasiveness and is found in numerous invasive herbaceous and woody plants (Castro-Díez et al., 2014; Jo et al., 2017; Kaushik et al., 2024). This ability enables invasive species to alter nitrogen cycling and availability, which may significantly impact the dynamics of ecosystems.

Invasive species possessing nitrogen-fixing abilities can influence nitrogen cycling by affecting nitrification rates, leading to alterations in soil nitrogen availability. This may create positive feedback loops that benefit invasive species' long-term survival compared to native species suited to low-nutrient habitats (Funk, 2013; Yelenik & D'Antonio, 2013). The strength of these loops often depends on interspecific interactions in the ecosystem (Lee et al., 2012). Further, these alterations in nitrogen availability can have broader biogeochemical implications, including differences in decomposition rates and release of N_2 between IAPs and native species (Ehrenfeld, 2010). Such changes in nutrient dynamics can create conditions conducive to secondary invasions by other exotic plants that persist in nitrogen-rich environments (Gaertner et al., 2014; Hughes & Denslow, 2005).

Overall, specific functional traits, such as nitrogen-fixing ability, can have cascading effects on ecosystem processes and community dynamics. By altering resource availability and biogeochemical cycles, invasive species with these traits can facilitate their own persistence and promote the establishment of other non-native species, potentially leading to further disruptions in ecosystem structure and function. Understanding these relationships is critical for successful prevention and reduction of the adverse effects of invasive plant species on native ecosystems.

2.6 Allelopathy and Invasion Success

Allelopathy is an intriguing biological phenomenon where an organism emits substances into its surroundings that have adverse effects on the development, viability, and procreation of other neighbouring organisms. In the case of invasive species, their capacity to generate specific compounds, termed allelopathy, plays a key role in their proliferation and establishment in new environments. These compounds can disrupt the beneficial relationships that native plants rely on; a concept termed the novel weapon hypothesis (Uddin et al., 2020). It is commonly observed that invasive plants produce allelopathic substances, research findings both supports and challenges the notion of their role in invasion dynamics. The total environmental and evolutionary impacts of allelopathic relationships between non-native and native plants are not yet completely known, resulting in numerous unresolved queries (Oduor et al., 2020; Yuan et al., 2021).

One intriguing aspect is that invasive species may increase their Allelochemical production in relation to strong competition for resources. This heightened production provides a competitive advantage over native plants, particularly as these chemicals often have a major effect on native species compared to exotic ones. However, whether this ability to produce allelopathic compounds can evolve in introduced environments, a theory called the evolution of the increased weaponry hypothesis has not been thoroughly investigated and lacks substantial evidence. Further, it is essential to recognize that allelochemical production is not static but is influenced by the availability and quality of resources in the environment. This underscores the intricate connection between allelopathy impacts on native communities and the dynamics of available resources in the ecosystems where these interactions occur (Zhang et al., 2021).

A recent study discovered a noteworthy trend: non-native species have a greater influence on indigenous species than do other indigenous species. On the other hand, even in cases when the non-native species do not share an evolutionary history, they have a lesser propensity to suppress other non-native species than do native species. This discovery highlights the potential role allelopathy may have in supporting the coexistence of closely related non-native species or in elevating invading species that are not closely related to non-native species. This research also reveals that the negative impacts of allelopathy are likely to increase with greater phylogenetic distance, implying that invasive species with fewer genetic ties to native species might gain an advantage from allelopathy. This highlights the importance of considering evolutionary relationships when trying to understand the dynamics of invasion and coexistence within ecosystems.

Local biotic and abiotic conditions in regions where species are introduced further underscores the significance of functional traits in determining invasiveness. The capacity of a species to adapt or evolve in response to the distinct conditions present in these new environments often dictates its progression along the invasion continuum. This highlights the dynamic nature of invasion processes and underscores the importance of considering both genetic and environmental factors when seeking to understand the success of invasive organisms.

2.7 Phenotypic Plasticity and Invasion Success

Phenotypic plasticity is the capacity of a specific genetic makeup to show a diverse range of physical characteristics in diverse environments. This is particularly significant in traits that are important for ecological interactions, as it has been linked to the successful invasion of species (Gómez et al., 2025; Sommer, 2020). The concept of the phenotypic plasticity hypothesis suggests that by being adaptable to diverse biotic and abiotic conditions, species can expand their ecological range and better adapt to the new environment. Furthermore, adaptability in characteristics that affect resource use and regeneration can be advantageous for adapting to climate change and promoting range expansions, allowing species with a wide variation in average trait values to potentially colonize diverse climatic conditions (Molina-Montenegro et al., 2018). Several theories have been proposed to explain the connection between invasion success and plasticity as invasive species may modify their morphology and physiology, they may be able to sustain high fitness over a wide range of environments, or they may experience greater fitness increases under favourable conditions than native species. Alternatively, invasive species may better maintain their fitness in unfavourable environments, or they may combine these strategies in both resource-rich and stressful environments. These theories suggest that phenotypic plasticity plays an important role in the success of invasive species (Drenovsky et al., 2012; Richards et al., 2006).

The phenomenon of plasticity in introduced ranges refers to a species' capacity to adjust to novel environmental circumstances outside of its natural range. This adaptation can occur through various mechanisms, such as changes in morphology, physiology, behaviour, or life history traits. Although the ability allows them to flourish in a variety of often challenging habitats, and plasticity is crucial to the success of invasive species. However, the research on plasticity in introduced ranges has produced conflicting results. Some previous studies have observed variations in plasticity among invasive and non-invasive species, indicating that some species may be more invasive due to plasticity. However, other research has not discovered any appreciable distinctions in the plasticity of invasive and non-invasive species (Davidson et al., 2011; Montesinos & Callaway, 2018; Zhang et al., 2022).

One complicating factor in these comparisons is the classification of species as invasive or non-invasive. Certain species may be considered non-invasive in their native range but can eventually become invasive when introduced to a new environment. This discrepancy in classification can affect the interpretation of results and make it challenging to generalize findings across different studies. Further, the plasticity behaviour may provide many benefits vary depending on the stage of invasion. Experiments such as Transplant Garden, which involve moving individuals of a species to different environments, have shown that effects of plasticity may be critical during the initial colonization stages of invasion. This suggests that plasticity may play a crucial role in facilitating the establishment of invasive populations in new environments. However, as established populations spread and become more genetically differentiated, genetic factors may become highly significant than plasticity in describing the species success.

For instance, a meta-analysis attempted to synthesize the existing empirical evidence on plasticity in introduced ranges (Montesinos & Callaway, 2018). However, the analysis concluded that the evidence remains inconclusive, highlighting the complexity of the phenomenon and the need for further research to comprehend plasticity role in invasion biology. The conflicting results in the literature suggest that additional research is required to elucidate the mechanisms underlying plasticity in introduced ranges and its implications for invasion biology (Van Kleunen et al., 2010). However, one key factor that can complicate our understanding is the potential for evolutionary changes in plasticity over time. While plasticity allows individuals to adapt to changing environmental scenarios, evolutionary changes in plasticity can occur as populations experience selection pressures in their new environments. These changes may not be immediately apparent and are often overlooked in studies of invasion biology (Matesanz et al., 2010).

Phenotypic plasticity can be beneficial if the phenotypes produced in response to environmental changes result in above-average fitness. For example, if a plant species exhibit plasticity traits that provide an advantage to grow well in a broad range of soil types, it may have a competitive benefit over species with less plasticity. High phenotypic plasticity, however, does not always equate to high fitness, and it is unclear how much it contributes to successful invasions in comparison to the mean species attributes or evolutionary changes (Hulme, 2008; Marchini et al., 2019). Few studies have evidence of increased plasticity in invasive populations, indicating that plasticity has a critical role in their success in new environments. However, other research have observed no significant differences in plasticity between native and invasive populations (Franks et al., 2014).

These discrepancies arise due to the costs associated with plasticity, which can influence whether the evolution of enhanced plasticity is beneficial. Plasticity requires energy and resources, and there may be trade-offs between investing in plasticity and other traits that contribute to fitness (Huang et al., 2015). Additionally, the presence of natural enemies in introduced ranges may influence the evolution of plasticity. Natural enemies can exert selection pressures on populations, favouring individuals with higher adaptive plasticity by decreasing the plasticity costs and enhancing the benefits.

3 Evolutionary Process and Invasion Success

Species invasion across different environments is rather related to higher evolutionary changes in various functional roles, driven by some uninvited selection pressures encountered in new introduced ranges (Colautti & Lau, 2016; Hierro et al., 2020; Molina-Montenegro et al., 2018). As non-native species confront different barriers along the invasion continuum, **every** stage of the invasion progression leaves genetic imprints on populations that invade, influencing their success in subsequent stages (Lachmuth et al., 2010). Various evolutionary mechanisms operate at different points in the invasion process (Zenni et al., 2017) (Fig. 2).

The evolutionary consequences stemming from pre-introduction history, such as adaptability to climate or adaptation to anthropogenic disturbances, can significantly influence a species' success in its new introduced environment. The term "anthropogenically induced adaptation to invade" was coined by Hufbauer et al. (2012) to describe scenarios where populations that have acclimatized to human disruptions in their native range may thrive in similarly disturbed situations in their new habitat. Furthermore, the evolutionary background of an introduced species in reaction to human-caused disruptions might offer a benefit in the face of ongoing significant global changes (Colautti & Lau, 2016). However, acclimatization might occur predominantly in the new introduced range, with the effectiveness of invasion being dependent on the introduction of very successful populations, which is observed as the invasive bridgehead effect (Lombaert et al., 2010).

Allee effects represent a phenomenon observed in many species, characterized by the positive dependence of densities at low population sizes. These effects often arise due to challenges in locating mates and bottlenecks related to the small initial populations. These impacts can significantly influence the naturalization of non-native plants and may even lead to stochastic extinction events (Simberloff, 2009). However, successive introductions of non-native species may help alleviate the impacts of demographic bottlenecks, which typically decrease genetic diversity in newer founding populations. Even though genetic bottlenecks are anticipated to lead to inbreeding depression and a higher fixation of harmful mutations due to genetic drift (referred to as drift load), many alien/ exotic plants have successfully become invasive. This phenomenon underscores that the negative consequences of genetic bottlenecks alone are insufficient to prevent non-native species from establishing and spreading. One hypothesis suggests that exotic plant species can adapt to their new environments through temporary or permanent relief from stressful conditions and through interactions with the local environment. For instance, they may exploit resources unavailable or underutilized in their original native range, allowing them to thrive in the new habitat (Schrieber & Lachmuth, 2017).

However, suboptimal conditions, competition amongst non-native and native species, and encounters with novel enemies can constrain the naturalization of alien plants. Despite these challenges, non-native species may experience changes related to escape from the specialist natural enemies, which can further facilitate their invasion success. For example, in the absence of their natural predators or pathogens, non-native species may experience reduced pressure, allowing them to proliferate more freely (Fridley et al. 2021; Gioria & Osborne, 2014).

During the initial invasion stage, many invasive species can reach high population densities, which can in turn promote demographic and evolutionary changes. These changes may include shifts in life history traits, reproductive strategies, or genetic composition that enhances the species' ability to persist and spread in changing environments. Ultimately, these demographic and evolutionary responses can help reduce extinction risks for non-native species, further bolstering their success in new introduced areas (Colautti et al., 2014; Mitchell et al., 2006).

3.1 Genetic Changes and Invasion Success

The success of invasions and the potential for invasibility are characterized by opportunities for both interspecific and intraspecific hybridization among those populations that were previously isolated. This leads to the formation of new adaptive gene combinations or an increase in genetic variation available for adaptive evolution. As biotic and abiotic conditions change over time and throughout the invasion process, new combinations of alleles and epigenetic changes may facilitate the success of invasions. (Mounger et al., 2021). A rising body of research indicates that genetic mixing, hybridization, and polyploidization, or enhanced ploidy, are linked to effective naturalisation and invasiveness (Zenni et al., 2017). Plant's genome size and its polyploidy status are the important attributes that rather associated to the naturalization and invasiveness of plant species. Studies conducted by Pandit et al. (2014) and Te Beest et al. (2012) have highlighted the significance of these traits in shaping the expression of plant characteristics. Polyploidy can arise through either autopolyploidy or allopolyploidy. Autopolyploidy occurs when a plant's genome duplicates within the same species, while allopolyploidy involves hybridization between different species, followed by genome duplication. Both types of polyploidies have significant ecological and evolutionary implications for introduced plant species (Rutland et al., 2021; Te Beest et al., 2012).

As observed by Mounger et al. (2021) and Te Beest et al. (2012), polyploidization triggers various genetic and epigenetic changes, resulting in altered functional traits that allow plants to better cope with environmental variability and stressful conditions, while exploiting new ecological niches. It is widely believed that polyploid plants have a greater chance of success as invasive species compared to diploid plants. This is because polyploidy enhances strength during the initial stages of invasion process and enhances the potential for adaptation due to the greater gene pool diversity it confers. As emphasized by research polyploidy contributes significantly to the evolution of the invasiveness trait in plant species (Rosche et al., 2017). In summary, genome size and polyploidy are crucial attributes that influence the success of plant species in newly introduced areas. Polyploidy enhances adaptation and provides a significant advantage for IAS, contributing for their establishment and spread in novel habitats.

Intra- and interspecific hybridization represent additional mechanisms triggering epigenetic and genetic changes, which may result in new functional traits and increased invasiveness. Hegarty et al. (2012) and Schierenbeck and Ellstrand (2009) showed that hybridization often results in trait variations favouring invasiveness, such as increased fecundity and size. However, the causal relationship between hybridization and invasiveness is not always straightforward, as highlighted by Hovick and Whitney (2014). Temporary interbreeding with local populations or previous invaders might assist in overcoming the limited genetic diversity caused by initial founding events. The impact of hybridization and polyploidization on invasion to successfully happen varies depending on factors such as introduction histories and whether hybrids emerge within founding populations or post-introduction. The enduring effects of hybridization, whether beneficial or detrimental, are still poorly understood and could potentially disrupt local adaptation, leading to genomic extinction.

4 Ecosystem Dynamics and Invasion Success

The analyses of invasion success can be framed within the parameters of the attributes of recipient ecosystems or the phylogenetic trajectories and functional contributions of their species (Bernard-Verdier & Hulme, 2015; Colautti et al., 2014) (Fig. 2). Invasibility, which is defined as the probability of successful establishment of species per introduced propagule, functions as a measurable indicator (Davis et al., 2005). It is imperative to distinguish invasibility from the degree of invasion, which refers to the magnitude of invasion within a system and varies temporally as an invasion progresses (Catford et al., 2012; Guo et al., 2015). Various research studies on the invasion of ecosystems tend to concentrate solely on the level of invasion, without considering the vulnerability of the ecosystem. This often leads to confusion, as such studies may wrongly associate vulnerability with the propagule pressure and the duration of the invader's residence time (Aikio et al., 2012). In the past, the concepts of invasiveness of species and ecosystem invasibility (Lonsdale, 1999), were independently examined (Richardson & Pyšek, 2012; Thuiller et al., 2010). Nevertheless, the significance of plant functional attributes for invasion success varies depending on recipient ecosystem characteristics (Hui et al., 2016), which impact invasibility and resilience against invasion (Elton, 2020; van Kleunen et al., 2010). As a result, certain species become invasive in specific ecosystems but not others, and their success eventually varies even within the same ecosystem type, though evidence for this is limited.

4.1 Edaphic Traits: Turmoil and Availability of Resources

Environmental conditions to which aliens are introduced can affect their ability to establish and spread. These characteristics include the availability of resources and how they vary over time, the frequency and intensity of disturbances, and changes in the environment. Additionally, the diversity of the existing plant community and the variety of species present can either resist invasions or create conditions that allow non-native species to thrive. Lastly, the presence or absence of predators and mutualistic relationships can also influence the invasibility of an ecosystem.

The impact of disturbances on recipient ecosystems has been closely linked to their susceptibility to invasion. Studies by Catford et al. (2012) and Davis et al. (2000) have found that natural or human-induced disturbances can significantly modify the resource availability, displace native plant species, and create gaps in vegetation which paves the way for the rapid establishment of exotic plants. This disturbance-mediated hypothesis suggests that by resetting the succession stages and increasing availability of resources, disruptions can favour the naturalization and full spread of alien plants that are better equipped to acquire resources than native species. Overall, disturbances have been found to enhance invasibility, making it convenient for exotic plants to thrive and even outcompete native species. Alien species are known to counter rapidly to a temporary increase in resources, giving them an unfair advantage over native species. Studies suggest that disturbances that facilitate such invasions occur when there are alterations in the turnover rates or resource fluctuations such as light, nutrients or space, due to changes in the historical disturbance regime. However, it should be noted that many invasion studies do not distinguish between the different components of ecosystem-level disturbance regimes.

According to scientific studies, naturalisation is positively dependent on ruderal and competitive tactics but negatively linked to stress tolerance (Grime, 2006), except for trees. In habitats with recurring and extreme disturbances, ruderal species outcompete indigenous rivals and even stress-tolerant plants during the initial phases of succession (Montesinos, 2022b). These ruderal species use disturbance-created apertures to avoid competition, yet they can also compete well (Grime, 2006). Although the existence of indigenous species which respond fast to various disturbances, provides resistance or barriers to invasions, indigenous ruderal species that are adapted to such disturbances may not colonise these disturbed areas as effectively as alien species (Gioria & Osborne, 2014; Tilman, 1997). Extreme perturbations that promote the colonisation of opportunistic non-native species (Davis, 2007) often impair native communities' competitiveness (Seabloom et al., 2003).

The fluctuations in resource availability hypothesis suggests that in productive and nutrient-rich ecosystems, fluctuations in resource availability, including disturbances such as pulse events, can alter plant-plant interactions (Houseman et al., 2014; Montesinos, 2022a). This disruption can have significant implications for the balance between invasive and native species within these ecosystems. In such environments, indigenous species are often well-acclimatized to low-resource or stressful conditions due to evolutionary pressures. However, when disturbances or fluctuations in resource availability occur, it can temporarily alter the competitive interactions between invasive and native species. Temporary disruption can pave the way for invasive species to fully establish and thrive. Neither the functional/phylogenetic parallels in resource utilization between native and non-native species nor the diversity of native communities are implied by the hypothesis. Instead, it focuses on the idea that changes in resource availability, whether increases or decreases, may impact the performance of invasive species. For example, in situations where there is an increase in resources, invasive species may have a competitive advantage over native species due to their ability to capitalize on the abundance of resources. Conversely, when resource availability decreases due to disturbances, native species may have adaptations that allow them to better withstand these conditions, potentially reducing the success of invasive species. However, there may be a complex and context-dependent relationship between invasibility, resource availability, and disturbances. Factors such as specific traits of invasive and native plant species, frequency and intensity of disturbances, and the dynamic interactions within the ecosystem all play a major role in establishing the result (Hobbs & Huenneke, 1992). Therefore, while the fluctuating resource availability hypothesis provides a framework for recognizing the relationships between invasive and native species, further research is needed to fully elucidate these relationships in different ecosystems.

According to the intermediate disturbance hypothesis, plant diversity in ecosystems can be maximized at moderate disturbance levels. However, the effects of such disturbance levels on the native and non-native species diversity can vary due to several interacting attributes (Catford et al., 2012). These factors include human-mediated dispersal, which can introduce non-native species into ecosystems at varying rates. Additionally, the non-native species pool may be overrepresented by early successional species, which are often opportunistic and quick to establish in disturbed environments. Furthermore, fast-growing species may benefit the most from the release of natural enemies in disturbed ecosystems, potentially enhancing the alien species success under intermediate disturbance levels. Human-altered habitats often experience increased disturbance levels, which can further influence the diversity of both alien and native species. These disturbances can disrupt natural ecosystems and create various opportunities for aliens to establish and proliferate (Funk, 2013; Waddell et al., 2020). Despite being traditionally considered less invasible, low-resource and stressful ecosystems have also been invaded by non-native plants. This unexpected invasion may be attributed to preadaptation, where non-native species possess traits that allow them to thrive in challenging environments, and rapid evolutionary changes that enable them to quickly adapt to new opportunities (Melbourne et al., 2007).

In response to great environmental variability, disruptions, and escape from pressure or natural predators, invasive plants flourish. Temporary nutrient increases brought about by either human or environmental disturbances may counteract the detrimental impacts of native species resource allocation competition in low-resource ecosystems. Non-native alien species can take over if resource fluctuations coincide with the availability of propagules. Previous findings include Waddell et al. (2020), Catford et al. (2009), Melbourne et al. (2007), and Davis et al. (2000) support these findings. We must be cautious and consider the complex factors contributing to their success when managing these invasive alien species.

The ecological balance may be severely impacted by non-native organisms, especially when they work together to establish themselves and cause harm. This phenomenon is known as invasional meltdown, and it can have disastrous consequences for ecosystems. Positive interactions among invasive species can amplify their impact or facilitate secondary invasions, leading to significant changes in the introduced habitat (Gaertner et al., 2014; Gioria et al., 2011). Differentiating between invasive species that are actively driving changes and those that are just passengers of broader environmental stresses is vital, as invasive species can cause changes that aid in their establishment and naturalization. By understanding these dynamics, we can better manage invasive species and protect our natural resources for generations to come (White et al., 2013).

4.2 Urban Ecosystems Prone to Invasion Success

Urban areas are particularly vulnerable to plant invasions due to their proximity to heavily modified ecosystems such as agricultural fields and forest plantations (Pati et al., 2023). These altered landscapes serve as significant sources of non-native seeds, which can easily spread to nearby urban environments (Pati et al., 2024). The strong propagule pressure and local introductions levels in urban areas is primarily driven by human activities such as construction, transportation, gardening, and other alterations to the environment (Potgieter & Cadotte, 2020). The introduction of alien plants in forestry and agriculture not only increases the availability of non-native seeds but also contributes to landscape fragmentation and degradation. These changes further disrupt ecosystems on a broader scale, facilitating the naturalization and spread of alien species. One significant outcome of the disruption is the expansion of numerous alien species into ecosystems that were traditionally considered less vulnerable. For example, forest remnants and other natural areas near urban centres are increasingly being invaded by non-native plants. This invasion is facilitated by the creation of habitat edges through fragmentation, which provides favourable conditions for colonization by alien species (Plenderleith et al., 2022). Habitat edges have different environmental conditions compared to the interior of natural ecosystems, making them more suitable for non-native species that may be less adapted to the native ecosystem's interior conditions. Additionally, urban areas tend to have higher disturbance levels, which can further promote the spread and colonisation of alien species. Overall, the proximity of urban areas to heavily altered ecosystems, coupled with the high level of human-induced changes and disturbance, creates ideal conditions for the expansion of plant invasions (Pati et al., 2024). This phenomenon not only threatens native biodiversity but also undermines the ecological integrity of both urban and adjacent natural environments (McDougall et al., 2018).

Transportation networks such as rivers and roads act as the routes for seed dissemination, enhancing ecosystem susceptibility to invasions (Waddell et al., 2020), even in biodiverse environments. Flood events promote invasions in riparian and flood-prone ecosystems by temporarily increasing nutrient availability and dispersing seeds. Roads and railways offer avenues for non-native plant establishment and spread, even in protected and species-rich habitats (Vorstenbosch et al., 2020). This process is particularly obvious in mountainous places around the world, where climate change, greater utilization of land by humans, along with the introduction of new propagules give chances for improved colonisation along elevation gradients (Pauchard et al., 2009). While it's typically observed that alien species occurrence diminishes as elevation rise, the warming climate is altering this pattern by allowing invasive plants that grow next to roadways in order to reach higher altitudes. Recent studies conducted on mountain roads worldwide have revealed that the traits of invasive species propagating into adjacent natural vegetation are distinct from those found colonizing roadsides.

Roadside invaders tend to be long-lived species that are not typically considered ruderal, i.e., they are not adapted to thrive in disturbed habitats. In contrast, species invading nearby natural vegetation are usually shade and moisture-tolerant, indicating adaptation to the specific conditions of those habitats. This contradictory result highlights the crucial role that roads play as conduits for the spread of IAS that could lack the traits requisite for long-distance spread (Foxcroft et al., 2019).

With tourism and recreational activities becoming increasingly important, recreational paths also serve as favourable environment and conduits for the spread of alien species. It poses a growing threat to the natural habitats and protected areas. Due to the rapid increase in human activities which continue to impact ecosystems, it's essential to understand the mechanism of roads and recreational paths in assisting non-native species' proliferation. Efforts to manage and mitigate the impact of invasive species should consider these pathways of spread to effectively protect native biodiversity and ecosystem integrity. Managing invasions in urban environments effectively requires an integrated approach that combines ecological, socio-economic, and policy perspectives. Recent advancements in remote sensing and citizen science have enhanced monitoring and management efforts (Young et al., 2021). Public awareness and community participation are crucial for implementing sustainable management practices (Richardson et al., 2022).

4.3 Propagule Pressure and Invasion Success

Assessing the risk posed by invasive species involves careful consideration of propagule pressure, describing the quantity and frequency of introduced propagules along with their survival rate (Lonsdale, 1999; Richardson & Pyšek, 2006). Propagule pressure stands out as the critical trait influencing the success rate of establishing alien species. This pressure affects invasive species establishment in various ways, including multiple introductions of alien species, preadaptation to introduced environments, human commensalism (where species benefit from close association with humans), high gene flow, and dispersal facilitated by human activities (Levin, 2021).

Protecting our ecosystems necessitates a thorough assessment of the potential risk such exotic species might pose, with propagule pressure serving as a key indicator (Cassey et al., 2018). By understanding and evaluating propagule pressure, we can better anticipate and mitigate the potential risks linked to invasive alien species introductions, thus safeguarding the integrity and biodiversity of our ecosystems.

Elevated propagule pressure significantly increases the possibility of alien species finding suitable conditions for establishment. This includes favourable microclimates, unoccupied ecological niches, or microhabitats lacking natural enemies. As a result, the process of range expansion is expedited for non-native species experiencing high propagule pressure (Rejmánek et al., 2013; Ricciardi & Jones, 2011). In low-stress areas with adequate availability of food and poor biotic resistance, this advantage is more prominent. A prime example is the annual ruderal *Arabidopsis thaliana*, which demonstrates how elevated propagule pressure facilitates colonization in such environments. This underscores the importance of managing propagule pressure to effectively mitigate the risks associated with invasive species introductions (Hovick & Whitney, 2014).

5 Climate Change and Invasion Success

There is significant concern regarding the impact of human-induced climate change, rising atmospheric CO_2 levels, and nitrogen deposition on the resilience of native populations, nutrient availability, and weak biotic resistance, as well as the vulnerability of ecosystems to invasion and the patterns of plant invasion. It is difficult to determine the long-term effects of global shifts on plant invasions owing to an array of unsolved issues. In response to warming climates, both native and non-native species may experience range changes, either diminishing or spreading into novel territories and habitats (Wan et al., 2021). This might result in the emergence of new species assemblages, modifying these communities' resilience to invasion and disturbing ecological relationships, which might benefit either native or alien species (Wallingford et al., 2020). It was found that human influences alter resource dynamics and disturbances (Baskin & Baskin, 2022; Dukes & Mooney, 1999), they also interfere with relationships between natural enemies and mutualists (Traveset &

Richardson, 2020), however, it is still unclear where these changes will ultimately end up.

Changes in global temperatures can also disrupt the timing of biological events in native species, potentially providing non-native species with windows of opportunity for the establishment with reduced competition (Gioria & Osborne, 2014; Howell et al., 2020). Studies on mountain ecosystems, which are particularly vulnerable to climate change and increased human activity, suggest that alien plant species are increasing their ranges at higher elevations due to warming temperatures (Vorstenbosch et al., 2020). According to previous studies non-native species thrive nearer to roadways and due to propagation over long distances, these species are propagating more rapidly than native species (Dainese et al., 2017). Further, findings from a study on clonal plants that are alien to protected areas, as highlighted by Wan et al. (2021), indicate that the risk of invasions may change with shifting climatic conditions. Specifically, there may be an increased invasion risk for biomes at high elevations and latitudes, while lower elevations and tropical/subtropical regions might experience a decrease. This shift can be attributed to the advantage of asexual reproduction in less favourable environments for sexual reproduction.

Additionally, invasive plants demonstrate rapid evolution and possess broad environmental tolerance, enabling them to quickly adapt to climate change. This adaptability enhances their competitiveness and facilitates their expansion into new habitats, presenting significant challenges for invasive species management and conservation efforts (Eyster & Wolkovich, 2021; Gentili et al., 2021). There is currently no conclusive evidence when it comes to the adaptation of resilience traits to climate change. However, demographic studies suggest that species with seeds capable of forming persistent soil seed banks, which are less vulnerable to temperature increases and related environmental changes, may be less affected by climate warming (Baskin & Baskin, 2022). The capacity to establish enduring seed banks is frequently linked to the processes of naturalization and becoming invasive. Consequently, numerous invasive species might demonstrate enhanced resistance and adaptability to climate change by dispersing their seeds over time. It is crucial to recognize that while seed banks may offer protection, this benefit might be short-lived. Additionally, data on how climate change affects plants is both scarce and unevenly distributed across regions. Research increasingly shows that invasive species are adapting to climate change by shortening seed dormancy, starting germination sooner, and boosting germination rates.

6 Conclusions and Future Prospects

In the last twenty years, there has been considerable advancement in comprehending the ecological and evolutionary factors that drive plant invasions. However, complex interactions among various factors make it challenging to explain invasion success with a single theory. Context dependencies play a crucial role, highlighting the need for tailored approaches focusing on specific taxonomic groups, functional groups, or habitats. The invasion syndromes approach offers a practical framework for achieving reasonable generalizations. Human activities have intensified invasion pressures across scales, challenging the resistance of ecosystems previously deemed invulnerable. By integrating these three components into ecological frameworks, researchers can gain a more thorough insight into the mechanisms and dynamics of invasions. This may assist in direct efforts meant to prevent, control, or reduce the effects of invasive species on ecosystems. For example: Identifying and prioritizing species with high invasiveness traits can help focus monitoring and management efforts on species most likely to cause ecological harm. Understanding community invasibility can guide efforts to restore or enhance native biodiversity and ecosystem resilience, making ecosystems more resistant to invasion. Considering the environmental context can help predict where invasive species are likely to spread and establish, allowing for targeted interventions in vulnerable areas. This approach promises to give a more thorough understanding of plant invasions and improve management efforts.

Acknowledgements The authors are thankful to the Head, Department of Botany, Dr. H.S. Gour University, Sagar, India, for providing facilities. Authors are also thankful to the contributors of all the research papers and information cited in this book chapter.

Competing Interests The authors declare no competing interests.

References

Aikio, S., Duncan, R. P., & Hulme, P. E. (2012). The vulnerability of habitats to plant invasion: Disentangling the roles of propagule pressure, time and sampling effort. *Global Ecology and Biogeography, 21*(8), 778–786.

Baker, H. G. (1974). The evolution of weeds. *Annual Review of Ecology and Systematics, 5*, 1–24.

Baker, H. G. (1965). Characteristics and modes of origin of weeds. In *Characteristics and modes of origin of weeds* (pp. 147–172).

Baruch, Z., & Goldstein, G. (1999). Leaf construction cost, nutrient concentration, and net CO_2 assimilation of native and invasive species in Hawaii. *Oecologia, 121*, 183–192.

Baskin, C. C., & Baskin, J. M. (Eds.). (2022). *Plant regeneration from seeds: a global warming perspective*. Academic Press.

Bernard-Verdier, M., & Hulme, P. E. (2015). Alien and native plant species play different roles in plant community structure. *Journal of Ecology, 103*(1), 143–152.

Blackburn, T. M., Pyšek, P., Bacher, S., Carlton, J. T., Duncan, R. P., Jarošík, V., Wilson, J. R., & Richardson, D. M. (2011). A proposed unified framework for biological invasions. *Trends in Ecology & Evolution, 26*(7), 333–339.

Blair, A. C., & Wolfe, L. M. (2004). The evolution of an invasive plant: An experimental study with Silene latifolia. *Ecology, 85*(11), 3035–3042.

Briski, E., Chan, F. T., Darling, J. A., Lauringson, V., MacIsaac, H. J., Zhan, A., & Bailey, S. A. (2018). Beyond propagule pressure: Importance of selection during the transport stage of biological invasions. *Frontiers in Ecology and the Environment, 16*(6), 345–353.

Burns, J. H. (2006). Relatedness and environment affect traits associated with invasive and noninvasive introduced Commelinaceae. *Ecological Applications, 16*(4), 1367–1376.

Cadotte, M. W., & Tucker, C. M. (2017). Should environmental filtering be abandoned? *Trends in Ecology & Evolution, 32*(6), 429–437.
Cadotte, M. W., Campbell, S. E., Li, S. P., Sodhi, D. S., & Mandrak, N. E. (2018). Preadaptation and naturalization of nonnative species: Darwin's two fundamental insights into species invasion. *Annual Review of Plant Biology, 69*, 661–684.
Cassey, P., Delean, S., Lockwood, J. L., Sadowski, J. S., & Blackburn, T. M. (2018). Dissecting the null model for biological invasions: A meta-analysis of the propagule pressure effect. *PLoS Biology, 16*(4), Article e2005987.
Castro-Díez, P., Godoy, O., Alonso, A., Gallardo, A., & Saldaña, A. (2014). What explains variation in the impacts of exotic plant invasions on the nitrogen cycle? A meta-analysis. *Ecology Letters, 17*(1), 1–12.
Catford, J. A., Jansson, R., & Nilsson, C. (2009). Reducing redundancy in invasion ecology by integrating hypotheses into a single theoretical framework. *Diversity and Distributions, 15*(1), 22–40.
Catford, J. A., Vesk, P. A., Richardson, D. M., & Pyšek, P. (2012). Quantifying levels of biological invasion: Towards the objective classification of invaded and invasible ecosystems. *Global Change Biology, 18*(1), 44–62.
Chauhan, S., & Verma, T. Invasive species management in forest ecosystems. In *Advances in forestry and agro-forestry* (p. 311).
Chrobock, T., Kempel, A., Fischer, M., & van Kleunen, M. (2011). Introduction bias: Cultivated alien plant species germinate faster and more abundantly than native species in Switzerland. *Basic and Applied Ecology, 12*(3), 244–250.
Colautti, R. I., & Lau, J. A. (2016). Contemporary evolution during invasion: evidence for differentiation, natural selection, and local adaptation. *Invasion Genetics: The Baker and Stebbins Legacy*, 101–121.
Colautti, R. I., Parker, J. D., Cadotte, M. W., Pysek, P., Brown, C. S., Sax, D. F., & Richardson, D. M. (2014). Quantifying the invasiveness of species.
Cornelissen, J. H., Lavorel, S., Garnier, E., Díaz, S., Buchmann, N., Gurvich, D. E., Reich, P. B., Ter Steege, H., Morgan, H. D., Van Der Heijden, M. G. A., Pausas, J. G., Poorter, H., et al. (2003). A handbook of protocols for standardised and easy measurement of plant functional traits worldwide. *Australian Journal of Botany, 51*(4), 335–380.
Craine, J. M., & Lee, W. (2003). Covariation in leaf and root traits for native and non-native grasses along an altitudinal gradient in New Zealand. *Oecologia, 134*, 471–478.
Dainese, M., Aikio, S., Hulme, P. E., Bertolli, A., Prosser, F., & Marini, L. (2017). Human disturbance and upward expansion of plants in a warming climate. *Nature Climate Change, 7*(8), 577–580.
Dainese, M., Aikio, S., Hulme, P. E., Bertolli, A., Prosser, F., & Marini, L. (2022). Human disturbance and upward expansion of plants in a warming climate. *Nature Climate Change, 12*(3), 264–270.
Davidson, A. M., Jennions, M., & Nicotra, A. B. (2011). Do invasive species show higher phenotypic plasticity than native species and if so, is it adaptive? A meta-analysis. *Ecology Letters, 14*(4), 419–431.
Davis, M. A., Lockwood, J. L., Hoopes, M. F., & Marchetti, M. P. (2007). Invasion ecology. *Journal of Vegetation Science, 18*(6), 925–927.
Davis, M. A., Grime, J. P., & Thompson, K. (2000). Fluctuating resources in plant communities: A general theory of invasibility. *Journal of Ecology, 88*(3), 528–534.
Davis, M. A., Thompson, K., Grime, J. P., & Liebhold, A. (2005). Invasibility: The local mechanism driving community assembly and species diversity. *Ecography, 28*(5), 696–704.
Daws, M. I., Hall, J., Flynn, S., & Pritchard, H. W. (2007). Do invasive species have bigger seeds? Evidence from intra-and inter-specific comparisons. *South African Journal of Botany, 73*(1), 138–143.
De Kroon, H., & Van Groenendael, J. (1997). *The ecology and evolution of clonal plants*. Backhuys.

Díaz, S. M., Settele, J., Brondízio, E., Ngo, H., Guèze, M., Agard, J., Arneth, A., Balvanera, P., Brauman, K., Butchart, S., Chan, K., Zayas, C., et al. (2019). The global assessment report on biodiversity and ecosystem services: Summary for policy makers.

Diez, J. M., Sullivan, J. J., Hulme, P. E., Edwards, G., & Duncan, R. P. (2008). Darwin's naturalization conundrum: Dissecting taxonomic patterns of species invasions. *Ecology Letters, 11*(7), 674–681.

Divíšek, J., Chytrý, M., Beckage, B., Gotelli, N. J., Lososová, Z., Pyšek, P., Richardson, D. M., & Molofsky, J (2018). Similarity of introduced plant species to native ones facilitates naturalization, but differences enhance invasion success. *Nature Communications, 9*(1), 4631.

Dostálek, T., Münzbergová, Z., Kladivová, A., & Macel, M. (2021). Invasive Lupinus polyphyllus benefits from high soil nutrient levels and competition release. *Journal of Ecology, 109*(2), 957–967.

Drenovsky, R. E., Grewell, B. J., D'antonio, C. M., Funk, J. L., James, J. J., Molinari, N., Parker, I. M., & Richards, C. L. (2012). A functional trait perspective on plant invasion. *Annals of Botany, 110*(1), 141–153.

Dukes, J. S., & Mooney, H. A. (1999). Does global change increase the success of biological invaders? *Trends in Ecology & Evolution, 14*(4), 135–139.

Durand, L. Z., & Goldstein, G. (2001). Photosynthesis, photoinhibition, and nitrogen use efficiency in native and invasive tree ferns in Hawaii. *Oecologia, 126*, 345–354.

Ehrenfeld, J. G. (2010). Ecosystem consequences of biological invasions. *Annual Review of Ecology, Evolution, and Systematics, 41*, 59–80.

Elton, C. S. (2020). *The ecology of invasions by animals and plants*. Springer Nature.

Epanchin-Niell, R., McAusland, C., Liebhold, A., Mwebaze, P., & Springborn, M. R. (2021). Biological invasions and international trade: Managing a moving target. *Review of Environmental Economics and Policy, 15*(1), 180–190.

Erfmeier, A., & Bruelheide, H. (2005). Invasive and native Rhododendron ponticum populations: Is there evidence for genotypic differences in germination and growth? *Ecography, 28*(4), 417–428.

Eyster, H. N., & Wolkovich, E. M. (2021). Comparisons in the native and introduced ranges reveal little evidence of climatic adaptation in germination traits. *Climate Change Ecology, 2*, Article 100023.

Finch-Savage, W. E., & Bassel, G. W. (2016). Seed vigour and crop establishment: Extending performance beyond adaptation. *Journal of Experimental Botany, 67*(3), 567–591.

Foxcroft, L. C., Spear, D., Van Wilgen, N. J., & McGeoch, M. A. (2019). Assessing the association between pathways of alien plant invaders and their impacts in protected areas. *NeoBiota, 43*, 1–25.

Franks, S. J., Weber, J. J., & Aitken, S. N. (2014). Evolutionary and plastic responses to climate change in terrestrial plant populations. *Evolutionary Applications, 7*(1), 123–139.

Fridley, J. D. (2013). Plant invasions across the Northern Hemisphere: A deep-time perspective. *Annals of the New York Academy of Sciences, 1293*(1), 8–17.

Fridley, J. D., Jo, I., Hulme, P. E., & Duncan, R. P. (2021). A habitat-based assessment of the role of competition in plant invasions. *Journal of Ecology, 109*(3), 1263–1274.

Fristoe, T. S., Chytrý, M., Dawson, W., Essl, F., Heleno, R., Kreft, H., Maurel, N., Pergl, J., Pyšek, P., Seebens, H., Weigelt, P., Van Kleunen, M., et al. (2021). Dimensions of invasiveness: Links between local abundance, geographic range size, and habitat breadth in Europe's alien and native floras. *Proceedings of the National Academy of Sciences, 118*(22), e2021173118.

Funk, J. L. (2013). The physiology of invasive plants in low-resource environments. *Conservation Physiology, 1*(1), cot026.

Gaertner, M., Biggs, R., Te Beest, M., Hui, C., Molofsky, J., & Richardson, D. M. (2014). Invasive plants as drivers of regime shifts: Identifying high-priority invaders that alter feedback relationships. *Diversity and Distributions, 20*(7), 733–744.

Gaertner, M., et al. (2017). Past, present, and future research on biological invasions in South Africa. *South African Journal of Science, 113*(11–12), 1–9.

Gallien, L., Thornhill, A. H., Zurell, D., Miller, J. T., & Richardson, D. M. (2019). Global predictors of alien plant establishment success: Combining niche and trait proxies. *Proceedings of the Royal Society B, 286*(1897), 20182477.

Gentili, R., Ambrosini, R., Augustinus, B.A., Caronni, S., Cardarelli, E., Montagnani, C., Müller-Schärer, H., Schaffner, U., & Citterio, S. (2021). High phenotypic plasticity in a prominent plant invader along altitudinal and temperature gradients. *Plants, 10*(10), 2144.

Gioria, M., Carta, A., Baskin, C. C., Dawson, W., Essl, F., Kreft, H., Pergl, J., van Kleunen, M., Weigelt, P., Winter, M., Pyšek, P. (2021). Persistent soil seed banks promote naturalisation and invasiveness in flowering plants. *Ecology Letters, 24*(8), 1655–1667.

Gioria, M., Dieterich, B., & Osborne, B. (2011). Battle of the giants: primary and secondary invasions by large herbaceous species. In *Biology and environment: proceedings of the Royal Irish academy* (pp. 177–193). Royal Irish Academy.

Gioria, M., Hulme, P. E., Richardson, D. M., & Pyšek, P. (2023). Why are invasive plants successful? *Annual Review of Plant Biology, 74*, 635–670.

Gioria, M., & Osborne, B. A. (2014). Resource competition in plant invasions: Emerging patterns and research needs. *Frontiers in Plant Science, 5*, 501.

Gioria, M., Pyšek, P., & Osborne, B. A. (2022). Resource competition in plant invasions: Emerging patterns and research needs. *Frontiers in Plant Science, 13*, Article 852147.

Gómez, J. M., Perfectti Álvarez, F., González Megías, A., & Armas, C. (2025). Quantifying phenotypic plasticity: A call for consistency.

Grime, J. P. (2006). *Plant strategies, vegetation processes, and ecosystem properties*. John Wiley & Sons.

Groeneveld, E., et al. (2018). Urban biodiversity and ecological design. *Urban Ecology, 23*, 315–330.

Grotkopp, E., Rejmánek, M., & Rost, T. L. (2002). Toward a causal explanation of plant invasiveness: Seedling growth and life-history strategies of 29 pine (Pinus) species. *The American Naturalist, 159*(4), 396–419.

Gulzar, R., Wani, S. A., Hassan, T., Reddy, C. S., Shrestha, B. B., Mukul, S. A., Shabbir, A., Iqbal, I. M., Ranwala, S. M., Dorjee, Sujanapal, P., Khuroo, A. A., et al. (2023). Looking beyond the political boundaries: an integrated inventory of invasive alien flora of South Asia. *Biological Invasions*, 1–22.

Guo, Q., Fei, S., Dukes, J. S., Oswalt, C. M., III, B. V. I., & Potter, K. M. (2015). A unified approach for quantifying invasibility and degree of invasion. *Ecology*, *96*(10), 2613–2621.

Hamilton, M. A., Murray, B. R., Cadotte, M. W., Hose, G. C., Baker, A. C., Harris, C. J., & Licari, D. (2005). Life-history correlates of plant invasiveness at regional and continental scales. *Ecology Letters, 8*(10), 1066–1074.

Hegarty, M. J., Abbott, R. J., & Hiscock, S. J. (2012). Allopolyploid speciation in action: the origins and evolution of *Senecio cambrensis*. In *Polyploidy and genome evolution* (pp. 245–270). Springer Berlin Heidelberg.

Hierro, J. L., Eren, Ö., Montesinos, D., Andonian, K., Kethsuriani, L., Özcan, R., Diaconu, A., Török, K., Cavieres, L., French, K., et al. (2020). Increments in weed seed size track global range expansion and contribute to colonization in a non-native region. *Biological Invasions*, *22*, 969–982.

Hobbs, R. J., & Huenneke, L. F. (1992). Disturbance, diversity, and invasion: Implications for conservation. *Conservation Biology, 6*(3), 324–337.

Houseman, G. R., Foster, B. L., & Brassil, C. E. (2014). Propagule pressure-invasibility relationships: Testing the influence of soil fertility and disturbance with Lespedeza cuneata. *Oecologia, 174*, 511–520.

Hovick, S. M., & Whitney, K. D. (2014). Hybridisation is associated with increased fecundity and size in invasive taxa: Meta-analytic support for the hybridisation-invasion hypothesis. *Ecology Letters, 17*(11), 1464–1477.

Howell, A., Winkler, D. E., Phillips, M. L., McNellis, B., & Reed, S. C. (2020). Experimental warming changes phenology and shortens growing season of the dominant invasive plant Bromus tectorum (cheatgrass). *Frontiers in Plant Science, 11*, Article 570001.

Huang, Q. Q., Pan, X. Y., Fan, Z. W., & Peng, S. L. (2015). Stress relief may promote the evolution of greater phenotypic plasticity in exotic invasive species: A hypothesis. *Ecology and Evolution, 5*(6), 1169–1177.

Hufbauer, R. A., Facon, B., Ravigné, V., Turgeon, J., Foucaud, J., Lee, C. E., Rey, O., Estoup, A., et al. (2012). Anthropogenically induced adaptation to invade (AIAI): contemporary adaptation to human-altered habitats within the native range can promote invasions. *Evolutionary Applications, 5*(1), 89–101.

Hughes, R. F., & Denslow, J. S. (2005). Invasion by a N_2-fixing tree alters function and structure in wet lowland forests of Hawaii. *Ecological Applications, 15*(5), 1615–1628.

Hui, C., Richardson, D. M., Landi, P., Minoarivelo, H. O., Garnas, J., & Roy, H. E. (2016). Defining invasiveness and invasibility in ecological networks. *Biological Invasions, 18*, 971–983.

Hulme, P. E. (2008). Phenotypic plasticity and plant invasions: Is it all Jack? *Functional Ecology, 22*(1), 3–7.

Iwanycki Ahlstrand, N., Gopalakrishnan, S., Vieira, F. G., Bieker, V. C., Meudt, H. M., Dunbar-Co, S., Rothfels, C. J., Martinez-Swatson, K. A., Maldonado, C., Hassemer, G., Shipunov, A., Rønsted, N., et al. (2022). Travel tales of a worldwide weed: genomic signatures of *Plantago major* L. reveal distinct genotypic groups with links to colonial trade routes. *Frontiers in Plant Science, 13*, 838166.

Jelbert, K., Stott, I., McDonald, R. A., & Hodgson, D. (2015). Invasiveness of plants is predicted by size and fecundity in the native range. *Ecology and Evolution, 5*(10), 1933–1943.

Jo, I., Fridley, J. D., & Frank, D. A. (2017). Invasive plants accelerate nitrogen cycling: Evidence from experimental woody monocultures. *Journal of Ecology, 105*(4), 1105–1110.

Jones, B. A. (2019). Tree shade, temperature, and human health: Evidence from invasive species-induced deforestation. *Ecological Economics, 156*, 12–23.

Jones, B. A., & McDermott, S. M. (2018). Health impacts of invasive species through an altered natural environment: Assessing air pollution sinks as a causal pathway. *Environmental and Resource Economics, 71*, 23–43.

Kaushik, P., Pati, P. K., Chadhar, B. L., Khan, M. L., & Khare, P. K. (2022a). Functional dispersion and variation in above-ground traits in tropical dry deciduous forests of Central India. *International Journal of Ecology and Environmental Sciences, 48*(2), 251–264.

Kaushik, P., Pati, P. K., Khan, M. L., & Khare, P. K. (2022b). Plant functional traits best explain invasive species' performance within a dynamic ecosystem-A review. *Trees, Forests and People, 8*, Article 100260.

Kaushik, P., Pati, P. K., Khan, M. L., & Khare, P. K. (2024). Effect of invasive species on forest composition in tropical dry forest: A case study of invasion by *Gliricidia sepium. Southern Forests: A Journal of Forest Science.*

Kleyer, M., Bekker, R. M., Knevel, I. C., Bakker, J. P., Thompson, K., Sonnenschein, M., et al. (2008). The LEDA Traitbase: A database of life-history traits of the Northwest European flora. *Journal of Ecology, 96*(6), 1266–1274.

Klisz, M., Chakraborty, D., Cvjetković, B., Grabner, M., Lintunen, A., Mayer, K., George, J. P., Rossi, S (2023). Functional traits of boreal species and adaptation to local conditions. In *Boreal forests in the face of climate change: sustainable management* (pp. 323–355). Springer International Publishing.

Kueffer, C. (2017). Plant invasions in the Anthropocene. *Science, 358*(6364), 724–725.

Kumschick, S., Journiac, L., Boulesnane-Genguant, O., Botella, C., Pouteau, R., & Rouget, M. (2025). Mapping potential environmental impacts of alien species in the face of climate change. *Biological Invasions, 27*(1), 1–14.

Lachmuth, S., Durka, W., & Schurr, F. M. (2010). The making of a rapid plant invader: Genetic diversity and differentiation in the native and invaded range of *Senecio inaequidens. Molecular Ecology, 19*(18), 3952–3967.

Lake, J. C., & Leishman, M. R. (2004). Invasion success of exotic plants in natural ecosystems: The role of disturbance, plant attributes and freedom from herbivores. *Biological Conservation, 117*(2), 215–226.

Lavoie, C., Joly, S., Bergeron, A., Guay, G., & Groeneveld, E. (2016). Explaining naturalization and invasiveness: New insights from historical ornamental plant catalogs. *Ecology and Evolution, 6*(20), 7188–7198.

Lee, M. R., Flory, S. L., & Phillips, R. P. (2012). Positive feedbacks to growth of an invasive grass through alteration of nitrogen cycling. *Oecologia, 170*, 457–465.

Levin, D. A. (2021). Propagule pressure and the establishment of emergent polyploid populations. *Annals of Botany, 127*(1), 1–5.

Liu, C., Wolter, C., Xian, W., & Jeschke, J. M. (2020). Most invasive species largely conserve their climatic niche. *Proceedings of the National Academy of Sciences, 117*(38), 23643–23651.

Liu, R. L., Yang, Y. B., Lee, B. R., Liu, G., Zhang, W. G., Chen, X. Y., Zhu, Z. H., et al. (2021). The dispersal-related traits of an invasive plant Galinsoga quadriradiata correlate with elevation during range expansion into mountain ranges. *AoB Plants, 13*(3), plab008.

Liu, X., Hejda, M., & Pyšek, P. (2021b). Traits associated with invasiveness in alien plants: Where do we stand? *Ecology Letters, 24*(6), 1167–1181.

Lombaert, E., Guillemaud, T., Cornuet, J. M., Malausa, T., Facon, B., & Estoup, A. (2010). Bridgehead effect in the worldwide invasion of the biocontrol harlequin ladybird. *PLoS ONE, 5*(3), Article e9743.

Lonsdale, W. M. (1999). Global patterns of plant invasions and the concept of invasibility. *Ecology, 80*(5), 1522–1536.

Lukács, B. A., Mesterházy, A., Vidéki, R., & Király, G. (2016). Alien aquatic vascular plants in Hungary (Pannonian ecoregion): Historical aspects, data set and trends. *Plant Biosystems-an International Journal Dealing with All Aspects of Plant Biology, 150*(3), 388–395.

Marchini, G. L., Maraist, C. A., & Cruzan, M. B. (2019). Trait divergence, not plasticity, determines the success of a newly invasive plant. *Annals of Botany, 123*(4), 667–679.

Marod, D., Duengkae, P., Kutintara, U., Sungkaew, S., Wachrinrat, C., Asanok, L., & Klomwattanakul, N. (2012). The influences of an invasive plant species (Leucaena leucocephala) on tree regeneration in Khao Phuluang Forest, Northeastern Thailand. *Agriculture and Natural Resources, 46*(1), 39–50.

Maron, J. L., Elmendorf, S. C., & Vilà, M. (2007). Contrasting plant physiological adaptation to climate in the native and introduced range of *Hypericum perforatum. Evolution, 61*(8), 1912–1924.

Marsh, A. S., Hayes, D. C., Klein, P. N., Zimmerman, N., Dalsimer, A., Burkett, D. A., Huebner, C. D., Rabaglia, R., Meyerson, L. A., Harper-Lore, B. L., Davidson, J. L., Lopez, V. M., et al. (2021). Sectoral impacts of invasive species in the United States and approaches to management. In *Invasive species in forests and rangelands of the United States: A comprehensive science synthesis for the United States forest sector* (pp. 203–229).

Mason, R. A., Cooke, J., Moles, A. T., & Leishman, M. R. (2008). Reproductive output of invasive versus native plants. *Global Ecology and Biogeography, 17*(5), 633–640.

Matesanz, S., Gianoli, E., & Valladares, F. (2010). Global change and the evolution of phenotypic plasticity in plants. *Annals of the New York Academy of Sciences, 1206*(1), 35–55.

Maurel, N., Hanspach, J., Kühn, I., Pyšek, P., & van Kleunen, M. (2016). Introduction bias affects relationships between the characteristics of ornamental alien plants and their naturalization success. *Global Ecology and Biogeography, 25*(12), 1500–1509.

McDougall, K. L., Lembrechts, J., Rew, L. J., Haider, S., Cavieres, L. A., Kueffer, C., Milbau, A., Naylor, B. J., Nuñez, M. A., Pauchard, A., Seipel, T., Alexander, J. M., et al. (2018). Running off the road: roadside non-native plants invading mountain vegetation. *Biological invasions, 20*, 3461–3473.

McKinney, M. L. (2002). Urbanization, biodiversity, and conservation. *BioScience, 52*(10), 883–890.

Melbourne, B. A., Cornell, H. V., Davies, K. F., Dugaw, C. J., Elmendorf, S., Freestone, A. L., Hall, R. J., Harrison, S., Hastings, A., Holland, M., Holyoak, M., Yokomizo, H., et al. (2007). Invasion in a heterogeneous world: resistance, coexistence or hostile takeover? *Ecology Letters, 10*(1), 77–94.

Mitchell, C. E., Agrawal, A. A., Bever, J. D., Gilbert, G. S., Hufbauer, R. A., Klironomos, J. N., Maron, J. L., Morris, W. F., Parker, I. M., Power, A. G., Seabloom, E. W., Vazquez, D. P., et al. (2006). Biotic interactions and plant invasions. *Ecology Letters, 9*(6), 726–740.

Molina-Montenegro, M. A., Acuña-Rodríguez, I. S., Flores, T. S., Hereme, R., Lafon, A., Atala, C., & Torres-Díaz, C. (2018). Is the success of plant invasions the result of rapid adaptive evolution in seed traits? Evidence from a latitudinal rainfall gradient. *Frontiers in Plant Science, 9*, 208.

Montesinos, D. (2022a). Fast invasives fastly become faster: Invasive plants align largely with the fast side of the plant economics spectrum. *Journal of Ecology, 110*(5), 1010–1014.

Montesinos, D. (2022b). Fast invasives slow down: Long-term growth declines in invasive trees. *Ecological Applications, 32*(1), Article e02489.

Montesinos, D., & Callaway, R. M. (2018). Traits correlate with invasive success more than plasticity: A comparison of three Centaurea congeners. *Ecology and Evolution, 8*(15), 7378–7385.

Moravcová, L., Pyšek, P., Jarošík, V., & Pergl, J. (2015). Getting the right traits: Reproductive and dispersal characteristics predict the invasiveness of herbaceous plant species. *PLoS ONE, 10*(4), Article e0123634.

Mounger, J., Ainouche, M. L., Bossdorf, O., Cavé-Radet, A., Li, B., Parepa, M., Salmon, A., Yang, J., & Richards, C. L. (2021). Epigenetics and the success of invasive plants. *Philosophical Transactions of the Royal Society B, 376*(1826), 20200117.

Oduor, A. M., van Kleunen, M., & Stift, M. (2020). Allelopathic effects of native and invasive *Brassica nigra* do not support the novel-weapons hypothesis. *American Journal of Botany, 107*(8), 1106–1113.

Osunkoya, O. O., Lock, C. B., Dhileepan, K., & Buru, J. C. (2021). Lag times and invasion dynamics of established and emerging weeds: Insights from herbarium records of Queensland, Australia. *Biological Invasions, 23*, 3383–3408.

Pandit, M. K., White, S. M., & Pocock, M. J. (2014). The contrasting effects of genome size, chromosome number and ploidy level on plant invasiveness: A global analysis. *New Phytologist, 203*(2), 697–703.

Pati, P. K., Kaushik, P., Khan, M. L., & Khare, P. K. (2022). Biodiversity and ecosystem services of trees outside forests: A case study from Dr. Harisingh Gour Vishwavidyalaya, Sagar, Central India. *Indian Journal of Ecology, 49*(2), 608–615.

Pati, P. K., Kaushik, P., Khan, M. L., & Khare, P. K. (2023). Effect of habitat specific wood specific gravity on biomass and carbon stock of trees in tropical dry deciduous forest of central India. *Tropical Ecology, 64*(3), 519–531.

Pati, P. K., Kaushik, P., Maurya, D., Adhikari, C., Bishwas, A. J., Khan, M. L., & Khare, P. K. (2024). Evaluating the role of roadside vegetation in atmospheric carbon dioxide mitigation: A case study. *Arboricultural Journal, 46*(2), 97–119.

Pattison, R. R., Goldstein, G., & Ares, A. (1998). Growth, biomass allocation and photosynthesis of invasive and native Hawaiian rainforest species. *Oecologia, 117*, 449–459.

Pauchard, A., Kueffer, C., Dietz, H., Daehler, C. C., Alexander, J., Edwards, P. J., Arévalo, J. R., Cavieres, L. A., Guisan, A., Haider, S., Jakobs, G., Seipel, T., et al. (2009). Ain't no mountain high enough: plant invasions reaching new elevations. *Frontiers in Ecology and the Environment, 7*(9), 479–486.

Perez-Harguindeguy, N., Diaz, S., Garnier, E., Lavorel, S., Poorter, H., Jaureguiberry, P., et al. (2013). New handbook for standardised measurement of plant functional traits worldwide. *Australian Journal of Botany, 61*(3), 167–234.

Plenderleith, F. A., Irrazabal, V. A., Burslem, D. F., Travis, J. M., & Powell, P. A. (2022). Predicting spatially heterogeneous invasive spread: Pyracantha angustifolia invading a dry Andean valley in northern Argentina. *Biological Invasions, 24*(7), 2201–2216.

Potgieter, L. J., & Cadotte, M. W. (2020). The application of selected invasion frameworks to urban ecosystems. *NeoBiota, 62*, 365–386.

Pyšek, P., Bacher, S., Kühn, I., Novoa, A., Catford, J. A., Hulme, P. E., Pergl, J., Richardson, D. M., Wilson, J. R., & Blackburn, T. M. (2020a). Macroecological Framework for Invasive Aliens (MAFIA): Disentangling large-scale context dependence in biological invasions

Pyšek, P., Jarošík, V., Pergl, J., Moravcová, L., Chytrý, M., & Kühn, I. (2014). Temperate trees and shrubs as global invaders: The relationship between invasiveness and native distribution depends on biological traits. *Biological Invasions, 16*, 577–589.

Pyšek, P., Manceur, A. M., Alba, C., McGregor, K. F., Pergl, J., Štajerová, K., Chytrý, M., Danihelka, J., Kartesz, J., Klimešová, J., Lučanová, M., Kühn, I., et al. (2015). Naturalization of central European plants in North America: species traits, habitats, propagule pressure, residence time. *Ecology, 96*(3), 762–774.

Pyšek, P., & Richardson, D. M. (2007). Traits associated with invasiveness in alien plants: where do we stand? *Biological Invasions*, 97–125.

Rai, P. K., & Singh, J. S. (2020). Invasive alien plant species: Their impact on environment, ecosystem services and human health. *Ecological Indicators, 111*, Article 106020.

Razanajatovo, M., van Kleunen, M., Kreft, H., Dawson, W., Essl, F., Pergl, J., Pyšek, P., Winter, M., & Weigelt, P. (2019). Autofertility and self-compatibility moderately benefit island colonization of plants. *Global Ecology and Biogeography, 28*(3), 341–352.

Rejmanek, M., & Richardson, D. M. (1996). What attributes make some plant species more invasive? *Ecology, 77*(6), 1655–1661.

Rejmánek, M., Richardson, D. M., Higgins, S. I., Pitcairn, M. J., & Grotkopp, E. (2005). Ecology of invasive plants: State of the art. *Scope-Scientific Committee on Problems of the Environment International Council of Scientific Unions, 63*, 104.

Rejmánek, M., Richardson, D. M., & Pyšek, P. (2013). Plant invasions and invasibility of plant communities. *Vegetation Ecology*, 387–424.

Ricciardi, A., & Jones, L. A. (2011). Åsa M. Kestrup and Jessica M. Ward. *Fifty Years of Invasion Ecology*, 225.

Richards, C. L., Bossdorf, O., Muth, N. Z., Gurevitch, J., & Pigliucci, M. (2006). Jack of all trades, master of some? On the role of phenotypic plasticity in plant invasions. *Ecology Letters, 9*(8), 981–993.

Richardson, D. M. (2013). *Plant invasions and invasibility of plant communities.*

Richardson, D. M., & Pyšek, P. (2006). Plant invasions: Merging the concepts of species invasiveness and community invasibility. *Progress in Physical Geography, 30*(3), 409–431.

Richardson, D. M., & Pyšek, P. (2012). Naturalization of introduced plants: Ecological drivers of biogeographical patterns. *New Phytologist, 196*(2), 383–396.

Richardson, D. M., et al. (2022). Challenges and opportunities for managing biological invasions in urban environments. *Global Ecology and Biogeography, 31*(1), 31–45.

Richardson, D. M., Pyšek, P., Rejmanek, M., Barbour, M. G., Panetta, F. D., & West, C. J. (2000). Naturalization and invasion of alien plants: Concepts and definitions. *Diversity and Distributions, 6*(2), 93–107.

Rosche, C., Hensen, I., Mraz, P., Durka, W., Hartmann, M., & Lachmuth, S. (2017). Invasion success in polyploids: The role of inbreeding in the contrasting colonization abilities of diploid versus tetraploid populations of *Centaurea stoebe* sl. *Journal of Ecology, 105*(2), 425–435.

Rutland, C. A., Hall, N. D., & McElroy, J. S. (2021). The impact of polyploidization on the evolution of weed species: Historical understanding and current limitations. *Frontiers in Agronomy, 3*, Article 626454.

Schierenbeck, K. A., & Ellstrand, N. C. (2009). Hybridization and the evolution of invasiveness in plants and other organisms. *Biological Invasions, 11*, 1093–1105.

Schrieber, K., & Lachmuth, S. (2017). The genetic paradox of invasions revisited: The potential role of inbreeding× environment interactions in invasion success. *Biological Reviews, 92*(2), 939–952.

Seabloom, E. W., Harpole, W. S., Reichman, O. J., & Tilman, D. (2003). Invasion, competitive dominance, and resource use by exotic and native California grassland species. *Proceedings of the National Academy of Sciences, 100*(23), 13384–13389.

Sharma, G. P., Raghubanshi, A. S., & Singh, J. S. (2019). Lantana invasion: An overview. *Weed Biology and Management, 5*(4), 157–167.
Simberloff, D. (2009). The role of propagule pressure in biological invasions. *Annual Review of Ecology, Evolution, and Systematics, 40*, 81–102.
Smith, A. L., Hodkinson, T. R., Villellas, J., Catford, J. A., Csergő, A. M., Blomberg, S. P., Crone, E. E., Ehrlén, J., Garcia, M. B., Laine, A. L., Roach, D. A., Buckley, Y. M., et al. (2020). Global gene flow releases invasive plants from environmental constraints on genetic diversity. *Proceedings of the National Academy of Sciences, 117*(8), 4218–4227.
Sodhi, D. S., Livingstone, S. W., Carboni, M., & Cadotte, M. W. (2019). Plant invasion alters trait composition and diversity across habitats. *Ecology and Evolution, 9*(11), 6199–6210.
Sommer, R. J. (2020). Phenotypic plasticity: From theory and genetics to current and future challenges. *Genetics, 215*(1), 1–13.
Te Beest, M., Le Roux, J. J., Richardson, D. M., Brysting, A. K., Suda, J., Kubešová, M., & Pyšek, P. (2012). The more the better? The role of polyploidy in facilitating plant invasions. *Annals of Botany, 109*(1), 19–45.
Thuiller, W., Gallien, L., Boulangeat, I., De Bello, F., Münkemüller, T., Roquet, C., & Lavergne, S. (2010). Resolving Darwin's naturalization conundrum: A quest for evidence. *Diversity and Distributions, 16*(3), 461–475.
Tilman, D. (1997). Community invasibility, recruitment limitation, and grassland biodiversity. *Ecology, 78*(1), 81–92.
Traveset, A., & Richardson, D. M. (Eds.). (2020). *Plant invasions: the role of biotic interactions* (Vol. 4). CABI.
Tripathi, S., Bhadouria, R., Srivastava, P., Singh, R., & Batish, D. R. (Eds.). (2023). *Plant invasions and global climate change.* Springer Nature.
Uddin, M. N., Asaeda, T., Shampa, S. H., & Robinson, R. W. (2020). Allelopathy and its coevolutionary implications between native and non-native neighbors of invasive Cynara cardunculus L. *Ecology and Evolution, 10*(14), 7463–7475.
Van Kleunen, M., Weber, E., & Fischer, M. (2010). A meta-analysis of trait differences between invasive and non-invasive plant species. *Ecology Letters, 13*(2), 235–245.
Vilà, M., et al. (2011). Ecological impacts of invasive alien plants: A meta-analysis of their effects on species, communities, and ecosystems. *Ecology Letters, 14*(7), 702–710.
Vorstenbosch, T., Essl, F., & Lenzner, B. (2020). An uphill battle? The elevational distribution of alien plant species along rivers and roads in the Austrian Alps. *NeoBiota, 63*, 1–24.
Waddell, E. H., Chapman, D. S., Hill, J. K., Hughes, M., Bin Sailim, A., Tangah, J., & Banin, L. F. (2020). Trait filtering during exotic plant invasion of tropical rainforest remnants along a disturbance gradient. *Functional Ecology, 34*(12), 2584–2597.
Wallingford, P. D., Morelli, T. L., Allen, J. M., Beaury, E. M., Blumenthal, D. M., Bradley, B. A., Dukes, J. S., Early, R., Fusco, E. J., Goldberg, D. E., Ibáñez, I., & Sorte, C. J. (2020). Adjusting the lens of invasion biology to focus on the impacts of climate-driven range shifts. *Nature Climate Change, 10*(5), 398–405.
Wan, J. Z., Wang, C. J., Zimmermann, N. E., Li, M. H., Pouteau, R., & Yu, F. H. (2021). Current and future plant invasions in protected areas: Does clonality matter? *Diversity and Distributions, 27*(12), 2465–2478.
Weiher, E., van der Werf, A., Thompson, K., Roderick, M., Garnier, E., & Eriksson, O. (1999). Chal-lenging Theophrastus. A common core list of plant traits for functional ecology. *Journal of Vegetation Science, 10*, 609–620.
Westoby, M., Falster, D. S., Moles, A. T., Vesk, P. A., & Wright, I. J. (2002). Plant ecological strategies: Some leading dimensions of variation between species. *Annual Review of Ecology and Systematics, 33*(1), 125–159.
White, S. R., Tannas, S., Bao, T., Bennett, J. A., Bork, E. W., & Cahill, J. F., Jr. (2013). Using structural equation modelling to test the passenger, driver and opportunist concepts in a Poa pratensis invasion. *Oikos, 122*(3), 377–384.

Williamson, M., & Fitter, M. (1996). The varying success of invaders. Advances in invasion ecology. *Ecology, 77*(6), 1661–1666.

Wyse, S. V., & Hulme, P. E. (2021). Dispersal potential rather than risk assessment scores predict the spread rate of non-native pines across New Zealand. *Journal of Applied Ecology, 58*(9), 1981–1992.

Xia, L., Geng, Q., & An, S. (2020). Rapid genetic divergence of an invasive species, *Spartina alterniflora*, in China. *Frontiers in Genetics, 11*, 284.

Xie, J., He, Z., Liu, X., & van Kleunen, M. (2022). Nitrogen deposition promotes invasive *Solidago canadensis* via enhanced nitrogen assimilation. *Journal of Plant Ecology, 15*(3), 456–468.

Ye, X., Zhang, Y., Liu, Z., & Li, Z. (2023). Clonality and root traits enhance invasiveness of *Alternanthera philoxeroides* under nutrient fluctuations. *Oecologia, 201*(1), 1–12.

Yelenik, S. G., & D'Antonio, C. M. (2013). Self-reinforcing impacts of plant invasions change over time. *Nature, 503*(7477), 517–520.

Young, A. M., et al. (2021). The role of citizen science in monitoring invasive species. *Conservation Biology, 35*(3), 678–687.

Yuan, L., Li, J. M., Yu, F. H., Oduor, A. M., & van Kleunen, M. (2021). Allelopathic and competitive interactions between native and alien plants. *Biological Invasions, 23*(10), 3077–3090.

Zenni, R. D., Dickie, I. A., Wingfield, M. J., Hirsch, H., Crous, C. J., Meyerson, L. A., Burgess, T. I., Zimmermann, T. G., Klock, M. M., Siemann, E., & Erfmeier, A. (2017). Evolutionary dynamics of tree invasions: Complementing the unified framework for biological invasions. *AoB Plants, 9*(1), plw085.

Zhang, L., Chen, A., Li, Y., Li, D., Cheng, S., Cheng, L., & Liu, Y. (2022). Differences in phenotypic plasticity between invasive and native plants responding to three environmental factors. *Life, 12*(12), 1970.

Zhang, Z., Liu, Y., Yuan, L., Weber, E., & van Kleunen, M. (2021). Effect of allelopathy on plant performance: A meta-analysis. *Ecology Letters, 24*(2), 348–362.

Zirbel, C. R., Bassett, T., Grman, E., & Brudvig, L. A. (2017). Plant functional traits and environmental conditions shape community assembly and ecosystem functioning during restoration. *Journal of Applied Ecology, 54*(4), 1070–1079.

Chapter 7
Ecosystem Services and Disservices Linked to Invasive Alien Plants

Raees Khan, Safia Belhadj, Amjad Khan, Sidra Batool, Syed Mukaram Shah, Mahrine Rashid, and Shazia Erum

Abstract Alien plant species have been introduced by humans or by natural forces globally, and many of them have become invasive. Invasive alien plant species (IAPS) have positive or normally detrimental effects on the ecosystem as they alter the ecosystem's properties, functions, and associated services. The ecosystem disservices (EDS) linked with IAPS include effects on human health, and threats to native biodiversity, soil, crops, horticulture, forestry, industry and the environment. The IAPS poses serious threats to native flora, especially endemic plants with a narrow distribution range. New pests and allergies become widespread along these IAPS damaging human health, crops and livestock. The ecosystem services (ES) linked to IAPS include the use of hundreds of alien woody species for timber, many species have ornamental, spiritual and aesthetic value, and several are cultivated for pharmaceutical and cosmetic compounds. Other ecosystem services (ES) of IAPS include ecosystem restoration, soil stabilisation, phytoremediation, erosion regulation and windbreaks. IAPS are among the top global environmental challenges and have large-scale environmental and socioeconomic effects. The chapter deconstructs the bifold roles played by invasive alien plants, describing their advantages (e.g., erosion control, pollination) and harm (e.g., invasive effect on native vegetation, economic loss), centered on context-dependent ecological and human trade-offs. IAPS have a major ecosystem disservices (EDS) and services (ES) that directly and indirectly affect human well-being, the environment, and socioeconomic sectors. This chapter

R. Khan (✉) · A. Khan · M. Rashid · S. Erum
National Herbarium of Pakistan (Stewart Collection), Plant Genetic Resources Institute, PARC-National Agricultural Research Centre, Islamabad, Pakistan
e-mail: raeeskhan@bs.qau.edu.pk

S. Belhadj
Faculty of Nature and Life Sciences, University of Djelfa, Djelfa, Algeria

A. Khan
Department of Plant Sciences, Quaid-I-Azam University—Islamabad, Islamabad, Pakistan

S. Batool
China University of Geosciences, Wuhan, Hubei, China

S. M. Shah
Centre of Plant Biodiversity, University of Peshawar, Peshawar, Pakistan

R. Singh et al. (eds.), *Invasive Alien Plants in Urban Ecosystems*,
https://doi.org/10.1007/978-3-032-00997-5_7

discusses the benefits (ecosystem services) and drawbacks (disservices) of alien invasive plants by exploring their ecological and socio-economic impacts. The main discussion includes their role in habitat protection, soil stabilization, nutrient cycling, and the negative effects like biodiversity loss and resource competition. Case studies display the impact of invasive species on ecosystems and human health.

Keywords Biodiversity · Climate change · Native flora · Threats · Invasive species

1 Introduction

Today, the world faces critical challenges, including climate change, biodiversity loss, environmental pollution, threatening biodiversity, deteriorating environmental health and food scarcity (Adigun & Odeleye, 2025; Khan et al., 2022). Biological invasions are among the central forces of change that influence many dimensions of life on Earth and impact ecosystem services (Bargali, 2024). Invasive alien plant species (IAPS) are one of the most significant causes of ecosystem degradation and loss of biodiversity (Gallardo et al., 2024; Junaedi et al., 2023; Lorena & Charles, 2008). The invasion of plant species is a global concern; their impact on ecosystems and native species are often irreversible (Gallardo et al., 2024; Kettunen et al., 2009). According to Shine et al. (2000), as reported by Sakhraoui et al. (2019) and Afreen et al. (2023), no area is effectively protected from this issue due to the difficulties encountered in effectively managing systems for plant material that is transported through several means of transportation.

According to the Biodiversity and Ecosystem Services (IPBES) projections, including global biodiversity hotspots, about one-fifth of the Earth's surface is at risk due to invasive species, which are one of the main direct drivers of biodiversity loss globally (IPBES, 2023). IAPS refer to species that are introduced into new areas, where they become established and pose a threat to the native ecosystems. These species can adapt and thrive, causing harm to the local environment through their physical characteristics, ecological interactions, and genetic traits (Khan et al., 2022). According to Shine et al. (2009) definition, as reported by Shiferaw et al. (2022), a species is considered as invasive alien species when it spreads beyond its native range. Invasion results when species are cut off from existing populations and inhabit novel geographic areas (Bargali, 2024; Gallardo et al., 2024). Some exotic species naturalize in their introduced range and may become invasive. Earlier, Shine et al. (2000), cited by Sakhraoui et al. (2019), reported that an invasive species, such as an exotic species that establishes itself in a natural or semi-natural habitat, induces changes and threatens indigenous biological diversity.

IAPS have a greater potential for invasion compared to native flora as they are usually more efficient competitors and with little pressure from local predators than native species (Ntalo et al., 2022). These species most often settle in disturbed environments, where they can serve as indicators of disturbance. Invasive species are a real danger to biodiversity via the reduction of species richness, genetic pollution and erosion. They are the second leading cause of biodiversity erosion after habitat destruction and fragmentation (Sakhraoui et al., 2019).

The global count of alien invasive species is on the rise, as established by several authors (Freer-Smith & Webber, 2017; Hanley & Roberts, 2019). Globally, more than 2500 species have become invasive. Asteraceae has the largest number of invasive species, as this family is cosmopolitan (Pagad et al., 2018). In Europe, nearly 1100 exotic species have been identified, with 10 to 15% showing negative impacts, resulting in annual expenses estimated at 12.5 billion euros, with nearly 77% related to damages and 23% to management (Kettunen et al., 2009). Climate change, trade, travel, tourism and accelerated anthropogenic activities have introduced and spread IAPS globally, causing severe impacts by modifying ecosystems, affecting the environment and human well-being (Hulme et al., 2018; Junaedi et al., 2023; Ljubojević, 2023; Rai & Singh, 2020). Botanical gardens, plant nurseries and imports of seeds have been the primary introductory Pathway for IAPS in the past (Hulme et al., 2018; Junaedi et al., 2023). Many ornamental, decorative and multipurpose IAPS are deliberately or accidentally introduced to new areas and can have unintended consequences and impacts on both human health and ecosystems. Correct taxonomy and accurate identification of IAPS are very significant for management policy-making and understanding their future spread (Khan et al., 2022). The invasive species can significantly alter the global environment and affect human health in very complex ways (Rai & Singh, 2020). South Asia has about 392 IAPS (Bhatta et al., 2023). *Rhododendron ponticum* was initially brought to the UK as an attractive ornamental plant and gained popularity among gardeners due to its low maintenance, evergreen characteristics, and vibrant blossoms. However, opinions about *R. ponticum* have shifted in recent times due to its rapid growth and its ability to outcompete native plant species. Consequently, initiatives to remove *R. ponticum* often encounter opposition from the public (Dehnen-Schmutz & Williamson, 2006).

IAPS are introduced intentionally or unintentionally and pose a threat to the environment by disrupting the delicate balance of biodiversity and ecosystems and outcompeting indigenous medicinal plants (Maema et al., 2016; Ul Abidin et al., 2023). For instance, Paper mulberry (*Broussonetia papyrifera* L.) is an Asian invasive species that was planted in the capital of Pakistan for greenery purposes, now triggers severe allergies and threatens native flora (Khan et al., 2022). Similarly, *Ailanthus altissima* is one of the very invasive species that was introduced as an ornamental plant in the eighteenth century in North Africa. It has rapidly spread due to its use for ornamental, fuel, and erosion control purposes. As an allelopathic species, *A. altissima* can negatively impact indigenous plant communities by suppressing resident species and altering competitive interactions (Kacimi & Sbabdji, 2022; Kasson et al., 2014; Montecchiari et al., 2020; Sladonja et al., 2015; Small et al., 2010). Even, it has the reputation as one of the best species for soil stabilization, *A. altissima* is known for its devastating invasive character across many countries because of its rapid multiplication (Sladonja et al., 2015; Wunder et al., 2014). As well, a survey conducted on the west coast of Algeria by Bachir et al. (2021) showed that the invasive alga *Caulerpa cylindracea* tends to colonize disturbed ecosystems, reflecting a reduction in native algal diversity, resulting in a severe impoverishment of the algal community. The study showed, as well, that the number of macroalgal species accompanying the invasive species has dropped by 52%.

Globalization facilitates species exchange and provides the foundation for promoting the exchange of species and agricultural seeds between various countries and regions. South Asia has a growing problem of invasive species, with more than half (53%) of local and regional inventories, with India having the highest number of invasive species (145) (Bhatta et al., 2023). While many of these species bring positive benefits and do not pose any harm to the native ecosystem, some can be detrimental when introduced to new areas. Trade, transportation of goods, tourism, the food industry, and travel are the primary factors that facilitate the introduction of IAPS, as highlighted by Copp et al. (2007) and Khan et al. (2022). Global trade of exotic plants will result in accelerated plant invasions in emerging economies under climate change (Seebens et al., 2015). In certain instances, these IAPS are introduced either unknowingly without understanding their harmful impacts or intentionally as biological weapons (bioterrorism) to disrupt agricultural practices in a country or region. Factors such as global trade, inadequate quarantine regulations, a scarcity of experts, lack of awareness within society, and global climate change create favourable conditions for invasive species to thrive, leading to detrimental effects on the local flora and fauna. Some common IAPS in various parts of Pakistan are shown in Plate 1. The society and the general public are often unaware of the situation and rarely concerned about the negative impacts associated with alien invasive species. The concept of IAPS is closely linked to other biodiversity-related issues, including habitat loss and climate change and these species have the potential to shift, expand, and establish themselves in new areas, which in turn has implications for local biodiversity and natural resources (Courchamp et al., 2017; Gallardo et al., 2024; Khan et al., 2022).

Species from all major botanical groups have become invasive (Downey & Richardson, 2016). Potentially, IAPS are essentially woody plants. Vegetative propagation allows the plant to become invasive if associated with rapid growth, as it is the case for *Arundo donax*. Exotic species that combine sexual and asexual reproduction can spread quickly to become invasive (Albert et al., 2015). This is particularly the case of *Acacia dealbata*, *Acacia horrida*, *Acacia saligna* and *Ailanthus altissima* (Sakhraoui et al., 2019). According to Di Castri (1990), several characteristics help species to become invasive. Morphological traits such as smaller size, lighter weight, the presence of winged or spiny seeds, and a high ability to move and disperse are the main features. Ecologically, invasive species exhibit a wide potential niche, lack specialized germination requirements, easily attract pollinators, possess dormancy capabilities, grow rapidly, and allocate significant resources to reproduce, produce long-lasting seeds and have edible fruits that attract animals for dispersal. Genetically, invasive species have high fecundity, rapid population growth, short and simple life cycles, high genetic variation, engage in uniparental reproduction, and may exhibit polyploidy (Skočajić & Nešić, 2021).

To achieve the Sustainable Development Goals (SDGs), proper management and control of invasive alien plant species is a vital goal (Rai & Singh, 2020). According

Plate 1 Invasive taxa: **a** *Alternanthera philoxeroides*, **b** *Asclepias curassavica*, **c** *Bidens pilosa*, **d** *Blumea sinuata*, **e** *Caesalpinia pulcherrima*, **f** *Cannabis sativa*, **g** *Carthamus oxyacantha*, **h** *Clerodendrum indicum*, **i** *Cousinia prolifera*, **j** *Eichhornia crassipes*, **k** *Emex spinosa*, **l** *Erigeron bonariensis*, **m** *Euphorbia heterophylla*, **n** *Euphorbia tirucalli*, **o** *Hydrocotyle umbellata*

to IUCN (2021), 33 plant species have been driven to extinction as a result of IAS (Hulme et al., 2018). The impacts associated with most IAPS have not been studied or are poorly understood or documented and little attention has been given to this one major global threat (Downey & Richardson, 2016; Khan et al., 2022). Here, this chapter discusses in detail Ecosystem services and disservices reported and linked to IAPS worldwide in Urban Ecosystems.

2 Ecosystem Services (ES)

IAPS can positively impact society by providing resources (Shackleton et al. 2018; Linders et al., 2020). Using the Common International Classification of Ecosystem Services (CICES; Haines-Young & Potschin, 2012), ES from IAPS can be categorized as: (i) provisioning services (including food, fiber, pharmaceuticals, water and others); (ii) regulation and maintenance services (climate, water and erosion regulation, nutrient cycling, pollination etc.); and (iii) cultural services (spiritual and aesthetic values as well as providing the foundation for tourism and recreation development). Here, the detailed services associated with IAPS directly or indirectly are outlined (Fig. 1).

i. **Provisioning Services**

IAPS increase food resources for wildlife and humans (Pejchar & Mooney, 2009). For example: *Lantana camara* (Plate 2) and *Prosopis* species are invasive but a good potential source of browse for livestock in Semi-Arid areas of South Africa (Ntalo et al., 2022; Ravhuhali et al., 2021). The IAPS helps to enhance the economy as it provides food, fuel, timber, and pollination services. IAPS provides an abundant food source for birds in the fall and winter, which increases their survival. IAPS provides ecological advantages, such as offering shelter and food resources to native species (Schlaepfer et al., 2011). Many invasive species are good sources of timber or fiber

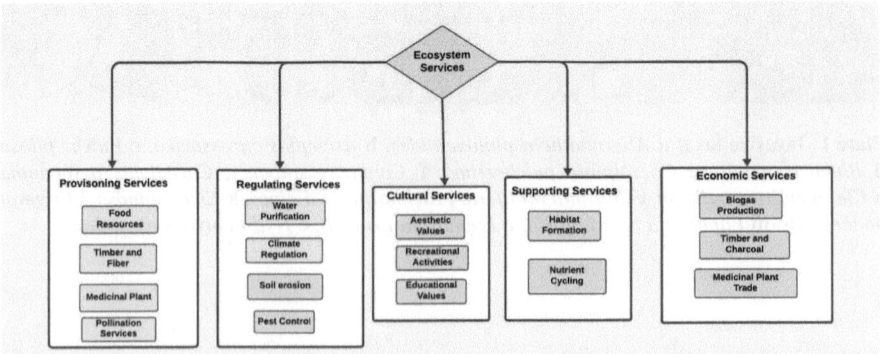

Fig. 1 Major ecosystem services (ES) associated with Invasive Alien Plant Species (IAPS)

production. *Eucalyptus* species were introduced in Pakistan and Algeria, for instance, to increase forest cover, horticulture and timber production (Chevalier, 1952; Khan et al., 2022). Many IAPS are used as medicinal plants with health benefits for humans and livestock (Adhikari et al., 2022; Khan et al., 2011; Maema et al., 2016).

Plate 2 Invasive taxa: **a** *Justicia brandegeeana*, **b** *Jacaranda mimosifolia*, **c** *Jatropha integerrima*, **d** *Lantana camara*, **e** *Leucaena leucocephala*, **f** *Martynia annua*, **g** *Mirabilis jalapa*, **h** *Nelumbo nucifera*, **i** *Nerium indicum*, **j** *Parthenium hysterophorus*, **k** *Passiflora caerulea*, **l** *Pithecellobium dulce*, **m** *Pittosporum tobira*, **n** *Prosopis juliflora*, **o** *Ricinus communis*

Khan et al. (2011) reported 39 invasive species from northeastern Bangladesh with medicinal uses for 37 diseases, while Adhikari et al. (2022) documented 24 invasive species for therapeutic uses in the Nepalese communities in Waterberg District, Limpopo Province, South Africa. Maema et al. (2016) reported invasive species (i.e. *Schinusmolle*, *Catharanthus roseus*, *Datura stramonium*, *Opuntia stricta*, *Opuntia ficus-indica*, *Sambucus canadensis*, *Ricinus communis* (Plate 3), *Melia azedarach*, *Argemone ochroleuca* and *Eriobotrya japónica*) are used as medicinal plants for the treatment of various diseases such as chest complaint, blood purification, asthma, hypertension and infertility in Waterberg District, Limpopo Province (South Africa). Williams et al. (2021) recorded 81 non-native taxa traded in South Africa of which 36 were alien plant taxa. Similarly, Khan et al. (2022) reported that *Broussonetia papyrifera*, *Eucalyptus camaldulensis*, *Prosopis julifora* and *Leucaena leucocephala* are associated in Pakistan for timber production, fuel wood and greenery on barren land (Plates 2, 3 and 4). Máximo et al. (2020) reported that many invasive species (i.e. *Carpobrotus edulis*, *Hakea salicifolia*, *Hakea sericea*, *Oxalis pes-caprae* and *Phytolacca americana*) which pose a threat to Portugal ecosystem are a potential source of bioactive metabolites. As well, *Paulownia* is considered an invasive plant species in Algeria. It is used in semi-arid and semi-desert areas to stabilize soils and as a source of fodder, in addition to the exploitation of its leaves and fibers in the medicinal field of the cosmetic and textile industries. *Paulownia* can over cope with temperatures over 49 °C, has rapid growth and may reach 7 m in height in only three years (APS, 2023).

ii. **Regulating Services**

Invasive alien plant species (IAPS) common regulating services include the provision of clean water and a stable climate. The most regulated services of invasive species are underappreciated, but they have widespread implications in water purification, pest control, pollination, fishery, agriculture and forestry, natural hazards and climate mitigation (Pejchar & Mooney, 2009). These species have the potential to modify the essential efficiency, hydrology, nutrient cycling (soil improvement), and decomposition in the ecosystem (Afreen et al., 2023; Ravhuhali et al., 2021; Shackleton et al., 2019). IAPS also contributes to climate regulation through carbon sequestration and helps in reducing air pollution (Nunes et al., 2021; Rai & Singh, 2020). Peltzer et al. (2010) reported that IAPS is associated with dramatically altering a range of aboveground and belowground ecosystem processes including carbon sequestration. In Portugal, *Acacia dealbata* Link is used for the carbon sequestration potential as a forest invasive species (Nunes et al., 2021). Many IAPS are responsible for water regulation by influencing hydrological cycles (Catford, 2017). *Eichhornia crassipes* (Mart) Solms. (water hyacinth), positively affects water quality by reducing arsenic contaminants from water and rendering it unobjectionable for human drinking (Kathryn, 2005). Several IAPS help to reduce and control soil erosion through root systems. *Prosopis* provides ecosystem services like reducing soil erosion (Ravhuhali et al., 2021). *Eucalyptus* species are associated with increased honey production (Pejchar & Mooney, 2009). Several IAPS e.g. *Cannabis* is used for the biological control of pests (Khan et al., 2022).

Plate 3 Invasive taxa: **a** *Volkameria heterophylla*, **b** *Youngia japonica*, **c** *Cestrum aurantiacum*, **d** *Holmskioldia sanguinea*, **e** *Heliotropium curassavicum*, **f** *Rondeletia odorata*, **g** *Phyllanthus niruri*, **h** *Broussonetia papyrifera*, **i** *Argemone mexicana*, **j** *Sambucus nigra*, **k** *Ipomea carnea*, **l** *Verbena tenuisecta*, **m** *Robinia pseudoacacia*, **n** *Dichondra repens*, **o** *Ruellia tuberosa*

Plate 4 Invasive taxa: **a** *Catharanthus roseus*, **b** *Calendula officinalis*, **c** *Combretum indicum*, **d** *Ixora coccinea*, **e** *Conocarpus erectus*, **f** *Adenium obesum*, **g** *Clerodendrum inerme*, **h** *Acacia salicina*, **i** *Guaiacum officinale*, **j** *Oxalis corymbosa*, **k** *Plumbago auriculata*, **l** *Tecomaria capensis*, **m** *Tabernaemontana divaricata*, **n** *Turnera ulmifolia*, **o** *Weigela florida*

iii. **Cultural Services**

Many IAPS have aesthetic value for recreational activities like gardening, beautification of roadsides, regulating the temperature of urban areas or nature appreciation. These invasive species serve as an environmental resource, improve socioeconomic health, reduce pollution, and provide beauty for local and regional travellers. Invasive species also provide more ornamental plants and shade trees. Many exotic species are planted in nurseries, conservatories, arboretums, and botanical gardens have educational value for scientific research, understanding phytogeography and environmental education. Invasive species play an important role in preserving the cultural heritage by providing more wood used for ceremonies, preventing the cutting of native sacred plants and less native flora for flower harvesting. Invasive species also provide recreation sites and boost tourism. Sales of invasive alien ornamental plants provide profits for nurseries (Pejchar & Mooney, 2009).

iv. **Supporting Services**

IAPS helps in the habitat formation of native species. Many invasive species provide nesting places for birds, and are home to many other wildlife. Invasive species help in nutrient cycling and community composition. Nutrient cycling is done through decomposition processes by invasive species. IAPS helps in the mitigation of floods by upholding soil on streams and rivers.

v. **Economic Services**

IAPS has a complex impact on the socio-economy with beneficial effects on low-income rural economies (Kaufmann, 2004). Many invasive species (e.g. *Pontederia crassipes*) are used for biogas, firewood and biofertilizer production (Bhatta et al., 2023). Several invasive species are used for medicine, charcoal, thatching, timber, and firewood in many rural communities, with the economic value of the firewood being US$ 2.8 million. *Melaleuca quinquenervia* is an invasive species that helps in the production of honey and benefits the economy by about US$ 15 million annually (Pejchar & Mooney, 2009). Medicinal plants, e.g. *Aspalathus linearis* used as Rooibos tea, export worth US$ 2.1 million (Pejchar & Mooney, 2009).

3 Ecosystem Disservices (EDS)

Global trade increased several-fold in the last few decades, resulting in the naturalization of a high number of invasive plants, and IAPS are a ticking time bomb for emerging economies (e.g. small African and Southeast Asian) (Seebens et al., 2015). IAPS are known for their negative ecological, economic, or social impacts (Fig. 2). By diverting resources more efficiently than native species, IAPS alter trophic dynamics, disturbance regimes, and ecosystem processes (Miniat et al., 2021). Production of goods and services can be greatly disrupted by IAPS (Hanley & Roberts, 2019). IAPS also serves as a vector of disease and is predicted to increase as a result of climate

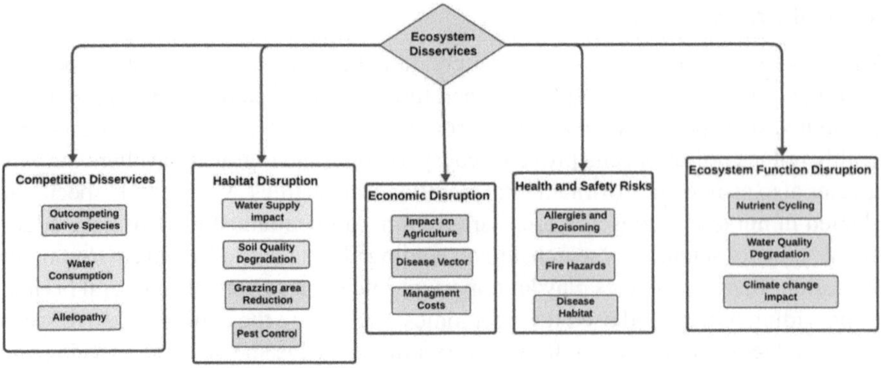

Fig. 2 Major ecosystem disservices (EDS) associated with Invasive alien plant species (IAPS)

change (Hanley & Roberts, 2019). Some major Ecosystem disservices associated with IAPS are outlined and discussed here (Fig. 2).

i. **Competition Disservices**

Chatterjee and Dewanji (2023) reported that alien invasive plants (AIPs) exert stronger effects on native flora, affecting their ability to survive in stressful environments with high temperatures, increased drought and widespread pollutants, or their vegetative or reproductive traits such as high specific leaf area and smaller seeds which help them to rapidly take up available resources, have high intrinsic growth rates allowing them to grow and extend their range rapidly, or their multiple modes of reproduction; or production of allele chemicals that suppress germination and growth of neighboring plants, soil amendments, or trophic imbalance and decreased species resilience—all of which can drastically reduce the biomass, coverage, and diversity of native communities (Kalusova et al. 2023).

Outcompeting native plant species for resources like water, nutrients, and sunlight. The IAPS have faster and more vigorous growth and usually use large amounts of water in their life cycle (Vergne et al., 2023). The IAPS, e.g. *Eucalyptus*, *Chromolaena* and *Lantana*, for instance, use Allelopathy (produce biochemicals) as a weapon to influence the growth, survival, and reproduction of native species. These allelochemicals also influence the nitrogen cycles and water quality (Chamier et al., 2012). The native vegetation and flora contribute significantly to carbon sequestration by acting as a sink for air pollutants (Shackleton et al., 2019). Disrupting the balance of native plant communities and reducing biodiversity can result in an increased concentration of hazardous pollutants in the air, causing cardiovascular/pulmonary problems (Rai & Singh, 2020). *Hydrocotyle vulgaris* is a potentially invasive species in the humid region of China which causes higher evapotranspiration in the native grassland community, resulting in a more significant increase in several invasive propagules (Tang et al., 2020).

ii. **Habitat Disruption**

The IAPS disrupts the habitat services, i.e. water supply, soil quality, and grazing areas (Ravhuhali et al., 2021). Altering or destroying native habitats leads to the displacement or extinction of native species (Máximo et al., 2020). Changing the structure and composition of ecosystems reduce overall ecological resilience. The hydrologic cycle is negatively affected by aquatic invasive species due to increasing temperatures. IAPS affect the transpiration properties and are the direct evidence for changes to the ecosystem hydrological cycle (Tang et al., 2020). These invasive species, through the increased use of water by non-native plants, impact the hydrology (Catford, 2017). IAPS are detrimental to the riparian ecosystem, affecting both native species and hydrology. Invasive woody shrubs and trees cost an estimated US$ 68 million annually in South Africa by reducing the water yield from upland fynbos ecosystems by 30% (Catford, 2017; Charles & Dukes, 2007). Many invasive species promote fire, severely damaging native ecosystems (Máximo et al., 2020). The invasive species, i.e. *Hedychium coronarium* can reduce evapotranspiration in invaded areas and increase it when immersed in the riparian forest (Vergne et al., 2023).

iii. **Economic Disruption**

IAPS have severe negative impacts on the economy by affecting human health, agriculture and forestry, and preserving the ecosystems and the services they provide (Pejchar & Mooney, 2009). Invasive species have the potential to harm food production (Engeman et al., 2010) and can also serve as carriers of diseases (Medlock & Leach, 2015). IAPS adversely affects crop growth and their ecosystems, resulting in decreased agricultural output and reduced yields (Vilà & Hulme, 2017). The control and management of IAPS come at huge economic costs, and also costly inputs such as pesticides to sustain production levels (Hanley & Roberts, 2019). The control and eradication of IAPS are very complex, time and resource-consuming (Nunes et al., 2021). The socioeconomic disturbances caused by plant invasions are accelerating on all continents (Bhatta et al., 2023). The invasive species has a global estimated cost of US$ 1.288 trillion between 1970 and 2017 to human societies. In India, the economic cost alone is estimated at US$ 176.7 billion annually (Diagne et al., 2021). *Xanthium strumarium* and *Verbesina encelioides* are the common urban invasive species in Pakistan (Plate 5). Several studies have been done to quantify the economic impact of IAPS. IAPS costs US$ 14.45 billion in China and the USA, from US$ 131 billion cumulative to US$ 128 billion annually. In the U.S., *Tamarix* (tamarisk) species are estimated to cost $53 million annually due to channel narrowing (Pejchar & Mooney, 2009). IAPS are also associated with disruption of services: In the UK, rail services spend money to control *R. ponticumas,* which causes interruptions to their operations (Dehnen-Schmutz & Williamson, 2006). *Centaurea solstialis* (yellow star thistle) is an unpalatable invasive species in California that costs about US$ 7.65 million annually (Pejchar & Mooney, 2009). Similarly, increasing costs for land management, invasive species control, and restoration efforts. More than $1.4 billion in water is lost to transpiration annually due to IAPS (Pejchar & Mooney, 2009).

Plate 5 Invasive taxa: **a** *Salvia coccinea*, **b** *Salvia reflexa*, **c** *Schinus molle*, **d** *Solanum elaeagnifolium*, **e** *Solanum nitidibaccatum*, **f** *Solanum pseudocapsicum*, **g** *Solanum rostratum*, **h** *Solanum sisymbriifolium*, **i** *Solidago canadensis*, **j** *Sonchus arvensis*, **k** *Sonchus oleraceus*, **l** *Symphyotrichum subulatum*, **m** *Tagetes minuta*, **n** *Verbesina encelioides*, **o** *Xanthium strumarium*

iv. **Health and Safety Risks**

Causing allergies or other health issues for humans and animals. Many invasive species are also poisonous species with detrimental impacts on humans and livestock (Khan et al., 2022). *Lantana camara* and *Miconia crenata* are reported as poisonous to livestock (Bhatta et al., 2023). Invasive species *Broussonetia papyrifera* and *Parthenium hysterophorus* are responsible for allergies and other skin-related diseases in Pakistan (Khan et al., 2022). *Chromolaena odorata* also has negative impacts on human health (Bhatta et al., 2023). Some invasive species, e.g. *Salvinia molesta* provide habitat to many diseases-carrying vectors like mosquitoes, causing dengue fever, elephantiasis, encephalitis and malaria (Chamier et al., 2012). IAPS are also responsible for creating fire hazards due to their flammability or altering fire regimes. *Acacia mangium* is an IAPS in Bangladesh reported to outcompete native species for water and light and change the fire regime. Similarly, *Cynodon dactylon, Ulex europaeus, Megathyrsus maximus* and *Lantana camara* are reported in South Asia to change the fire regime (Bhatta et al., 2023).

v. **Ecosystem Function Disruption**

By disrupting nutrient cycling and soil fertility, many invasive species release allelochemicals, disturbing the soil microorganisms. These invasive species damage the quality of water, causing water quality problems globally. Eutrophication of water bodies is caused by invasive species as they increase the metabolism rate, contributing to the increase of biomass (Chamier et al., 2012). IAPS alters the water availability, distribution and quality, affecting aquatic ecosystems. IAPS degrades the hydrological function of waterbodies by changing their physiology, morphology, behaviour and interactions with other species. Alien invasive species like tamarisk, *Sagittaria* and *Mimosa* invade wetlands, lakes, and rivers and can trap sediment, narrow flow channels and reduce flood attenuation (Catford, 2017). *Eichhornia crassipes, Salvinia molesta, Pistia stratiotes, Myriophyllum aquaticum* and *Azolla filiculoides* are some of the major aquatic invasive species that impact water quality and the aquatic ecosystem (Chamier et al., 2012). Climate change mitigation efforts are negatively impacted by invasive plant species. Similarly, IAPS reduce the resilience of green infrastructure against flooding and sea-level rise. The introduction of extensive alien tree plantations can alter regional precipitation patterns by modifying: (1) atmospheric water storage, (2) surface roughness, and (3) concentrations of condensation nuclei aerosols (Catford, 2017). These invasive species significantly impact ecosystem hydrology through competitive water uptake from native vegetation (van Wilgen et al., 2008). IAPS negatively affect both forestry operations and overall ecosystem health (Vilà & Hulme, 2017). Through vegetation and land cover modifications, IAPS alter key microclimatic factors including local temperatures, evapotranspiration rates, surface roughness, and aerosol particle concentrations (Catford, 2017).

Kalusova et al. (2023) conceptualized habitat invasion by alien plants through the metric "level of invasion," quantifying alien species presence and abundance at the site, community, and habitat scales. Their research identified that invasion levels peak in frequently disturbed habitats with high or variable resource availability, correlating strongly with the size of the introduced species pools. A clear gradient emerges, with invasion levels declining progressively from anthropogenically modified habitats to undisturbed natural ecosystems.

Urbanization drives land transformations that frequently reduce biodiversity and compromise ecosystem services, though these impacts vary geographically due to divergent geo-climatic and socio-economic contexts (Szabó et al., 2023). Population growth typically accelerates both economic development and international trade—key vectors for alien invasive plant (AIP) introductions. Intensive urbanization has precipitated approximately 50% decline in local species richness and a 38% reduction in total species abundance (McKinney, 2006). Climate change exacerbates invasion risks by facilitating range expansions, thereby threatening ecosystem integrity through functional, structural, and service-level degradation (Chatterjee & Dewanji, 2023).

Many IAPS have transformed the structure and function of ecosystems by changing the nutrient cycling regimes (Castro-Díez & Alonso, 2017; Raheem et al., 2024). Impacts have economic implications, because of reduced stream flow from watersheds in South African fynbos following alien tree invasion, or through disruption to fishing and navigation after invasion of aquatic plants such as *Eichhornia crassipes* (Rejmánek et al., 2013). Sakhraoui et al. (2019) reported the case of *Oxalis pescaprae,* which constitutes a good example of escapes with harmful effects. This species, introduced into Algeria as an ornamental plant, was reported like escaped from a garden, and 41 years later, Battandier and Trabut (1888) reported it as being very common and invasive in Algiers. Finally, after a century of expansion, Quezel and Santa (1963) declared this species as very common throughout the Algerian Tell. Similar ecological threats were documented by Shiferaw et al. (2022) in Northeast Ethiopia, where *Prosopis juliflora* was deliberately introduced to the Afar region during the late 1970s for rehabilitating degraded lands. Native to South America, the Caribbean, Mexico, and Central America, *P. juliflora* represents a classic case of intentional introduction with unintended consequences. While providing limited benefits as fuelwood and dry-season fodder in rural communities, its aggressive invasion of agricultural lands and associated biodiversity loss now present substantial ecological risks.

This case reflects a broader pattern in plant invasions. While invasive plants typically show less severe biodiversity impacts compared to introduced pathogens or predators (Rejmánek et al., 2013), their cumulative effects remain significant. Notably, of approximately 3000 naturalized plant species in North America (including over 1000 invasives), most show minimal detectable impacts on biotic communities. Nevertheless, many alien plants substantially reduce native plant abundance, fitness, and diversity. Importantly, no native plant extinctions have been conclusively attributed solely to competition with invasive plants (Rejmánek et al., 2013).

After Sakhraoui et al. (2020), 11 naturalized plant species are considered as invasive in the Mediterranean, as *Acacia saligna*, *Ailanthus altissima*, *Arundo donax*,

Austrocylindropuntia subulata, Carpobrotus edulis, Nicotiana glauca, Opuntia ficus-indica, Opuntia stricta, Oxalis pes-capreae, Ricinus *communis* and *Tropaeolum majus.*

In a literature review, Meddour and El Mokni (2016) reported that a large portion (45.29%) of plants introduced into Algeria are naturalized, with approximately 18% of them considered invasive or potentially invasive. In North-eastern Algeria, in the Skikda district, *Acacia saligna, Carpobrotus edulis* and *Optuntia ficus-indica* were the main threatening species of the biological richness (Sakhraoui et al., 2020). The coastal areas are the most affected as the introduced species are concerned with the naturalization. *A. saligna* and *O. ficus-indica* were recorded inside the open scrubland with trees while *C. edulis* was observed on the beaches and the coastal dunes. *A. donax* and *A. altissima* were both recorded at the edge of the creeks.

The proliferation of *A. saligna* is of particular concern as the plant is advancing inland, endangering the local biodiversity. Jeanmonod (1998) reported that the penetration of an allochtone species into natural environments is an indication of the threat to native flora. The invasive nature of this plant is visible, as it establishes itself among *Myrtus communis* and *Olea europaea,* in very dense populations, showing a rapid extension dynamic. This plant induces the diminution of the specific richness, diversity and abundance of the native flora, the alteration of the floristic composition and changes in the soil, the structure, the diversity and the function of the microbial communities. Despite its negative impact, this plant is used in the reforestation programs (Sakhraoui et al., 2019). *Arundo donax* is another invasive plant that dominates the ground cover at the edges of creeks and water bodies (Khan et al., 2017a, 2017b). This plant is usually associated with *Ailanthus altissima.* The riparian vegetation is disappearing in some places to the benefit of these two invasive plants. The same scheme is noticed nearby the coastal dunes, where *Carpobrotus edulis* is present. The endemic flora of this fragile area is endangered by the proliferation of this plant. Although appreciated, *O. ficus-indica* is responsible for several harmful effects (competition with the autochthonous vegetation and the expulsion of the latter). The proliferation of these species, considered as invasive in the natural Mediterranean habitats, makes the conservation process complicated, as these areas are already under threat (Sakhraoui et al., 2020; Toubal et al., 2014).

4 Conclusions and Future Prospects

The studies of IAPS unequivocally highlight the importance of integrating the subject of biological invasions into trade laws, climate change and promoting global partnerships to identify emerging invasive species. However, the magnitude of invasion and impacts are often context-dependent and vary by invading species, habitats, and regions. It is crucial to prioritize regions at an elevated risk of invasive species and involve them in international policy discussions. Therefore, identifying successful invaders and vulnerable habitats are crucial for the best decision-making and management. However, it is important to have inventories of AIPS in order to understand the dynamics of this invasion.

Although the ecosystem disservices (EDS) and services (ES) are reported, still detailed studies are needed to understand the economic quantification, livelihood considerations, biotechnological prospects (phytoremediation, bioenergy, phytosynthesis of nanoparticles, biomedical, industrial applications, etc.) and human health risk assessments of current and potential IAPS. Regardless of their effects, plant invasions provide chances to understand some evolutionary processes that are beyond the ethics of standard ecological experiments.

References

Adhikari, P., Acharya, S., Sigdel, S., & Kandel, G. (2022). Ethno-medicinal usage ethno-medicinal usage of invasive plants in traditional health care practices: A review. *International Journal of Applied Biology, 6*(2), 65–78.

Adigun, O. J., & Odeleye, D. A. (2025). The devastating consequences of environmental pollution on human health. *British Journal of Multidisciplinary and Advanced Studies, 6*(1), 37–46.

Afreen, T., Kumari, S., Bhadouria, R., Devi, R. S., Singh, S., & Tripathi, S. (2023). Plant invasion and soil processes: A mechanistic understanding. In S. Tripathi, R. Bhadouria, P. Srivastava, R. Singh, & D. R. Batish (Eds.), *Plant invasions and global climate change*. Springer. https://doi.org/10.1007/978-981-99-5910-5_10

Albert, A., Brisson, J., Belzile, F., Turgeon, J., & Lavoie, C. (2015). Strategies for a successful plant invasion: The reproduction of Phragmites australis in north-eastern North America. *Journal of Ecology, 103*(6), 1529–1537

APS (Algerian Press service). (2023). Introduction de nouvelles espèces floristiques: le paulownia à travers 3 wilayas dès octobre. https://www.aps.dz/regions/159248

Bachir, B. B., Ghellai, M., Daoudi, M., Behmene, I. E., & Bachir Bouiadjra, M. E. A. (2021). Impacts of the invasive species *Caulerpa cylindracea* Sonder, 1845 on the algae flora of the west coast of Algeria. *Biodiversity Data Journal, 9*, Article e64535. https://doi.org/10.3897/BDJ.9.e64535

Bargali, S. S. (2024). Biological invasion, biodiversity and people. *Journal of Environmental Biology, 45*(1), III–VI.

Battandier, M., & Trabut. (1888). Excursion botanique dans le sud de la province d'Oran. *Bulletin de la Société botanique de France, 35*(6), 338–348.

Bhatta, S., Shrestha, B. B., & Pyšek, P. (2023). Invasive alien plants in South Asia: Impacts and management. *NeoBiota, 88*, 135–167.

Castro-Díez, P., & Alonso, Á. 2017. Alteration of nitrogen cycling as a result of invasion. In *Impact of biological invasions on ecosystem services* (pp. 49–62).

Catford, J. A. (2017). Hydrological impacts of biological invasions. In *Impact of biological invasions on ecosystem services* (pp. 63–80).

Chamier, J., Schachtschneider, K., Le Maitre, D. C., Ashton, P. J., & Van Wilgen, B. W. (2012). Impacts of invasive alien plants on water quality, with particular emphasis on South Africa. *Water Sa, 8*(2), 345–356.

Charles, H., & Dukes, J. S. (2007). Impacts of invasive species on ecosystem services. In *Biological invasions* (pp. 217–237).

Chatterjee, S., & Dewanji, A. (2023). The distribution of invasive alien plant species in peri-urban areas: A case study from the city of Kolkata. *Community Ecology*. https://doi.org/10.1007/s42974-023-00169-z

Chevalier, A. (1952). Recherches sur les Phoenix africains. *Journal d'agriculture traditionnelle et de botanique appliquée, 32*(355), 205–236.
Copp, G. H., Templeton, M., & Gozlan, R. E. (2007). Propagule pressure and the invasion risks of non-native freshwater fishes: A case study in England. *Journal of Fish Biology, 71*, 148–159.
Courchamp, F., Fournier, A., Bellard, C., Bertelsmeier, C., Bonnaud, E., Jeschke, J. M., & Russell, J. C. (2017). Invasion biology: Specific problems and possible solutions. *Trends in Ecology & Evolution, 32*(1), 13–22.
Dehnen-Schmutz, K., & Williamson, M. (2006). *Rhododendron ponticum* in Britain and Ireland: Social, economic and ecological factors in its successful invasion. *Environment and History., 12*(3), 325–350.
Di Castri, F. (1990). On invading species and invaded ecosystems: The interplay of historical chance and biological necessity. *Biological invasions in Europe and the Mediterranean Basin* (pp. 3–16). Springer, Netherlands.
Diagne, C., Leroy, B., Vaissière, A. C., Gozlan, R. E., Roiz, D., Jarić, I., Salles, J. M., Bradshaw, C. J., & Courchamp, F. (2021). High and rising economic costs of biological invasions worldwide. *Nature, 592*(7855), 571–576.
Downey, P. O., & Richardson, D. M. 2016. Alien plant invasions and native plant extinctions: a six-threshold framework. *AoB Plants, 8*, plw047.
Engeman, R. M., Laborde, J. E., Constantin, B. U., Shwiff, S. A., Hall, P., Duffiney, A., & Luciano, F. (2010). The economic impacts to commercial farms from invasive monkeys in Puerto Rico. *Crop Protection, 29*(4), 401–405.
Freer-Smith, P. H., & Webber, J. F. (2017). Tree pests and diseases: The threat to biodiversity and the delivery of ecosystem services. *Biodiversity and Conservation, 26*(13), 3167–3181.
Gallardo, B., Bacher, S., Barbosa, A. M., Gallien, L., González-Moreno, P., Martínez-Bolea, V., Sorte, C., Vimercati, G., & Vilà, M. (2024). Risks posed by invasive species to the provision of ecosystem services in Europe. *Nature Communications, 15*(1), 2631.
Haines-Young, R., & Potschin, M. (2012). Common international classification of ecosystem services (CICES, Version 4.1). *European Environment Agency, 33*, 107.
Hanley, N., & Roberts, M. (2019). The economic benefits of invasive species management. *People and Nature, 1*(2), 124–137.
Hulme, P. E., Brundu, G., Carboni, M., Dehnen-Schmutz, K., Dullinger, S., Early, R., Essl, F., González-Moreno, P., Groom, Q. J., Kueffer, C., Kühn, I., & Verbrugge, L. N. (2018). Integrating invasive species policies across ornamental horticulture supply chains to prevent plant invasions. *Journal of Applied Ecology, 55*(1), 92–98.
IPBES. (2023). In H. E. Roy, A. Pauchard, P. Stoett, T. Renard Truong, S. Bacher, B. S. Galil, P. E. Hulme, T. Ikeda, K. V. Sankaran, M. A. McGeoch, L. A. Meyerson, M. A. Nuñez, A. Ordonez, S. J. Rahlao, E. Schwindt, H. Seebens, A. W. Sheppard, & V. Vandvik (Eds.), *Summary for policymakers of the thematic assessment report on invasive alien species and their control of the Intergovernmental science-policy platform on biodiversity and ecosystem services.* IPBES secretariat. https://doi.org/10.5281/zenodo.7430692
IUCN, S. (2021). Amphibian Specialist Group (2020). *The Amphibian Conservation Action Plan (ACAP): A status review and roadmap for global amphibian conservation.*
Jeanmonod, D. (1998). Les plantes introduites en Corse: Impacts, menaces et propositions de protection de la flore insulaire. *Biocosme Mésogéen, 15*(1), 45–68.
Junaedi, D. I., Nasution, T., Putri, D. M., Iryadi, R., Lestari, R., Kurniawan, V., Pratiwi, R. A., & Handayani, A. (2023). Threatened exotic species of botanical gardens: Application of trait-based naturalized species risk scoring assessment. *South African Journal of Botany, 152*, 321–331.
Kacimi Y., & Sbabdji M. (2022). Risque de l'invasion biologique de l'ailante glanduleux (*Ailanthus altissima* (MILL.) SWINGLE) au niveau du parc national de chréa, dynamique de propagation et essai den lutte. *Revue Agrobiologia 12*(2), 3063–3076.
Kalusova, V., Chytry, M., Vecera, M., Svenning, J. C., Biurrum, I., Kintrova, K., Agrillo, E., Carli, E., Ecker, K., Garbolino, E., Sibikova, M., Silc, U., & Axmanova, I. (2023). Neophyte invasions in European heathlands and scrub. *Biological Invasions, 25*, 1739–1765.

Kasson, M. T., Short, D. P. G., O'Neal, E. S., Subbarao, K. V., & Davis, D. D. (2014). Comparative pathogenicity, biocontrol efficacy, and multilocus sequence typing of *Verticillium nonalfalfae* from the invasive *Ailanthus altissima* and other hosts. *Phytopathology, 104*(3), 282–292.

Kathryn, L. (2005). *Getting to the root of the arsenic problem.* Chemistry World. http://www.rsc.org/chemistryworld/Issues/2005/May/arsenic_root.asp

Kaufmann, J. C. (2004). Prickly pear cactus and pastoralism in southwest Madagascar. *Ethnology, 1*, 345–361.

Kettunen, M., Mappendem, C., Herkenrath, A., Silvestri, P., Ten Brink, S. (2009). Technical support to EU strategy on invasive species (IAS)—Analysis of the impacts of policy options/measures to address IAS. Final report for the European Commission. UNEP-WCMC (Cambridge). Institute for European Environmental Policy (IEEP). Brussels, Belgium.

Khan, M. A. S. A., Sultana, F., Rahman, M. H., Roy, B., & Anik, S. I. (2011). Status and ethno-medicinal usage of invasive plants in traditional health care practices: A case study from northeastern Bangladesh. *Journal of Forestry Research, 22*, 649–658.

Khan, R., Ahmad, M., Zafar, M., & Ullah, A. (2017a). Scanning electron and light microscopy of foliar epidermal characters: A tool for plant taxonomists in the identification of grasses. *Microscopy Research and Technique, 80*(10), 1123–1140.

Khan, R., Iqbal, I. M., Ullah, A., Ullah, Z., & Khan, S. M. (2022). Invasive alien species: An emerging challenge for the biodiversity. *Biodiversity, conservation and sustainability in Asia: Volume 2: Prospects and challenges in South and Middle Asia* (pp. 459–471). Springer International Publishing.

Khan, R., Ullah, A. S. A. D., Khan, B., Khan, S. M., & Rashid, A. (2017b). Comparative foliar micro morphological studies of tribe Arundinacea, Aristideae and Chlorideae from Malakand Agency, Khyber Pakhtunkhwa. *Pakistan Journal of Botany, 49*(SI), 33–42.

Linders, T. E., Bekele, K., Schaffner, U., Allan, E., Alamirew, T., Choge, S. K., Eckert, S., Haji, J., Muturi, G., Mbaabu, P. R., & Shiferaw, H. (2020). The impact of invasive species on social-ecological systems: relating supply and use of selected provisioning ecosystem services. *Ecosystem Services, 41*, 101055.

Ljubojević, M. (2023). Invasive alien plant species: From the molecular to the economic approach. *Frontiers in Plant Science, 14*, 1185567.

Lorena, G.-A., & Charles, D. (2008). Neighbourhood analyses of the allelopathic effects of the invasive tree *Ailanthus altissima* in temperate forests. *The Journal of Ecology, 96*, 447–458.

Maema, L. P., Potgieter, M., & Mahlo, S. M. (2016). Invasive alien plant species used for the treatment of various diseases in Limpopo Province, South Africa. *African Journal of Traditional, Complementary and Alternative Medicines, 13*(4), 223–231.

Máximo, P., Ferreira, L. M., Branco, P. S., & Lourenço, A. (2020). Invasive plants: Turning enemies into value. *Molecules, 25*(15), 3529.

McKinney, M. L. (2006). Urbanization as a major cause of biotic homogenization. *Biological Conservation, 127*(3), 247–260.

Meddour, R., & El Mokni, R. (2016). State of the art on invasive or invasive plants introduced in Algeria and Tunisia. In *Proceedings of the XV OPTIMA meeting, June 6–11, 2016,* Montpellier, France.

Medlock, J. M., & Leach, S. A. (2015). Effect of climate change on vector-borne disease risk in the UK. *The Lancet Infectious Diseases, 15*(6), 721–730.

Miniat, C. F., Fraterrigo, J. M., Brantley, S. T., Callaham Jr, M. A., Cordell, S., Dukes, J. S., Giardina, C. P., Jose, S., & Lovett, G. (2021). Impacts of invasive species on forest and grassland ecosystem processes in the United States. In *Invasive species in forests and rangelands of the United States: A comprehensive science synthesis for the United States forest sector* (pp. 41–55).

Montecchiari, S., Allegrezza, M., Pelliccia, V., & Tesei, G. (2020). First syntaxonomical contribution to the invasive Ailanthus altissima (Mill.) Swingle forest communities at its southern limit in Europe. *Plant Sociology, 57*, 145–160.

Ntalo, M., Ravhuhali, K. E., Moyo, B., Hawu, O., & Msiza, N. H. (2022). Lantana camara: Poisonous species and a potential browse species for goats in southern Africa—A review. *Sustainability, 14*(2), 751.

Nunes, L. J., Raposo, M. A., Meireles, C. I., Pinto Gomes, C. J., & Almeida Ribeiro, N. M. (2021). Carbon sequestration potential of forest invasive species: A case study with Acacia dealbata Link. *Resources, 10*(5), 51.

Pagad, S., Genovesi, P., Carnevali, L., Schigel, D., & McGeoch, M. A. (2018). Introducing the global register of introduced and invasive species. *Scientific Data, 5*(1), 1–12.

Pejchar, L., & Mooney, H. A. (2009). Invasive species, ecosystem services and human well-being. *Trends in Ecology & Evolution, 24*(9), 497–504.

Peltzer, D. A., Allen, R. B., Lovett, G. M., Whitehead, D., & Wardle, D. A. (2010). Effects of biological invasions on forest carbon sequestration. *Global Change Biology, 16*(2), 732–746.

Raheem, A., Yohanna, P., Li, G., Noh, N. J., Iqbal, B., Tang, J., Du, D., Alahmadi, T.A., Ansari, M. J., Zhan, A., & Son, Y. (2024). Unraveling the ecological threads: How invasive alien plants influence soil carbon dynamics. *Journal of Environmental Management, 356*, 120556.

Rai, P. K., & Singh, J. S. (2020). Invasive alien plant species: Their impact on environment, ecosystem services and human health. *Ecological Indicators, 111*, Article 106020.

Ravhuhali, K. E., Mudau, H. S., Moyo, B., Hawu, O., & Msiza, N. H. (2021). *Prosopis* species—An invasive species and a potential source of browse for livestock in semi-arid areas of South Africa. *Sustainability, 13*(13), 7369.

Rejmánek, M., Richardson, D. M., & Pyšek, P. 2013. Plant invasions and invasibility of plant communities. *Vegetation Ecology,* 387–424.

Sakhraoui N, Boussouak R., Metallaoui S., Chefrour A., & Hadef A. (2020). La flore endémique du Nord-Est algérien face à la menace des espèces envahissantes. *Acta Botanica Malacitana, 45*, 67–79. https://doi.org/10.24310/abm.v45io.138

Sakhraoui, N., Metallaoui, S., Chefrour, A., & Hadef, A. (2019). The potentially invasive exotic flora of Algeria: first description of species cultivated in nurseries and gardens. *BASE.*

Schlaepfer, M. A., Sax, D. F., & Olden, J. D. (2011). The potential conservation value of non-native species. *Conservation Biology, 25*(3), 428–437.

Seebens, H., Essl, F., Dawson, W., Fuentes, N., Moser, D., Pergl, J., Pyšek, P., van Kleunen, M., Weber, E., Winter, M., & Blasius, B. (2015). Global trade will accelerate plant invasions in emerging economies under climate change. *Global Change Biology, 21*(11), 4128–4140.

Shackleton, R. T., Shackleton, C. M., & Kull, C. A. (2019). The role of invasive alien species in shaping local livelihoods and human well-being: A review. *Journal of Environmental Management, 229*, 145–157.

Shiferaw, W., Demissew, S., Bekele, T., & Aynekulu, E. (2022). Effects of Prosopis juliflora invasion on native species diversity and woody species regenerations in rangelands of Afar National Regional State, Northeast Ethiopia. *Journal of Resources and Ecology, 14*(1), 35–45.

Shine, C., Kettunen, M., & Ten Brink, P. (2009). *Technical support to EU strategy on invasive species (IAS)—Recommendations on policy options to control the negative impacts of IAS on biodiversity in Europe and the EU*. Final report for the European Commission, Brussels, Belgium.

Shine, C., Williams, N., & Gündling, L. (2000). *A guide to designing legal and institutional frameworks on alien invasive species*. Gland, Switzerland; Cambridge, UK; and Bonn, Germany: IUCN/IUCN Environmental Law Centre

Skočajić, D., & Nešić, M. (2021). Invasive species: Routes of introduction, establishment, and expansion. *Life on Land*, 571–582.

Sladonja, B., Susek, M., & Guillermic, J. (2015). Review on invasive tree of heaven (*Ailanthus altissima* (Mill.) Swingle) conflicting values: Assessment of its ecosystem services and potential biological threat. In *Environmental management*. Springer Science.

Small, C. J., White, D. C., & Hargbol, B. (2010). Allelopathic influences of the invasive Ailanthus altissima on a native and a non-native herb1, 2. *The Journal of the Torrey Botanical Society, 137*(4), 366–372.

Szabó, B., Korányi, D., Gallé, R., Lövei, G. L., Bakonyi, G., & Batáry, P. (2023). Urbanization decreases species richness and increases abundance in dry climates whereas decreases in wet climates: A global meta-analysis. *Science of the Total Environment, 859*, Article 160145.

Tang, L., Mo, K., Chen, Q., Zhang, J., Xia, J., & Lin, Y. (2020). An experimental study on potential changes in plant community evapotranspiration due to the invasion of *Hydrocotyle vulgaris. Journal of Hydro-Environment Research, 30*, 63–70.

Toubal, O., Boussehaba, A., Toubal, A., & Samraoui, B. (2014). Biodiversité méditerranéenne et changements globaux: cas du complexe de zones humides de Guerbès-Senhadja (Algérie). *Physio-Géo. Géographie physique et environnement, 8*, 273–295.

Ul Abidin, S. Z., Khan, R., Ahmad, M., Cuerrier, A., Zafar, M., Ullah, A., Khan, J., Saeed, A., Al-Qahtani, W. H., & Kazi, M. (2023). Wild edible fruits as a source of food and medicine: a study among tribal communities of Southern Khyber Pakhtunkhwa. *Plants, 13*(1), 39.

van Wilgen, B. W., Reyers, B., Le Maitre, D. C., Richardson, D. M., & Schonegevel, L. (2008). A biome-scale assessment of the impact of invasive alien plants on ecosystem services in South Africa. *Journal of Environmental Management, 89*(4), 336–349.

Vergne, D. D. C., Rosalem, L. M. P., Wendland, E. C., Anache, J. A. A., Martins da Silva, M. C., Boschi, R. S., & Silva Matos, D. M. D. (2023). Experimental study on potential influence of the invasive Hedychium coronarium J. König on the evapotranspiration of riparian plant community. *Plants, 12*(9), 1746.

Vilà, M., & Hulme, P. E. (2017). Non-native species, ecosystem services, and human well-being. In *Impact of biological invasions on ecosystem services* (pp. 1–14). Springer International Publishing.

Williams, V. L., Wojtasik, E. M., & Byrne, M. J. (2021). A chronicle of alien medicinal plants used as traditional medicine in South Africa, and their status as invasive species. *South African Journal of Botany, 142*, 63–72.

Wunder, J., Nobis, M., Conedera, M., & Gay, J. L. (2014). The place of tree of heaven in the forest remains to be studied. *La Forêt, 11*, 14.

Chapter 8
Impacts of Invasive Plant Species on Fishes, Birds, and Animals

Shalini Ojha, Simarpreet Kaur, Rishikesh Singh, Shalinder Kaur, Daizy R. Batish, and Harminder Pal Singh

Abstract Invasive species are a global threat to ecosystems and trigger a chain of the ecological and economic consequences. They disrupt biodiversity, destabilize ecosystems, and reduce agricultural productivity by replacing native vegetation, altering ecological interactions and habitats, as well as altering food webs, eventually declining fish, birds, and animal populations. For fish, these plants destroy their aquatic habitat and lower the water quality. Also, the disruptions harm birds and animal populations by changing food chains, behaviour, and disproportionately affecting native wildlife. The decline of keystone species and degradation of ecosystem services like water filtration and carbon sequestration make these impacts even worse. In this chapter, emphasis has been given on the impacts of invasive alien plants on the fishes, birds, and animal species present in the urban ecosystems. Management of these plants is necessary to prevent destructive ecological, economic, and social impacts. Therefore, in the latter half of the chapter, appropriate management approaches and scenarios have been highlighted. There is a need for a holistic approach that combines mechanical, biological, and chemical control with restoration, good governance, and community engagement. By doing so, we can conserve biodiversity, restore balance, and sustain our ecosystems for future generations.

Keywords Biodiversity · Cascading effects · Ecosystems · Food web disruption · Mitigation · Native species · Restoration

S. Ojha · S. Kaur · S. Kaur · D. R. Batish (✉)
Department of Botany, Panjab University, Chandigarh, India
e-mail: daizybatish@yahoo.com

H. P. Singh
Department of Environment Studies, Panjab University, Chandigarh, India

R. Singh
Amity School of Earth and Environmental Sciences, Amity University Punjab, Mohali, Punjab, India

R. Singh et al. (eds.), *Invasive Alien Plants in Urban Ecosystems*,
https://doi.org/10.1007/978-3-032-00997-5_8

1 Introduction

Invasive plant species are a mounting challenge globally, as they harm native ecosystems and animals residing in those disturbed ecosystems (Rai & Singh, 2020). They disrupt habitats, outcompete native flora, and alter food webs, leading to declines in fish, birds, and other animal populations (Gallardo et al., 2024; NWF, 2025; Mollot et al., 2017; Sharma et al., 2023). The introduction and spread of invasive plants, often mediated by human activities, can provoke and escalate cascading effects that compromise ecosystem services and resilience, so we need to have effective management and conservation strategies to mitigate the impact (Gallardo et al., 2024; NOAA Fisheries, 2025). These noxious plants are one of the major reasons for 40% of endangered species listings and one of the main causes of global biodiversity loss (UNEP, 2023). The impact of these is reflected in the changes in the flora structure, complexity, and plant community, and the large-scale impact on ecosystems and individual species (Gioria et al., 2023; Pyšek & Richardson, 2010; Rai & Singh, 2020). They achieve this through several mechanisms, such as:

- **Competition**: Invasive plants compete with native species for resources such as sunlight, water, and nutrients, often outcompeting them because of their aggressive growth and lack of natural predators or diseases (Omondi & Merceline, 2023; Zheng et al., 2015).
- **Habitat Alteration**: Invasive plants can change the physical structure and composition of habitats, forming dense monocultures or altering soil chemistry, making habitats unsuitable for native animals (Künzi et al., 2015).
- **Food Web Disruption**: By replacing native food sources or providing little to no nutritional value, invasive plants can disrupt food webs and negatively impact animals that rely on native plants for food (Carvalheiro et al., 2010; McCary et al., 2016).
- **Allelopathy**: Some invasive plants release chemicals into the soil that inhibit the growth of other plants, further reducing biodiversity and impacting animals that depend on multiple plant species (Kalisz et al., 2021; Singh et al., 2003).

The impact of invasive plants goes beyond terrestrial ecosystems, affecting aquatic environments and the fish and wildlife that live in them (Emery-Butcher et al., 2020; NOAA Fisheries, 2025). Aquatic invasive plants block waterways, reduce the dissolved oxygen levels in the water, and change the water quality, harming fish and disrupting aquatic food chains (Hassan & Nawchoo, 2020). We need to understand the many ways these plant species affect fish, birds, and animals to develop effective management and conservation strategies to protect native ecosystems and biodiversity (Gallardo et al., 2024; Rai & Singh, 2020). In this chapter, we emphasize on exploring the impacts of invasive alien plants on the fish, birds, and animal species present in the urban ecosystems. In addition, the effective management of invasive weeds and conservation strategies of the urban ecosystems have also been given equal attention at the end of the chapter.

2 Impact on Fishes

Invasive plants are a major threat to aquatic ecosystems, directly and indirectly affecting fish populations (Gallardo et al., 2016). Invasive aquatic plants can cause habitat alteration, breeding disruption, food web disruption, reduced biodiversity, and even the extinction of native species, and ultimately biodiversity and habitat destruction (Schultz & Dibble, 2012; WildFish, 2021). Some of the key mechanisms or processes are presented as follows:

2.1 Habitat Alteration

Invasive aquatic plants may change the physical structure of aquatic habitats (Havel et al., 2015). For example, Eurasian watermilfoil (*Myriophyllum spicatum*) forms a dense surface mat that reduces light penetration and can reduce native aquatic plant diversity (Eiswerth et al., 2000). Similarly, dense stands of hydrilla (*Hydrilla verticillata*) can change the composition of invertebrate assemblages compared with macrophyte multi-species treatments, making conditions unsuitable for native plants and animals to live (Brown & Maceina, 2002; Havel et al., 2015). High densities of plants can reduce the growth and health of largemouth bass and black and white crappie, most likely by reducing foraging efficiency (Hoyle et al., 2012; Wildhaber et al., 2023).

2.2 Disruption of Breeding

Invasive plants may disrupt the breeding cycles of fish. Physical obstruction from dense plant growth can prevent fish from accessing spawning grounds or reduce spawning success. Alterations in water quality, such as decreased dissolved oxygen levels caused by the decomposition of invasive plants, can also negatively impact fish reproduction (Sandilyan et al., 2018; Schultz & Dibble, 2012).

2.3 Feeding/Food Web Disruption

Invasive plants can disrupt aquatic food webs by altering the availability and composition of food sources (Kuehne et al., 2016; Vashishtha, 2017). Changes in the plant community can result in poorer food resources for macroinvertebrates and higher-level consumers (Gallardo et al., 2016; Kuehne et al., 2016). Invasive fish and invertebrate species can dramatically alter food web structures, decreasing the food available for native species and leading to direct competition that results in population declines and loss of biodiversity (Vashishtha, 2017).

The replacement of native kelp forests by invasive green macroalgae modifies the quality of macroalgal detritus transported to deeper ocean environments, impacting the abundance and diversity of macrofaunal species, while invasive terrestrial plants that displace native forest vegetation alter pond food web dynamics by introducing novel allelochemicals that leach into ponds, reducing zooplankton growth rates and affecting aquatic food web structures (Peller & Altermatt, 2024).

2.4 Reduced Biodiversity

The combined effects of habitat alteration, breeding disruption, and food web disruption can lead to a reduction in fish biodiversity (WildFish, 2021). As invasive plants spread and take over aquatic ecosystems, they can outcompete native plants and lead to a decline in marine habitats. This, in turn, can impact fish species that use different habitats at different stages of their life cycle (Omondi & Merceline, 2023).

2.5 Case Studies

- In India, water hyacinth is invasive in aquatic ecosystems and reduces water flow in irrigation channels by 40–95%, causing huge economic losses in West Bengal through crop suppression, poor rice germination and harvesting, and disrupted inland water transport in Kerala's backwaters (ISC, 2025).
- A striking example is the impact of hydrilla (*Hydrilla verticillata*) in Lake Seminole, Georgia, where dense hydrilla altered the habitat and led to slower growth rates and lower fecundity in largemouth bass, reduced foraging efficiency, and impacted the overall health and population dynamics of the bass. It shows how invasive aquatic plants disrupt native ecosystems, the need for management strategies to control hydrilla, and requirement of maintaining the optimal vegetation levels to restore ecological balance and enhance fish populations (Brown & Maceina, 2002; Maceina & Reeves, 1996).

These examples illustrate the varied and far-reaching impacts of invasive plant species on fish populations and aquatic ecosystems, and warrants effective management and conservation strategies to mitigate these impacts and protect native fish stocks (Havel et al., 2015) (Table 1).

Table 1 Economic and environmental impacts of invasive plant species on fish populations

S. No.	Invasive plant	Economic impact	Environmental impact	Native species affected	References
1	Water hyacinth	Impedes water transport, reduces crop yields	Decreases biodiversity, alters ecosystem processes	Fish and invertebrates	Villamagna and Murphy (2010), ISC (2025)
2	Hydrilla	Increases maintenance costs for waterways	Reduces dissolved oxygen, impacts fish reproduction	Largemouth Bass, Crappie	Maceina and Reeves (1996), Brown and Maceina (2002)
3	Eurasian watermilfoil	Affects recreational activities, reduces property values	Alter habitat structure impacts native plant diversity	Various fish species	Eiswerth et al. (2000)

3 Impact on Birds

Invasive plant species are a serious concern for birds in terms of altering their habitat and foraging habits (Fletcher et al., 2019). Invasive plants spread fast, displace native vegetation, and reduce the food and shelter needed by birds to survive, which can lead to birds moving to new areas, changing their migration patterns, or experiencing population decline (Schirmel et al., 2016).

3.1 Loss of Nesting and Foraging Habitats

Invasive plant species are the main cause of habitat loss for birds, altering landscapes and reducing biodiversity, especially in riparian ecosystems, where they simplify habitat structures, leading to bird population decline and diversity (Schirmel et al., 2016). According to a study conducted by Reynolds et al. (2019) in southeastern Poland, invasive Caucasian hogweeds (*Heracleum sosnowskyi*) significantly impacted bird communities by altering nesting and foraging dynamics, especially in fragmented young forests during early succession stages. These towering herbaceous plants reduced alpha diversity by lowering species richness in both forest and non-forest bird communities. Reynolds et al. (2019) concluded that in areas with a lot of human-made habitats that are severely invaded, you see a marked decline in the number of forest birds, and that decline highlights just how bad the consequences of these invasive plants can be. In tropical dry forests, non-native grasses and woody legumes have disrupted fire cycles and nutrient availability, further degrading the bird habitats and leading to population declines (Hardy et al., 2025). In places where invasive plants took over, there were fewer kinds of birds, whether they lived in forests or open lands (Grzędzicka, 2022). In places where invasive plants took over, there were fewer kinds of birds, whether they lived in forests or open lands (Grzędzicka, 2022).

Exotic plantations dominated by species like *Gliricidia sepium* and *Leucaena leucocephala* creates dense tree cover, eliminating the grassy understory vital for savanna birds (Choudaj & Wankhade, 2021). These trees release chemicals that suppress native vegetation, further reducing habitat quality (Schirmel et al., 2016). Birds and insects struggle to find resources in such environments, leading to declining populations (Choudaj & Wankhade, 2021).

3.2 Reduced Food Availability

As invasive species outcompete native plants, they reduce insect populations and other food sources for birds (Tallamy & Shriver, 2021). Narango et al. (2018) looked at how non-native plants affect Carolina chickadees (*Poecile carolinensis*) in suburban areas and found that yards with these plants had fewer insects, particularly caterpillars, a key food source for these birds. So, chickadees in areas with more non-native plants had to adjust their diet and eat less nutritious prey like spiders, which had serious consequences for the birds—reduced reproduction and unsustainable population growth. The researchers found that when non-native plants make up more than 30% of the vegetation, the area is no longer suitable for a sustainable chickadee population. Instead, it becomes a "population sink" where more birds die than are born (Narango et al., 2018).

The decline in native insect populations due to invasive plants also affects insect-eating birds, forcing them to change their feeding behaviours. This shift can influence migration patterns and overall bird fitness (Tallamy & Shriver, 2021).

Choudaj and Wankhade (2021) reported that exotic plantations support omnivorous birds and led to decline in predator and herbivore species by causing lesser availability of specialized food sources like seeds and insects. This affects ground-foraging birds the most as there is no food on the forest floor dominated by invasive species. Moreover, the disruption of the predator–prey relationship has a broader impact on the ecosystem (Choudaj & Wankhade, 2021).

3.3 Behavioural and Reproductive Changes

Invasive plants not only change the landscape but also affect native wildlife. For example, Chen et al. (2022) found that the spread of smooth cordgrass (*Spartina alterniflora*) in China's coastal wetlands changed the singing habits of plain prinia birds (*Prinia inornata*). In areas covered with this invasive plant, the birds sang shorter and less varied songs compared to birds living in natural habitats, which is likely to have happened because the thick, uniform structure of the smooth cordgrass affected sound transmission and communication. Since birds rely on songs to attract mates and defend their territory, these changes might negatively affect their populations in the longer run (Chen et al., 2022).

Invasive plants also change the feeding and nesting patterns of birds. For instance, in some cities, non-native trees like *Bombax ceiba* attract nectar-eating birds, but in the scarcity of food availability, they shift their usual feeding behaviour and start depending more on invasive trees (Silva et al., 2023). Additionally, some birds adjust their diets to eat more fruits from invasive plants when native food is lacking. Although these new habitats may seem suitable, they often do not provide the necessary resources birds need to nest and raise their chicks. Over the time, this can cause bird populations to decline (Nelson et al., 2017).

3.4 Case Studies

3.4.1 The Impact of *Broussonetia Papyrifera* on Local Vegetation and Wildlife of Chandigarh

Invasive plant species replace native vegetation and alter habitat conditions (Gioria et al., 2023). A study by Maan et al. (2021) in Chandigarh shows that the invasive tree *Broussonetia papyrifera* plays a major role as a food source for different bird and mammal species. Its negative impact on understorey vegetation is mainly due to its ability to spread easily. The tree produces attractive fruits with many seeds that are spread by birds and other animals. Although *B. papyrifera* benefits some species temporarily, its widespread presence threatens native biodiversity by outcompeting native plants and spreading its seeds through animal interaction (Maan et al. 2021).

3.4.2 The Impact of *Lantana Camara* on Local Wildlife in Nanded, Maharashtra

Lantana camara, a fast-spreading shrub, has affected biodiversity in many parts of India (Negi et al., 2019). A study in Nanded, Maharashtra found that despite being invasive *L. camara* supports around 33 native species by providing food and shelter, with birds like Red-vented Bulbul, Indian White-eye, and Orange-headed Thrush which were seen consuming its berries. At the same time, Ashy Prinia and Indian Silverbill used the branches of *L. camara* for nesting and insects visited for berries and nectar (Jadhav et al., 2023). *L. camara* helps some species for a while, but its spread threatens native biodiversity, which can lead to ecosystem imbalance and decline of native species, so we need to control *L. camara* and promote native plants for a stable ecosystem (Jadhav et al., 2023) (Table 2).

In short, invasive plant species affect bird populations by changing habitats, reducing food, and altering bird behaviour, thus threatening bird diversity and survival. To address this, conservation strategies should focus on managing invasive

Table 2 Species dependent on *Lantana camara*

S. No.	Category	Examples	Dependency on *Lantana camara*	References
1	Butterflies	Common Mormon, Blue Tiger, Lemon Emigrant	Nectar source for feeding	Mukherjee et al. (2015), Jadhav et al. (2023)
2	Insects	Honeybees, Wasp Moth, Blue Banded Bee	Feeding on nectar, pollination	Priyanka and Joshi (2013), Jadhav et al. (2023)
3	Birds	Red-vented Bulbul, Indian White-eye, Ashy Prinia	Feeding on berries, nesting in branches	Sharma et al. (2005), Jadhav et al. (2023)
4	Mammals	Squirrel	Consuming berries as a food source	Priyanka and Joshi (2013), Negi et al. (2019), Jadhav et al. (2023)

species, reviving native plant communities, and safeguarding diverse ecosystems for bird populations.

4 Impact on Animals

Invasive plant species are a problem everywhere in the world, disrupting habitats, outcompeting native vegetation, and altering food chains with greater impacts on animals, including reducing numbers, changing behaviour, and disrupting entire ecosystems (Schirmel et al., 2016). Some of the impacts of invasive plants on mammalian species are as following.

4.1 Reduction in Animal Population Size

One of the most common effects of invasive plants is a decline in native animal populations (Stewart et al., 2021). Enhancing the diversity, coverage, or density of native plants in urban green spaces has consistently been associated with an increase in animal biodiversity (Parsons et al., 2006; Wilkinson, 2006). Research on species preferences also indicates that native animals are more likely to use native plants, whereas non-native animals tend to favor exotic plant species (Fletcher et al., 2019). Schirmel et al. (2016) found that invasive plants had a strong negative effect on animal numbers and diversity in coastal and grassland ecosystems. The impact was smaller but still noticeable in scrublands, cities (affecting population size), and marine habitats (affecting animal health and fitness).

4.2 Disruption of Food Chain and Ecosystem

Invasive plants affect not only individual species but the entire food chain (McCary et al., 2016). When plants replace native vegetation, they change the availability of food for herbivores, which in turn affects carnivores (Engelkes et al., 2012). Schirmel et al. (2016) found that in areas heavily invaded by non-native plants, the number of predators like birds and mammals decrease due to a reduction in prey species, triggering cascading effects throughout the ecosystem and resulting in long-term biodiversity loss. Reduced prey density and increased habitat complexity reduce predator efficiency and reshape the ecosystem (Schirmel et al., 2016).

In addition, invasive plants can alter the natural decomposition process by changing the composition of leaf litter (Kuglerová et al., 2017). This affects decomposers like insects and fungi, which are key to nutrient cycling. As a result, the entire ecosystem can become less stable over time (Schirmel et al., 2016).

4.3 Changes in Animal Behaviour

Apart from reducing population sizes, invasive plants can also change how animals behave (Mooney & Cleland, 2001). Some animals get caught in "evolutionary traps" where their instincts lead them to make bad survival decisions (Robertson & Chalfoun, 2016). Invasive plants alter animal behavior by changing their foraging patterns and habitat use, often forcing them to adapt to new food sources and nesting sites, and these shifts can lead to increased competition, altered predator–prey dynamics, and changes in movement patterns (Stewart et al., 2021). The invasive shrub *Rhododendron ponticum* creates dense vegetation that protects wood mice (*Apodemus sylvaticus*) but at the same time reduces the growth of native food plants so the mice cannot find food and end up in a paradoxical situation where they are safe from predators but can't survive due to food scarcity (Stewart et al., 2021). Invasive plants can also alter the chemical cues in the environment and confuse animals that rely on smell or taste to find food or avoid predators (Kelley et al., 2018; Price et al., 2024). These changes in behaviour can have long-term consequences and affect reproduction and survival rates (Stewart et al., 2021).

4.4 Native vs. Exotic Animal Responses

Interestingly, not all animals respond to plant invasions uniformly. Studies indicates that invasive plants tend to harm native species more than exotic ones, with native animals which have evolved in stable environments, struggling to adapt to rapid environmental changes, whereas exotic animals, possibly having co-evolved with similar invasive plants, often exhibit little to no negative effects (Fletcher et al., 2019). A meta-analysis by Fletcher et al. (2019) revealed that invasive plants decrease native animal abundance, especially annelids, and chordates, but have minimal impact on exotic animals because native species are not adapted to rapid environmental change since they evolved in stable environments, whereas exotic species have broader ecological tolerance.

4.5 *Increased Human-Wildlife Conflict (WHC)*

Human-wildlife conflict is getting complicated due to the spread of invasive plant species, which disrupts natural habitats and leads to food scarcity for wildlife (Nyhus, 2016). This makes animals search for alternative food sources, and they often end up in human-dominated landscapes, causing crop damage, livestock predation, and sometimes human casualties (IUCN, 2022). For example, according to an article published on *scroll.in* by Sibi Arasu (2017), in Nilgiris, elephants raid crops due to the lack of natural fodder caused by the presence of an invasive plant of the family Fabaceae, that is, *Prosopis juliflora* (synonym: *Neltuma juliflora*).

4.6 *Case Study: The Impact of Invasive Plant Amur Honeysuckle on White-Footed Mice in Urban Forest Patches*

Urbanization and invasive plant species are reshaping forest ecosystems, often in unexpected ways. One such example is the impact of invasive shrubs like Amur honeysuckle (*Lonicera maackii*) on the habitat and population dynamics of small mammals. Busala et al. (2024) investigated how these invasive plants influence the habitat quality for white-footed mice (*Peromyscus leucopus*), a species known for its role in Lyme disease transmission. The study highlighted how invasive plant species can indirectly affect wildlife by modifying habitat conditions, while Amur honeysuckle provides additional shelter for white-footed mice, it also contributes to their population growth, which could have broader ecological and public health impacts. Managing invasive plants in urban forests may therefore be an important step in controlling small mammal populations and reducing disease risks.

In short, invasive plants greatly threaten animal populations by reducing their food, shelter, and overall survival chances. Some animals, like exotic species, can adapt better, but most local wildlife suffers. Invasive plants also disrupt the food chain, leading to fewer predators and changes in animal behaviour. Some animals get trapped in environments that seem safe but do not have enough food. To save wildlife, we need to control invasive plants and let native plants grow back. This will create a balanced nature and support different animal species.

5 Cascading Effects Across Ecosystems

Invasive plant species trigger a series of cascading effects that extend throughout ecosystems, impacting keystone species, leading to a decline in biodiversity and ecosystem services, and creating feedback loops that accelerate ecosystem degradation (Fig. 1) (Sage, 2020). Introductory invasions by exotic plants usually start quite like dominoes across an ecosystem, disrupting the equilibrium of ways things normally work and kicking off a series of results that impact lots of layers and ecosystem processes (Ellis et al., 2011; Sage, 2020).

5.1 Impacts on Keystone Species

Keystone species are superstars of ecosystem health, crucial and responsible for defining how different parts of ecosystems work together and the bigger picture they create together, and they get a lot of attention in the literature because their absence can change things dramatically, especially by overturning the balance of the ecosystem (Hale & Koprowski, 2018). They often have a disproportionately large impact on their environment relative to their abundance. Invasive plants can negatively impact the keystone species by altering and disturbing their habitat, which can further trigger cascading effects and destabilize the entire ecosystem (Hale & Koprowski, 2018). The plants can also reduce food supplies and directly compete with keystone species for important things to survive (Mouquet et al., 2013; Siegert et al., 2023).

5.2 Deline in Biodiversity and Ecosystem Services

Invasive plants outcompete native plants for light, water, and nutrients and thus decline native plant diversity (Rai & Singh, 2020). Native plants displaced by invasive species disrupt ecological interactions like pollination and seed dispersal and the special relationships between native plants and other organisms which can compromise ecosystem services like carbon sequestration, soil stabilization, and water purification and ultimately have big implications for the benefits we get from ecosystems (Gallardo et al., 2019).

5.3 Feedback Loop in Ecosystem Degradation

Invasive plants can create a positive feedback loop that promotes their dominance (Gaertner et al., 2014). As they establish and spread, invasive species can change

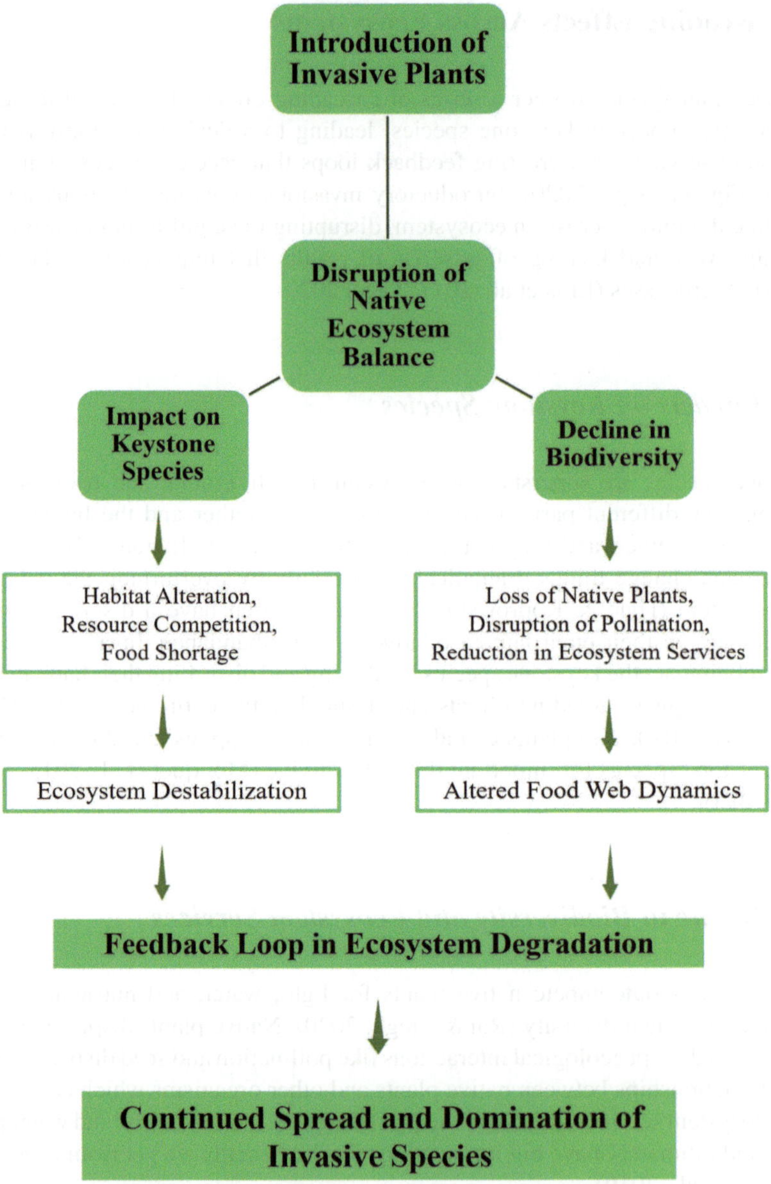

Fig. 1 Cascading effects of invasive plant species across urban ecosystems

ecosystem processes and conditions to make the environment more suitable for themselves and less for native plants (Fig. 1) (Gioria et al., 2023). For example, some invasive plants can alter soil structure, nutrient cycling, hydrological regimes, and ecosystem processes (Eviner et al., 2012). These alterations can cascade down the food chain and affect the abundance and distribution of other organisms in the ecosystems (Ellis et al., 2011).

According to a study conducted by Crawford and Knight (2016), the invasive species *Lespedeza cuneata* exemplifies feedback loops that drive ecosystem degradation, particularly in urban areas, by altering soil properties and competing with native plants. As a leguminous forb, it influences both abiotic and biotic soil conditions, including microbial communities, which enhance its competitive advantage. Its ability to fix nitrogen allows it to thrive in nutrient-poor soils, while allelopathic effects suppress native species either directly or indirectly by modifying soil chemistry. Additionally, *L. cuneata* outcompetes native plants through rapid growth, shading, and high seed production, forming dense monocultures that reduce biodiversity. These mechanisms create reinforcing feedback loops that favor its dominance and disrupt native ecosystems, highlighting the importance of understanding plant-soil interactions in managing invasive species and mitigating the urban ecosystem degradation (Crawford & Knight, 2016).

The chain reactions of invasive plant species need to be understood to manage and conserve these species effectively (Dimitriadis et al., 2021). To effectively manage invasive species and for proficient ecosystem restoration, one must mainly focus on improving native plants' reproductive success and reducing negative effects from invasives (Weidlich et al., 2020). By tackling the root causes of degradation in ecosystems and increasing their resilience, invasive plant impacts in the long run could be minimized (Rai, 2022).

6 Mitigation and Management Strategies

Managing invasive plants is key to protecting native ecosystems and the residing animals, including fish, birds, and other wildlife (Blossey, 1999). Effective mitigation strategies for invasive plants involve an integrated approach that combines mechanical, biological, chemical, and community-driven methods and habitat restoration, policy frameworks, community engagement, and case studies to reduce the impact of invasive plants and restore native ecosystems (Hussner et al., 2017; Meyerson et al., 2022; Pyšek & Richardson, 2010).

6.1 Invasive Alien Plants Control Methods

- Mechanical Control: It involves physically removing invasive plants by hand-pulling, cutting, and mowing (Hussain et al., 2018; Sobuj & Byun, 2023). Hand-pulling is labour-intensive but effective for small infestations, especially before plants go to set seed; cutting and mowing can manage larger populations but often need to be repeated to prevent re-growth (Lommen et al., 2018). Girdling is another technique used on woody invasives where the bark is removed to kill the plant while leaving it standing as a dead snag that can be a habitat for wildlife (Merceron et al., 2016).
- Biological Control: It involves using natural enemies or predators to manage invasive plant populations by introducing herbivorous insects or pathogens that specifically target invasive species without harming native flora (Culliney, 2005). For example, the introduction of the leaf-eating beetle (*Galerucella calmariensis*) has been used to control purple loosestrife (*Lythrum salicaria*) and has reduced its spread in wetland areas (Humphrey, 2024; Malecki et al., 1993).
- Chemical Control: Under this method, herbicides are used to target invasive plants selectively. Foliar applications involve spraying herbicides directly onto the plant leaves, while soil-applied herbicides stop root growth, also, the timing and herbicide choices are key factors in minimizing collateral damage to native species (Culliney, 2005; Gunstone et al., 2021; Netherland et al., 2005).
- Integrated Weed Management (IWM): It is a combination of multiple approaches for creating a whole approach to control invasive plants. This considers the life cycle of the invasive species, ecological interactions, and site-specific conditions (Gage & Schwartz-Lazar, 2019). By integrating mechanical removal, biological controls, chemical applications, and habitat restoration, IWM aims to achieve long-term management objectives while minimizing impact on native ecosystems (Pavlović et al., 2022).

6.2 Habitat Restoration Efforts

After the removal of invasive plants, restoration is critical to re-establishing native plant communities and building ecosystem resilience (Schaefer, 2011). Successful habitat restoration includes conducting **site assessments** to identify invasive species present and evaluate ecosystem conditions, **re-establishment of plants,** including replanting with native species selected based on local ecological conditions, so restored habitats support wildlife, and **monitoring and maintenance,** which is essential to track the progress of restoration and to quickly address re-emerging invasive populations (Méndez-Toribio et al., 2021).

6.3 Policy and Community Engagement

Managing invasive plants often requires policy frameworks at the local, state, and national levels, and these should focus on preventing the introduction of these plants and propagules through trade and transport regulations and public awareness campaigns about invasive plants (Epanchin-Niell, 2017). Community engagement is key to successful management, local stewardship, and participation in restoration projects that can increase public awareness and commitment to the local ecosystem through monitoring or volunteering days (Cerveny et al., 2022).

6.4 Ongoing Management Efforts

Keeping track of and managing the present state is important with invasive plants. This includes **education and outreach** by providing resources and training for landowners and land managers on proper methods of identification and management of these plants (Zhang et al., 2023), **research projects,** funding for new control methods (van Rees et al., 2022), and government agencies, conservation organizations, researchers and local communities working together, in **collaboration,** to tackle invasive plant issues (Graham, 2019).

6.5 Success Stories

Several successful stories demonstrate effective management of invasive species:

- The Mexican beetle, *Zygogramma bicolorata,* was brought to India in 1982 as a biocontrol agent to tackle the invasive Congress grass (*Parthenium hysterophorus*). Tests to ensure the beetle wouldn't harm native or cultivated plants came to be positive. Overall, the introduction of *Zygogramma bicolorata* as a biocontrol agent in India has been a positive step in the integrated management of Congress grass (Dhileepan & Wilmot Senaratne, 2009). However, its release was to be withdrawn as the beetle was not specific to *Parthenium* and it started damaging the sunflower crop also.
- In the San Francisco Bay Area of California, government agencies, non-profits, and local communities have been working together to reduce the spread of invasive yellow starthistle (*Centaurea solstitialis*). This has resulted in a better habitat for native wildlife (Epanchin-Niell et al., 2009).
- Parts of Europe have seen the re-establishment of native riparian flora and habitats for fish and bird populations following the removal of Japanese knotweed (*Fallopia japonica*) along riverbanks (Rotherham, 2021).

- In New Zealand, targeted biological control of invasive weeds like gorse (*Ulex europaeus*) has improved pasture for livestock and native birds (Roberts & Florentine, 2021).

To conclude, the combined use of multifaceted methods, such as control measures, habitat restoration, policy support, community participation, and examples of success and ongoing management will help to minimize the impact of invasive plant species on fish, birds, and animals. By prioritizing these strategies, we can protect biodiversity and restore our natural environments.

7 Conclusions and Future Prospects

Invasive plant species are causing multifaceted threats to ecosystems around the world setting off a chain of adverse effects, also posing a significant threat to urban ecosystems. These plants are a real menace to biodiversity, disrupting ecosystem stability and green infrastructure. They out-compete local vegetation, disturb the ecological interactions, and modify the natural habitats, ultimately leading to worsened habitats and economic losses. Such a disruption affects the population of animals as well, by altering food chains, changing behaviours, and disproportionately affecting indigenous faunal species. Further down the consequences chain is the decline in the number of some keystone species that provide stability for ecosystems, along with the degradation of vital ecosystem services, including water filtration and carbon sequestration. To address these challenges, cities need a comprehensive management strategy utilizing mechanical, biological, and chemical control with restoration efforts, good governance, and community participation to address the impacts. By focusing on these strategies, we can work toward conserving biodiversity, restoring environmental equilibrium, and sustaining the unspoiled state of our urban ecosystems for the generation yet to come.

Going forward, we need to improve our understanding and quantification of invasive species' ecological impacts, longer-term studies across larger scales with multiple species, evolutionary relationships, and a wider range of taxa. By getting a better understanding of these complex interactions, we can develop enhanced and effective management strategies for invasive plants and mitigate their impacts on fish, birds, and animals.

Acknowledgements Shalini Ojha acknowledges the Council of Scientific & Industrial Research (CSIR), New Delhi, for Research Funding. Daizy R. Batish, Shalinder Kaur, and Rishikesh Singh are thankful to the National Mission on Himalayan Studies (NMHS), Ministry of Environment, Forest and Climate Change (MoEFCC), New Delhi, for the Financial Assistance in the form of a Research Project (NMHS2024-25/SC-XII/MG95-200/SL06).

References

Arasu, S. (2017, March 14). In the Nilgiris, invasive plant species are driving animals into conflict with humans. Wildlife conservation. *Scroll.in*. https://scroll.in/article/830777/in-the-nilgiris-invasive-plant-species-are-taking-over-forests-driving-human-animal-conflict

Blossey, B. (1999). Before, during and after: The need for long-term monitoring in invasive plant species management. *Biological Invasions, 1*, 301–311.

Brown, S. J., & Maceina, M. J. (2002). The influence of disparate levels of submersed aquatic vegetation on largemouth bass population characteristics in a Georgia reservoir. *Journal of Aquatic Plant Management, 40*, 28–35.

Busala, G. M., Phillips, P. M., & Behm, J. E. (2024). Variation in forest patch habitat quality for white-footed mice driven by invasive plants in an urbanizing landscape. *Ecosphere, 15*, 70068.

Carvalheiro, L. G., Buckley, Y. M., & Memmott, J. (2010). Diet breadth influences how the impact of invasive plants is propagated through food webs. *Ecology, 91*, 1063–1074.

Cerveny, L. K., Derrien, M. M., Miller, A. B., & Meyer, C. (2022). Partnership and community engagement models for stewarding national scenic trails: A social-ecological systems perspective. *Tourism Planning & Development, 19*, 204–226.

Chen, P., Chen, T., Liu, B., Zhang, M., & Lu, C. (2022). Song variation of a native songbird in a modified habitat by invasive plant. *Integrative Zoology, 17*, 93–104.

Choudaj, K., & Wankhade, V. (2021). Reduction in avian diversity due to exotic tree plantations on the native savannas of Pune City, India. *Tropical Ecology, 62*, 499–507.

Crawford, K. M., & Knight, T. M. (2016). Competition overwhelms the positive plant–soil feedback generated by an invasive plant. *Oecologia, 183*, 211–220.

Culliney, T. W. (2005). Benefits of classical biological control for managing invasive plants. *Critical Reviews in Plant Sciences, 24*, 131–150.

Dhileepan, K., & Wilmot Senaratne, K. A. D. (2009). How widespread is Parthenium hysterophorus and its biological control agent Zygogramma bicolorata in South Asia? *Weed Research, 49*, 557–562.

Dimitriadis, C., Fournari-Konstantinidou, I., Sourbès, L., Koutsoubas, D., & Katsanevakis, S. (2021). Long term interactions of native and invasive species in a marine protected area suggest complex cascading effects challenging conservation outcomes. *Diversity, 13*, 71.

Eiswerth, M. E., Donaldson, S. G., & Johnson, W. S. (2000). Potential environmental impacts and economic damages of Eurasian watermilfoil (Myriophyllum spicatum) in western Nevada and northeastern California. *Weed Technology, 14*, 511–518.

Ellis, B. K., Stanford, J. A., Goodman, D., Stafford, C. P., Gustafson, D. L., Beauchamp, D. A., Chess, D. W., Craft, J. A., Deleray, M. A., & Hansen, B. S. (2011). Long-term effects of a trophic cascade in a large lake ecosystem. *Proceedings of the National Academy of Sciences, 108*, 1070–1075.

Emery-Butcher, H. E., Beatty, S. J., & Robson, B. J. (2020). The impacts of invasive ecosystem engineers in freshwaters: A review. *Freshwater Biology, 65*, 999–1015.

Engelkes, T., Wouters, B., Bezemer, T. M., Harvey, J. A., & van der Putten, W. H. (2012). Contrasting patterns of herbivore and predator pressure on invasive and native plants. *Basic and Applied Ecology, 13*, 725–734.

Epanchin-Niell, R., Hufford, M., Aslan, C., Sexton, J., Port, J., & Waring, T. (2009). Controlling invasive species in complex social landscapes. *Frontiers in Ecology and the Environment, 8*, 210–216.

Epanchin-Niell, R. S. (2017). Economics of invasive species policy and management. *Biological Invasions, 19*, 3333–3354.

Eviner, V. T., Garbach, K., Baty, J. H., & Hoskinson, S. A. (2012). Measuring the effects of invasive plants on ecosystem services: Challenges and prospects. *Invasive Plant Science and Management, 5*, 125–136.

Fletcher, R. A., Brooks, R. K., Lakoba, V. T., Sharma, G., Heminger, A. R., Dickinson, C. C., & Barney, J. N. (2019). Invasive plants negatively impact native, but not exotic, animals. *Global Change Biology, 25*, 3694–3705.

Gaertner, M., Biggs, R., Te Beest, M., Hui, C., Molofsky, J., & Richardson, D. M. (2014). Invasive plants as drivers of regime shifts: Identifying high-priority invaders that alter feedback relationships. *Diversity and Distributions, 20*, 733–744.

Gage, K. L., & Schwartz-Lazaro, L. M. (2019). Shifting the paradigm: An ecological systems approach to weed management. *Agriculture, 9*, 179.

Gallardo, B., Bacher, S., Barbosa, A. M., Gallien, L., González-Moreno, P., Martínez-Bolea, V., Sorte, C., Vimercati, G., & Vilà, M. (2024). Risks posed by invasive species to the provision of ecosystem services in Europe. *Nature Communications, 15*, 2631.

Gallardo, B., Bacher, S., Bradley, B., Comín, F. A., Gallien, L., Jeschke, J. M., Sorte, C. J., & Vilà, M. (2019). InvasiBES: Understanding and managing the impacts of Invasive alien species on biodiversity and ecosystem services. *NeoBiota, 50*, 109–122.

Gallardo, B., Clavero, M., Sánchez, M. I., & Vilà, M. (2016). Global ecological impacts of invasive species in aquatic ecosystems. *Global Change Biology, 22*, 151–163.

Gioria, M., Hulme, P. E., Richardson, D. M., & Pyšek, P. (2023). Why are invasive plants successful? *Annual Review of Plant Biology, 74*, 635–670.

Graham, S. (2019). Coordinating invasive plant management among conservation and rural stakeholders. *Land Use Policy, 81*, 247–255.

Grzędzicka, E. (2022). Impact of invasive weeds on the diversity and dissimilarity of bird communities in forested areas. *Diversity, 14*, 229.

Gunstone, T., Cornelisse, T., Klein, K., Dubey, A., & Donley, N. (2021). Pesticides and soil invertebrates: A hazard assessment. *Frontiers in Environmental Science, 9*, Article 643847.

Hale, S. L., & Koprowski, J. L. (2018). Ecosystem-level effects of keystone species reintroduction: A literature review. *Restoration Ecology, 26*, 439–445.

Hardy, N. G., Kuebbing, S. E., Duguid, M. C., Ashton, M. S., Sheban, K. C., Inman, S. E., & Martin, M. P. (2025). Non-native invasive plants in tropical dry forests: A global review of presence, impacts, and management. *Restoration Ecology, 33*, Article e14288.

Hassan, A., & Nawchoo, I. A. (2020). Impact of invasive plants in aquatic ecosystems. In K. Hakeem, R. Bhat, & H. Qadri (Eds.), *Bioremediation and biotechnology* (pp. 55–73). Springer.

Havel, J. E., Kovalenko, K. E., Thomaz, S. M., Amalfitano, S., & Kats, L. B. (2015). Aquatic invasive species: Challenges for the future. *Hydrobiologia, 750*, 147–170.

Hoyle, J. A., Bowlby, J. N., Brousseau, C. M., Johnson, T. B., Morrison, B. J., & Randall, R. G. (2012). Fish community structure in the Bay of Quinte, Lake Ontario: The influence of nutrient levels and invasive species. *Aquatic Ecosystem Health & Management, 15*, 370–384.

Humphrey, J. (2024). History, impacts, and control of purple loosestrife (Lythrum salicaria) in Minnesota. *Scholarly Horizons: University of Minnesota, Morris Undergraduate Journal, 11*, 5.

Hussain, M., Farooq, S., Merfield, C., & Jabran, K. (2018). Mechanical weed control. In K. Jabran & B. S. Chauhan (Eds.), *Non-chemical weed control* (pp. 133–155). Academic Press.

Hussner, A., Stiers, I., Verhofstad, M. J. J. M., Bakker, E. S., Grutters, B. M. C., Haury, J., Van Valkenburg, J. L. C. H., Brundu, G., Newman, J., Clayton, J. S., & Anderson, L. W. J. (2017). Management and control methods of invasive alien freshwater aquatic plants: A review. *Aquatic Botany, 136*, 112–137.

International Union for Conservation of Nature (IUCN) (2022). *Human-wildlife conflict.* https://iucn.org/resources/issues-brief/human-wildlife-conflict

Invasive Species Centre (ISC), Canada. (2025). *Invasive fish and invertebrates.* https://www.invasivespeciescentre.ca/invasive-species/meet-the-species/fish-and-invertebrates/

Jadhav, P. L., Shinde, A. C., Patil, P. V., & Mali, R. P. (2023). Ecological dependency on invasive species Lantana camara by native organisms in the ecosystem of Nanded, Maharashtra (India). *International Journal of Researches in Biosciences, Agriculture and Technology, Special Issue (XI)*, 106–113.

Kalisz, S., Kivlin, S. N., & Bialic-Murphy, L. (2021). Allelopathy is pervasive in invasive plants. *Biological Invasions, 23*, 367–371.
Kelley, J. L., Chapuis, L., Davies, W. I., & Collin, S. P. (2018). Sensory system responses to human-induced environmental change. *Frontiers in Ecology and Evolution, 6*, 95.
Kuehne, L. M., Olden, J. D., & Rubenson, E. S. (2016). Multi-trophic impacts of an invasive aquatic plant. *Freshwater Biology, 61*, 1846–1861.
Kuglerová, L., García, L., Pardo, I., Mottiar, Y., & Richardson, J. S. (2017). Does leaf litter from invasive plants contribute the same support of a stream ecosystem function as native vegetation? *Ecosphere, 8*, Article e01779.
Künzi, Y., Prati, D., Fischer, M., & Boch, S. (2015). Reduction of native diversity by invasive plants depends on habitat conditions. *American Journal of Plant Sciences, 6*, 2718–2733.
Lommen, S. T., Jongejans, E., Leitsch-Vitalos, M., Tokarska-Guzik, B., Zalai, M., Müller-Schärer, H., & Karrer, G. (2018). Time to cut: Population models reveal how to mow invasive common ragweed cost-effectively. *NeoBiota, 39*, 53–78.
Maan, I., Kaur, A., Singh, H. P., Batish, D. R., & Kohli, R. K. (2021). Exotic avenue plantations turning foe: Invasive potential, distribution and impact of Broussonetia papyrifera in Chandigarh, India. *Urban Forestry & Urban Greening, 59*, 127010.
Maceina, M. J., & Reeves, W. C. (1996). Relationships between Largemouth Bass and Aquatic Plants in. *American Fisheries Society Symposium, 16*, 382–393.
Malecki, R. A., Blossey, B., Hight, S. D., Schroeder, D., Kok, L. T., & Coulson, J. R. (1993). Biological control of purple loosestrife. *BioScience, 43*, 680–686.
McCary, M. A., Mores, R., Farfan, M. A., & Wise, D. H. (2016). Invasive plants have different effects on trophic structure of green and brown food webs in terrestrial ecosystems: A meta-analysis. *Ecology Letters, 19*, 328–335.
Méndez-Toribio, M., Martínez-Garza, C., & Ceccon, E. (2021). Challenges during the execution, results, and monitoring phases of ecological restoration: Learning from a country-wide assessment. *PLoS ONE, 16*, Article e0249573.
Merceron, N. R., Lamarque, L. J., Delzon, S., & Porté, A. J. (2016). Killing it softly: Girdling as an efficient eco-friendly method to locally remove invasive Acer negundo. *Ecological Restoration, 34*, 297–305.
Meyerson, L. A., Pauchard, A., Brundu, G., Carlton, J. T., Hierro, J. L., Kueffer, C., Pandit, M. K., Pyšek, P., Richardson, D. M., & Packer, J. G. (2022). Moving toward global strategies for managing invasive alien species. In D. R. Clements, M. K. Upadhyaya, S. Joshi, & A. Shrestha (Eds.), *Global plant invasions* (pp. 331–360). Springer.
Mollot, G., Pantel, J. H., & Romanuk, T. N. (2017). The effects of invasive species on the decline in species richness: A global meta-analysis. *Advances in Ecological Research, 56*, 61–83.
Mooney, H. A., & Cleland, E. E. (2001). The evolutionary impact of invasive species. *Proceedings of the National Academy of Sciences, 98*, 5446–5451.
Mouquet, N., Gravel, D., Massol, F., & Calcagno, V. (2013). Extending the concept of keystone species to communities and ecosystems. *Ecology Letters, 16*, 1–8.
Mukherjee, S., Banerjee, S., Basu, P., Saha, G. K., & Aditya, G. (2015). Lantana camara and butterfly abundance in an urban landscape: Benefits for conservation or species invasion? *Ekologia, 34*, 309.
Narango, D. L., Tallamy, D. W., & Marra, P. P. (2018). Nonnative plants reduce population growth of an insectivorous bird. *Proceedings of the National Academy of Sciences, 115*, 11549–11554.
National Wildlife Federation (NWF). (2025). *Invasive species.* https://www.nwf.org/Educational-Resources/Wildlife-Guide/Threats-to-Wildlife/Invasive-Species
Negi, G. C., Sharma, S., Vishvakarma, S. C., Samant, S. S., Maikhuri, R. K., Prasad, R. C., & Palni, L. M. (2019). Ecology and use of Lantana camara in India. *The Botanical Review, 85*, 109–130.
Nelson, S. B., Coon, J. J., Duchardt, C. J., Fischer, J. D., Halsey, S. J., Kranz, A. J., Parker, C. M., Schneider, S. C., Swartz, T. M., & Miller, J. R. (2017). Patterns and mechanisms of invasive plant impacts on North American birds: A systematic review. *Biological Invasions, 19*, 1547–1563.

Netherland, M. D., Getsinger, K. D., & Stubbs, D. R. (2005). Aquatic plant management: Invasive species and chemical control. *Outlooks on Pest Management, 16*, 100.

NOAA Fisheries. (2025). *Invasive and exotic marine species*. https://fisheries.noaa.gov/insight/invasive-and-exotic-marine-species

Nyhus, P. J. (2016). Human–wildlife conflict and coexistence. *Annual Review of Environment and Resources, 41*, 143–171.

Omondi, A., & Merceline, A. (2023). Invasive species and their impacts on the ecology of Lake Victoria: A rapid review. *Pan Africa Science Journal, 2*.

Parsons, H., Major, R. E., & French, K. (2006). Species interactions and habitat associations of birds inhabiting urban areas of Sydney, Australia. *Austral Ecology, 31*, 217–227.

Pavlović, D., Vrbničanin, S., Anđelković, A., Božić, D., Rajković, M., & Malidža, G. (2022). Non-chemical weed control for plant health and environment: Ecological integrated weed management (EIWM). *Agronomy, 12*, 1091.

Peller, T., & Altermatt, F. (2024). Invasive species drive cross-ecosystem effects worldwide. *Nature Ecology & Evolution, 8*, 1087–1097.

Price, C. J., Bytheway, J. P., Finnerty, P. B., Grant, L. S., Masani, S., Orlando, C. G., Parker, F. C., Possell, M., McArthur, C., & Banks, P. B. (2024). Altering reality—Sensory tactics to manage wildlife and conserve threatened species. *Australian Zoologist, 43*, 510–517.

Priyanka, N., & Joshi, P. K. (2013). A review of Lantana camara studies in India. *International Journal of Scientific and Research Publications, 3*, 1–11.

Pyšek, P., & Richardson, D. M. (2010). Invasive species, environmental change and management, and health. *Annual Review of Environment and Resources, 35*, 25–55.

Rai, P. K. (2022). Environmental degradation by invasive alien plants in the anthropocene: Challenges and prospects for sustainable restoration. *Anthropocene Science, 1*, 5–28.

Rai, P. K., & Singh, J. S. (2020). Invasive alien plant species: Their impact on environment, ecosystem services and human health. *Ecological Indicators, 111*, Article 106020.

Reynolds, S. J., Ibáñez-Álamo, J. D., Sumasgutner, P., & Mainwaring, M. C. (2019). Urbanisation and nest building in birds: A review of threats and opportunities. *Journal of Ornithology, 160*, 841–860.

Roberts, J., & Florentine, S. (2021). Biology, distribution and control of the invasive species Ulex europaeus (Gorse): A global synthesis of current and future management challenges and research gaps. *Weed Research, 61*, 272–281.

Robertson, B. A., & Chalfoun, A. D. (2016). Evolutionary traps as keys to understanding behavioral maladapation. *Current Opinion in Behavioral Sciences, 12*, 12–17.

Rotherham, I. D. (2021). The impacts of recolonisation of an urbanised river by native and non-native species. *Frontiers in Ecology and Evolution, 9*, Article 618371.

Sage, R. F. (2020). Global change biology: A primer. *Global Change Biology, 26*, 3–30.

Sandilyan, S., Meenakumari, B., Biju Kumar, A., & Mandal, R. (2018). A review on impacts of invasive alien species on Indian inland aquatic ecosystems. *National Biodiversity Authority*, 88.

Schaefer, V. H. (2011). Remembering our roots: A possible connection between loss of ecological memory, alien invasions and ecological restoration. *Urban Ecosystems, 14*, 35–44.

Schirmel, J., Bundschuh, M., Entling, M. H., Kowarik, I., & Buchholz, S. (2016). Impacts of invasive plants on resident animals across ecosystems, taxa, and feeding types: A global assessment. *Global Change Biology, 22*, 594–603.

Schultz, R., & Dibble, E. (2012). Effects of invasive macrophytes on freshwater fish and macroinvertebrate communities: The role of invasive plant traits. *Hydrobiologia, 684*, 1–14.

Sharma, A., Kaur, A., Kaur, S., Kohli, R. K., & Batish, D. R. (2023). Plant invasion and climate change: A global overview. In S. Tripathi, R. Bhadouria, P. Srivastava, R. Singh, & D. R. Batish (Eds.), *Plant invasions and global climate change* (pp. 3–30). Springer.

Sharma, G. P., Raghubanshi, A. S., & Singh, J. S. (2005). Lantana invasion: An overview. *Weed Biology and Management, 5*, 157–165.

Siegert, N. W., McCullough, D. G., Luther, T., Benedict, L., Crocker, S., Church, K., & Banks, J. (2023). Biological invasion threatens keystone species indelibly entwined with Indigenous cultures. *Frontiers in Ecology and the Environment, 21*, 310–316.

Silva, P. A., Cherutte, A. G., Gomes, A. C. S., Silva, L. L., Brito, L., Rodrigues, B. M., & Maruyama, P. K. (2023). The ecological role and potential impact of an alien tree highly attractive to native nectar-feeding birds in urban areas. *Urban Ecosystems, 26*, 1029–1040.

Singh, H. P., Batish, D. R., & Kohli, R. K. (2003). Allelopathic interactions and allelochemicals: New possibilities for sustainable weed management. *Critical Reviews in Plant Sciences, 22*, 239–311.

Sobuj, N., & Byun, C. (2023). A synthesis of plant invasion control: Important factors to consider when choosing a control method. *Écoscience, 30*, 234–246.

Stewart, P. S., Hill, R. A., Stephens, P. A., Whittingham, M. J., & Dawson, W. (2021). Impacts of invasive plants on animal behaviour. *Ecology Letters, 24*, 891–907.

Tallamy, D. W., & Shriver, W. G. (2021). Are declines in insects and insectivorous birds related? *The Condor, 123*, 1–8.

United Nations Environment Programme (UNEP). (2023). *Five drivers of the nature crisis.* https://www.unep.org/news-and-stories/story/five-drivers-nature-crisis#:~:text=Invasive%20species,-Invasive%20species%20pose&text=The%20negative%20effects%20of%20these,where%20the%20cause%20is%20known

van Rees, C. B., Hand, B. K., Carter, S. C., Bargeron, C., Cline, T. J., Daniel, W., Ferrante, J. A., Gaddis, K., Hunter, M. E., Jarnevich, C. S., & McGeoch, M. A. (2022). A framework to integrate innovations in invasion science for proactive management. *Biological Reviews, 97*, 1712–1735.

Vashishtha, A. (2017). Assessing the impact of invasive plant species on native plant communities and ecosystems. *International Journal of Research in Engineering and Applied Sciences, 7*, 172–181.

Villamagna, A. M., & Murphy, B. R. (2010). Ecological and socio-economic impacts of invasive water hyacinth (Eichhornia crassipes): A review. *Freshwater biology, 55*(2).

Weidlich, E. W., Flórido, F. G., Sorrini, T. B., & Brancalion, P. H. (2020). Controlling invasive plant species in ecological restoration: A global review. *Journal of Applied Ecology, 57*, 1806–1817.

WildFish Organization. (2021). *The impact of invasive species on the aquatic environment.* https://wildfish.org/latest-news/invasive-species/

Wildhaber, M. L., West, B. M., Ditter, K. K., Moore, A. P., & Peterson, A. S. (2023). A review of grass carp and related species literature on diet, behavior, toxicology, and physiology focused on informing development of controls for invasive grass carp populations in North America. *Fishes, 8*, 547.

Wilkinson, N. (2006). Factors influencing the small-scale distribution of House Sparrows Passer domesticus in a suburban environment. *Bird Study, 53*, 39–46.

Zhang, B., Zhai, L., Ervin, G. N., & Coyle, D. R. (2023). Effective and timely use of models to inform on-the-ground management of invasive plants. *Biological Invasions, 25*, 2089–2102.

Zheng, Y., Feng, Y., Valiente-Banuet, A., Li, Y., Liao, Z., Zhang, J., & Chen, Y. (2015). Are invasive plants more competitive than native conspecifics? Patterns vary with competitors. *Scientific Reports, 5*, 15622.

Part III
Socio-ecological Impacts of Invasive Alien Plant Species in Urban Ecosystems

Chapter 9
Economic Impacts of Invasive Alien Plant Species on Urban Ecosystems

Prakash Rajak, Ritturaj Singh, Sara Fatma, Anindya Shankar Nandi, and Hema Singh

Abstract Urban ecosystems are dynamic environments where human infrastructure and natural features coexist and interact. However, these habitats are becoming more and more threatened by the invasive plant species' quick proliferation. These alien plants spread swiftly, frequently pushing out local plants and upsetting the natural equilibrium. They can cause major maintenance and repair expenses by causing damage to buildings, roadways, and drainage systems. Additionally, invasive plants diminish the aesthetic and biological significance of urban green spaces, which detracts from their allure for tourists and leisure. Invasive species in cities do more than just physical harm to the economy. Reduced property prices in the impacted areas and higher medical costs as a result of allergies or pests linked to these plants are examples of indirect costs. The cost of controlling infestations, from mechanical removal and chemical treatments to public awareness campaigns, is frequently high for municipalities. This chapter presents an in-depth exploration of the economic implications of invasive plant species in urban settings. It highlights the ways in which these plants affect city budgets, recreational opportunities, urban aesthetics, and public services. We highlight the financial load and offer insights into efficient, long-term control solutions using real-world examples and management techniques. Reducing long-term costs and safeguarding urban ecosystems require early discovery, quick response, and logical, species-specific management strategies. The chapter's ultimate goal is to assist environmental managers, city planners, and legislators by providing workable answers and highlighting the significance of proactive, economical actions to maintain the resilience and health of urban ecosystems.

Keywords Climate change · Economic cost · Management · Plant invasion · Urbanization

---

P. Rajak (✉) · R. Singh · S. Fatma · A. S. Nandi · H. Singh
Ecosystem Analysis Laboratory, Department of Botany, Banaras Hindu University, Varanasi, India
e-mail: prakashrajak95@gmail.com

R. Singh et al. (eds.), *Invasive Alien Plants in Urban Ecosystems*,
https://doi.org/10.1007/978-3-032-00997-5_9

1 Introduction

Urbanization is one of the most pervasive forms of land-use change worldwide and plays a crucial role in shaping ecological processes across landscapes. It is widely recognized as a key driver in the introduction, establishment, and spread of alien species (Alston & Richardson, 2006; Pyšek, 1998). The transformation of natural landscapes into urbanized environments leads to habitat fragmentation, the loss of native vegetation, and the creation of new ecological niches (Kowarik, 2011). These conditions facilitate the movement and proliferation of non-native species, many of which are capable of becoming invasive (Aronson et al., 2014). Cities are characterized by several unique environmental features, such as the prevalence of impermeable surfaces, altered hydrology, novel soil conditions, and increased pollution levels (Imhoff et al., 2010). These changes contribute to the formation of distinct urban microclimates-typically warmer, more nutrient-rich, and more disturbed than surrounding natural habitats, an effect commonly referred to as the urban heat island phenomenon (Imhoff et al., 2010; McKinney, 2006). While some pockets of low-nutrient or unimproved habitats persist in urban areas, the overall high fertility and disturbance levels, combined with significant within-habitat variation, make urban ecosystems especially conducive to the establishment of ruderal species, including many alien plants (Aronson et al., 2014; Pyšek & Richardson, 2010). Invasive alien species (IAS)-often labeled as exotic, non-native, or non-indigenous—are species introduced outside their natural geographic ranges, either intentionally or accidentally (Elton & Elton, 1958; Vilà et al., 2011). These species have the potential to cause severe ecological disruption by displacing native flora and fauna, altering ecosystem services, and impacting biodiversity at multiple scales (Blackburn et al., 2014; Pimentel et al., 2000). Urban ecosystems, with their dynamic history, diverse land use, and frequent disturbances, serve as epicenters for biological invasions. Invasive alien plants (IAPs), in particular, are now a common feature of cityscapes around the world, presenting growing challenges to both ecological integrity and human well-being (Padayachee et al., 2017). A comprehensive understanding of biological patterns within urban landscapes is therefore essential. This includes recognizing how urbanization shapes the distribution and behavior of IAPs, and how these species interact with both natural and human-modified environments (Aronson, 2015). It is within this complex ecological and socio-economic context that the role of IAPs must be examined more critically. In this chapter, we present an in-depth analysis of the economic consequences of invasive alien plant species in urban ecosystems. We explore the direct and indirect financial impacts these species impose on infrastructure, public health, property values, and city management. Furthermore, we highlight sustainable, cost-effective strategies for controlling IAPs, aiming to support evidence-based policymaking and long-term urban ecosystem resilience.

2 The Changing Dynamics of Urban Landscapes

Urban landscapes are diverse, with linear axes facilitating the flow of people and goods between various purposes, such as parks, leisure areas, commercial and industrial zones, and residential districts (Larkham, 2006). These scenes are dynamic, with new neighborhoods emerging on the outskirts of cities. To understand the biological processes and patterns in cities, it is crucial to capture this diversity of spatial and temporal urban usage. However, ecological research often offers limited descriptions of urban environments, based on broad land use classifications like "urban," "peri-urban," and "rural" (Fratarcangeli et al., 2019). This categorization overlooks the history and use of different regions, as well as significant structural and social components that influence ecological patterns (Ariori et al., 2017; Aronson et al., 2015a, 2015b). A more comprehensive depiction of urban environments is needed to enhance our understanding of biological processes in cities. To capture urban heterogeneity, ecological patterns on relevant spatial scales must be analyzed in light of the diversity of urban land uses (Cadotte et al., 2017; Ramalho & Hobbs, 2012). Large spatial resolutions ($\geq 1\ km^2$) can help study plant patterns in metropolitan settings, while smaller areas (1 ha or less) can show variations in alpha diversity or community features. A multiscale approach could help better understand the mechanisms behind urban ecological patterns (Fratarcangeli et al., 2019; Godefroid & Koedam, 2007; Johnson et al., 2018; Lososová et al., 2011).

3 Invasive Alien Plant Species (IAPS): A Growing Threat to Global Biodiversity

A serious and expanding danger to biodiversity worldwide is posed by IAPS. Intentionally or inadvertently, these plant species have been brought outside of their original environments, where they establish, spread, and damage native ecosystems (Early et al., 2016; Mehraj et al., 2018; Vilà & Hulme, 2017; Vilà et al., 2010). They have significant and varied ecological effects, including displacing native species, changing habitats, and lowering the amount of vital nutrients available to local plants and animals. The introduction and naturalization of IAPS in new areas has been made much easier by the growth of global commerce and transportation networks. Frequent disruptions, environmental stressors, and increased propagule pressure—the flow of seeds or reproductive material—make urban regions particularly susceptible to these incursions (Pyšek et al., 2020). These elements provide the perfect environment for IAPS to grow and flourish. High reproductive output, quick growth, and good environmental adaptations are some of the characteristics that make IAPS particularly competitive with native species. As a result, the demise of several native species that are vulnerable or endangered is accelerating. The problem of biological invasions has drawn more attention due to these effects on a global and regional level, notably on the Indian subcontinent (van Kleunen et al., 2015). Invasive alien species are the second biggest danger to biodiversity after habitat loss, according to the Convention on Biological Diversity, which acknowledges the seriousness of this problem. The Convention urges swift action to stop their introduction and, if required, to manage or eliminate species that threaten natural habitats, ecosystems, or species.

4 Infiltration and Expansion of IAPS in Urban Ecosystems

IAPS expansion into urban ecosystems is a complicated process that is fueled by a variety of environmental changes and human-mediated activities. There are many ways that alien species are introduced, both purposefully and accidentally. These encompass intentional introductions for ecological, agricultural, or ornamental purposes (Pyšek & Hulme, 2005) as well as inadvertent pathways like evasion from cultivation, contaminating products, dispersal through man-made passageways, and secondary introductions from previously infected areas (Blackburn et al., 2011; Puth & Post, 2005).

The main entrance sites for terrestrial alien plants in urban environments are often parks, gardens, and residential neighborhoods. The species may progressively spread into peri-urban, rural, and even protected natural regions from these original locations in a "stepping-stone" pattern. This process is greatly aided by transportation corridors including roads, railroads, and sidewalks, which provide continuous, disturbed environments that support the growth and spread of plants (Blackburn et al., 2011; Puth & Post, 2005; Rajak et al., 2021). The migration of IAPS is also effectively facilitated by urban aquatic settings. Invasive species may spread quickly both within and outside of urban bounds because of the long-distance plant propagules that streams, drainage channels, and urban rivers can convey (Catford et al., 2012; Dyderski & Jagodziński, 2016). Because of its fragmented landscapes and diverse microhabitats, the urban matrix generates special ecological niches that support the survival and growth of adaptive IAPS. Apart from these vectors, the physical layout of the city itself makes invasion easier. According to Tallent-Halsell and Watt (2009), roadsides, railway embankments, riverbanks, and disturbed urban soils often provide the perfect environment for the development of pioneer invasive species because they provide open space, light, and less competition. In addition to serving as invasion routes, these regions serve as reservoirs or sources for further dissemination. In addition, because of their high levels of connection and human migration, metropolitan environments are key hubs for the long-distance dissemination of alien plants (Hodkinson & Thompson, 1997; Von der Lippe & Kowarik, 2008). Invasive species may spread from these hubs into nearby ecosystems, having an influence at a larger landscape level. Therefore, it is essential to comprehend how human activity and structural elements interact at various spatial scales in order to forecast and control IAPS dynamics in urban environments. Säumel and Kowarik (2013) have emphasized that successful management and mitigation methods need a holistic strategy that incorporates biological, geographical, and societal variables.

5 Climate Change and Human Activity in Urban Plant Invasions

Human disturbances and climate change are the main causes of plant invasions, which are a worldwide problem (Capellini et al., 2015; Colautti & MacIsaac, 2004). Due to resource competition and environmental compatibility, several plant species have been introduced outside of their natural ranges (Holmes et al., 2009; Rajak et al., 2025). Invasive species spread more easily due to rapid climate change, which also contributes to changing invasion patterns (Catford et al., 2009; Marco et al., 2010). Because of their resistance and quick growth, ornamental plants are often invasive in urban environments (Bellard et al., 2013, 2018). Invasion hazards are influenced by variables including neighborhood age, plant characteristics, and community planting methods. Global invasive species distribution is predicted to shift due to climate change, presenting major risks to agriculture, human health, ecosystems, native biodiversity, and economies (van Kleunen et al., 2015; Early et al., 2016; Matthews et al., 2018; Vila and Hulme, 2017; Mehraj et al., 2018).

6 Sociocultural and Economic Perceptions of IAPS

In many regions, invasive species play a complex role by providing essential resources such as food, medicine, income, and items of sociocultural importance (Bellard et al., 2018). Recent research has examined how community views of invasive species—specifically, their effects on livelihoods and traditional practices—are influenced by local ecological knowledge. Siges, for example, looked into how *Piper aduncum* was used in rural African communities, showing how perceptions differed by gender and age. While males pointed out that other traditional crops had been displaced by sweet potatoes, which were growing alongside *P. aduncum*, older participants voiced worry about the invasion's impact on the loss of forest goods. Conversely, women bemoaned the loss of mushrooms that were previously prevalent at the borders of forests. It's interesting to note that some people valued how the plant simplified the management of agricultural fields.

Divergent opinions were also expressed on the effect on food production; some said that tuber yields had decreased, while others thought that they had increased. According to local populations in Kenya, *Prosopis juliflora* had no discernible impact on subsistence farming (Hulme, 2017; Rajak et al., 2025). Similarly, *Mikania micrantha* is still largely considered to be environmentally detrimental even though it is utilized for a variety of purposes in rural Nepal. These variations imply that local perceptions of invasive species may change in response to invasion severity and the immediate advantages or disadvantages encountered. Notably, some communities were unable to appropriately estimate long-term changes in forest product production, despite a reported increase in resource availability. This suggests that perception and ecological reality are not aligned (Dyderski & Jagodziński, 2016).

7 IAPS Introduced in Indian Urban Ecosystem

A large number of alien plant species are purposefully introduced in India. However, there are serious hazards associated with introducing new species. *Ageratum conyzoides, Asclepias curassavica, Cryptostegia grandiflora, Cytisus scoparius, Eichhornia crassipes, Lantana camara,* and other exotic plant species were purposefully brought to India for their aesthetic appeal (Reddy, 2008). Unintentional propagules of foreign plants being carried by ballast water, ocean rafting, etc., is another significant issue. As Table 1 illustrates, invasive plants that have devastated terrestrial ecosystems include *Ageratum conyzoides, Eupatorium adenophorum, Eupatorium odoratum, Lantana camara, Leucaena leucocephala; Mikania micrantha, Parthenium hysterophorus,* and *Prosopis juliflora;* and *Eichhornia crassipes, Ipomoea* spp., and *Salvinia molesta* in aquatic environments. Due to its invasion of urban ecosystems and natural places, including the Andhra University campus, the introduction of species including *Chromolaena odorata, Antigonon leptopus,* and *Lantana camara* has become troublesome.

Only after a careful evaluation of the risks involved can biological control be used. The majority of IAPS in India are neo-tropical (South American). This is similar to Andhra University Campus, which has a majority contribution of 62% and has been promoted across India via a number of means. The Delhi Ridge Forest has been supplanted by invasive species such as *Prosopis juliflora* (Rajak et al., 2025), *Leucaena leucocephala* (Mishra et al., 2024), and *Lantana camara.* Invasive species, exotics, and agricultural escapes now make up a significant amount of the vegetation in addition to native Aravalli species. The Delhi Ridge Forest is made up of the Northern, Central, South-Central, and Southern Delhi Ridges, which are portions of the original Aravalli vegetation that has withstood these alterations. Silk Road merchants introduced several exotic plant species to India from countries in Africa, the Mediterranean, and Eurasia.

Prosopis juliflora was initially brought to Delhi in the 1900s by skilled gardener William R. Mustoe as part of the colonial afforestation initiative. Even after independence, it remained a major player in green efforts. Because of its rapid growth and resistance to drought, this plant was widely introduced in many Indian states, including Gujarat, Madhya Pradesh, Rajasthan, and Delhi. After then, it extended to agricultural areas, pasturelands, and woods. It is one of the most aggressive plant invaders and is now considered a nuisance in many arid and semi-arid parts of Asia, Africa, and North America. *P. juliflora* has been shown to alter the physical and chemical properties of soils and to have traits that prevent other plants from growing underneath it. Experts attribute the disappearance of native tree species such as *Prosopis cineraria* on Delhi Ridge to *P. juliflora.* It is believed that the African tamarind, *Tamarindus indica,* was initially introduced to India by Ethiopian traders (Shah, 2014).

Parthenium hysterophorus, an invasive weed of tropical America, entered India and remained obscure till 1965, when it was noticed first time in Pune. It has caused serious health problems in many tropical and subtropical regions of the world

Table 1 Globally, invasive alien plant species (IAPS) have been recorded for their detrimental effects on the economy

S. No	Species name	Region	Economic loss	References
1	*Opuntia stricta*	Africa	US$ 500–1000 per household per year	Shackleton et al. (2017)
2	*Baccharis halimifolia*	France and Australia	US$ 0.1 billion from 1960 to 2020	Fernandez et al. (2023)
3	*Euphorbia esula*	USA	Approx. US$ 1.03 billion from 1960 to 2020	Fantle-Lepczyk et al. (2022)
4	*Tamarix ramosissima*	USA	US$ 26.3–67.8 million	Pejchar and Mooney (2009)
5	*Hedychium coronarium*	Argentina	US$ 44–215 million per year	Zilio et al. (2025)
6	*Tamarix* spp., *Ligustrum lucidum,* and *Tithonia tubaeformis*	Argentina	US$ 255,000 million per year	Zilio (2019)
7	*Parthenium hysterophorus*	Australia	Approx. US$16.8 million per year	Chippendale and Panetta (1994)
8	*Eichhornia crassipes*	China	Approx. US$1.46 million per year	Wei (2009)
9	*Bidens pilosa*	Russia	Approx. 91 billion rubles	Magomedov et al. (2013)
10	*Centaurea solsitialis*	USA	Approx. US$12.7 million per year	Julia et al. (2007)
11	*Parthenium hysterophorus, Leucaena leucocephala, Oreochromis mossambicus,* etc.	India	US$ 127.3 billion to 182.6 billion over 1960–2020	Bang et al. (2022)

(Reddy & Bagyanarayana, 2007). The plant's introduction into India is debatable; evidence points to an unintentional introduction around 1810. William Roxburgh's 1814 book Hortus bengalensis has the earliest known mention of Parthenium (Kohli et al., 2006). Since then, the plant has expanded to Jammu and Kashmir, naturalized in Delhi, and is currently found up to 1600 m above sea level across India.

Another invasive species, *Lantana camara*, was brought to India as an ornamental plant in the 1800s because of its eye-catching blooms. It may spread vegetatively, yields a large number of seeds, and grows quickly (Parsons & Cuthbertson, 2001). Lantana can withstand a broad range of climatic conditions and grows dense thickets that push out native seedlings. It is the most invasive weed in India's national parks and natural reserves, spreading to disturbed regions, the edges of forests, and gaps in the forest (Palmer & Pullen, 1995). Belonging to the Verbenaceae family, lantana has an impact on native woods and grasslands in more than 60 countries throughout the world. It affects 14 key crops and infests millions of hectares of grazing pasture in a variety of habitats and soil types. It tends to decay in swampy or hydromorphic soils and is vulnerable to frosts, low temperatures, and salty soils (Sharma et al., 2005).

8 Economic Impacts of the IAPS

At least 434 invasive alien tree species have been linked to detrimental ecological and economic impacts globally, according to the InvaCost database. However, financial estimates are available for less than 20% of these species. The predicted global costs range from $23.9 billion to $64.7 billion. The Acacia species (*Acacia* spp., *Tamarix* spp., *Ligustrum* spp., *Prosopis* spp., and *Leucaena* spp.) and the Fabaceae family suffer the highest costs due to their widespread introduction for forestry and ornamental purposes. Several IAPs that were put in place for human welfare have a history of having a disastrous impact on the economy and environment, claim Souza et al. (2018). Therefore, using people's perceptions of the occurrence and their local ecological knowledge is a viable way to categorize the consequences of IAPS. *Acacia mangium*, an IAPS in the northern Brazilian Amazon, is known to negatively affect the environment, the economy, and indigenous people by altering the water's quality (Souza et al., 2018). The invasion of aquatic macrophytes like *Eichhornia crassipes* (water hyacinth) in Lake Victoria has been damaging to human welfare since it reduces fish productivity and the potential for ecotourism (Kasulo, 2000; Pejchar & Mooney, 2009). Additionally, ecological niche models and global climate models predict that water hyacinth would move toward European and Mediterranean regions under a regime of climate change. This implies that there would be significant economic consequences from such an invasion (Kriticos & Brunel, 2016; Rajak et al., 2025).

The *Tamarix ramosissima* invasion in the USA has caused significant water loss, valued at $26.3–67.8 million (Pejchar & Mooney, 2009; Reid et al., 2005; Zavaleta, 2000). *Melaleuca quinquenervia* and Eucalyptus species also use a lot of groundwater in Florida and California. *Myriophyllum spicatum* and other aquatic macrophytes in Lake Tahoe are responsible for 1% of recreational losses annually, or US$500,000 (Eiswerth et al., 2005; Schmitz et al., 1997). Every year, pathogenic invasive species such as *Euphorbia esula* and *Xanthomonas campestris* cause economic losses of over $200 million USD. Between 1906 and 1991, 79 invasive species caused losses of US$97 billion, according to the Office of Technology Assessment (Andersen et al., 2004). The yearly harm caused by pathogenic invaders was US$1.5 trillion, with costs in the US alone, amounting to $120–138 billion (Rajak et al., 2025; Xu et al., 2006). 283 flora invaders cost China $14.45 billion in lost revenue. *Opuntia stricta* and other high-risk invasive species cause $500 to $1000 in yearly economic losses for each family in Africa (Shackleton et al., 2017). In Southeast Asia, disease-carrying foreign invaders cost the human health sector US$1.85 billion, according to Nghiem et al. (2013). Invaders caused more economic harm to the agricultural industry than to the human health sector. Many invasive plants have socioeconomic consequences, including increased maintenance costs for urban green areas, health problems, and impacts on regional agriculture, in addition to posing a threat to urban biodiversity. Effective management requires concerted efforts that include ecological restoration initiatives, the implementation of policies, and public education.

9 Global Economic Impact of IAPS

IAPS cause significant economic losses worldwide, impacting agriculture, forestry, infrastructure, tourism, and public health. These losses can be categorized into direct and indirect economic damages. IAPS pose significant economic challenges to urban ecosystems, affecting infrastructure, biodiversity, public health, and ecosystem services (Panzavolta et al., 2021). Their economic impact can be broadly categorized into direct costs, indirect costs, and opportunity costs:

9.1 Direct Economic Losses Caused by IAPS

Direct economic losses occur when invasive plant species directly impact agricultural yields, livestock productivity, forestry resources, and infrastructure (Eschen et al., 2021). These losses need quick financial outlays for things like damage repair, weed management, and decreased output. IAPS mainly impact land management, forestry, and agriculture, resulting in significant direct economic losses on a global scale. Reduced wood output, damaged pasturelands, higher management costs, and lower agricultural yields are the causes of these losses. The costs associated with managing, controlling, and eliminating invasive plants are considered direct economic losses. Large quantities of money are spent by governments and municipalities on biological control initiatives, pesticides, and mechanical removal. Furthermore, certain invasive species, such Japanese Knotweed (*Reynoutria japonica*), cause substantial maintenance costs because they grow aggressively and harm infrastructure, such as sidewalks, roads, drainage systems, and building foundations (Docking, 2024; Mishra et al., 2024). Water bodies are also clogged by invasive aquatic plants, which lowers water quality and raises the cost of filtering drinking water (Fig. 1).

9.2 Agricultural Losses

Invasive plant species compete with crops for essential resources such as nutrients, water, and sunlight, leading to lower agricultural productivity. Farmers are forced to

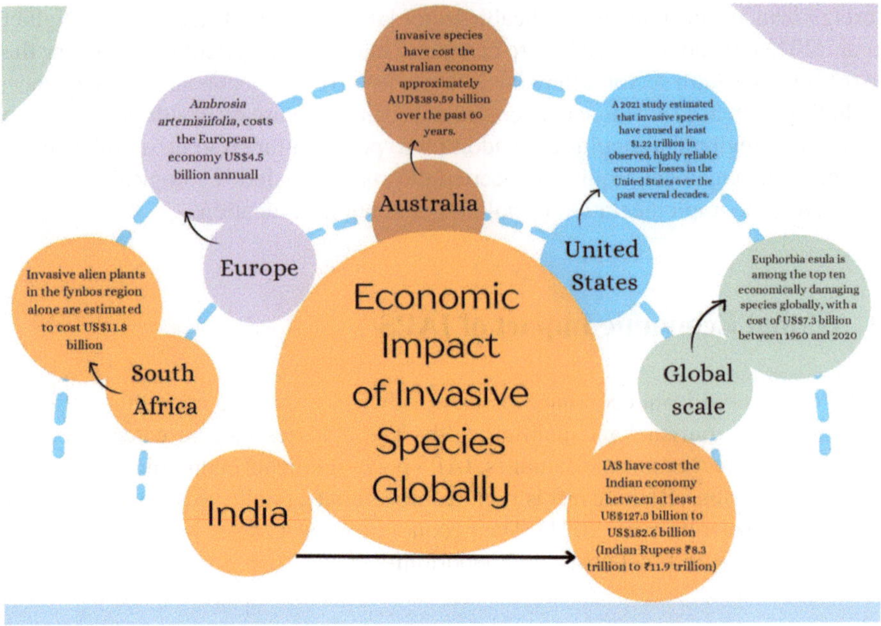

Fig. 1 Economic loss resulting from socio-ecological and economic factors affecting human well-being in various nations is a measure of the effects of IAPS

invest heavily in herbicides, manual removal, and other weed management strategies (Moss, 2019). Some key examples include: *Parthenium hysterophorus* (Parthenium weed): Causes severe yield reductions in cereal crops like wheat and maize in India, Australia, and Africa. *Striga* spp. (Witchweed): A parasitic plant that significantly reduces yields of staple crops like sorghum, maize, and millet in sub-Saharan Africa. *Eupatorium adenophorum* (Crofton weed): Reduces tea, coffee, and spice production in South Asia.

9.3 Livestock and Pasture Degradation

Invasive plants often invade grazing lands, reducing the availability of nutritious fodder, and sometimes causing toxicity in livestock. The economic losses in the livestock industry arise from reduced weight gain, lower milk production, and increased veterinary costs (Ruiz-Nieto et al., 2020; Rajak et al., 2021). Key examples include: *Prosopis juliflora* (Mesquite): Forms dense thickets in pasturelands across Africa, India, and South America, displacing native grasses and reducing available forage. *Lantana camara* (Lantana): Toxic to cattle and other livestock, leading to poisoning and mortality. *Centaurea* spp. (Knapweed): Reduces grazing capacity in North America, leading to significant economic losses for ranchers.

9.4 Increased Costs for Weed Control and Management

Governments, farmers, and industries spend billions of dollars annually on invasive plant control, including herbicide applications, mechanical removal, and biological control programs. Some costly control measures include:

The U.S. spends over $11 billion annually on invasive weed control in agriculture (Pimentel et al., 2005). Australia invests heavily in weed eradication programs, with annual costs exceeding $5 billion due to species like *Opuntia stricta* (Prickly Pear).

9.5 Reduced Timber and Forestry Output

Invasive plant species disrupt forest ecosystems, affecting timber production and biodiversity. Fast-growing invasive trees suppress the regeneration of commercially valuable native tree species. *Acacia mearnsii* (Black Wattle): Invades forests in South Africa and South America, reducing the availability of high-value timber species (Juba 2020). *Ailanthus altissima* (Tree of Heaven): Spreads aggressively in Europe and North America, damaging commercial forestry plantations. *Leucaena leucocephala* (*Leucaena*): Rapidly colonizes tropical forests, reducing biodiversity and timber yields.

9.6 Infrastructure Damage

The aggressive root systems of some invasive plants damage roads, buildings, and irrigation systems, leading to high maintenance costs (Shi et al., 2023). *Ailanthus altissima* (Tree of Heaven): Causes structural damage to roads, sidewalks, and pipelines. *Tamarix* spp. (Saltcedar): Clogs water channels and irrigation systems, increasing costs for water management authorities.

Agriculture is one of the industries most impacted, as invasive plants compete for vital nutrients, water, and sunshine, lowering agricultural production (Fried et al., 2017). Farmers incur higher expenses as a result of this competition, as they have to spend more on hand removal, pesticides, and other management methods. For example, in Africa, India, and Australia, *Parthenium hysterophorus*, often known as Parthenium weed, drastically lowers crop output. Invasive plants also have a detrimental effect on pasturelands and cattle. In Africa and Asia, grazing grounds are degraded by toxic species like *Prosopis juliflora* (mesquite), which results in lower production of meat and dairy products (Mishra et al., 2024; Rajak et al., 2025; Ruiz-Nieto et al., 2020). Because fast-growing alien species inhibit native forest regeneration and reduce wood yields, the forestry industry also suffers. For instance, the lumber sector has been disrupted by the invasion of South African forests by *Acacia mearnsii*, sometimes known as Black Wattle.

9.7 Indirect Economic Losses

The ecological and environmental effects of invasive plants result in indirect economic costs. Because these plants outcompete native vegetation, biodiversity is lost, and ecosystem functions, including pollination, carbon sequestration, and air purification impacted adversely (Daba & Dejene, 2018). Certain invasive species, including *Salvinia molesta* and *Prosopis juliflora,* change hydrological systems by obstructing streams and raising the possibility of urban floods. Because plants like *Parthenium hysterophorus* cause allergies, respiratory problems, and skin illnesses; they also have an impact on public health and raise healthcare expenses (Mishra et al., 2024; Shabbir et al., 2024). Secondary impacts of invasive plant species, such as higher public health expenses, environmental deterioration, biodiversity loss, and trade disruption, lead to indirect economic losses. These losses have long-term economic repercussions even if they might not be quantifiable right away. Invasive plant species result in significant indirect economic effects in addition to direct losses. Due to the invasive root systems of species like *Ailanthus altissima* (Tree of Heaven), which harm roads, bridges, and pipelines, and raise maintenance costs, infrastructure is especially vulnerable (Sladonja et al., 2015). Invasive aquatic plants, such as *Eichhornia crassipes* (Water Hyacinth), also cause financial losses to the tourist and leisure sectors by clogging lakes and rivers, lowering the attractiveness of national parks and protected areas, and restricting prospects for water-based tourism. Furthermore, allergenic plants like *Ambrosia artemisiifolia* (ragweed), which cause respiratory illnesses and raise medical bills, contribute to an increase in public health expenditures.

9.8 Increased Public Health Costs

Some invasive plants release allergens that cause respiratory diseases, leading to higher healthcare costs (Lazzaro et al., 2018). Example: *Ambrosia artemisiifolia* (Ragweed) is a major cause of seasonal allergies in Europe and North America.

9.9 Tourism and Recreational Losses

Invasive plants reduce the aesthetic and ecological value of national parks, forests, and water bodies, leading to a decline in tourism revenue (Charles & Dukes, 2007). Example: *Eichhornia crassipes* (Water Hyacinth) clogs lakes and rivers, restricting boating, fishing, and water sports.

10 Ecosystem Degradation and Biodiversity Loss

Invasive plants outcompete native species, leading to biodiversity loss and ecological imbalances (Linders et al., 2019). Example: *Lantana camara* disrupts native forests in India, affecting wildlife and ecosystem services.

Case Studies

International Trade and Economic Losses

Invasive plants have an effect on global supply networks and international trade. Strict biosecurity regulations are enforced in many nations, which raises the price of quarantine facilities and customs checks. Export restrictions resulting from invasive species can also keep impacted agricultural goods from accessing international markets (Perrault et al., 2003). Importing nations frequently reject contaminated products or crops, which causes financial losses for exporters and trade difficulties. In order to prevent economic harm and preserve biodiversity, governments and businesses throughout the world make significant investments in invasive plant species control (Pickett & Grove, 2009; Pickett et al., 2001, 2011; Rebele, 1994).

Impact of IAPS in Indian Cities

IAPS pose significant hazards to society, the economy, and the environment on a global scale. Among its negative impacts are ecological degradation, higher water demand, fire hazards, and biodiversity loss. This page provides a comprehensive analysis of the effects of IAPS with the assistance of several studies and references. India, which is categorized as mega-biodiverse and has a rapidly expanding economy, is home to four of the world's biodiversity hotspots (Myers et al., 2000). Despite this, there is currently no national invasive alien species (IAS) synthesis. India now has thousands of non-native species (Rajak et al., 2021; Singh et al., 2023). There are 173 invasive plant species (Reddy, 2008; Botanical Survey of India) and 157 invasive animal species (Government of India 2018), according to cautious estimates. Due to habitat changes brought on by India's fast urbanization, invasive plant species are more likely to proliferate (Afreen et al., 2023; Mallick et al., 2019). Urban environments are significantly impacted by these invasive species, which have an impact on infrastructure, public health, biodiversity, and the economy. Concern over the unchecked spread of invasive species in parks, roadsides, water bodies, and vacant spaces is growing among urban planners and environmentalists (Mishra et al., 2024; Rajak et al., 2025).

Economic Impact of *Prosopis juliflora* in Urban ecosystem of India

One of the most aggressive invasive plant species in India is *Prosopis juliflora*, sometimes referred to as Mesquite or Vilayati Babul. Although it was first brought in to help with afforestation and to fight desertification, it has now gone out of control. It is seriously harming the environment and the economy in a number of different areas (Hussain et al., 2021).

Economic Losses in Water Management

Groundwater Depletion: The deep root system of *Prosopis juliflora* extracts excessive amounts of groundwater, causing water scarcity in the arid and semi-arid regions of Rajasthan, Gujarat, and Maharashtra. This forces farmers and local communities to invest in deeper wells and water transport, increasing their financial burden.

Clogging of Water Bodies: The species aggressively colonizes riverbanks, lakes, and reservoirs, reducing water availability for drinking and irrigation purposes. This has led to expensive removal projects by local governments to clear water sources.

Infrastructure and Urban Economic Losses

Damage to Roads and Buildings: The aggressive root system of *Prosopis juliflora* causes cracks in pavements, roads, and drainage systems, leading to increased maintenance costs for municipalities and urban planners (Esquivel & Quijas, 2021).

Invasion of Urban Green Spaces: Public parks and vacant lands in cities like Delhi and Chennai have been overrun by *Prosopis juliflora*, reducing the aesthetic and recreational value of urban areas.

Cost of Eradication and Control Measures

The removal of *Prosopis juliflora* requires heavy machinery, manual labour, and herbicide treatments, making control efforts expensive and time-consuming. Government agencies and NGOs spend crores of rupees annually in eradication and afforestation programs to restore degraded lands (Dagar & Gupta, 2022). Despite efforts, the plant's rapid regeneration capacity makes complete eradication nearly impossible, leading to recurring control costs. Although *Prosopis juliflora* has some economic benefits, such as providing fuelwood and charcoal for rural populations, its negative impacts far outweigh the advantages (Eshetu, 2024). The widespread invasion of this species in India has led to significant economic losses in agriculture, water management, biodiversity conservation, infrastructure maintenance, and public health. Urgent action is needed to implement sustainable management practices, promote the use of native vegetation, and control the spread of this invasive species to minimize long-term economic impacts.

Economic Impact of *Leucaena leucocephala* in Urban Ecosystem

The time needed to locate and treat dispersed plants is a major factor in the \$5.9–\$14.7 million per hectare cost of leucaena infestations in Queensland. There haven't been any direct research on Leucaena's detrimental effects on primary output, however primary farmers in Queensland see it as a beneficial species. With the exception of low-risk elements like water and animals, non-graziers are unlikely to have plants sprout on their land. Nevertheless, two farmers reported that flood-waters had brought leucaena to their land. Leucaena can be controlled, and infected locations can be controlled (Shelton et al., 2001). Currently, little money is needed to regulate undesirable places, but if they expand, more will be needed. For dense infestations, using registered herbicides helps lower control expenses.

Management of IAPS

IAPS pose severe environmental, economic, and social challenges, requiring structured International frameworks like the Convention on Biological Diversity and the Sustainable Development Goals emphasize the importance of preventing the introduction of IAPS and controlling their impacts. Effective management requires a comprehensive approach that includes prevention, early detection, rapid response, containment, eradication, and impact mitigation. Control methods include mechanical, chemical, biological, and cultural or control by utilization techniques (Fig. 2), each tailored to specific species and environmental conditions (Simberloff et al., 2013).

Integrated pest management (IPM) is recommended, combining manual, mechanical, chemical, biological, and cultural control techniques. IAPS management requires cost-effective, evidence-based, and targeted strategies, integrating ecological restoration, public engagement, policy regulation, and interdisciplinary research (Shackleton et al., 2019). Challenges like, a lack of coordinated policies, research consistency, and adequate funding hinder effective management. An "adaptive collaborative restoration" approach aims to address these gaps by integrating continuous monitoring, risk assessment, stakeholder participation, and long-term ecosystem restoration. Rapid Response (EDRR) programs and Integrated Vegetation Management (IVM) are essential for managing invasive species like Japanese knotweed, which can impact urban ecology, infrastructure, biodiversity, and ecosystem services (Wagner et al., 2022). Public and private sector collaboration

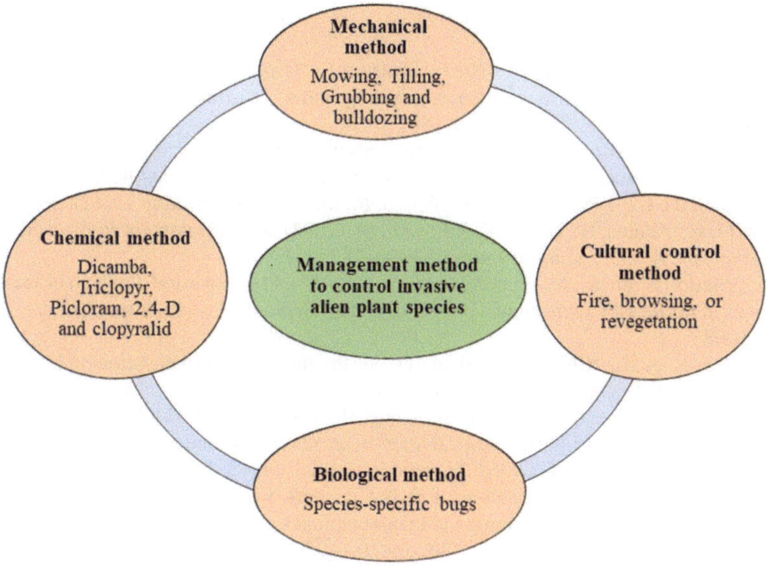

Fig. 2 Management of IAPS includes mechanical, chemical, biological and cultural methods

is crucial, with policies and legislation establishing regulations and guidelines to prevent and manage invasive species, promoting public health and safety.

11 Conclusions and Future Prospects

Invasive Alien Plant Species pose a significant threat to urban ecosystems in India, causing economic, environmental, and social impacts. These species disrupt native biodiversity, degrade infrastructure, and impose financial burdens on the city management. They also contribute to biodiversity loss, alter ecosystem services, and negatively affect human health. Their proliferation in urban areas leads to increased maintenance costs for green spaces and infrastructure, and decreased property values. Effective management of IAPS requires a multifaceted approach, including prevention, early detection, rapid response, and public awareness campaigns. Integrated Pest Management (IPM) strategies, which combine mechanical, chemical, biological, and cultural control methods, are recommended to address the complexities of IAPS management. Strategies include policy development, public engagement, research and monitoring, and international collaboration. By adopting these measures, urban areas can enhance their resilience against IAPS threats, ensuring the preservation of biodiversity and the well-being of urban communities.

References

Afreen, T., Rajak, P., & Singh, H. (2023). Invasion by *Hyptis suaveolens* modifies the effects of altered rainfall variability on nutrient cycling across seasons in a dry tropical grassland experiment. *Ecoscience, 30*(3–4), 170–186.

Alston, K. P., & Richardson, D. M. (2006). The roles of habitat features, disturbance, and distance from putative source populations in structuring alien plant invasions at the urban/wildland interface on the Cape Peninsula, South Africa. *Biological Conservation, 132*(2), 183–198.

Andersen, M. C., Adams, H., Hope, B., & Powell, M. (2004). Risk assessment for invasive species. *Risk Analysis: An International Journal, 24*(4), 787–793.

Ariori, C., Aiello-Lammens, M. E., & Silander, J. A. (2017). Plant invasion along an urban-to-rural gradient in northeast Connecticut. *Journal of Urban Ecology, 3*, 1–13. https://doi.org/10.1093/jue/jux008

Aronson, M. F. J., et al. (2014). A global analysis of the impacts of urbanization on bird and plant diversity reveals key anthropogenic drivers. *Proceedings of the Royal Society B: Biological Sciences, 281*(1780), 20133330. https://doi.org/10.1098/rspb.2013.3330

Aronson, M. F. J., et al. (2015a). Hierarchical filters determine community assembly of urban species pools. *Ecology, 97*(11), 2952–2963. https://doi.org/10.1890/15-0825.1

Aronson, M. F. J., Handel, S. N., La Puma, I. P., & Clemants, S. E. (2015b). Urbanization promotes non-native woody species and diverse plant assemblages in the New York metropolitan region. *Urban Ecosystems, 18*, 31–45. https://doi.org/10.1007/s11252-014-0382-z1

Bellard, C., Jeschke, J. M., Leroy, B., & Mace, G. M. (2018). Insights from modeling studies on how climate change affects invasive alien species geography. *Ecology and Evolution, 8*(11), 5688–5700.

Bellard, C., Thuiller, W., Leroy, B., Genovesi, P., Bakkenes, M., & Courchamp, F. (2013). Will climate change promote future invasions? *Global Change Biology, 19*(12), 3740–3748.

Blackburn, T. M., Essl, F., Evans, T., Hulme, P. E., Jeschke, J. M., Kühn, I., & Bacher, S. (2014). A unified classification of alien species based on the magnitude of their environmental impacts. *PLoS Biology, 12*(5), Article e1001850.

Blackburn, T. M., Pyšek, P., Bacher, S., Carlton, J. T., Duncan, R. P., Jarošík, V., & Richardson, D. M. (2011). A proposed unified framework for biological invasions. *Trends in Ecology & Evolution, 26*(7), 333–339.

Cadotte, M. W., Yasui, S. L. E., Livingstone, S., & MacIvor, J. S. (2017). Are urban systems beneficial, detrimental, or indifferent for biological invasion? *Biological Invasions, 19*, 3489–3503. https://doi.org/10.1007/s10530-017-1586-y

Capellini, I., Baker, J., Allen, W. L., Street, S. E., & Venditti, C. (2015). The role of life history traits in mammalian invasion success. *Ecology Letters, 18*(10), 1099–1107.

Catford, J. A., Daehler, C. C., Murphy, H. T., et al. (2012). The intermediate disturbance hypothesis and plant invasions: Implications for species richness and management. *Perspectives in Plant Ecology, Evolution and Systematics, 14*, 231–241. https://doi.org/10.1016/j.ppees.2011.12.002

Catford, J. A., Jansson, R., & Nilsson, C. (2009). Reducing redundancy in invasion ecology by integrating hypotheses into a single theoretical framework. *Diversity and Distributions, 15*, 22–40. https://doi.org/10.1111/j.1472-4642.2008.00521.x

Charles, H., & Dukes, J. S. (2007). Impacts of invasive species on ecosystem services. *Biological Invasions*, 217–237.

Colautti, R. I., & MacIsaac, H. J. (2004). A neutral terminology to define 'invasive' species. *Diversity and Distributions, 10*(2), 135–141.

Daba, M. H., & Dejene, S. W. (2018). The role of biodiversity and ecosystem services in carbon sequestration and its implication for climate change mitigation. *Environmental Sciences and Natural Resources, 11*(2), 1–10.

Dagar, J. C., & Gupta, S. R. (2022). Ecological restoration for achieving the goals of land degradation neutrality with special reference to India. *Journal of Soil Salinity and Water Quality, 14*(2), 161–174.

de Souza, T. A. F., de Andrade, L. A., Freitas, H., & da Silva, S. A. (2018). Biological invasion influences the outcome of plant-soil feedback in the invasive plant species from the Brazilian semi-arid. *Microbial Ecology, 76*, 102–112.

Diagne, C., Leroy, B., Vaissière, A. C., Gozlan, R. E., Roiz, D., Jarić, I., & Courchamp, F. (2021). High and rising economic costs of biological invasions worldwide. *Nature, 592*(7855), 571–576.

Docking, D. (2024). Invasive invaders: Japanese knotweed and non-native species in our homes. *Journal of Building Survey, Appraisal & Valuation, 13*(1), 46–56.

Dyderski, M. K., & Jagodziński, A. M. (2016). Patterns of plant invasions at small spatial scale correspond with that at the whole country scale. *Urban Ecosystems, 19*, 983–998. https://doi.org/10.1007/s11252-015-0524-y

Early, R., Bradley, B. A., Dukes, J. S., Lawler, J. J., Olden, J. D., Blumenthal, D. M., & Tatem, A. J. (2016). Global threats from invasive alien species in the twenty-first century and national response capacities. *Nature Communications, 7*(1), 12485.

Eiswerth, M. E., Darden, T. D., Johnson, W. S., Agapoff, J., & Harris, T. R. (2005). Input–output modeling, outdoor recreation, and the economic impacts of weeds. *Weed Science, 53*(1), 130–137.

Elton, C. S., & Elton, C. S. (1958). *The reasons for conservation* (pp. 143–153). Springer.

Eschen, R., Beale, T., Bonnin, J. M., Constantine, K. L., Duah, S., Finch, E. A., & Taylor, B. (2021). Towards estimating the economic cost of invasive alien species to African crop and livestock production. *CABI Agriculture and Bioscience, 2*, 1–18.

Eshetu, A. A. (2024). A valuable or a curse resource? A systematic review on expansion, perception of local community, benefits and side effects of Prosopis juliflora. *Frontiers in Conservation Science, 5*, 1491618.

Esquivel, T., & Quijas, S. (2021). Damage to infrastructure (pavement) caused by urban trees in Puerto Vallarta, Jalisco. *Revista Mexicana de Ciencias Forestales, 12*(68), 178–201.

Fratarcangeli, C., Fanelli, G., Franceschini, S., et al. (2019). Beyond the urban-rural gradient: Selforganizing map detects the nine landscape types of the city of Rome. *Urban for Urban Green, 38*, 354–370. https://doi.org/10.1016/j.ufug.2019.01.012

Fried, G., Chauvel, B., Reynaud, P., & Sache, I. (2017). Decreases in crop production by non-native weeds, pests, and pathogens. In *Impact of biological invasions on ecosystem services* (pp. 83–101).

Godefroid, S., & Koedam, N. (2007). Urban plant species patterns are highly driven by density and function of built-up areas. *Landscape Ecology, 22*, 1227–1239. https://doi.org/10.1007/s10980-007-9102-x

Hodkinson, D. J., & Thompson, K. (1997). Plant dispersal: The role of man. *Journal of Applied Ecology*, 1484–1496.

Holmes, T. P., Aukema, J. E., Von Holle, B., Liebhold, A., & Sills, E. (2009). Economic impacts of invasive species in forests: Past, present, and future. *Annals of the New York Academy of Sciences, 1162*(1), 18–38.

Hulme, P. E. (2017). Climate change and biological invasions: Evidence, expectations, and response options. *Biological Reviews, 92*(3), 1297–1313.

Hussain, M. I., Shackleton, R., El-Keblawy, A., González, L., & Trigo, M. M. (2021). Impact of the invasive Prosopis juliflora on terrestrial ecosystems. *Sustainable Agriculture Reviews, 52*, 223–278.

Imhoff, M. L., Zhang, P., Wolfe, R. E., & Bounoua, L. (2010). Remote sensing of the urban heat island effect across biomes in the continental USA. *Remote Sensing of Environment, 114*(3), 504–513. https://doi.org/10.1016/j.rse.2009.10.008

Johnson, A. L., Borowy, D., & Swan, C. M. (2018). Land use history and seed dispersal drive divergent plant community assembly patterns in urban vacant lots. *Journal of Applied Ecology, 55*, 451–460. https://doi.org/10.1111/1365-2664.12958

Kasulo, V. (2000). The impact of invasive species in African lakes. *The Economics of Biological Invasions*, 183.

Kohli, R. K., Batish, D. R., Singh, H. P., & Dogra, K. S. (2006). Status, invasiveness and environmental threats of three tropical American invasive weeds (*Parthenium hysterophorus* L., *Ageratum conyzoides* L., *Lantana camara* L.) in India. *Biological Invasions, 8*, 1501–1510.

Kowarik, I. (2011). Novel urban ecosystems, biodiversity, and conservation. *Environmental Pollution, 159*(8–9), 1974–1983.

Kriticos, D. J., & Brunel, S. (2016). Assessing and managing the current and future pest risk from water hyacinth, (*Eichhornia crassipes*), an invasive aquatic plant threatening the environment and water security. *PLoS ONE, 11*(8), Article e0120054.

Larkham, P. J. (2006). The study of urban form in Great Britain. *Urban Morphology, 10*, 117–141.

Lazzaro, L., Essl, F., Lugliè, A., Padedda, B. M., Pyšek, P., & Brundu, G. (2018). *Invasive alien plant impacts on human health and well-being.*

Linders, T. E. W., Schaffner, U., Eschen, R., Abebe, A., Choge, S. K., Nigatu, L., & Allan, E. (2019). Direct and indirect effects of invasive species: Biodiversity loss is a major mechanism by which an invasive tree affects ecosystem functioning. *Journal of Ecology, 107*(6), 2660–2672.

Lososová, Z., Horsák, M., Chytrý, M., et al. (2011). Diversity of Central European urban biota: Effects of human-made habitat types on plants and land snails. *Journal of Biogeography, 38*, 1152–1163. https://doi.org/10.1111/j.1365-2699.2011.02475.x

Mallick, S. N., Ekka, N. X., Kumar, S., & Sahu, S. C. (2019). Invasive alien flora in and around an urban area of India. In *Diversity and ecology of invasive plants*. IntechOpen.

Marco, A., Lavergne, S., Dutoit, T., & Bertaudiere-Montes, V. (2010). From the backyard to the backcountry: How ecological and biological traits explain the escape of garden plants into Mediterranean old fields. *Biological Invasions, 12*, 761–779. https://doi.org/10.1007/s10530-009-9

Matthews, H. D., Zickfeld, K., Knutti, R., & Allen, M. R. (2018). Focus on cumulative emissions, global carbon budgets and the implications for climate mitigation targets. *Environmental Research Letters, 13*(1), Article 010201.

McKinney, M. L. (2006). Urbanization as a major cause of biotic homogenization. *Biological Conservation, 127*(3), 247–260. https://doi.org/10.1016/j.biocon.2005.09.005

Mehraj, G., Khuroo, A. A., Qureshi, S., Muzafar, I., Friedman, C. R., & Rashid, I. (2018). Patterns of alien plant diversity in the urban landscapes of global biodiversity hotspots: A case study from the Himalayas. *Biodiversity and Conservation, 27*, 1055–1072.

Mishra, A., Rajak, P., Raghubanshi, A. S., & Singh, H. (2024). Understanding competition between two invasive woody plants of India under an altered rainfall regime. *Plant Ecology,* 1–19.

Moss, S. (2019). Integrated weed management (IWM): Why are farmers reluctant to adopt non-chemical alternatives to herbicides? *Pest Management Science, 75*(5), 1205–1211.

Myers, N., Mittermeier, R. A., Mittermeier, C. G., Da Fonseca, G. A., & Kent, J. (2000). Biodiversity hotspots for conservation priorities. *Nature, 403*(6772), 853–858.

Nghiem, L. T., Soliman, T., Yeo, D. C., Tan, H. T., Evans, T. A., Mumford, J. D., & Carrasco, L. R. (2013). Economic and environmental impacts of harmful non-indigenous species in Southeast Asia. *PLoS ONE, 8*(8), Article e71255.

Padayachee, A. L., Irlich, U. M., Faulkner, K. T., Gaertner, M., Procheş, Ş, Wilson, J. R., & Rouget, M. (2017). How do invasive species travel to and through urban environments? *Biological Invasions, 19*, 3557–3570.

Palmer, W. A., & Pullen, K. R. (1995). The phytophagous arthropods associated with Lantana camara, L. hirsuta, L. urticifolia, and L. urticoides (Verbenaceae) in North America. *Biological Control, 5*(1), 54–72.

Panzavolta, T., Bracalini, M., Benigno, A., & Moricca, S. (2021). Alien invasive pathogens and pests harming trees, forests, and plantations: Pathways, global consequences and management. *Forests, 12*(10), 1364.

Parsons, W. T., & Cuthbertson, E. G. (2001). *Noxious weeds of Australia.* CSIRO publishing.

Pejchar, L., & Mooney, H. A. (2009). Invasive species, ecosystem services and human well-being. *Trends in Ecology & Evolution, 24*(9), 497–504.

Perrault, A., Bennett, M., Burgiel, S., Delach, A., & Muffett, C. (2003). *Invasive species, agriculture and trade: Case studies from the NAFTA context.* North American Commission for Environmental Cooperation Working Paper.

Pickett, S. T., Cadenasso, M. L., Grove, J. M., Boone, C. G., Groffman, P. M., Irwin, E., & Warren, P. (2011). Urban ecological systems: Scientific foundations and a decade of progress. *Journal of Environmental Management, 92*(3), 331–362.

Pickett, S. T., Cadenasso, M. L., Grove, J. M., Nilon, C. H., Pouyat, R. V., Zipperer, W. C., & Costanza, R. (2001). Urban ecological systems: Linking terrestrial ecological, physical, and socioeconomic components of metropolitan areas. *Annual Review of Ecology and Systematics, 32*(1), 127–157.

Pickett, S. T., & Grove, J. M. (2009). Urban ecosystems: What would Tansley do? *Urban Ecosystems, 12*, 1–8.

Pimentel, D., Lach, L., Zuniga, R., & Morrison, D. (2000). Environmental and economic costs of nonindigenous species in the United States. *BioScience, 50*(1), 53–65.

Pimentel, D., Zuniga, R., & Morrison, D. (2005). Update on the environmental and economic costs associated with alien-invasive species in the United States. *Ecological Economics, 52*(3), 273–288.

Puth, L. M., & Post, D. M. (2005). Studying invasion: Have we missed the boat? *Ecology Letters, 8*(7), 715–721.

Pyšek, P. (1998). Alien and native species in Central European urban floras: A quantitative comparison. *Journal of Biogeography, 25*(1), 155–163.

Pyšek, P., & Hulme, P. E. (2005). Spatio-temporal dynamics of plant invasions: Linking pattern to process. *Ecoscience, 12*(3), 302–315.

Pyšek, P., Hulme, P. E., Simberloff, D., Bacher, S., Blackburn, T. M., Carlton, J. T., & Richardson, D. M. (2020). Scientists' warning on invasive alien species. *Biological Reviews, 95*(6), 1511–1534.
Pyšek, P., & Richardson, D. M. (2010). Invasive species, environmental change and management, and health. *Annual Review of Environment and Resources, 35*, 25–55. https://doi.org/10.1146/annurev-environ-033009-095548
Rajak, P., Afreen, T., Raghubanshi, A. S., & Singh, H. (2025). Rainfall fluctuation causes the invasive plant Prosopis juliflora to adapt ecophysiologically and change phenotypically. *Environmental Monitoring and Assessment, 197*(1), 1–17.
Rajak, P., Afreen, T., & Singh, H. (2021). Assessing the impact of invasive alien plant species on environment, ecosystem services and human health. *Plant Archives, 21*(1), 2155–2163.
Ramalho, C. E., & Hobbs, R. J. (2012). Time for a change: Dynamic urban ecology. *Trends in Ecology & Evolution, 27*(3), 179–188.
Rebele, F. (1994). Urban ecology and special features of urban ecosystems. *Global Ecology and Biogeography Letters,* 173–187.
Reddy, C. S. (2008). Catalogue of invasive alien flora of India. *Life Science Journal, 5*(2), 84–89.
Reddy, C. S., & Bagyanarayana, G. (2007). Invasion of exotic species: Threat to the biodiversity. In *Proceedings of the National Seminar on Conservation of Eastern Ghats* (pp. 40–44).
Reid, W. V., Mooney, H. A., Cropper, A., Capistrano, D., Carpenter, S. R., Chopra, K., & Zurek, M. B. (2005). *Ecosystems and human well-being-Synthesis: A report of the millennium ecosystem assessment*. Island Press.
Ruiz-Nieto, J. E., Hernández-Ruiz, J., Hernández-Marín, J., Mendoza-Carrillo, J., Abraham-Juárez, M., Isiordia-Lachica, P. M., & Mireles-Arriaga, A. I. (2020). Mesquite (Prosopis spp.) tree as a feed resource for animal growth. *Agroforestry Systems, 94*(4), 1139–1149.
Säumel, I., & Kowarik, I. (2013). Propagule morphology and river characteristics shape secondary water dispersal in tree species. *Plant Ecology, 214*, 1257–1272.
Schmitz, O. J., Beckerman, A. P., & O'Brien, K. M. (1997). Behaviorally mediated trophic cascades: Effects of predation risk on food web interactions. *Ecology, 78*(5), 1388–1399.
Shabbir, A., Bajwa, A. A., Mao, R., Kezar, S., Dorji, S., & Adkins, S. W. (2024). Biology of invasive plants 6. *Parthenium hysterophorus* L. *Invasive Plant Science and Management, 17*(3), 129–156.
Shackleton, R. T., Richardson, D. M., Shackleton, C. M., Bennett, B., Crowley, S. L., Dehnen-Schmutz, K., & Larson, B. M. (2019). Explaining people's perceptions of invasive alien species: A conceptual framework. *Journal of Environmental Management, 229*, 10–26. https://doi.org/10.1016/j.jenvman.2018.04.045
Shackleton, R. T., Witt, A. B., Piroris, F. M., & van Wilgen, B. W. (2017). Distribution and socio-ecological impacts of the invasive alien cactus Opuntia stricta in eastern Africa. *Biological Invasions, 19*, 2427–2441.
Shah, N. C. (2014). Tamarindus indica: Introduction in India and culinary, medicinal, and industrial uses. *Asian Agrihist, 18*(4), 343–355.
Sharma, G. P., Raghubanshi, A. S., & Singh, J. S. (2005). Lantana invasion: An overview. *Weed Biology and Management, 5*(4), 157–165.
Shelton, M., McNeill, F., Dalzell, S., & Graham, G. (2001). *An assessment of the weed status and management of Leucaena leucocephala (Lam.) de Wit in Queensland*. Report provided to Queensland Department of Natural Resources and Mines, Queensland DPI. Environmental Protection Agency, Queensland, by The University of Queensland, Brisbane.
Shi, F., Meng, Q., Pan, L., & Wang, J. (2023). Root damage of street trees in urban environments: An overview of its hazards, causes, and prevention and control measures. *Science of the Total Environment, 904*, Article 166728.
Simberloff, D., Martin, J. L., Genovesi, P., Maris, V., Wardle, D. A., Aronson, J., & Vilà, M. (2013). Impacts of biological invasions: What's what and the way forward. *Trends in Ecology & Evolution, 28*(1), 58–66. https://doi.org/10.1016/j.tree.2012.07.013

Singh, R., Singh, A. K., Rajak, P., & Singh, H. (2023). Plant invasion and climate change: An overview on history, impacts, and management practices. *Plant Invasions and Global Climate Change*, 343–365.

Sladonja, B., Sušek, M., & Guillermic, J. (2015). Review on invasive tree of heaven (*Ailanthus altissima* (Mill.) Swingle) conflicting values: Assessment of its ecosystem services and potential biological threat. *Environmental Management, 56*, 1009–1034.

Tallent-Halsell, N. G., & Watt, M. S. (2009). The invasive *Buddleja davidii* (butterfly bush). *The Botanical Review, 75*, 292–325.

Van Ham, C., Genovesi, P., & Scalera, R. (2013). *Invasive alien species: The urban dimension. Case studies on strengthening local action in Europe.* IUCN European Union Representative Office.

Van Kleunen, M., Dawson, W., Essl, F., Pergl, J., Winter, M., Weber, E., & Pyšek, P. (2015). Global exchange and accumulation of non-native plants. *Nature, 525*(7567), 100–103.

van Rees, C. B., Hand, B. K., Carter, S. C., Bargeron, C., Cline, T. J., Daniel, W., & Luikart, G. (2022). A framework to integrate innovations in invasion science for proactive management. *Biological Reviews, 97*(4), 1712–1735. https://doi.org/10.1002/eap.1511

Vilà, M., et al. (2011). Ecological impacts of invasive alien plants: A meta-analysis of their effects on species, communities and ecosystems. *Ecology Letters, 14*(7), 702–708. https://doi.org/10.1111/j.1461-0248.2011.01628.x

Vilà, M., Basnou, C., Pyšek, P., Josefsson, M., Genovesi, P., Gollasch, S., DAISIE partners. (2010). How well do we understand the impacts of alien species on ecosystem services? A pan-European, cross-taxa assessment. *Frontiers in Ecology and the Environment, 8*(3), 135–144.

Vilà, M., & Hulme, P. E. (Eds.). (2017). *Impact of biological invasions on ecosystem services* (Vol. 12). Springer.

Von der Lippe, M., & Kowarik, I. (2008). Do cities export biodiversity? Traffic as dispersal vector across urban–rural gradients. *Diversity and Distributions, 14*(1), 18–25.

Xu, H., Ding, H., Li, M., Qiang, S., Guo, J., Han, Z., & Wan, F. (2006). The distribution and economic losses of alien species invasion to China. *Biological Invasions, 8*, 1495–1500.

Zavaleta, E. (2000). The economic value of controlling an invasive shrub. *AMBIO: A Journal of the Human Environment, 29*(8), 462–467.

Chapter 10
Invasive Alien Weeds in the Urban Environment: A Case Study of Chandigarh

Shalini Ojha, Pooja Rani, Bhupender Singh, Rajat Kumar, Khushwinder Singh, Amrit Pal, Rishikesh Singh, Shalinder Kaur, and Daizy R. Batish

Abstract Urban ecosystems, though hubs of human activity and biodiversity, are increasingly threatened by the proliferation of invasive plant species, particularly in fully urbanized regions like Chandigarh, India. This chapter delves into the complex ecological, socio-economic, and health-related impacts posed by these species in an urban context. Chandigarh, renowned for its planned urban landscape and green infrastructure, has paradoxically become a hotspot for numerous invasive species, including *Ageratum conyzoides, Broussonetia papyrifera, Lantana camara, Leucaena leucocephala, Mesosphaerum suaveolens, Neltuma juliflora, Parthenium hysterophorus, Pontederia crassipes, Ricinus communis,* etc. These species not only outcompete native flora through aggressive reproduction, allelopathy, and ecological plasticity, but also alter soil composition, disrupt trophic interactions, and impair urban water systems. The study documents the patterns of invasion, ecological traits, and dispersal mechanisms of these species, with a focus on localized case studies within Chandigarh. It further assesses the profound implications for urban biodiversity, public health, agriculture, and infrastructure, while evaluating current control measures including biological, chemical, and mechanical interventions. Emphasis is placed on the necessity of integrated weed management, restoration ecology, interdisciplinary research, and policy implementation to mitigate threats of invasive plant species and preserve the ecological integrity of urban landscapes.

Pooja Rani, Bhupender Singh, Rajat Kumar, Khushwinder Singh and Amrit Pal these authors contributed equally.

S. Ojha · P. Rani · B. Singh · K. Singh · A. Pal · S. Kaur · D. R. Batish (✉)
Department of Botany, Panjab University, Chandigarh, India
e-mail: daizybatish@yahoo.com

R. Kumar · R. Singh
Amity School of Earth and Environmental Sciences, Amity University Punjab, Mohali, Punjab, India

R. Singh et al. (eds.), *Invasive Alien Plants in Urban Ecosystems*,
https://doi.org/10.1007/978-3-032-00997-5_10

Keywords Biological invasion · Ecological impacts · Green spaces · Invasive plants · Urban landscape · Union Territory

1 Introduction

Plant invasions, primarily driven by human-mediated introductions of non-native species, pose significant ecological, economic, and social threats worldwide (Gallardo & Aldridge, 2013). With Urban areas currently home to over 55% of the world's population, a number set to reach ~ 68% by 2050 (WHO, 2025), this has led to land-use changes including habitat loss and fragmentation (Reynolds et al., 2019). These urban areas create novel disturbance regimes and favourable conditions for non-native species to establish (Gaertner et al., 2017). As a result, non-native plants are now an integral part of the urban vegetation all over the world, often spreading into adjacent natural ecosystems where they become invasive and ecologically disruptive.

The expansion of invasive alien vegetation in urban environments not only affects the viability of indigenous vegetation but also greatly influences the ecological environment and human life in local areas (Potgieter et al., 2022). Although much research has focused on invasions in forest and agricultural ecosystems, urban areas, where the majority of plants originate, remain comparatively understudied (Staudhammer et al., 2015). Urban green spaces such as parks, gardens, and natural areas are key hotspots for invasive plants (Gaertner et al., 2017). Urbanization-driven land use changes, such as the development of green belts and landscape gardens, often inadvertently introduce exotic species. Additionally, extensive-use transport and logistics systems in cities and night-time light pollution due to the installation of streetlights during urbanization further aid in the dispersal of exotic species (Murphy et al., 2022). Warmer climates, ample nutrient availability, and limited natural predators in urban man-made habitats are also favorable to the growth and proliferation of invasive alien vegetation (Gaertner et al., 2017). Thus, it is necessary to understand the distribution patterns and ecological impacts in complex urban settings for informed decision-making and effective invasion management.

In India, the invasive alien plants have gained more attention over the last few decades, especially concerning biodiversity loss, ecosystem degradation, and socio-economic effects (Kohli et al., 2006a). While a lot of research has been done on forested and agricultural ecosystems, urban invasive ecology is still a bit of a blind spot. Cities like Chandigarh, lauded for their green infrastructure and extensive urban forestry, are quietly grappling with a growing threat from urban weeds, many of which are invasive alien species (Fig. 1) (Batish et al., 2004a, 2004b).

Chandigarh, the joint capital of Punjab and Haryana, is a meticulously planned city, known for its lush green spaces, including parks, gardens, roadside plantations, and forest areas (Chalana, 2015). While, these green spots are crucial for the city's resilience and beauty, but they also unintentionally become hotspots for invasions, hosting a range of alien plant species such as *Ageratum conyzoides, Broussonetia papyrifera, Lantana camara, Leucaena leucocephala, Mesosphaerum suaveolens,*

(a)

(b)

Fig. 1 Chandigarh's contrasting landscapes: **a** a symbol of Green Oasis **b** invasive plant outgrowth

Neltuma juliflora, Parthenium hysterophorus, and Pontederia crassipes (Fig. 2) (Batish et al., 2004a, 2004b; Maan et al., 2021; Sharma et al., 2017). These plants thrive in disturbed, nutrient-rich urban environments, giving them a competitive edge over native or less aggressive species.

Urban plant invasions are caused by a combination of factors, like global trade, ornamental horticulture, changes in landscaping, and climate change (Essl et al., 2011). In Chandigarh, for instance, roadside plantations and public gardens often feature ornamental exotics, some of which manage to escape cultivation and spread into natural or semi-natural areas (Maan et al., 2021). Moreover, frequent urban disturbances, such as road widening, construction, and soil excavation, further facilitate these invasions.

Fig. 2 Chandigarh's prominent invasive weeds: an illustrated depiction

The impact of urban plant invasions on the ecosystem is significant. Invasive weeds can outcompete native species for essential resources like light, nutrients, and space; they can also change soil chemistry, affect water flow, and disrupt trophic interactions (Ehrenfeld, 2010). For instance, *Parthenium hysterophorus* releases allelochemicals that inhibit the growth of native plants, while *Lantana camara* creates dense patches that diminish the diversity of the understory (Kohli et al., 2006a). Beyond ecological issues, many invasive plants in urban areas are associated with health problems, damage to infrastructure, and rising costs for weed management (Mazza et al., 2014). Despite these challenges, the study of urban invasion ecology in India is still quite limited, particularly when it comes to city-specific data (Gaertner et al., 2017). Localized studies that document the composition, spread, and ecological impacts of invasive alien plants are crucial for developing early warning systems, control strategies, and conservation efforts for urban biodiversity (Maan et al., 2020). Given its organized layout, rich vegetation, and urban biodiversity initiatives, Chandigarh presents a unique model to examine urban weed dynamics and develop effective management strategies.

This chapter intends to address this gap by studying diversity, ecological traits, introduction pathways, and the spread of invasive plant species in Chandigarh. It also outlines key ecological challenges of urban invasives and offers strategies to support resilient, evidence-based urban management.

2 Urban Ecology and Biodiversity of Chandigarh

2.1 The Urban Landscape

Chandigarh city is located in the foothills of the Shivalik range and was established as India's Union Territory (UT) region. Known as the 'City Beautiful', Chandigarh was planned by French Architect Le Corbusier, and is admired for its unique concept, modern architecture, and presence of 1400 greenbelts/gardens/parks (TERI, 2024). The city's total area is 114 sq. km, and there are about 22 villages that fall under its jurisdiction. It is famous for wide tree-lined roads, well-maintained gardens, unique architecture, greenery, the Rock Garden, Sukhna Lake, and the quality of life for its citizens. Chandigarh has 35 km^2 of recorded forest area, out of which 3 km^2 is Unclassed Forests and a 32 km^2 is Reserved Forest. The urban forests in the UT are of utmost importance to maintain the ecological balance of the region (TERI, 2024).

A well-developed urban forestry, strategic tree plantations, and its proximity to the Shivalik Hills and lower Himalayas have contributed to Chandigarh's rich biodiversity (TERI, 2024). Chandigarh was planned to have open spaces, landscaped areas, recreational areas, and tree-lined roads, avenues, and gardens spread all over the city (Anonymous, 2025). Tree plantations along roads, open spaces, green belts, and around building complexes are among the most captivating features of the City's landscaping. These features not only enhance the aesthetic value but also serve ecological functions.

However, the recent trends and analysis indicate that Chandigarh is struggling to balance economic, environmental, and social sustainability. Today, the periphery of Chandigarh is characterized by unregulated construction and rapid urbanization, which leads to habitat fragmentation. The introduction of invasive plant species has significantly altered its ecological dynamics (Bedi et al., 2020). Urbanization also reduces the richness of plant species and pollinators' availability, which homogenizes the environment, promotes the invasion of non-native species, causes changes in phenological events, and presents difficulties for plant growth (de Barros Ruas et al., 2022).

2.2 Role of Green Spaces in Supporting Urban Biodiversity

Green spaces in Chandigarh, such as the Sukhna Wildlife Sanctuary, the Rose Garden, and the Rock Garden, provide habitats for both native and exotic species. These areas help maintain ecological balance and support biodiversity within a densely populated urban landscape and serve as havens where native and exotic species can coexist in harmony. According to Gao et al. (2021), they function as interconnecting corridors that facilitate wildlife movement between fragmented habitats, fostering gene flow that is crucial for maintaining genetic variety and adaptation. Diversity in plant and animal species enhance environmental health by enhancing biodiversity and fostering a more sustainable urban landscape (Edeigba et al., 2024). Native plant species are commonly found in the microenvironments of these green spaces and have adapted to the local climate, soil conditions, and ecosystems over time. In addition, they serve as refuges in the city, providing both biodiversity and pollution mitigation and preventing the introduction of non-native species (Theodorou, 2022).

2.3 Key Challenges to Native Flora

Despite its rich urban flora, Chandigarh faces significant challenges in conserving native plant species. The fast-growing weeds like *Parthenium hysterophorus* and *Ageratum conyzoides* outcompete native plants, reducing biodiversity (Batish et al., 2012; Kohli et al., 2006a). Furthermore, construction and road expansions have destroyed natural vegetation, pushing native species to the brink. As a result, changes in habitat structure, soil, and microclimate conditions induced by invasion and/or fragmentation alter habitat quality for native flora and, through bottom-up mechanisms, higher trophic levels, and modify biotic interactions (Pehle & Schirmel, 2015).

3 Prominent Invasive Weeds in Chandigarh: Diversity and Ecology

3.1 Terrestrial Herbaceous Weeds

3.1.1 *Parthenium hysterophorus* L.

- **Origin and Spread**

Parthenium hysterophorus L. is a highly invasive alien plant species commonly known as parthenium weed, Congress grass, or Carrot grass, that presents a significant danger to biodiversity, agriculture, and human health worldwide (Adkins & Shabbir, 2014). It is one of the fiercest invasive alien species due to factors like impressive reproductive capacity, allelopathic properties, and ability to colonize disturbed habitats (Kohli et al., 2006b). This species of *Parthenium* is native to the American tropics and subtropics, but has spread to over 40 countries largely through international trade practices (Shabbir et al., 2024).

It is guessed that in India, the introduction *of Parthenium hysterophorus* is through the PL 480 grain import program from the United States during the 1950s, with contaminated wheat consignments (Kumar & Aggarwal, 2024). Thereafter, it has established itself in almost all Indian states and is one of the worst alien weeds in the nation (Kaur et al., 2014). In urban settlements like Chandigarh, it has been reported to proliferate in vacant plots, roadside and sub-urban areas, competing with cultivated as well as wild flora. This species was first described in India by botanist William Roxburgh in *Hortus Bengalensis* in 1814, but it got recognition in the 1950s when it started to spread invasively (Batish et al., 2012).

- **Distribution (Global, India, Chandigarh)**

Parthenium hysterophorus, which is native to the Gulf of Mexico, Central America, and the Caribbean, has now expanded across Africa, Asia, and Australia (Fig. 3a) (Adkins et al., 1996). It has expanded to over 20 countries and continues to do so because of its properties like tolerance to different soil types, high seed viability, and resistance to drought and heat (Shabbir et al., 2024). In India (Fig. 4a), the weed has established itself widely, with significant infestations reported in states such as Punjab, Haryana, Himachal Pradesh, Karnataka, Tamil Nadu, and Uttar Pradesh (Kohli et al., 2006c).

Field surveys in the city of Chandigarh further confirm *Parthenium hysterophorus* as one of the most prominent invasive weeds in open urban spaces, specifically vacant plots, where it competes with native flora (Batish et al., 2002).

- **Mechanisms of Invasion**

Parthenium hysterophorus displays an exceptional ability to grow rapidly across different types of habitats. It engages several seed dispersal mechanisms such as wind-borne seeds, water runoff, animal fur, human clothing, and contaminated agricultural machinery (Kohli et al., 2006c).

- *Seed Production and Dispersal*: Under favourable conditions, a single *Parthenium* plant has the capacity to produce 15,000–20,000 seeds per growing season, with some exceptional individuals producing close to 100,000 seeds (Adkins & Shabbir, 2014). These seeds are small (~ 2 mm), lightweight (0.04 mg), and non-dormant, allowing them to germinate quickly under favourable conditions.
- *Human-Mediated Dispersal*
  - Contaminated Agricultural Produce: The introduction of this invasive weed is linked to wheat imports under the PL 480 program (Kumar & Aggarwal, 2024). Today, crop seeds, fodder, and soil transported from affected areas often contain *Parthenium* seeds (Kaur et al., 2014).
  - Construction Activities: At infrastructure development sites, regular movement of soil, debris, and construction material provide the ideal condition for the dispersal of *Parthenium* seeds to new locations.
  - Transportation Corridors: *Parthenium* has been found proliferating along national and state highways, railway tracks, and pedestrian trails, with vehicular movement facilitating seed dispersal.

- *Soil Seed Bank Persistence: Parthenium hysterophorus* seeds can show viability up to 6 years under natural conditions, establishing a tenacious soil seed bank (Adkins & Shabbir, 2014). Studies in India indicate that over 80% of seeds show viability for the first two years, supporting regrowth after monsoon or disturbance.

Even after complete mechanical removal in a few urban parks, seedlings regenerated within 2–3 weeks following rainfall, confirming a robust seed bank (Batish et al., 2002).

- *Climatic adaptability: Parthenium hysterophorus* is a highly tolerant and adaptable species thriving across a wide range of climatic conditions, temperatures from 20 to 35 °C, and resists both drought and mild flooding. *Parthenium* has adapted well in the subtropical climate of Chandigarh (average rainfall ~ 1100 mm and temperatures from 4 to 45 °C), showing year-round presence, with peak growth during monsoon (July to September) and flowering extending to late autumn.
- *Absence of Natural Enemies*: In India, the lack of natural predators or pathogens allows *Parthenium* populations to grow unchecked. Unlike in its native range, natural herbivores regulate its spread. To address this, four Indian research institutes have initiated Integrated Pest Management (IPM) programs against *Parthenium hysterophorus.* While chemical and myco-herbicides based on indigenous diseases have been examined, but their efficacy has been limited due to the economic constraints and a lack of suitable exploitable fungi (Kaur et al., 2014).
- **Control Efforts and Case Studies**

  - Eucalypt oil derived from *Eucalyptus globulus* and *E. citriodora* has shown inhibiting effects on the germination of *Parthenium hysterophorus.* Also, it significantly reduced weeds' cellular respiration and chlorophyll content, indicating their potential as natural herbicides (Kohli et al., 1998).
  - A study conducted in Gorakhpur, Uttar Pradesh, examined the distribution and management strategies of *Parthenium hysterophorus*, and it was concluded that manual uprooting before the beginning of the reproductive phase of the plant is one of the effective methods for the management of this weed. As *Parthenium* readily colonizes both managed and natural grasslands and germinates throughout the year, this approach is particularly relevant in urban and semi-urban regions where chemical control may not be feasible (Tripathi et al., 2023).
  - In another study conducted along the roadsides in an area of 5 km^2 near Attibele–Sarjapura road (948A) in Anekal taluk of Bengaluru Urban district, Karnataka, a 70–80% disease occurrence by *Puccinia abrupta* var. *partheniicola* (winter rust) in large tracts of *Parthenium hysterophorus* was found, suggesting a great addition to the biocontrol agent's armoury for *Parthenium* weed (Kumar, 2024).

3.1.2 *Ageratum conyzoides* L.

- **Origin and Spread**

Ageratum conyzoides L. is a tenacious invasive weed originally found in Central and South America and the Caribbean. It is commonly called Billygoat weed, Goatweed, or Whiteweed. It has now proliferated to Asia, Africa, and the Pacific Islands (Vélez-Gavilán, 2016). It exhibits properties such as high ecological plasticity, allelopathy, rapid growth, and prolific seed production, aiding its success as an invader (Batish et al., 2004a, 2004b). It was identified as an invasive weed in the early 20th century in India. Originally found along forest edges and wastelands, its spread was promoted due to anthropogenic activities. Since then, it has spread rapidly in agricultural and urban areas.

- **Distribution (Global, India, Chandigarh)**

Ageratum conyzoides, which is native to Central and South America, has now spread to almost 70 countries across the tropics and subtropics (Fig. 3b). In India, it has spread to almost all agriculturally fitting climatic zones, particularly in the Himalayan Foothills, northeastern states, and the Indo-Gangetic plains. A prominent presence has been observed in Himachal Pradesh, Uttarakhand, and Punjab (Fig. 4b), where it is found in degraded forest lands, urban parks, agricultural and roadside margins (Vélez-Gavilán, 2016).

The plant is found in food gardens, forest/plantation areas, water channels, and wastelands in Chandigarh, as well as in the ratoon sugarcane crop fields, maize fields, and wastelands of Himachal Pradesh, India (Kaur et al., 2012).

- **Mechanisms of Invasion**

  Ageratum conyzoides propagates rapidly due to its properties like ecological plasticity, efficient seed dispersal, and a combination of sexual and asexual reproductive strategies:

  - *Seed Production and Longevity*: Seeds of *Ageratum conyzoides* show high germination rates and viability for multiple years in the soil seed bank, aiding in regermination under favourable conditions. A single plant can produce more than 40,000 small and lightweight seeds in a single season (Singh et al., 2002).
  - *Wind and Water Dispersal*: *Ageratum conyzoides* produces seeds with a pappus-like structure, which aids in dispersal by wind up to significant distances. During the rainy season, runoff water carries these seeds into agricultural fields and water bodies (Sharma & Raghubanshi, 2007).
  - *Human Mediated Spread*: Agricultural equipment and machinery, construction vehicles, and other infrastructure-related equipment significantly aid in its seed dispersal. In urban and sub-urban areas, its seeds are frequently dispersed due to traffic along railway tracks and the roadside.
  - *Livestock Movement*: Seeds are dispersed by sticking to animal fur and hooves, further aiding their spread in grazing lands and forests (Kaur et al., 2012).
  - *Disturbed Habitat Colonization*: *Ageratum conyzoides* prospers in disturbed habitats like construction sites, overgrazed grasslands, etc. The weed spreads rapidly in such sites and can establish dominance in a single growing season (Batish et al., 2004a, 2004b).

- **Control Measures**

  - Various methods have been proposed and practiced by agronomists and natural resource professionals for weed control. It is possible to physically control

plants through uprooting, burning, or cutting, but these methods may not be useful if the area is large, and seeds have already been planted (Batish et al., 2004a, 2004b).
- Furthermore, it can be reduced by using conservation tillage systems, flooding the field, and hand-weeding (Basavaraj & Reddy, 2017).
- To control *A. conyzoides*, agroecosystems employ preemergence herbicides (oxadiazon, atrazine, oxyfluorfen, diuron, methazole, simazine, etc.) and postemergence herbicides (glyphosate, 2, 4-D, etc.). However, these herbicides pose environmental risks and may result in the production of resistant species (Batish et al., 2004a, 2004b).
- Monoterpenes, such as cineole and citronellol, as well as natural plant products, such as parthenin and volatile essential oils from *Callistemon viminalis*, are effective against *A. conyzoides* (Motmainna et al., 2021).

3.1.3 *Mesosphaerum suaveolens* (L.) Poit.

- **Origin and Spread**

Mesosphaerum suaveolens (L.) Poit., synonym *Hyptis suaveolens,* commonly known as Vilayti Tulsi, is a genus of Lamiaceae (mint family) with about 400 species, and most of the species are native to tropical America (Raizada, 2006). Although *Hyptis* has several medicinal properties (Jesus et al., 2013) and is used in folklore remedies, its spread is so fast that over a lapse of time after its establishment, it upsets the natural ecological patterns in the areas of invasion (Padalia et al., 2013).

- **Distribution (Global, India, Chandigarh)**

It is a pan-tropical weed that has spread from the tropics to the sub-tropics. The widespread invasion of *Hyptis* was reported from the Northern Territory in Australia (Fig. 3c), Vidhyans in Central India (Sharma et al., 2017), and the Himalayan foothills (Padalia et al., 2013). It is a ruderal weed capable of colonizing disturbed lands. Although *Hyptis* has several medicinal properties and is used in folklore remedies, its rapid spread over time significantly upsets the natural ecological pattern in the invaded areas. In India, *Hyptis* occurrence is reported from North-East India, Vindhyans, Deccan Peninsula, and Andaman and Nicobar Islands (Fig. 4c) (Padalia et al., 2013).

Chandigarh possesses a favourable climate for this species, and due to urban sprawl and expansion of satellite towns, the peri-urban areas around Chandigarh are degrading fast, thereby paving the way for *H. suaveolens* invasion (Sharma et al., 2017).

- **Mechanisms of Invasion**

The plant forms large thickets and is believed to produce allelochemicals, which impede the seed germination of native species. It is threatening the course of natural succession, the existence of vulnerable taxa, and wild animals in the invaded areas (Padalia et al., 2013).

- **Negative Impacts**
  - It displaces useful species like *Abutilon indicum, Justicia adhatoda, Anisomeles indica, Barleria cristata, Carissa carandas, Dioscorea deltoidea, Murraya koenigii*, and *Paspalidium flavidum*, which also affect the local people (Sharma et al., 2017).
  - It negatively affects plant diversity and soil properties of a grassland, and severely reduces species number, richness, diversity, and evenness (Afreen et al., 2018).
  - The phytotoxicity of *Hyptis* restricts the growth and establishment of other plants near it (Raizada, 2006).
- **Control Measures**
  - It has been stated by many researchers that the use of chemicals provides the most effective method to control *H. suaveolens*.
  - Pandey and Pandey (2009) isolated a fungus, *Phoma herbarium*, from diseased *H. suaveolens*, which proved to be a potential mycoherbicide against this weed.
  - Suaveolens (2009) stated that early dry-season or pre-dispersal fires are effective due to the weed's high flammability and germination sensitivity to fire-induced soil temperatures. Crops harvested mid-to-late in the rainy season can limit seed bank replenishment.

3.2 Invasive Shrub

3.2.1 *Lantana camara* L.

- **Origin and Spread**

Lantana camara L., commonly known as Red Sage, belonging to the family Verbenaceae, is an aromatic invasive weed, native to tropical America. After being introduced into various tropical and subtropical regions of the world for ornamental hedging, it eventually became a serious weed in forests, agricultural land, pastures, and disturbed habitats (Kohli et al., 2006b). Over time, it has spread to open areas along roadsides, railway tracks, crop fields, and open forests. In India, seven or eight species, such as *L. camara, L. indica, L. veronicifoila,* and *L. trifolia,* have been reported, besides species of horticultural value (Negi et al., 2019).

- **Distribution (Global, India, Chandigarh)**

Globally, *L. camara* has been found in a wide range of habitats, including Africa, Southern Europe (such as Spain and Portugal), the Middle East, India, tropical Asia, Australia, New Zealand, as well as many islands in the Atlantic, Pacific, and Indian

Oceans (Fig. 3d). In India, *Lantana* was initially introduced in 1807 at the National Botanical Garden, and as an ornamental hedge plant in Calcutta in the early nineteenth century (Hakimuddin, 1929). In Indian forests, lantana invades 86,806 km^2 (38.8%), particularly in areas that are warm and humid. The species is most widely distributed in the Southern Western Ghats, the Shivalik highlands, and Central India (Fig. 4d) (Mungi et al., 2020).

In Chandigarh, the Shivalik Hills and the forests around the city are abundant with this weed to a moderate to heavy extent, and thus adversely affecting forest ecology and biodiversity. In Sukhna Wildlife Sanctuary, the weed was found in more than 50% of the forest area (Grewal, 2014).

- **Mechanisms of Invasion**

The primary method of reproduction is by seeds, although vegetative propagation (layering), when a stem is rooted after coming into contact with soil, is also possible. As a result, it heavily infests wastelands, edges of forests, urban parks, and forests in Chandigarh, particularly in the Sukhna Wildlife Sanctuary. Mature plants can yield as many as 12,000 seeds annually (Negi et al., 2019).

- **Negative Impacts**
  - The *Lantana* plant has been estimated to have caused the extinction of many endemic medicinal plants and trees in the Sukhna Wildlife Sanctuary (TERI, 2024).
  - Kato-Noguchi and Kurniadie (2021) stated that it forms dense monospecific stands by reducing the germination and seedling growth of native plant species, as well as by increasing seedling mortality.
  - *Lantana* releases a variety of chemicals (volatile and non-volatile) from its fresh parts, residues, and essential oils, which affect other species and are toxic to grazing animals (Kato-Noguchi & Kurniadie, 2021). *Lantana* invasion is reported to have negative impacts on biodiversity and all four categories of ecosystem services.

- **Control Measures**
  - According to historical records, the removal of *Lantana* took a lot of effort and resources. The combination of fire, mechanical, and biological control methods dates back to 1921 in India (Bhagwat et al., 2012).
  - Various approaches, such as stem clipping, bulldozing, ploughing, and grubbing, can be employed for lantana removal. Usually, Herbicides are often applied before or after this activity to increase their effectiveness (Bhutia et al., 2021).
  - Various Bankar et al. (2023) reported that ~ 95 percent of *Lantana* fruit was destroyed as a consequence of the accidentally imported *Epinotia lantana* and *Ophiomyia lantanae*.

- ***Control of Lantana and Restoration of Biodiversity in Reserve Forests of Chandigarh: A Case Study (By*** Singh, 2012*)*

In 2001, the Chandigarh Forest Department launched an integrated project to control *Lantana camara* infestation, which was hindering natural regeneration and degrading grassland ecosystems. Initial removal involved manual extraction of clumps with minimal disturbance to fauna, followed by measures to prevent reseeding. Despite efforts, a 25% regrowth was observed in the first year, necessitating further control. *Arundo donax* was introduced to suppress regrowth, followed by native species like bamboo and jungle jalebi to aid habitat recovery. Over time, native flora re-established itself naturally, enhancing biodiversity and restoring key habitats such as Sukhna Wildlife Sanctuary. By 2009–2010, the project successfully eradicated *Lantana*, demonstrating the effectiveness of strategic, long-term ecological restoration and offering a replicable model for invasive species management.

3.3 *Invasive Tree Species*

3.3.1 *Leucaena leucocephala* (Lam.) de Wit

- **Origin and spread**

In disturbed regions like roadsides, forest margins, and semi-urban habitats, *Leucaena leucocephala* (Lam.) de Wit is known to be an urban weed, typically reaching a height of 18 m. *Leucaena* is an evergreen and thornless leguminous species, belonging to the family Fabaceae. It is one of the worst invaders in the world and spans over Africa, Asia, Europe, and America. It creates dense thickets, lowers biodiversity, and endangers endemic species (Campbell et al., 2019). This species holds a tag for being one of the top five terrestrial invasive species, with an extensive global presence (Sharma et al., 2022). The plant is of Central American origin, which distributed in tropical and subtropical regions globally. In 1990, it was introduced to China to reverse desertification (Ramírez-Bahena et al., 2019). It continues to grow naturally and is a fast-growing species in disturbed habitats. It is identified as a weed in over 20 countries, dispersed across every continent (Bakewell-Stone, 2023).

- **Distribution (Global, India, Chandigarh)**

Globally, *Leucaena leucocephala* is found in North America (12 countries), South America (7), Oceania (23), Africa (17), Asia (17), and Europe (1) (Fig. 3e). It is regarded as one of the top five invasive terrestrial plant species in the world (Sharma et al., 2022). In India (Fig. 4e), *Leucaena* was intentionally planted throughout the country for afforestation purposes, and at present, its severe infestations can be observed in north-eastern regions and western Himalayas of India (Debnath & Debnath, 2017). It continues to grow naturally and is a fast-growing species in disturbed habitats. According to Ghate (1991), it was brought to India's Western Ghats in the late 1800s and is currently spreading quickly in all habitats.

In Chandigarh, the Planting of *Leucaena leucocephala* along with other species is done to cope with cascading sedimentation in Sukhna Lake, Chandigarh, as suggested by the team of Scientists under the Central Soil and Water Conservation Research and Training Institute, led by PR Mishra (Chaturvedi et al., 2018).

- **Mechanisms of Invasion**

Leucaena's invasive strategies acquired from its self-compatibility, large number of propagule production, having a persistent seed bank, flowering throughout the year, and being environmentally stress-tolerant established its presence in non-native habitats (Marler, 2020).

- **Prominent Negative Impact**

Alopecia, poor body condition, infertility, low birth weight, thyroid malfunction, and organ toxicity from Mimosine (a toxic amino acid) are some of the bad effects that *Leucaena leucocephala* dietary interaction produces.

- **Control Measures**
  - Mechanical methods such as cutting the young trees and shading their trunks with sheets of plastic to inhibit the sprouting for further emergence.
  - Chemical methods include repetitive application of specific herbicides, as this species is resistant to the most common herbicides. Chemicals, i.e., Glyphosate, Triclopyr, Picloram, and metsulfuron-methyl, have shown better efficacy in this species.
  - *Sesbania virgata* (Cav). Pers native to South American origin showed phytotoxic effect on *Leucaena* through its seed leachates (Mignoni et al., 2017).
  - Timely application, suitable target selection, and perfect timing for application lead to successful weed management. A study by Hwang et al. (2010) reported that the summer season is the ideal condition for species control, as the seedling requirements occur during this period.
  - *Leucaena* is used as fodder, timber, and soil amelioration by enhancing inter-cropping systems, especially in semi-arid regions. This species is also used in biomass production and as a potential raw material for industrial applications (Prasad et al., 2010).

3.3.2 *Neltuma juliflora* (Sw.) Raf.

- **Origin and Spread**

Neltuma juliflora (Sw.) Raf., synonym *Prosopis juliflora,* is a dominant weed in semi-arid urban areas of India, especially along highways and bare ground. It flourishes in soils that are enriched with metals and have the ability to accumulate and detoxify contaminants, which helps with phytoremediation efforts (Saini et al., 2012). The plant is indigenous to the Northern region of South America and Central America, and is expanded from southern Mexico to Panama, then from the Caribbean Islands to Venezuela and Northern Peru (Dakhil et al., 2021). Presently, it covers the Arabic peninsula, the Indian subcontinent, Sub-Saharan African nations, the Pacific region, and the Middle East's Semi-arid and Arid zones (Abedelnoor et al., 2009). The main purpose of introducing this species is to fetch solid fuelwood, provide fodder and shade, and strengthen the sand dune in deserts (Table 1).

- **Distribution (Global, India, Chandigarh)**

Many *Neltuma* species were introduced to the Caribbean, Chile, Peru, and the Pacific coast of Central America, most likely during the period when human settlements expanded over the American continent (Fig. 3f). By the early 1900s, *N. juliflora* had become prevalent in Egypt. In 1917, an official from the Egyptian Department of Agriculture in Giza and South Africa introduced it to Sudan (Pasiecznik et al., 2001). The exact origins of *N. juliflora* and its subsequent imports in East Africa are unknown, however, it may belong to plants brought to the region in the 1930s by traders from India or Southern Africa, or animals from Sudan or Southern Africa. It seems to be planted along the railway route that runs from Mombasa to Nairobi and beyond (Maundu et al., 2009). In India, *Neltuma juliflora* was initially planted in Jodhpur, Rajasthan, in 1913. Due to the plant's amazing growth rates, the state ruler at the time designated it a "Royal Plant" in 1940. Successive introductions were progressively made into the states of Tamil Nadu, Madhya Pradesh, Maharashtra, Haryana, Punjab, and Uttar Pradesh (Fig. 4f) (Sivakumar et al., 2018). Even though

Table 1 Global distribution of *Neltuma juliflora* with regions and purpose of introduction

Region	Purpose of the introduction	Reference
African Continent		
Kenya	Fuelwood, fodder	Choge et al. (2022)
Somalia	Fuelwood and dune stabilization	FAO (2020)
Sudan	Reverse desertification, fuelwood, and timber	Mishra et al. (2003)
Djibouti	Along the street for shade	Eshetu (2024)
South Africa	Erosion control, wind fence, timber	Maitre et al. (2002)
Arab Peninsula		
UAE	Restoration of degraded desert lands	El-Keblawy and Al-Rawai (2006)
Oman	Aesthetic, productivity (land)	Al-Rawahy et al. (2003)
Indian Sub-Continent		
India	Soil reclamation, wood fuel species, fiber	Elfadl and Luukkanen (2005)
Pakistan	Sand dune stabilization and poverty alleviation (fuel wood collection)	Essa et al. (2017)

the main purpose for the establishment of India's first plantations was conservation, *Prosopis* has emerged as the primary fuel source in rural regions and, to a greater extent, in urban and semi-urban areas today. In India's tropical dry and semi-arid regions, it supplies over 70% of the firewood required for rural residents.

In Chandigarh, *Neltuma juliflora,* along with *Acacia modesta* and other species, was planted to combat siltation due to heavy soil erosion. The study conducted during 2008–09 in the catchment area confirmed the presence of *Neltuma juliflora* in this area.

- **Negative Impacts**

This species depletes the soil moisture content, hence limits the availability of water to the local plants. Also, it favours breeding and spreading malaria-causing mosquitoes (Kishoin et al., 2024). Additionally, it frequently infiltrates dry riverbeds and other waterways, clogging them and sending flood rivers into a state of chaos. *Neltuma juliflora* has demonstrated its auto-toxicity by the suppression of its seed germination by its leachates (Khan & Shaukat, 2006). *Neltuma juliflora* is a major cause of food insecurity and Political Instability in Somalia, as reported by a report published by the FAO (2020).

- **Control Measures**
  - Mechanical methods like slashing, burning, and uprooting of seedlings or young *Neltuma* plants are the most common control measures (Seid et al., 2020).
  - Mera-71,2,4-D ester with Paraquat is formulated to affect Weed Recovery. *N. juliflora* regrowth also ceases by application of Diuron, Mera-71 (Glyphosate), and 2,4- equivalent to paraquat, diuron.
  - Evippe moth species established effective against *N. juliflora.*
  - UAV systems for early weed detection for auto identification and removal of *Neltuma* seedlings exhibit the pros of Integrated Weed Management.
  - Community Awareness: The *Neltuma juliflora* is a blessing in disguise. People in the Arid and semi-arid tropics are somewhat aware of the *N. juliflora* species as it adversely affects crop production, and it becomes a noxious weed in the near future (Balcha, 2022).

3.3.3 *Broussonetia papyrifera* (L.) Vent.

- **Origin and Spread**

Broussonetia papyrifera (L.) Vent., commonly known as paper mulberry, belongs to the Moraceae family, and is an exotic tree that is found all over the world for its economic, aesthetic, and medicinal benefits (Seth, 2003). This tree hails from southeastern and eastern Asia, as well as the Pacific Islands, and is used to be quite popular as a versatile tree, valued for its timber, paper production, tapa cloth, and ornamental uses. However, it has now been labelled as an invasive species in several countries, including India. The tree spreads mainly through seeds, often carried by birds, and through root suckers that it produces in abundance from its shallow roots (Whistler & Elevitch, 2006). The genus *Broussonetia* includes four species, with *B. papyrifera* being one of the most recognized due to its significant role in papermaking and its potential to become invasive (POWO, 2025). *B. papyrifera* boasts a fascinating history that stretches back thousands of years. This plant is present in the subtropical areas of Asia, including places like mainland China, Taiwan, Japan, Korea, and Southeast Asia. It was highly prized for its bark, which was transformed into paper and cloth, especially in ancient China, Japan, and Korea (Maan et al., 2021).

- **Distribution (Global, India, Chandigarh)**

Broussonetia papyrifera is indigenous to East Asia, specifically China, Japan, Korea, and Taiwan. It is widely naturalized across Asia, such as India, Pakistan, Thailand, Malaysia, and Vietnam. Beyond Asia, it has been naturalized and introduced into North America, Africa, Europe, and the Pacific Islands (Fig. 3g) (CABI 2019). In India, *B. papyrifera* has a wide distribution in different states. It was introduced in 1880 and has since spread to areas such as Meghalaya, Punjab, and Uttar Pradesh (Fig. 4g) (Ali et al., 2022). The species grows well in varied habitats, ranging from sea level to altitudes of up to 1000 m (Olupot & Isabirye-Basuta, 2022).

Its arrival in Chandigarh likely took place during the city's development in the mid-twentieth century as part of urban landscaping initiatives (Maan et al., 2020). In Chandigarh, *B. papyrifera* occurs in urban and peri-urban regions, frequently in disturbed environments like roadsides, in abandoned lots, and in managed green spaces. Its presence is significant because of its invasive potential (Gupta et al., 2024). *B. papyrifera* shows a high level of invasiveness due to its fast growth, adaptability in reproduction, and ecological effects.

- **Key Features of Distribution**
  - **Native Range:** East Asia, including China, Japan, Korea, and Taiwan
  - **Introduced Regions:** North America, Africa, Europe, and the Pacific Islands
  - **Indian Distribution:** Meghalaya, Punjab, Uttar Pradesh, and other regions
  - **Chandigarh Distribution:** Urban and peri-urban areas, particularly in disturbed habitats.

- **Mechanisms of Invasion**

Broussonetia papyrifera employs several strategies to spread and establish itself in new areas:

- **Seed Dispersal**

The tree produces fruits that are consumed by birds, bats, and other animals. These animals then deposit the seeds in their droppings, often in new locations, facilitating long-distance dispersal (Whistler & Elevitch, 2006). Seeds can also be spread through moving water and dumped garden waste, further expanding their range.

- **Vegetative Propagation**

Broussonetia papyrifera can spread locally through root suckers. When the stems get damaged, these suckers pop up, allowing the tree to regenerate and form dense thickets (Apetorgbor & Bosu, 2011). Broken stems and roots can also grow into new plants if they are transported to suitable environments, such as through dumped garden waste or moving water.

- **Human Activity**

People often plant this tree for its quick growth and attractive appearance. This intentional planting has played a big role in its spread, especially in urban areas like Chandigarh (Olivares et al., 2019).

- **Negative Impacts**

Broussonetia papyrifera poses several prospective impacts in Chandigarh. Its invasive tendencies often push out native plant species, which diminishes biodiversity and disrupts the ecosystem's natural balance (Maan et al., 2022). The tree also has allelopathic traits that can hinder the growth of other plants, making the loss of biodiversity even worse (Negi et al., 2016). On top of that, its pollen can cause respiratory allergies, which concerns public health (Wu et al., 2019). From an economic standpoint, the rapid growth and spread of this tree can lead to higher maintenance costs for urban infrastructure, and its ecological effects can harm soil quality and upset nutrient cycles. In summary, effectively managing *B. papyrifera* is essential to decrease these adverse impacts and maintain the ecological harmony of Chandigarh's urban landscape.

- **Control Efforts**

Efforts to tackle the invasion of *Broussonetia papyrifera* in Chandigarh are mainly focused on understanding its ecological effects and crafting effective management strategies (Gupta et al., 2024). Although there are not many large-scale control programs documented, various research initiatives have been undertaken to evaluate their invasive potential and effects on local ecosystems. Here are some of the strategies and efforts being considered:

- **Phenological Management**: Studies have emphasized the significance of managing *B. papyrifera* based on its life cycle. This means targeting the plant during specific stages of its life cycle, like before it flowers, to prevent seed dispersal and reduce its spread (Maan et al., 2020).
- **Biochar Production**: Research has explored converting *B. papyrifera* biomass into biochar as a method to manage its spread while also potentially enhancing soil fertility. This could be a sustainable method to make use of invasive plant material (Gupta et al., 2024).
- **Ecological Research**: Scientists from institutions such as Panjab University, Chandigarh, have carried out in-depth studies on the allelopathic effects of

B. papyrifera, which sheds light on how it outcompetes native species. This understanding can guide control strategies.
- **Public Awareness**: There's an increasing need for campaigns to raise public awareness about the invasive nature of *B. papyrifera* and the importance of not planting it as an ornamental tree
- **Mechanical Removal**: Local authorities and environmental groups sometimes engage in the mechanical removal of invasive species, although this can be tricky due to the tree's ability to regrow from its roots.

While these initiatives are in progress, there's a pressing need for more comprehensive and coordinated management plans to effectively control the spread of *B. papyrifera* in Chandigarh.

3.4 Invasive Aquatic Weed

***Pontederia crassipes* (Mart.) Solms**

- **Origin and Spread**

Commonly referred to as common water hyacinth or jal kumbhi, *Pontederia crassipes* (Mart.) Solms (previously *Eichhornia crassipes*) is a South American aquatic plant that has spread throughout the world and is frequently invasive outside of its natural area. It originates from Amazonia, Brazil, and probably Argentina (Barrett & Forno, 1982). It has spread to all tropical and sub-tropical nations and is currently found on every continent, except Antarctica. It is the only species of the subgenus *Oshunae* within the genus *Pontederia* (Leandro et al., 2021). Introduced over a century ago, as an ornamental species to decorate the water bodies of numerous nations, this fast-growing, free-floating, perennial aquatic plant has since spread throughout Europe, Africa, Asia, and North America (Patel, 2012). If growth circumstances are favourable, this invasive macrophyte can treble its biomass in six to fourteen days.

- **Distribution (Global, India, Chandigarh)**

It originates from Amazonia, Brazil, and probably Argentina. The plant has spread to all tropical and sub-tropical nations and is currently found on every continent, except Antarctica. Introduced over a century ago as an ornamental species to decorate the water bodies of numerous nations, this fast-growing, free-floating, perennial aquatic plant has since spread throughout Europe, Africa, Asia, and North America (Fig. 3h) (Patel, 2012). In India (Fig. 4h), it is widely distributed across many states due to its tropical and subtropical climate, providing ideal conditions for its growth (Rojas-Sandoval & Acevedo-Rodríguez, 2013).

In Chandigarh, Sukhna Lake is the most prominent water body and is frequently infested with water hyacinth, especially during the monsoon and post-monsoon months (Sharma, 2020).

- **Mechanisms of Invasion**
  - The water hyacinth is a fast-growing plant that replaces native vegetation on water bodies during the rainy season. The plant produces a large amount of seeds that are viable for up to 30 years (Dechassa, 2020).
  - It also spreads through fragmentation of established plants and may resprout from rhizomes or germinate from seeds. This aggressive weed is challenging the ecological stability and posing a threat to biodiversity. It can further alter the structure and function of the ecosystem, disrupting the food chain and nutrient cycling (Khanna et al., 2011).
- **Negative Impacts**
  - Expansion of the invasive water hyacinth plant, along with intensive boating, has led to fewer migratory birds arriving at Sukhna Lake in Chandigarh (Sharma, 2020). *Pontederia crassipes* grows very fast and covers ponds, lakes, and rivers. Mats block light from penetrating water bodies, depleting oxygen rapidly, and inducing changes in the flora and fauna underneath (Fig. 5) (Chandam, 2017).
  - If growth circumstances are favourable, this invasive macrophyte can treble its biomass within six to fourteen days. It has an effect index of 4 and a spread index of 3, making it one of the worst weeds in the world (Huang et al., 2008).
- **Control Measures**
  - Mechanical removal of water hyacinth is the best short-term solution for controlling the proliferation of the weed in small areas. Manual removal of water hyacinth is also suitable for extremely small areas, but ineffective for large infestations control (Dechassa, 2020).
  - Biocontrol agents have been widely used in many parts of the world, and improved monitoring capability will help to choose the most appropriate control method.
  - Chemical herbicides such as 2,4-D, endothall dipotassium salt, endothall dimethyl alkylamine salts, and glyphosate are effective against water hyacinth, but long-term use degrades water quality and puts aquatic life at risk (Malik, 2007).
  - Biocontrol by weevils, fungi, and allelopathic plant extracts has been met with partial success only (Tessmann et al., 2008). However, an integrated approach using insects and plant pathogens synergistically has been successful in controlling *P. crassipes*. Biocontrol agents have proved futile in controlling weeds, and integration of management modalities seems to be the only option to battle against the resilient weed menace (Patel, 2012).

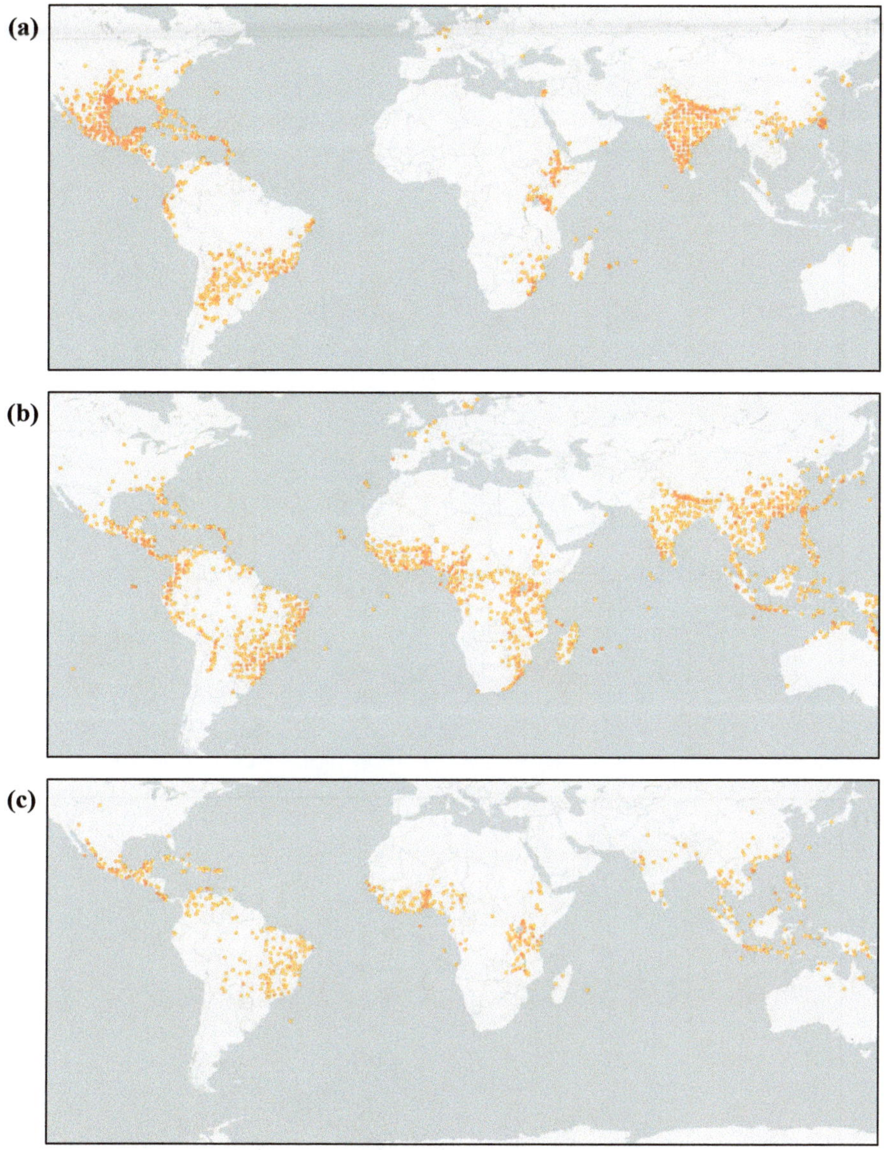

Fig. 3 Global distribution maps of invasive plant species **a** *Parthenium hysterophorus* **b** *Ageratum conyzoides* **c** *Mesosphaerum suaveolens* **d** *Lantana camara* **e** *Leucaena leucocephala* **f** *Neltuma juliflora* **g** *Broussonetia papyrifera* **h** *Pontederia crassipes*. Accessed from: Global Biodiversity Information Facility (GBIF) portal (GBIF, 2025)

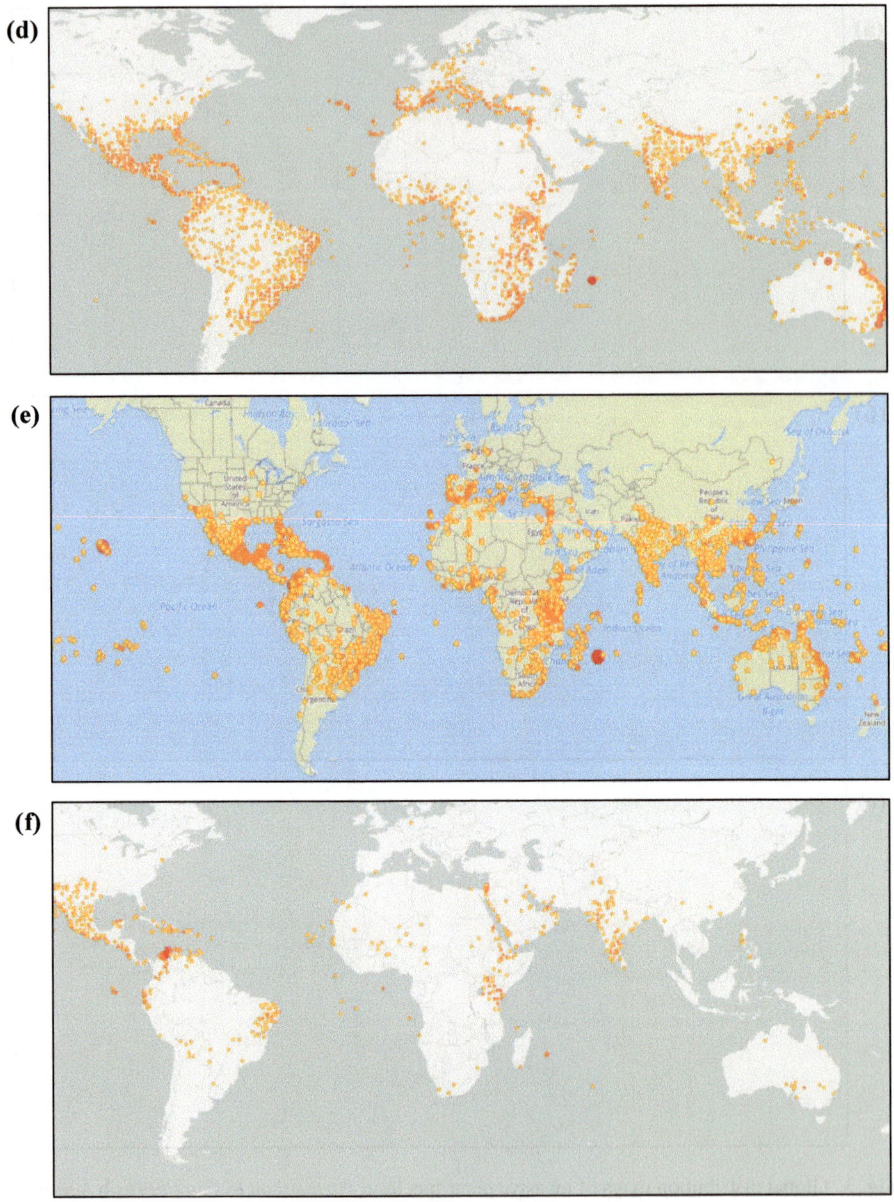

Fig. 3 (continued)

(g)

(h)

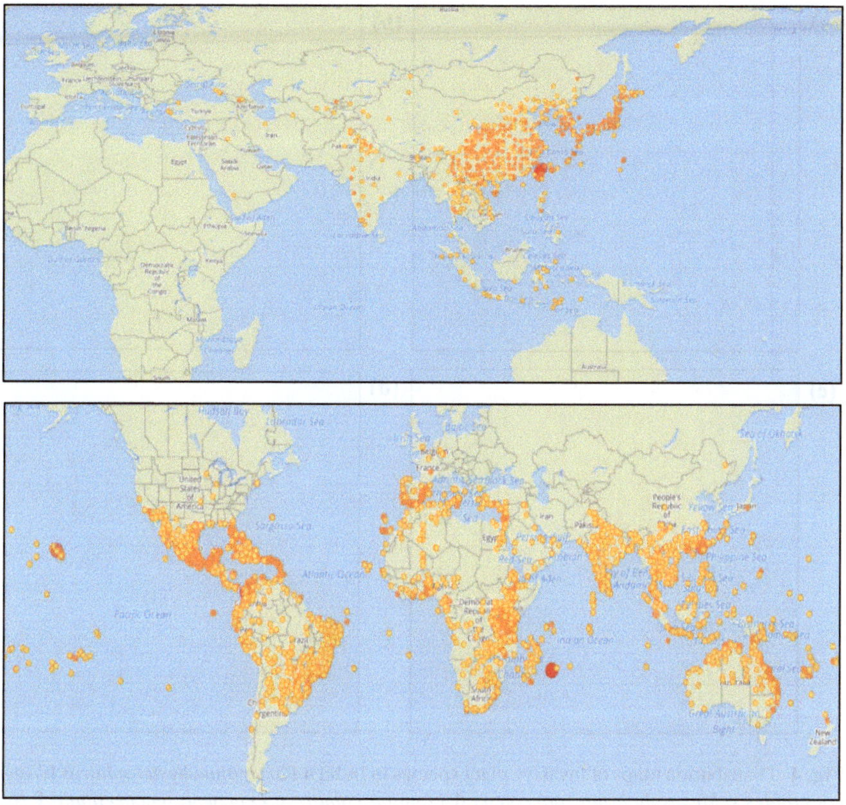

Fig. 3 (continued)

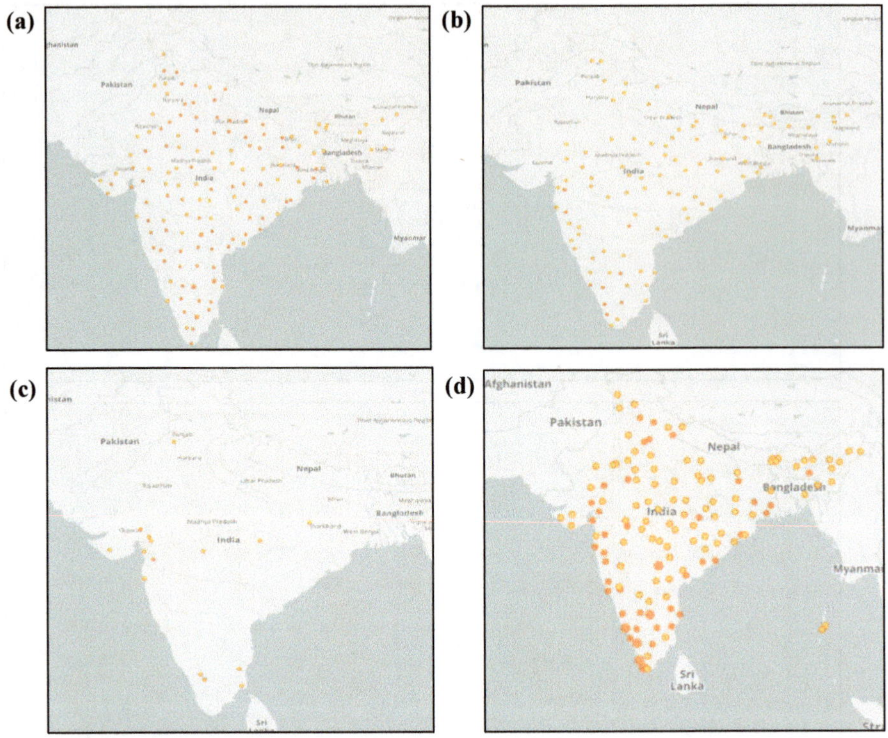

Fig. 4 Distribution maps of invasive plant species in India **a** *Parthenium hysterophorus* **b** *Ageratum conyzoides* **c** *Mesosphaerum suaveolens* **d** *Lantana camara* **e** *Leucaena leucocephala* **f** *Neltuma juliflora* **g** *Broussonetia papyrifera* **h** *Pontederia crassipes*. Accessed from: Global Biodiversity Information Facility (GBIF) portal (GBIF, 2025)

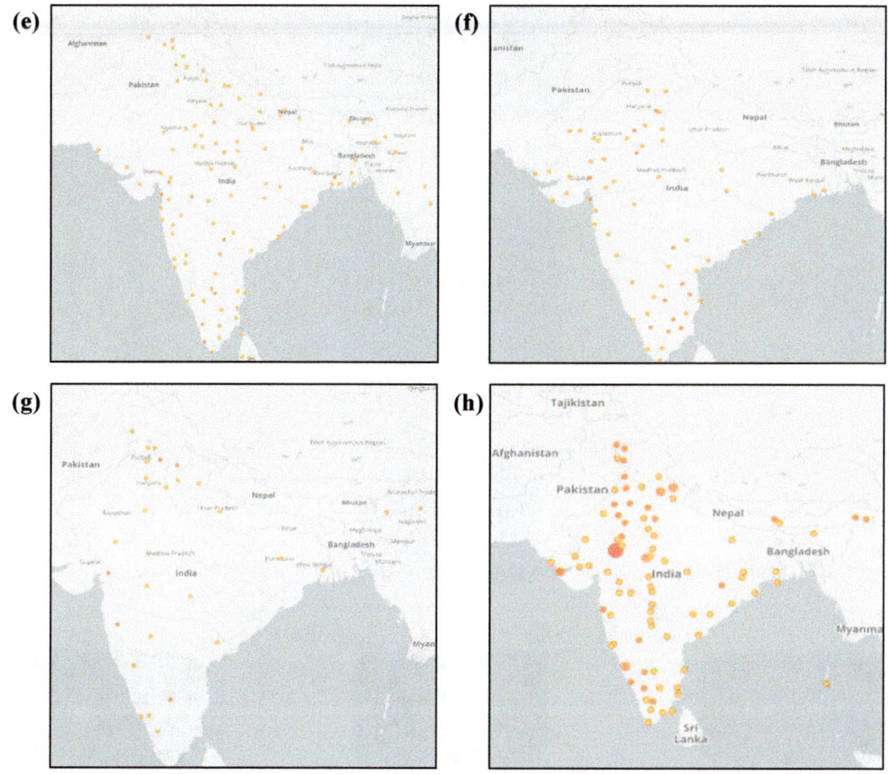

Fig. 4 (continued)

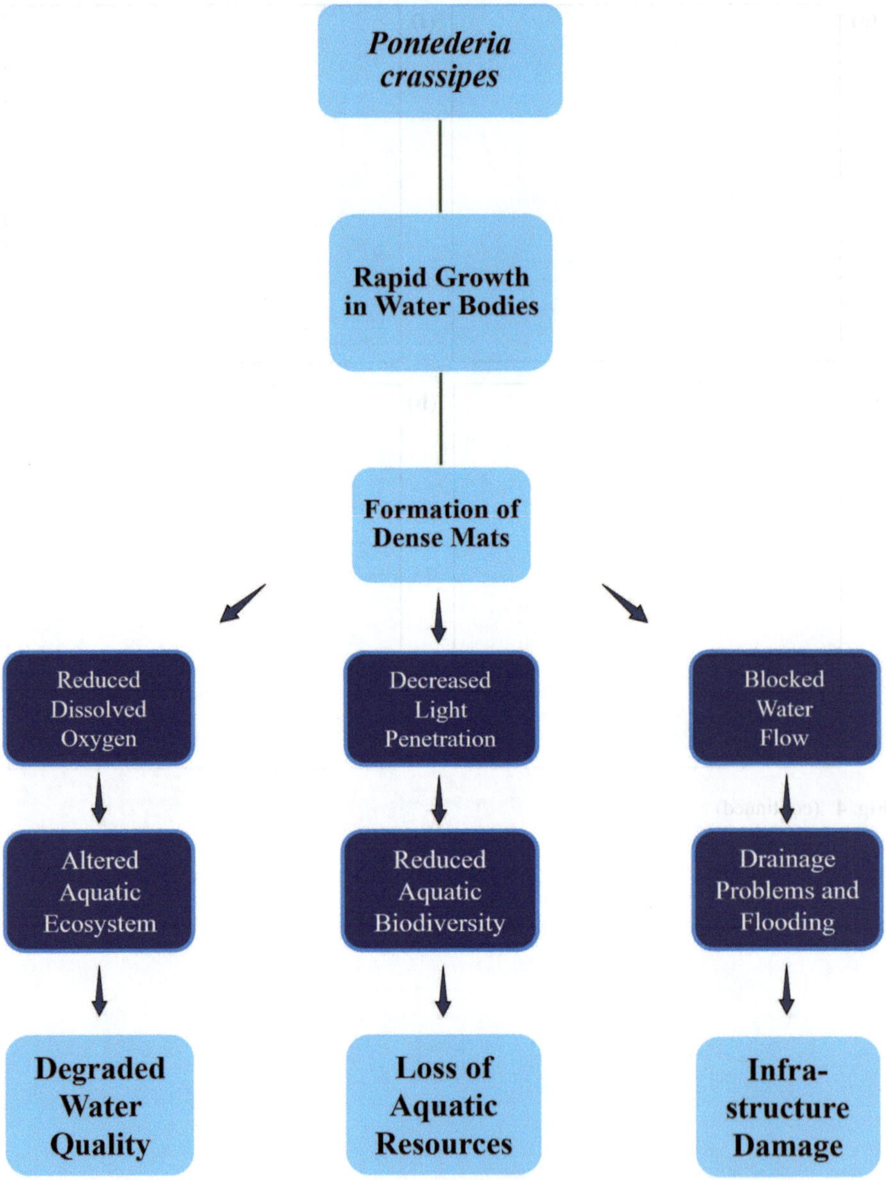

Fig. 5 Impact of *Pontederia crassipes*, an aquatic invasive plant, on urban ecosystems

4 Ecological and Socio-economic Impacts of Invasive Plant Species

Invasive alien plant species have been one of the most disruptive forces in nature, disturbing native ecosystems, reducing agricultural yield, and affecting human health. Invasive species affect biodiversity, disturb ecological processes, and cause a substantial monetary cost to agriculture, forestry, and urban environments. Invasive alien species affect the natural resources of economically and ecologically valuable ecosystems. These species bring structural and functional changes to these ecosystems (Shackleton et al., 2019). The impacts of invasive alien species on native ecosystems depend on several ecological, environmental, and climatic factors; thus, it is a complex phenomenon.

4.1 Impact on Native Flora and Fauna

4.1.1 Displacement of Native Plant Species

The ability of invasive alien species to outcompete the native flora, and consume a major proportion of the essential resources, such as water, nutrients, etc., has been one of the prominent concerns. Invasive alien species like *Parthenium hysterophorus* and *Ageratum conyzoides* are capable of inhibiting the germination and growth of native plant species due to the presence of allelopathic properties, wherein they release phytotoxins (Gnanavel & Natarajan, 2013). These invasive species are responsible for the homogenization of native plant ecosystems, thus reducing the plant diversity in all types of ecosystems. *Lantana camara*, a woody shrub, replaces native undergrowth vegetation and alters forest succession patterns; the food chain is also affected due to the replacement of native plant species, further limiting resources for herbivores and ultimately affecting the predator–prey dynamics (Negi et al., 2019).

4.1.2 Impact on Wildlife and Pollinators

The loss of native flora has a direct effect on the availability of food and habitat for native fauna. The rapid proliferation of *Prosopis juliflora* in semi-arid grasslands has reduced fodder availability for herbivores like nilgai and blackbuck (Kumar & Mathur, 2014). Invasive plant species displace nectar-rich native flower species, thereby directly altering the pollinator networks (Pathak et al., 2019). In Sukhna Wildlife Sanctuary (Chandigarh), a decrease in butterfly species richness was recorded in areas significantly invaded by *Lantana camara* as compared to less invaded ones (Singh, 2012).

4.2 Economic Consequences for Agricultural and Urban Areas

4.2.1 Agricultural Losses

Invasive weeds significantly reduce agricultural productivity by competing for valuable resources like water, nutrients, and minerals. As *Parthenium hysterophorus* invades major crops such as wheat and maize, further reducing production and increasing the cost of weed control (Kaur et al., 2014). *Parthenium hysterophorus* and *Leucaena leucocephala* have been found responsible for a loss in production in maize, sugarcane, and potato, and increased weed management costs. Weeds cause 10–80% crop yield losses besides impairing product quality and causing health and environmental hazards (Rao & Chauhan, 2015). In vegetables, *Ageratum conyzoides* reportedly has a negative effect on root development and reduces production (Syngkli & Rai, 2024).

4.2.2 Infrastructure Damage and Management Costs

Urban settlements face infrastructure difficulties due to invasive alien plants. The proliferation of *Eichhornia crassipes* blocks drainage systems and pollutes urban water bodies, thereby leading to flooding and inflating the water treatment costs (Degaga, 2018). The Chandigarh Municipal Corporation reported spending over 30 lakh rupees yearly on eradicating *Eichhornia crassipes* from Sukhna Lake and regulating *Parthenium* growth in public parks (TERI, 2024).

4.3 Health Hazards and Allergies

4.3.1 Respiratory and Dermatological Effects

Many invasive alien plants produce allergenic pollen and toxic compounds. In the majority of urban and rural settlements, *Parthenium hysterophorus* is a prominent allergenic plant causing asthma, allergic rhinitis, and contact dermatitis (Allan et al., 2019). In Chandigarh, *Broussonetia papyrifera* is reported as a contributor to seasonal allergic rhinitis (Ravindra et al., 2023).

4.3.2 Impacts on Livestock

Toxic compounds present in invasive alien plants present a health risk to livestock animals. *Neltuma juliflora* pods have alkaloids that can cause gut health and dental issues in ruminants (William & Jafari, 2015). Herders reported increased cases of infirmity and gut issues in goats grazing in proximity to *Neltuma* patches in the peripheral regions of Chandigarh. *Lantana camara* ingestions lead to liver toxicity, radiant sensitivity, and in some cases death in cattle and sheep (Mossie & Yirdaw, 2023).

4.4 Soil Erosion and Water Quality Issues

4.4.1 Hydrological Degradation

Aquatic invasive alien plants like *Pontederia crassipes* degrade water quality by reducing DO (Dissolved oxygen), water clarity, and nitrogen and phosphorus. These alterations disrupt aquatic ecosystems (Villamagna & Murphy, 2010). Zones with dense growth of *Eichhornia* in Sukhna Lake have shown a reduction in OD levels during seasonal monitoring (TERI, 2024).

4.4.2 Soil Degradation

Invasive alien species like *Lantana camara* and *Leucaena leucocephala* impact the soil nutrient profiles and decrease microbial diversity in soil (Sharma & Raghubanshi, 2009). Their thick root mats make the soil firm and reduce penetration, promoting erosion and nutrient runoff during precipitation (Sharma et al., 2016). Loss of grassland due to *Neltuma juliflora* further exacerbates these problems, degrading rangelands and reducing soil quality.

5 Management and Mitigation Strategies

- **Biological Control and Mechanical Removal**

An integrated weed management approach is a prerequisite for the mitigation of invasive alien species. Biological control agents like *Zygogramma bicolorata* have demonstrated success in reducing *Parthenium hysterophorus* density (Bashar et al., 2021). *Puccinia lantanae* fungi have displayed the ability to infect *Lantana camara* (Thomas et al., 2021). In Sukhna Wildlife Sanctuary and Nek Chand Rock Garden, manual uprooting and manual extraction, followed by replanting native flora, have yielded positive results in the mitigation of invasive plant species such as *Lantana* (TERI, 2024).

- **Chemical Control**

Chemical herbicides from chloroacetanilide, imidazolinone, sulfonylurea, and pyrimidinyl-carboxy groups are employed for weed management practices worldwide, although overuse presents the risk of resistance development and non-target species damage (Bamal et al., 2024). In Chandigarh, selective spraying has been combined with physical removal to reduce dependence on chemical agents.

- **Ecological Restoration and Phytoremediation**

Reinstatement of native vegetation is the way to prevent reinvasion. Species like *Desmodium trifolium, Vetiveria zizanioides,* and *Cymbopogon martinii* have shown the ability to stabilise degraded soils (Mandal et al., 2017). In wetlands invaded by *Eichhornia*, phytoremediation using *Typha latifolia* and *Cyperus rotundus* has improved nutrient cycling (Shingare et al., 2017).

In summary, the ecological and socioeconomic impacts of invasive alien species in Chandigarh, Punjab, and peripheral regions are complex and growing critically. These species disrupt native ecosystems, disturb plant-pollinator relations, impact biodiversity, and simultaneously inflict major monetary burdens through reduced agricultural productivity, high weed management expenses, and enhanced risk to human and animal health. The rapid proliferation of invasive plants like *Parthenium hysterophorus, Ageratum conyzoides, Broussonetia papyrifera, Lantana camara, Leucaena leucocephala, Neltuma juliflora,* and *Pontederia crassipes* emphasizes the need for extensive, region-specific strategies that integrate scientific findings, policy intervention, and community participation. Effective management requires prioritizing invasive plant control based on impact severity, ensuring comprehensive implementation (as shown in Fig. 6). Efficacious mitigation must combine early detection, biological control, mechanical removal, and reintroduction of native flora to curb the ecological decline and promote sustainable land use.

6 Conclusions and Future Prospects

The well-planned and architecturally significant urban landscapes of Chandigarh are not only immune to the slow functioning and persistent entry of the invasive plant species, but also serve as active grounds for ecological resilience and innovative urban biodiversity management. Through this exploration of the city's invasive flora, it becomes evident that these initially unnoticeable weeds are far more than botanical oddities; they are living agents of environmental transformation. Species like *Ageratum conyzoides, Broussonetia papyrifera, Lantana camara, Leucaena leucocephala, Mesosphaerum suaveolens, Neltuma juliflora, Parthenium hysterophorus, and Pontederia crassipes* not only flourish in the urban habitat but also actively modify it, displacing native flora, changing soil chemistry, and altering local biodiversity. This chapter has brought to the fore the aesthetic, ecological, and socioeconomic dimensions of these urban invaders. They are highly successful due to

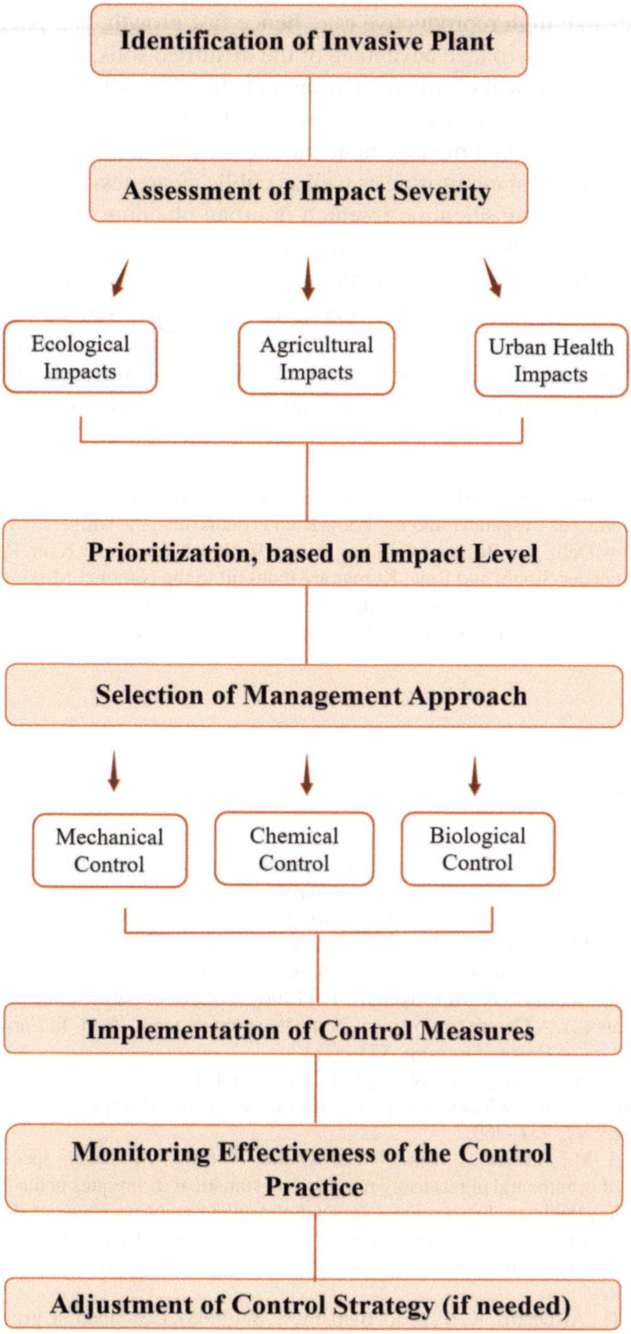

Fig. 6 Flow chart showing management (control) options for invasive plants

characteristics like high reproductive rate, hence fast growth, and phenotypic plasticity, supporting them to take advantage of the disturbed soils, fragmented habitats, and microclimates typical of urban environments like Chandigarh.

As this city continues to develop and grow, its urban biodiversity should develop accordingly. Invasive plant management and the maintenance of native biodiversity are dependent on robust monitoring as well as public awareness, community engagement, and incorporating scientific research in urban planning. Moreover, exploring the impact of climate change in modifying invasion dynamics and fostering interdisciplinary collaborations can lead to better insights and more sustainable strategies in conserving urban biodiversity. The investigations of urban weeds not only hold the answer to the resilience and persistence of invasive plant species but also may mirror the well-being of an urban ecosystem. Acknowledging and addressing the concerning issues of invasive vegetation is the critical next step toward more sustainable, greener cities.

Acknowledgements Shalini Ojha acknowledges the Council of Scientific & Industrial Research (CSIR), New Delhi, for Research Funding. Pooja Rani is thankful to the University Grants Commission (UGC), New Delhi, for Research Funding. Daizy R. Batish, Shalinder Kaur, Rishikesh Singh, Amrit Pal, Bhupender Singh, and Rajat Kumar are thankful to the National Mission on Himalayan Studies (NMHS), Ministry of Environment, Forest and Climate Change (MoEFCC), New Delhi, for the Financial Assistance in the form of a Research Project (NMHS2024-25/SC-XII/MG95-200/SL06).

References

Abedelnoor, T. M., Talib, N. H., Mabrouk, A. A., Mohamed, M. A., El-Mahi, M. I., Abu-Eisa, H. H., & Bokrezion, H. (2009). *The use of alternative animal feeds to enhance food security and environmental protection in the Sudan (The case for Prosopis Juliflora)* (pp. 38). PENHA-APRC.

Adkins, S., & Shabbir, A. (2014). Biology, ecology and management of the invasive parthenium weed (*Parthenium hysterophorus* L.). *Pest Management Science, 70*, 1023–1029.

Adkins, S. W., Navie, S. C., & McFadyen, R. E. (1996). Control of parthenium weed (Parthenium hysterophorus L.): A Centre for Tropical Pest Management team effort. In *Eleventh Australian Weeds Conference Proceedings* (pp. 573–578).

Afreen, T., Srivastava, P., Singh, H., & Singh, J. S. (2018). Effect of invasion by *Hyptis suaveolens* on plant diversity and selected soil properties of a constructed tropical grassland. *Journal of Plant Ecology, 11*, 751–760.

Ali, S., Khan, S. M., Ahmad, Z., Ejaz, U., & Khalid, N. (2022). Indicator species analysis and assessment of ornamental plants using multivariate statistical techniques in the home gardens of Rustam Valley, Pakistan. *International Journal of Applied and Experimental Biology, 1*, 87–95.

Allan, S., Shi BoYang, S. B., & Adkins, S. W. (2019). Impact of parthenium weed on human and animal health. *Parthenium weed: Biology, ecology and management* (pp. 105–130). CAB International.

Al-Rawahy, S. H., Al-Dhafri, K. S., & Al-Bahlany, S. S. (2003). Germination, growth and drought resistance of native and alien plant species of the genus Prosopis in the sultanate of Oman. *Asian Journal of Plant Sciences, 2*, 1020–1023.

Anonymous. (2025). Ecology and environment. In *Chandigarh master plan—2031* (pp. 463–474). Available online at: https://chandigarh.gov.in/chandigarh-master-plan-2031. Accessed on April 6, 2025.

Apetorgbor, M. M., & Bosu, P. P. (2011). Occurre CE AD CO Trol of paper mulberry (*Brousso etia* papyrifera) I souther GHA a. *Ghana Journal of Forestry, 27*, 40–51.

Bakewell-Stone, P. (2023). CABI compendium, *Leucaena leucocephala (leucaena)*. Available online at: https://doi.org/10.1079/cabicompendium.31634. Accessed on: April 11, 2025.

Balcha, L. D. (2022). Prosopis juliflora distribution, impacts, and control methods available in Ethiopia. *International Journal of Natural Resource Ecology and Management, 7*, 132–144.

Bamal, D., Duhan, A., Pal, A., Beniwal, R. K., Kumawat, P., Dhanda, S., Goyat, A., Hooda, V. S., & Yadav, R. (2024). Herbicide risks to non-target species and the environment: A review. *Environmental Chemistry Letters, 22*, 2977–3032.

Bankar, D. R., Renuka, S., Mahajan, P. M., & Priti, K. (2023). Biological control of weeds. *Emerging Trends in Entomology*, 15.

Barrett, S. C. H., & Forno, I. W. (1982). Style morph distribution in new world populations of *Eichhornia crassipes* (Mart.) Solms-Laubach (water hyacinth). *Aquatic Botany, 13*, 299–306.

Basavaraj, P., & Reddy, V. C. (2017). Influence of non chemical weed management practices on weed composition and diversity in transplanted organic finger millet. *International Journal of Agriculture Sciences, 9*, 3808–3811.

Bashar, H. K., Juraimi, A. S., Ahmad-Hamdani, M. S., Uddin, M. K., Asib, N., Anwar, M. P., & Rahaman, F. (2021). A mystic weed, Parthenium hysterophorus: Threats, potentials and management. *Agronomy, 11*, Article e1514.

Batish, D. R., Kohli, R. K., Singh, H. P., & Kaur, G. (2012). Biology, ecology and spread of the invasive weed *Parthenium hysterophorus* in India. *Invasive alien plants: An ecological appraisal for the Indian subcontinent* (pp. 10–18). CABI.

Batish, D. R., Singh, H. P., Dogra, K. S., Kaur, S., & Kohli, R. K. (2004a). Allelopathic interference of *Ageratum conyzoides* with growth and yield of soybean (*Glycine max*). *Weed Biology and Management, 4*, 153–158.

Batish, D. R., Singh, H. P., Kohli, R. K., Johar, V., & Yadav, S. (2004b). Management of invasive exotic weeds requires community participation1. *Weed Technology, 18*, 1445–1448.

Batish, D. R., Singh, H. P., Pandher, J. K., Arora, V., & Kohli, R. K. (2002). Phytotoxic effect of Parthenium residues on the selected soil properties and growth of chickpea and radish. *Weed Biology and Management, 2*, 73–78.

Bedi, P., Mahavir, & Tripathi, N. G. (2020). Smart urban green spaces for smart Chandigarh. In Vinod Kumar, T. (Eds.), *Smart environment for smart cities. Advances in 21st century human settlements* (pp. 111–148). Springer.

Bhagwat, S. A., Breman, E., Thekaekara, T., Thornton, T. F., & Willis, K. J. (2012). A battle lost? Report on two centuries of invasion and management of *Lantana camara* L. in Australia, India and South Africa. *PLoS One, 7*, e32407.

Bhutia, P., Kumar, M., & Khola, O. P. S. (2021). Control of lantana invasion in rangeland. *Agriculture & Food: E-Newsletter, 3*, 362–364.

CABI Compendium. (2019). *Broussonetia papyrifera (paper mulberry)*. Available online at: https://doi.org/10.1079/cabicompendium.10017. Accessed on April 12, 2025.

Campbell, S., Vogler, W., Brazier, D., Vitelli, J., & Brooks, S. (2019). Weed leucaena and its significance, implications and control. *Tropical Grasslands—Forrajes Tropicales, 7*, 280–289.

Chalana, M. (2015). Chandigarh: City and periphery. *Journal of Planning History, 14*, 62–84.

Chandam, C. D. (2017). *Assessment of human health in the neighbourhood of chemical and petrochemical industries: A case study of Vadodara City, Gujarat*. Doctoral dissertation, Maharaja Sayajirao University of Baroda (India).

Chaturvedi, O. P., Dagar, J. C., Handa, A. K., Kaushal, R., & Pandey, V. C. (2018). Agroforestry potential for higher productivity from degraded ravine watersheds. In J. Dagar & A. Singh (Eds.), *Ravine lands: Greening for livelihood and environmental security* (pp. 335–360). Springer.

Choge, S., Mbaabu, P. R., & Muturi, G. M. (2022). Management and control of the invasive *Prosopis juliflora* tree species in Africa with a focus on Kenya. In *Prosopis as a heat tolerant nitrogen fixing desert food legume* (pp. 67–81). Academic Press.

Dakhil, M. A., El-Keblawy, A., El-Sheikh, M. A., Halmy, M. W. A., Ksiksi, T., & Hassan, W. A. (2021). Global invasion risk assessment of *Prosopis juliflora* at biome level: Does soil matter? *Biology, 10*, Article e203.

de Barros Ruas, R., Costa, L. M. S., & Bered, F. (2022). Urbanization driving changes in plant species and communities–A global view. *Global Ecology and Conservation, 38*, Article e02243.

Debnath, A., & Debnath, B. (2017). Diversity, invasion status and usages of alien plant species in northeastern Hilly state of Tripura: A confluence of Indo-Barman hotspot. *American Journal of Plant Sciences, 08*, 212–235.

Dechassa, N. (2020). Origin, distribution, impact and management of water hyacinth (*Eichhornia crassipes* (Martius) Solm): A review. *Journal of Geography Environment and Earth Science International, 10*, 13–18.

Degaga, A. H. (2018). Water hyacinth (*Eichhornia crassipes*) biology and its impacts on ecosystem, biodiversity, economy and human well-being. *Journal of Life Science and Biomedicine, 8*, 94–100.

Edeigba, B. A., Ashinze, U. K., Umoh, A. A., Biu, P. W., Daraojimba, A. I. (2024). Urban green spaces and their impact on environmental health: A global review. *World Journal of Advanced Research and Reviews, 21*, 917–927.

Ehrenfeld, J. G. (2010). Ecosystem consequences of biological invasions. *Annual Review of Ecology, Evolution, and Systematics, 41*, 59–80.

Elfadl, M., & Luukkanen, O. (2005). Field studies on the ecological strategies of *Prosopis juliflora* in a dryland ecosystem. *Journal of Arid Environments, 66*, 1–15.

El-Keblawy, A., & Al-Rawai, A. (2006). Impacts of the invasive exotic *Prosopis juliflora* (Sw.) D.C. on the native flora and soils of the UAE. *Plant Ecology, 190*, 23–35.

Eshetu, A. A. (2024). A valuable or a curse resource? A systematic review on expansion, perception of local community, benefits and side effects of *Prosopis juliflora*. *Frontiers in Conservation Science, 5*, Article e1491618.

Essa, M., Babar, S., & Uddin, Z. (2017). Mesquite (*Prosopis juliflora*) plant prevalence in different districts of Balochistan & antibacterial activity of its pods. *Pure and Applied Biology (PAB), 6*, 1503–1515.

Essl, F., Dullinger, S., Rabitsch, W., Hulme, P. E., Hülber, K., Jarošík, V., Kleinbauer, I., Krausmann, F., Kühn, I., Nentwig, W., Vilà, M., Genovesi, P., Gherardi, F., Desprez-Loustau, M.-L., Roques, A., & Pyšek, P. (2011). Socioeconomic legacy yields an invasion debt. *Proceedings of the National Academy of Sciences, 108*, 203–207.

Food and Agriculture Organization of the United Nations (FAO). (2020). *Innovative ways of managing Prosopis juliflora trees in Somalia*. Available online at: https://openknowledge.fao.org/items/33b4b12f-6cf3-497a-ac5b-8882be2abbd0. Accessed on: April 15, 2025.

Gaertner, M., Wilson, J. R., Cadotte, M. W., MacIvor, J. S., Zenni, R. D., & Richardson, D. M. (2017). Non-native species in urban environments: Patterns, processes, impacts and challenges. *Biological Invasions, 19*, 3461–3469.

Gallardo, B., & Aldridge, D. C. (2013). The 'dirty dozen': Socio-economic factors amplify the invasion potential of 12 high-risk aquatic invasive species in Great Britain and Ireland. *Journal of Applied Ecology, 50*, 757–766.

Gao, Z., Song, K., Pan, Y., Malkinson, D., Zhang, X., Jia, B., Xia, T., Guo, X., Liang, H., Huang, S., Da, L., Van Bodegom, P. M. & Cieraad, E. (2021). Drivers of spontaneous plant richness patterns in urban green space within a biodiversity hotspot. *Urban Forestry & Urban Greening, 61*, e127098.

Ghate, V. S. (1991). CABI databases. In *Noteworthy plant invasions in the flora of Western Ghats of Maharashtra*. Available online at: https://doi.org/10.5555/19922326398. Accessed on: April 11, 2025.

Global Biodiversity Information Facility (GBIF). (2025). *Open Access*. Available online at: https://www.gbif.org/. Accessed on: April 12, 2025.

Gnanavel, I., & Natarajan, S. K. (2013). *Parthenium hysterophorus* L.: A major threat to natural and agro eco-systems in India. *Science International, 1*, 124–131.

Grewal, S. S. (2014). Impact of soil conservation measures on vegetation recovery in the catchment of a rapidly silting recreational lake at Chandigarh. *Indian Journal of Soil Conservation, 42*, 305–313.

Gupta, I., Singh, R., Kaushik, A., Singh, H. P., & Batish, D. R. (2024). Comparative investigation of powder and extract of biochar from *Broussonetia papyrifera* on the growth and eco-physiological attributes of *Vigna radiata*. *Carbon Research, 3*, Article e31.

Hakimuddin, M. (1929). Lantana in northern India as a pest and its probable utility in solving the cowdung problem. *Indian Forester, 56*, 405–410.

Huang, L., Zhao, D., Wang, J., Zhu, J., & Li, J. (2008). Scale impacts of land cover and vegetation corridors on urban thermal behavior in Nanjing, China. *Theoretical and Applied Climatology, 94*, 241–257.

Hwang, C., Hsu, L., Liou, Y., & Wang, C. (2010). Distribution, growth, and seed germination ability of lead tree *(Leucaena leucocephala)* plants in Penghu Islands, Taiwan. *Weed Technology, 24*, 574–582.

Jesus, N. Z. T., Falcão, H. S., Lima, G. R. M., Caldas Filho, M. R. D., Sales, I. R. P., Gomes, I. F., Santos, S. G., Tavares, J. F., Barbosa-Filho, J. M., & Batista, L. M. (2013). *Hyptis suaveolens* (L.) Poit (Lamiaceae), a medicinal plant protects the stomach against several gastric ulcer models. *Journal of Ethnopharmacology, 150*, 982–988.

Kato-Noguchi, H., & Kurniadie, D. (2021). Allelopathy of Lantana camara as an invasive plant. *Plants, 10*, Article e1028.

Kaur, M., Aggarwal, N. K., Kumar, V., & Dhiman, R. (2014). Effects and management of *Parthenium hysterophorus*: A weed of global significance. *International Scholarly Research Notices, 2014*, Article e368647.

Kaur, S., Batish, D. R., Kohli, R. K., & Singh, H. P. (2012). Ageratum conyzoides: An alien invasive weed in India. *Invasive alien plants: An ecological appraisal for the Indian subcontinent* (pp. 57–76). CABI.

Khan, D., & Shaukat, S. S. (2006). Phytotoxic effects of *Prosopis juliflora* Swartz. DC. Against some of its field associates and the cultivated species. *International Journal of Biology and Biotechnology, 3*, 353–366.

Khanna, S., Santos, M. J., Ustin, S. L., & Haverkamp, P. J. (2011). An integrated approach to a biophysiologically based classification of floating aquatic macrophytes. *International Journal of Remote Sensing, 32*, 1067–1094.

Kishoin, V., Tumwesigye, W., Turyasingura, B., Wilber, W., Chavula, P., Gweyi-onyango, J. P., ... & Jaufer, L. (2024). The negative and positive impacts of Prosopis juliflora on the Kenyan and Ethiopian ecosystems: A review study. *Notulae Scientia Biologicae, 16*, 11832–11832.

Kohli, R. K., Batish, D. R., & Singh, H. P. (1998). Eucalypt oils for the control of parthenium (*Parthenium hysterophorus* L.). *Crop Protection, 17*, 119–122.

Kohli, R. K., Batish, D. R., & Singh, H. P. (2006a). Allelopathic interactions in agroecosystems. *Allelopathy: A physiological process with ecological implications* (pp. 465–493). Springer, Netherlands.

Kohli, R. K., Batish, D. R., Singh, H. P., & Dogra, K. S. (2006b). Status, invasiveness and environmental threats of three tropical American invasive weeds (*Parthenium hysterophorus* L., *Ageratum conyzoides* L., *Lantana camara* L.) in India. *Biological Invasions, 8*, 1501–1510.

Kohli, R. K., Batish, D. R., Singh, H. P., & Dogra, K. S. (2006c). Status, invasiveness and environmental threats of *Parthenium hysterophorus* in India: A review. *International Journal of Agriculture and Biology, 8*, 2–12.

Kumar, N., & Aggarwal, N. K. (2024). A review on valorization, management, and applications of the hazardous weed *Parthenium hysterophorus*. *Systems Microbiology and Biomanufacturing, 4*, 607–619.

Kumar, P. S. (2024). The exotic rust fungus *Puccinia abrupta* var. partheniicola on the invasive alien weed *Parthenium hysterophorus* in India: Rediscovery and first report of an epiphytotic. *CABI Agriculture and Bioscience, 5*, e37.

Kumar, S., & Mathur, M. (2014). Impact of invasion by *Prosopis juliflora* on plant communities in arid grazing lands. *Tropical Ecology, 55*, 33–46.

Leandro, T. D., Holsback, Z. D. R., & Scremin-Dias, E. (2021). The aquatic species *Pontederia azurea* and *P. crassipes* (Pontederiaceae) in the Pantanal, Brazil: Evidence of how plant structure can simultaneously reflect phylogeny and ecology. *Acta Botanica Brasilica, 35*, 79–91.

Maan, I., Kaur, A., Sharma, A., Singh, H. P., Batish, D. R., Kohli, R. K., & Arora, N. K. (2022). Variations in leaf litter decomposition explain invasion success of *Broussonetia papyrifera* over confamilial non-invasive Morus alba in urban habitats. *Urban Forestry & Urban Greening, 67*, Article e127408.

Maan, I., Kaur, A., Singh, H. P., Batish, D. R., & Kohli, R. K. (2020). Evaluating the role of phenology in managing urban invasions: A case study of *Broussonetia papyrifera. Urban Forestry & Urban Greening, 48*, Article e126583.

Maan, I., Kaur, A., Singh, H. P., Batish, D. R., & Kohli, R. K. (2021). Exotic avenue plantations turning foe: Invasive potential, distribution and impact of *Broussonetia papyrifera* in Chandigarh, India. *Urban Forestry & Urban Greening, 59*, Article e127010.

Maitre, D. L., Van Wilgen, B., Gelderblom, C., Bailey, C., Chapman, R., & Nel, J. (2002). Invasive alien trees and water resources in South Africa: Case studies of the costs and benefits of management. *Forest Ecology and Management, 160*, 143–159.

Malik, A. (2007). Environmental challenge vis a vis opportunity: The case of water hyacinth. *Environment International, 33*, 122–138.

Mandal, D., Srivastava, P., Giri, N., Kaushal, R., Cerda, A., & Alam, N. M. (2017). Reversing land degradation through grasses: A systematic meta-analysis in the Indian tropics. *Solid Earth, 8*, 217–233.

Marler, T. E. (2020). Three invasive tree species change soil chemistry in Guam forests. *Forests, 11*, Article e279.

Maundu, P., Kibet, S., Morimoto, Y., Imbumi, M., & Adeka, R. (2009). Impact of *Prosopis juliflora* on Kenya's semi-arid and arid ecosystems and local livelihoods. *Biodiversity, 10*, 33–50.

Mazza, G., Tricarico, E., Genovesi, P., & Gherardi, F. (2014). Biological invaders are threats to human health: An overview. *Ethology Ecology & Evolution, 26*, 112–129.

Mignoni, D. S. B., Simões, K., & Braga, M. R. (2017). Potential allelopathic effects of the tropical legume *Sesbania virgata* on the alien *Leucaena leucocephala* related to seed carbohydrate metabolism. *Biological Invasions, 20*, 165–180.

Mishra, A., Sharma, S. D., & Gupta, M. K. (2003). Soil Rehabilitation through afforestation: Evaluation of the performance of *Prosopis juliflora*, *Dalbergia sissoo* and *Eucalyptus tereticornis* plantations in a sodic environment. *Arid Land Research and Management, 17*, 257–269.

Mossie, T., & Yirdaw, B. (2023). Documentation of major poisonous plants and their toxic effects on livestock: A review. *American Journal of Bioscience and Bioengineering, 11*, 47–55.

Motmainna, M., Juraimi, A. S., Uddin, M. K., Asib, N. B., Islam, A. M., Ahmad-Hamdani, M. S., Berahim, Z., & Hasan, M. (2021). Physiological and biochemical responses of *Ageratum conyzoides*, *Oryza sativa* f. spontanea (weedy rice) and *Cyperus iria* to *Parthenium hysterophorus* methanol extract. *Plants, 10*, e1205.

Mungi, N. A., Qureshi, Q., & Jhala, Y. V. (2020). Expanding niche and degrading forests: Key to the successful global invasion of *Lantana camara* (sensu lato). *Global Ecology and Conservation, 23*, Article e01080.

Murphy, S. M., Vyas, D. K., Sher, A. A., & Grenis, K. (2022). Light pollution affects invasive and native plant traits important to plant competition and herbivorous insects. *Biological Invasions, 24*, 599–602.

Negi, A., Batish, D. R., Singh, H. P., & Kohli, R. K. (2016). Allelopathic effect of leaves of invasive tree *Broussonetia papyrifera* against some crop plants. *Annals of Plant Sciences, 5*, 1261–1264.

Negi, G. C., Sharma, S., Vishvakarma, S. C., Samant, S. S., Maikhuri, R. K., Prasad, R. C., & Palni, L. M. (2019). Ecology and use of *Lantana camara* in India. *The Botanical Review, 85*, 109–130.

Olivares, G., Peña-Ahumada, B., Peñailillo, J., Payacán, C., Moncada, X., Saldarriaga-Córdoba, M., Matisoo-Smith, E., Chung, K. F., Seelenfreund, D. & Seelenfreund, A. (2019). Human mediated translocation of Pacific paper mulberry [*Broussonetia papyrifera* (L.) L'Hér. ex Vent.(Moraceae)]: Genetic evidence of dispersal routes in Remote Oceania. *Plos One, 14*, e0217107.

Olupot, W., & Isabirye-Basuta, G. (2022). Nucleation by isolated emergent native trees can push back the shade-intolerant invader *Broussonetia papyrifera* (L.) L'Hér. ex Vent. *Trees, Forests and People, 8*, e100220.

Padalia, H., Kudrat, M., & Sharma, K. P. (2013). Mapping sub-pixel occurrence of an alien invasive *Hyptis suaveolens* (L.) Poit. using spectral unmixing technique. *International Journal of Remote Sensing, 34*, 325–340.

Pandey, A., & Pandey, A. K. (2009). Preliminary assessment of Phoma herbarum as a potential mycoherbicide against *Hyptis suaveolens. Annals of Plant Protection Sciences, 17*, 270–271.

Pasiecznik, N. M., Felker, P., Harris, P. J. C., Harsh, L. N., Cruz, G., Tewari, J. C., Cadoret, K., & Maldonado, L. J. (2001). *The Prosopis juliflora—Prosopis pallida complex: A monograph* (pp. 172). HDRA.

Patel, S. (2012). Threats, management and envisaged utilizations of aquatic weed *Eichhornia crassipes*: An overview. *Reviews in Environmental Science and Bio/technology, 11*, 249–259.

Pathak, R., Negi, V. S., Rawal, R. S., & Bhatt, I. D. (2019). Alien plant invasion in the Indian Himalayan Region: State of knowledge and research priorities. *Biodiversity and Conservation, 28*, 3073–3102.

Pehle, A., & Schirmel, J. (2015). Moss invasion in a dune ecosystem influences ground-dwelling arthropod community structure and reduces soil biological activity. *Biological Invasions, 17*, 3467–3477.

Plants of the World Online (POWO). (2025). *Broussonetia L'Hér. ex Vent.* Available online on: https://powo.science.kew.org/taxon/urn:lsid:ipni.org:names:331428-2. Accessed on: April 12, 2025.

Potgieter, L. J., Shrestha, N., & Cadotte, M. W. (2022). Prioritizing terrestrial invasive alien plant species for management in urban ecosystems. *Journal of Applied Ecology, 59*, 872–883.

Prasad, J. V. N. S., Korwar, G. R., Rao, K. V., Srinivas, K., Srinivasarao, C., Pedababu, B., Venkateswarlu, B., Rao, S. N., & Kulkarni, H. D. (2010). On-farm evaluation of two fast growing trees for biomass production for industrial use in Andhra Pradesh, Southern India. *New Forests, 42*, 51–61.

Raizada, P. (2006). Ecological and vegetative characteristics of a potent invader, *Hyptis suaveolens* Poit from India. *Lyonia, 11*, 115–120.

Ramírez-Bahena, M. H., Flores-Félix, J. D., Velázquez, E., & Peix, Á. (2019). The Mimosoid tree *Leucaena leucocephala* can be nodulated by the symbiovar genistearum of Bradyrhizobium canariense. *Systematic and Applied Microbiology, 43*, Article e126041.

Rao, A. N., & Chauhan, B. S. (2015). Weeds and weed management in India—A review. In *Weed science in the Asian Pacific Region* (pp. 87–118). Indian Society of Weed Science.

Ravindra, K., Goyal, A., & Mor, S. (2023). Identification of prominent airborne pollen in a city situated in foot-hills of Himalayas, Chandigarh, India. *Environmental Science and Pollution Research, 30*, 15775–15788.

Reynolds, J. S., Ibáñez-Álamo, J. D., Sumasgutner, P., & Mainwaring, M. C. (2019). Urbanisation and nest building in birds: A review of threats and opportunities. *Journal of Ornithology, 160*, 841–860.

Rojas-Sandoval, J., & Acevedo-Rodríguez, P. (2013). CABI compendium, *Eichhornia crassipes (water hyacinth).* Available online at: https://doi.org/10.1079/cabicompendium.20544#REF-DDB-200685. Accessed on: April 14, 2025.

Saini, P., Khan, S., Baunthiyal, M., & Sharma, V. (2012). Organ-wise accumulation of fluoride in *Prosopis juliflora* and its potential for phytoremediation of fluoride contaminated soil. *Chemosphere, 89*, 633–635.

Seid, O., Haji, J., & Legesse, B. (2020). Rural households' perception on the effects of *Prosopis juliflora* invasion: The case of Amibara District of Afar National Regional State, Ethiopia. *Pastoralism, 10*, Article e21.

Seth, M. K. (2003). Trees and their economic importance. *The Botanical Review, 69*, 321–376.

Shabbir, A., Bajwa, A. A., Mao, R., Kezar, S., Dorji, S., & Adkins, S. W. (2024). Biology of invasive plants 6. *Parthenium hysterophorus* L. *Invasive Plant Science and Management, 17*, 129–156.

Shackleton, R. T., Shackleton, C. M., & Kull, C. A. (2019). The role of invasive alien species in shaping local livelihoods and human well-being: A review. *Journal of Environmental Management, 229*, 145–157.

Sharma, A., Batish, D. R., Singh, H. P., Jaryan, V., & Kohli, R. K. (2017). The impact of invasive *Hyptis suaveolens* on the floristic composition of the periurban ecosystems of Chandigarh, Northwestern India. *Flora, 233*, 156–162.

Sharma, G. P., & Raghubanshi, A. S. (2007). Effect of *Lantana camara* L. cover on local depletion of tree population in a tropical dry deciduous forest of India. *Applied Ecology and Environmental Research, 5*, 109–121.

Sharma, G. P., & Raghubanshi, A. S. (2009). Lantana invasion alters soil nitrogen pools and processes in the tropical dry deciduous forest of India. *Applied Soil Ecology, 42*, 134–140.

Sharma, P., Kaur, A., Batish, D. R., Kaur, S., & Chauhan, B. S. (2022). Critical insights into the ecological and invasive attributes of *Leucaena leucocephala*, a tropical agroforestry species. *Frontiers in Agronomy, 4*, Article e890992.

Sharma, P., Tiwari, A. K., Bhatt, V. K., & Sathiya, K. (2016). Changes in vegetation cover and soil erosion in a forest watershed on removal of weed Lantana camara in lower Shivalik region of Himalayas. *Journal of Soil and Water Conservation, 15*, 193–198.

Sharma, S. (2020, December 15). Chandigarh's famous Sukhna Lake losing its migratory birds. *NewsClick*. Available online at: https://www.newsclick.in/Chandigarh-Famous-Sukhna-Lake-Losing-Migratory-Birds

Shingare, R. P., Nanekar, S. V., Thawale, P. R., Karthik, R., & Juwarkar, A. A. (2017). Comparative study on removal of enteric pathogens from domestic wastewater using *Typha latifolia* and *Cyperus rotundus* along with different substrates. *International Journal of Phytoremediation, 19*, 899–908.

Singh, H. P., Batish, D. R., & Kohli, R. K. (2002). Allelopathic effect of two volatile monoterpenes against bill goat weed (*Ageratum conyzoides* L.). *Crop Protection, 21*, 347–350.

Singh, I. (2012). Control of Lantana and restoration of biodiversity in reserve forests of Chandigarh: A case study. *Invasive alien plants: An ecological appraisal for the Indian subcontinent* (pp. 292–298). CABI.

Sivakumar, K., Rawat, G. S., Badola, R., Adhikari, B. S., & Kamalakannan, B. (2018). *Study on ecological and socio-economic impact of invasive species, Prosopis juliflora and Lantana Camara, and their removal from forest, common and fallow land of Tamil Nadu.* Technical Report, Wildlife Institute of India, Dehradun.

Staudhammer, C. L., Escobedo, F. J., Holt, N., Young, L. J., Brandeis, T. J., & Zipperer, W. (2015). Predictors, spatial distribution, and occurrence of woody invasive plants in subtropical urban ecosystems. *Journal of Environmental Management, 155*, 97–105.

Suaveolens, U. (2009). A matrix model for the population dynamics of *Hyptis suaveolens*, an annual weed. *Ecotrópicos, 22*, 23–36.

Syngkli, R. B., & Rai, P. K. (2024). Allelopathic effects of *Ageratum conyzoides* L. on the germination and growth of *Zea mays* L., *Lactuca sativa* L. and *Solanum lycopersicum* L. *Allelopathy Journal, 62*, 193–204.

TERI. (2024). *State of environment report for Chandigarh UT–2021.* The Energy and Resources Institute.

Tessmann, D. J., Charudattan, R., & Preston, J. F. (2008). Variability in aggressiveness, cultural characteristics, cercosporin production and fatty acid profile of *Cercospora piaropi*, a biocontrol agent of water hyacinth. *Plant Pathology, 57*(5), 957–966.

Theodorou, P. (2022). The effects of urbanisation on ecological interactions. *Current Opinion in Insect Science, 52*, Article e100922.

Thomas, S. E., Evans, H. C., Cortat, G., Koutsidou, C., Day, M. D., & Ellison, C. A. (2021). Assessment of the microcyclic rust *Puccinia lantanae* as a classical biological control agent of the pantropical weed *Lantana camara*. *Biological Control, 160*, Article e104688.

Tripathi, S., Rai, S. K., & Pandey, S. K. (2023). Distribution pattern and management strategy of invasive weed *Parthenium hysterophorus* L. in Gorakhpur, India. *Asian Journal of Biological and Life Sciences, 12*, 138–143.

Vélez-Gavilán, J. (2016). *CABI compendium, Ageratum conyzoides (billy goat weed)*. Available online at: https://doi.org/10.1079/cabicompendium.3572. Accessed on: April 14, 2025.

Villamagna, A. M., & Murphy, B. R. (2010). Ecological and socio-economic impacts of invasive water hyacinth (*Eichhornia crassipes*): A review. *Freshwater Biology, 55*, 282–298.

Whistler, W. A., & Elevitch, C. R. (2006). *Broussonetia papyrifera* (paper mulberry). *Species Profiles for Pacific Island Agroforestry, 9*, 1–3.

William, K., & Jafri, L. (2015). Mesquite (*Prosopis juliflora*): Livestock grazing, its toxicity and management. *Journal of Bioresource Management, 2*, 49–58.

World Health Organization (WHO). (2025). Available online at: https://www.who.int/news-room/fact-sheets/detail/urban-health. Accessed on: March 28, 2025.

Wu, P. C., Su, H. J., Lung, S. C. C., Chen, M. J., & Lin, W. P. (2019). Pollen of *Broussonetia papyrifera*: An emerging aeroallergen associated with allergic illness in Taiwan. *Science of the Total Environment, 657*, 804–810.

Chapter 11
Some Approaches of Indigenous Peoples and Local Communities to Invasive Alien Species with Remarks on Urban Communities

Wycliffe Wanzala and Sheila I. Minyoso

Abstract Indigenous Peoples and Local Communities (IPLC) have profound ethnoknowledge of the environment and are able to diagnose their respective ecological niches and detect any changes ranging from biotic to abiotic with precision. The IPLC, therefore, are able to detect new species in the environment, together with those species, that are threatened and becoming extinct, including their impacts on the ecosystem as a result of their emergence and/or extinction. This is particularly common with indigenous people who have stayed in a particular ecosystem for over 70 years to have the opportunity to witness the changing nature in time and space, a scenario that should not be interpreted to mean evolution. Today, no country in the world exists without experiences of the emergence of new species and the extinction of the existing ones. These two antagonistic processes have devastating effects on the sustainable management of the ecosystems. The emergency of the Invasive Alien Species (IAS) is problematic worldwide, leading to biodiversity loss, and Indigenous Peoples and Local Communities have responded to this problem in some myriad ways in different ecosystems, henceforth, the need to increase their collaboration and partnerships. This Chapter outlines some of the responses and approaches to IAS by IPLC with remarks on urban communities since they provide the fulcrum point for trade in live organisms from many distant and different agroecological zones, supporting the livelihoods of very unique organisms.

Keywords Alien species · Biological invasions · Ecosystems · Ethnoknowledge · Old generation · Species extinction · Urbanization

---

W. Wanzala (✉)
Department of Biological Sciences, School of Pure and Applied Sciences, Maasai Mara University, Narok, Kenya
e-mail: wanzala@mmarau.ac.ke

W. Wanzala · S. I. Minyoso
Matungu Rural Herbal and Research Centre, Mumias, Kenya

S. I. Minyoso
The Scholarship Network Centre, Education and Research Division, Nairobi, Kenya

R. Singh et al. (eds.), *Invasive Alien Plants in Urban Ecosystems*,
https://doi.org/10.1007/978-3-032-00997-5_11

1 Introduction

The alien species are organisms existing outside their natural geographical range following their deliberate and/or accidental introduction (Polce et al., 2023; Tripathi et al., 2023). The invasion of Invasive Alien Species (IAS) in world ecosystems is unique and dynamic in nature, threatening biodiversity, ecosystem resilience, policy-making processes and socioeconomic stability of livelihoods (Eschen et al., 2021; Polce et al., 2023; Wehi et al., 2023). In those ecosystems, the IAS causes devastating environmental, socio-economic, agricultural, health and well-being, and political impacts on livelihoods in many human societies (Bacher et el., 2018; Barbet-Massin et al., 2020; Bartz et al., 2010; Shackleton et al., 2019; Shine, 2005). This therefore implies that IAS risks achieving UN Sustainable Development Goals (ICIPE, 2018; IUCN, 2020) if the mitigations put in place continually remain elusive. Indigenous Peoples and Local Communities (IPLC) around the world have therefore developed myriad ways of dealing with these impacts to some extent to their advantage (Bartz & Kowarik, 2019). The IPLC holds and is the custodian of very important and useful Indigenous and Local Knowledge (ILK) that are critical to the scientific research and sustainable management of IAS and their disastrous and beneficial effects (Hill et al., 2020; Sankaran et al., 2023). However, IAS are perplexing to understand and manage due to their high reproduction rates, lack of natural predators/enemies, lack of competition (for space, food and mates) and their adaptability characteristic that allows them to get established very fast and well under varying weather, climatic and habitat conditions (Heuzé et al., 2015; Polce et al., 2023). Evolutionary wise, these characteristics make IAS very adaptive, more competitive and henceforth successful organisms with respect to the natural selection theory of Charles Darwin (Kimura, 1968) as drivers of species extinction and biodiversity degradation (Mulema et al., 2022). With respect to this law, the IAS therefore initiates the process of replacing the native and local species, changing the profile of ecosystems, change communal structural well-being and affecting the food-web associations and ecosystem evolutions a great deal, thus destabilizing the entire environment (Lavergne et al., 1999). The situation becomes worse if the invasive alien species is a potential cyclical and/or mechanical transmitter of deadly dangerous disease(s) as a vector (Polce et al., 2023). However, urban centres provide hot spots for the evolution of Invasive Alien Species due to their dynamism in all their operations, cultural mix and evolution and finally, a high level of anthropogenic activities (drastic changes in trade patterns, human behaviours and environmental demands due to climate change phenomena), which provide opportunities and a conducive environment for the much-needed bio-interactions and henceforth, evolution. We therefore focus on the interactions of indigenous peoples and local communities (IPLC) versus Invasive Alien Species (IAS) with remarks on urban communities.

2 The Indigenous Peoples and Local Communities in IAS in Urban Communities

2.1 General Entry of IAS into an Environment

Entry of Invasive Alien Species into an environment is by many different ways, such as during the human and animal migration process, zoopharmacognosy and colonialism mechanism as a way of providing alternatives to native species (Reo & Ogden, 2018). Some flora and fauna species are introduced into an environment as ornamentals due to their beauty, the case of the common water hyacinth in Lake Victoria in East Africa (United Nations News Highlights, 2008). In some circumstances, weather and climatic changes dictate the emergence of new plant and animal species as well as the natural dispersal mechanisms of flora and fauna species by wind/air, animals and water (Heuzé et al., 2015). During seasonal festivals and ceremonies, a lot of plant and animal species are exchanged in form of batter trade and presents and/or gifts and by this mechanism, new species are found in new environments and deliberately introduced. Some flora and fauna species found themselves in new environments as species for research and studies in the laboratories, herbaria, aquaculture, mariculture, restocking for commercial and recreational fisheries, biological weapons, arboreta, home gardens and botanical gardens, as well as the development of new breeds in the bioengineering laboratories and released into the environment. Thus businesses involving animals, microorganisms, plants and their products coupled with transport mechanisms, tourism and travel, largely driven by globalization increasingly spreading IAS (ICIPE, 2018).

2.2 The Analysis of Entry of IAS in Urban Centres

Urban communities are indeed very dynamic in their evolution at all levels due to the high level of competitiveness for limited resources in time and space (Lawson, 2016). Urban communities involve a high level of to-and-fro migrations, which provide a significant avenue and basis for the entry of new species, which may easily turn invasive in nature. Additionally, the dynamically changing environment and the evolution of ecological niches of various organisms, provide opportunities for biointeractions whose resultant effect is the emergence of new species to fit in the circumstances and scenarios of the mechanisms of natural selection theory. The dominant migration pattern within countries involving rural–urban movements comes along with the migration of species from one region to another, as humans want to maintain their original environmental orientation and patterns in terms of plant and animal species (particularly pets) surrounding them. Parasitic organisms found on and inside bodies of humans, pets, garden plant species and tools/objects humans use on a daily basis in life are accidentally and deliberately introduced in these urban environments, and unfortunately, some species become shockingly invasive in nature for unknown reasons (Mazaleyrat et al., 2023; McDonald et al., 2023). Some of the notable entry avenues of new species in urban environments include, but are not limited to, the following 20 different ways:

1. ***Slaughterhouses built in urban centres***: Many slaughterhouses are built-in urban environments due to readily available markets and sustainable network of transport systems. Many herbivores, both imported and from distant geographical locations are slaughtered for beef while the contents of the gastrointestinal tract, particularly the digesta is disposed off in the urban environments containing viable seeds from a variety of eaten plant species during grazing and a large biodiversity of endoparasites, ranging from bacteria, protozoa, fungi, archaea, helminths to viruses (Fig. 1). While hides/skins carry with them all types of ectoparasites, particularly ticks and microorganisms in the ears, around the bases of hornes, and under the tail and noses (Wanzala et al., 2004). Humans use digesta as organic fertilizers in the market gardening fields existing in the urban environments, thus spreading the new species in the ecosystem. Humans become infected with the above endoparasites and ectoparasites, passing some together with faeces into the sewage systems of urban environments as mainly larvae and cysts that start surveying the new environment for habitation.
2. ***Supply Chain of Foodstuffs in Urban Centres***: The ever-growing population of urban centres are fed on a variety of foodstuffs supplied from distant ecological zones in rural farms. The plant species coming in the urban environments through the supply chain of foodstuffs are in the form of viable seeds and vegetative stems and leaves from which new plant species can emerge and grow in the new urban environment (Fig. 2). Contamination with a wide range of microorganisms in the native original environment is inevitable, acting as mechanical vectors to enable them to reach the new urban environment.
3. ***Private and Public Service Vehicles (PSV) Coming in Urban Centres***: Vehicles moving from distant geographical locations into the urban environments carry with them many arthropod species, such as mosquitoes, ticks and other flies, which become introduced into the new ecological systems to start the process of establishment (Fig. 3). Some establish and become invasive, e.g., the case of mosquitoes that emerge from vehicles moving from malaria-endemic lowlands to uplands when such vehicles stop a long the way at a roadside and passengers open its doors and windows that allow the mosquitoes to fly out of the vehicle such as this colourful minibus.
4. ***Human being Behaviours in Urban Centres***: Human beings themselves serve a function similar to the one mentioned above of slaughterhouses in urban environments, mainly through their faeces, footwear, clothing, nails, hairs etc., moving in and out of urban centres as visitors, tourists, job seekers, workers and businessmen.
5. ***Livestock Farming and Pet keeping in Urban Centres***: Livestock farming impacts to livelihoods, both positively and negatively at all levels in life. Livestock farming in urban centres create unique ecological niches from decomposing

Fig. 1 The scenario at the abattoir waste disposal composite, which comprises contents of the gastrointestinal tract, rotting flesh and bones particularly the digesta, containing many different viable seeds from a variety of eaten plant species during grazing and a large biodiversity of endoparasites, ranging from bacteria, protozoa, fungi, archaea, helminthes to viruses. The germinating seed (top left) and a patch of lush green Kikuyu grasss (top right) both emerging from the rotting organic matter of the deposited digesta as abattoir waste. A pile of manure from the digesta (bottom left) contains viable seeds and vegetations, which germinate in arable farms when used as raw organic manure while the spread animal hides surrounded by grass and dirt (bottom right), from the slaughtered animals provide a source of new and unknown ectoparasites in the area

organic manure for potentially unique and new species in squeezed and limited environmental spaces to start the process of establishment.

Here in Kenya, pets of all kinds are found, ranging from the traditional cats, birds, fish, rabbits, guinea pigs and dogs to more exotic animals such as snakes, tortoises, monkeys, mongoose and lions (Fig. 4). Additionally, a culture of rearing pets is increasingly becoming common and migrating with pets into new urban environments, indeed pose great environmental challenges for the introduction of new species as well as the endoparasite and ectoparasite species that come along with these pets from their original habitats into the new urban centres, so they may in future become invasive, just like the issue of herbivores being brought into the urban centres to be slaughtered for the beef supply.

Fig. 2 Foodstuffs in a market in Narok, Kenya, showing a variety of fruits, particularly mangoes, bananas, oranges and other foodstuffs coming from different distant and varied ecological zones are all now introduced at one point in the urban center

Fig. 3 Organisms such as mosquitoes or tse tse flies can move from one region to another by easily becoming stuck in either a van, bus or train and be able to transmit malaria, henceforth, the resurgence of malaria in Nairobi and some highland regions in Kenya and many parts of Africa. Prospective data suggest that persons returning from the malaria-endemic lowlands such as Kisumu (Lake Victoria Basin of Kenya) transport malaria parasites up the mountain to Kericho highlands in Kenya (Ernst et al., 2006; Shanks et al., 2005). While evidence exists that residents living near train and bus entry and exit points are at a slight risk of being bitten by infected *Anopheles* mosquitoes (Centre for Tropical and Travel Medicine)

Fig. 4 All these four animals (a white rabbit on the left, standing on dry grass; a brown heifer in the center standing on the green grass; a German Shepherd dog lying on a tiled floor and a coiled snake with a patterned body on the right, resting on a wooden surface) are pets and of different ecological niches associated with different risks. To provide a sustainable preventive, control and management levels in life, different animal species are supposed to be regularly screened, particularly before moving from their original habitats (Hufnagel, 2004). Apart from them being new species, they carry with them endoparasites (intestinal and blood/tissue) and ectoparasites, which are introduced at urban market centres. In addition, their cow dung may contain many different viable seeds that passed through their gastrointestinal tract undigested and may be introduced at the urban centres as new plant species, which may grow to become invasive species in future. Sustainable prevention, control and management of pathogenic agents associated with these pets should be dealt with at the point of origin rather than when movement has already occurred, supplemented by an efficient and effective mechanism of surveillance for a specified period of time

6. ***Market Gardening, Botanical Development and Agriculture in Urban Centres***: Urban centres have cleared all their lands and paved the way for the development of mainly social amenities and buildings. House plants, garden flowers around buildings and on roadsides and small arable and livestock farming gardens, all originating from distant and different rural areas with different ecological conditions (Fig. 5). This is the case of *Lantana camara* and common water hyacinth (*Eichhornia crassipes*) in sub-Saharan Africa (SSA), which were all introduced as ornamental flower garden plant species and house plants, respectively (ICIPE, 2018). However, here in Kenya, these species have ever since become invasive, negatively impacting to the ecosystem and affecting human livelihoods in many different ways, thus becoming a socio-economic concern (Kedera & Kuria, 2005; Mulema et al., 2022).
7. ***The Practice of Ethnomedicines in Urban Centres***: This involves the utilization and application of herbalism and zoopharmacognosy, which uses different species of plants and animals, respectively, as curative agents of a wide range of human, plant and animal ill-health conditions. Some of these species are brought to the urban centres in the form of viable seeds and vegetative stems and leaves from which new plant species can emerge and grow. From the environments where the ethnomedicines are identified, harvested, packaged, stored and carried to the urban centres for dispensing, they are bound to carry with them a wide range of bacteria, protozoa, fungi, archaea, helminthes, viruses etc. (Fig. 6). All these species are introduced in the new ecological zones as new species in the urban centres and may in the future become invasive.

Fig. 5 Planting of new species of plants in the urban botanical garden of Maasai Mara University located in Narok town in Kenya. The left image shows the site in the Botanical Garden for planting aquatic plant species whereas the right image shows the site in the Botanical Garden for planting terrestrial plant species If care is not taken, these new plant species normally introduced in the new area, get out of control from their confinement framework into the uncontrolled environments in the neighbourhoods and may become invasive. Alternatively, they attract new species in form of pests, both below-ground and above-ground species from their original environments

8. ***Disposal Sites of Industrial and House Wastes in Urban Centres***: This is indeed problematic in many urban centres worldwide and creates opportunities for the existence of very unique microecological zones and niches that attract and support the emergence of a wide range of new species that may become invasive in future.
9. ***Urban Flooding due to nearby Streams/rivers and Poor Harvesting of Rain-water***: During rains, runoff water and that from poor urban drainage systems, increase the spread of species from their original location to new ones with a view to establishing themselves in the new environments.
10. ***Building and Construction in Urban Centres***: Building materials used are brought from different geographical locations, including timber from forests and sand harvested in rivers and streams, around coming with them seeds dispersed by water. Building materials carry with them a wide range of bacteria, protozoa, fungi, archaea, helminthes, viruses etc. All these species are introduced in the new ecological zones as new species in the urban centres, and may in the future become invasive.
11. ***Refugees in Urban Centres due to Civil Wars and Unrest***: Refugees present a very unique scenario in urban centres where most of them prefer to settle as they come along with them a wide range of bacteria, protozoa, fungi, archaea, helminthes, viruses etc., in all sorts of manner together with their foodstuffs, which may have viable seeds and some vegetative stems and leaves to grow into new plants are now in a new environment. All these species are introduced

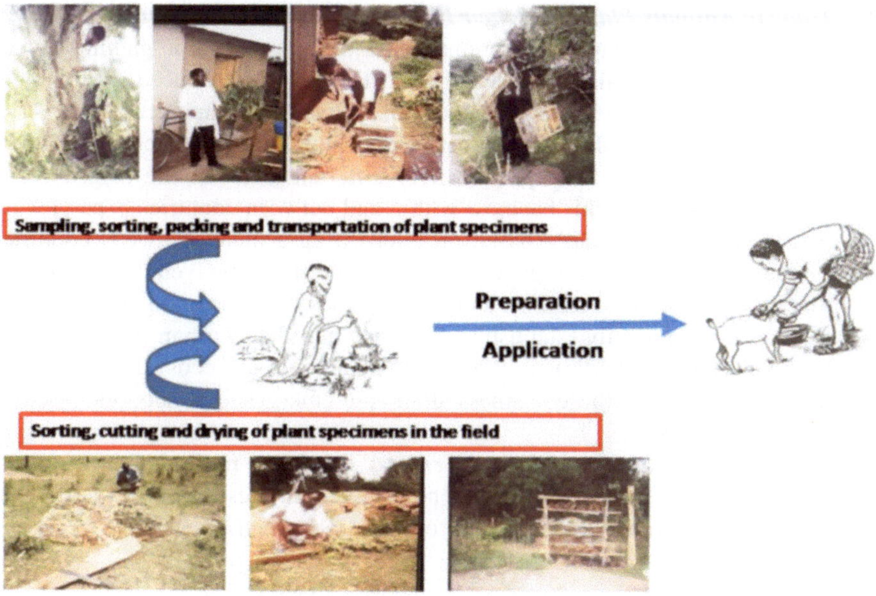

Fig. 6 The flow chart visually representing the steps from collection, sorting, preparation to application of ethnomedicines prepared from plant species. The sampling and collection of useful plant materials from the field for use as ethnomedicines is a demanding process conducted in many distant geographical locations and brought at the urban centres in most cases for confirmation (in case of uncertainties) in respective herbaria and subsequent preparation and application. Some plant materials may contain viable seeds and vegetative parts, which may grow and develop into new plant species in the new environments when the plant materials used are not well deposited

in the new ecological zones as new species in the urban centres and may in the future become invasive.

12. ***Catastrophic Wind Blowing across Urban Centres***: This particularly applies to organisms existing in the form of spores and cysts; and light hairy seeds, which are normally blown by strong winds and land on tall buildings in urban centres. Additionally, urban people inhale these organisms (in the form of spores and cysts), thereby causing infections and henceforth, establishing themselves as new organisms in the environment.
13. ***Emergency Relief and Aid***: Foodstuffs for emergency relief and aid have been the drivers for introducing new species in urban centres, particularly for refugees, who in most cases are always found residing in urban centres. These new species may become invasive with disastrous effects on livelihoods and ecosystems.
14. ***Biological Control Agents***: These are organisms, usually former enemies in the native environment, introduced to feed on other disastrous pest organisms in the new environment. The newly introduced biological agents may become invasive with disastrous effects to livelihoods and ecosystems.

15. ***Trade in Animals/Plants and their Introduction for Gaming and/or Trophies/ Gifts***: The newly introduced animals and plant species for gaming and/or trophies may gradually be established in the newly introduced environment to become invasive with disastrous effects to livelihoods and ecosystems. The profit-making businesses in livestock, plants, wildlife and associated products are huge and multifaceted and involves different levels in life (Fig. 7). The legislative control of this business within and between countries is perplexing and just complex, usually based on bilateral agreements between countries (Fèvre et al., 2006) to mitigate and manage the risks involved (Kellar, 1993). The newly introduced animals and plant species for trade and gaming, and/or trophies may gradually establish in the newly introduced urban environment to become invasive with disastrous effects to livelihoods and ecosystems, the most important output of these business affairs are emergence of zoonotic diseases (Thiermann, 2004).
16. ***Research Laboratories in Urban Centres***: This is the most challenging entry point of new species into the environment, as many research laboratories fail to contain research materials in the form of viable seeds, vegetative stems and

Fig. 7 A bustling outdoor livestock at the urban market for sale, with animals and people from distant geographical locations. Apart from the animals being m being new species introduced into the new environment, they carry with them endoparasites (intestinal and blood/tissue) and ectoparasites, which are introduced at urban market centres. In addition, their cow dung may contain many different viable seeds that passed through their gastrointestinal tract undigested and may be introduced at the urban centres as new plant species, which may grow to become invasive species in future

leaves and live animals; these materials escape into the neighborhood environments to start the process of establishment gradually. Following establishment, they may become invasive with disastrous effects on livelihoods and ecosystems.

17. ***Migratory Birds and other Organisms***: Due to the presentation of different conditions for different seasons and erratic changes in weather and climate in various ecosystems worldwide, birds and other organisms as well as cattle for pastoralists (Fig. 8) tend to move from one habitat to another many thousands of kilometers away, such as famous Kenya tourism hotspot of the Great Wildebeest Migration in the Maasai Mara National Reserve, the journey that begins from northern Tanzania, crossing the border to southern Kenya. Other organisms such as birds, endoparasites, ectoparasites, arthropods etc. migrate with these wildebeest and permanently establish themselves in a new environment together with the undigested and viable seeds and other vegetable materials eaten from northern Tanzania to southern Kenya (Wehi et al., 2023). Moving from distant geographical locations into the urban environments in search of suitable ecological niches, carry with them many arthropod species, such as the hardy ticks and many endoparasites, ranging from bacteria, protozoa, fungi, archaea, helminthes to viruses. The organisms become introduced into the new ecological systems to start the process of establishment. Some establish and become invasive.
18. ***Traditional Regalia used in Ceremonies***: The traditional ceremonies take place both in urban and rural environments. Many of the regalia used in these ceremonies are made from animal skins and hides, tails, horns, hooves and bones (Fig. 9). Some may involve the use of parts of the gastrointestinal tract. They carry with them many arthropod species, such as the hardy ticks and sometimes, the larvae and cysts of endoparasites, ranging from bacteria, protozoa, fungi,

Fig. 8 Cattle move many kilometers in search of good water and pastures as well as market centres for sale. Just like migrating wild animals, pastoralists with their animals move from one place to another looking for green pastures and water. In the course of their movement, the animals carry with them endoparasites (intestinal and blood/tissue) and ectoparasites, which are introduced at urban market centres and in the new environments where they have gone to look for clean water and good pastures. In addition, their cow dung may contain many different viable seeds that are passed through their gastrointestinal tract undigested and may be introduced at the urban centres as new plant species, which may grow to become invasive species in future

Fig. 9 Traditional regalia worn by humans—a biointeractions with nature: On the left, a person stands outdoors on grass wearing a traditional garment made from the hide of the Gazalle—*Likutu lie Swara* (Left), a hat from the skin of black and white colobus monkey (in the fore background) and on the right, the same person on different occasion, is wearing a traditional garment made from the Goat—*Likutu lie Mbusii* (right). While on the hat (right side), there is an Ostrich feather. All these items, if care is not taken, carry with them ectoparasites and cysts of endoparasites together with viable seeds of some plant species, which are most likely to be introduced in a new environment

archaea, helminthes to viruses. Those made from wild animals may carry with them the spores and seeds from plant species. These organisms become introduced into the new ecological systems to start the process of establishment. Some establish and become invasive.

19. ***Ecotourism and Camping in Urban Centres***: Just like elsewhere in the world, ecotourism and camping present lifetime travel moments in Kenya, ranging from walking safaris, camping, the big five game drives, village bike tours to yoga sunset wholesome experiences. As humans go for mesmerizing sceneries around the world, they come along with them a wide range of bacteria, protozoa, fungi, archaea, helminthes, viruses etc., in all sorts of manner together with their foodstuffs, which may have viable seeds and some, vegetative stems and leaves to grow into new plants now in a new environment. All these species are introduced in the new ecological zones as new species in the urban centres and may in future become invasive.
20. ***Urban Trade in Second-hand Clothing (locally known as* Mitumba*in Kenya)***: This business presents a lot of challenges as it involves the transit of second-hand items around the world in an unchecked manner. They carry with them many

anthropogenic-based pathogens in the form of viable eggs and cysts of microorganisms. Some clothing materials contain spores and seeds from distant plant species. These organisms become introduced into the new ecological systems to start the process of establishment. Some establish and become invasive.

All these entry avenues of new species into urban centres, present significant scenarios that surpass any existing legislative regulatory mechanisms put in place to prevent, control and manage the invasive alien species through monitoring and evaluation tools may not be possible to deal with every item at all the entry points in today's dynamic environments of urban centres coupled with corruption, tribalism, favouritism and nepotism in many circumstances of operations. Under the prevailing circumstances, therefore, prevention, control, and management of IAS can be achieved, if there must be a consensus agreed upon as a personal commitment and collective responsibilities for every individual in the world to be involved at all level of operations in the human societies.

2.3 The Identification of IAS by ILPC

Indigenous and local people who have lived in any one ecosystem for considerably many years, spanning several decades, while interacting and sustainably utilizing the ecological niches to their advantage and that of nature, are able to detect any changes in species diversity of that geographical location with precision (Ellis et al., 2021; IPBES, 2019). This change in species is normally realized by the following undertakings:

1. experiencing the long-term and the short-term disastrous effects of the newly introduced and emerging species,
2. gradual and drastic reduction in population of the original species, more particularly useful ones,
3. emergence of new species,
4. deliberate extinction of rare species and
5. experiencing new and unique food-web patterns in different ecological zones.

By realizing these changes in the acquainted ecosystem, new species are detected and invasive ones are determined (Wehi et al., 2021) and consequently, their use is uniquely sought if need be in relationship with the existing species and previous holistic experience of ethnoknowledge of the entire ecosystem. These ethnoknowledge-based strategic approaches of indigenous and local people appear realistic in developing long-term and sustainable mitigations against the bad effects of the invasive alien species. Therefore, the role of indigenous people and local communities in IAS in urban communities is of paramount importance in the prevention, control and management of IAS at all levels, including their disastrous and beneficial effects to the affected ecosystem. Nevertheless, while indigenous and local people understand the dynamic nature of their ecosystems very well, and are able to detect new species and their effects, their valuable ethnoknowledge, however, remains elusively underutilized in many different ways in managing the problems facing environments as a result of IAS (Wehi et al., 2023).

3 How Some IPLC Determine the Impacts of IAS

Indigenous Peoples and Local Communities (IPLC) have made use of their Indigenous and Local Knowledge (ILK) through observations and record keeping by memory for many years (mainly over 70 years old) in a given environment to be able to know the presence of Invasive Alien Species (IAS) (Sankaran et al., 2023). In most cases, the IAS threatens the existence and survival of the biodiversity of flora and fauna throughout the ecosystems in the world (Gopal, 2018; Heuzé et al., 2015; Rangi, 2018). The emergency and invasion of IAS destroy habitats and create problems for other native species, respectively, thus causing species endangerment and extinction, henceforth accruing irretrievable loss to the world (Rangi, 2018; Thapa et al., 2014). Additionally, the IAS causes tremendous socio-economic losses in livestock, forest, aquatic and arable farming productivity in the world (Heuzé et al., 2015; Wilt, 2018), thus threatening the survival of livelihoods. Therefore, by being problematic to the world ecosystems and IPLC, the case of Larger grain borer (*Prostephanus truncates*), Nile perch (*Lates niloticus*), House sparrow (*Passer domesticus*), Crown gall (*Agrobacterium tumefaciens*), Black Sigatoka (*Mycosphaerella fijiensis*), armyworm (moth), potato cyst nematode, fruit flies species (*Bactrocera zonata, B. dorsalis* and *B.* latifrons), *Aedes albopictus*, *Citrus* psyllid (*Diaphorina citri*), *Tuta absoluta* (moth), *Cypress* aphid (*Cinara cupressivora*), Louisiana crayfish (*Procambarus clarkia*), *Striga hermonthica, Chromolaena odorata, Mimosa diplotricha, Xanthium strumarium, Tagetes minuta, Parthenium hysterophorus, Mathenge* plant (*Prosopis juliflora*), *Eucalyptus* spp., *Lantana camara* and common water hyacinth (*Eichhornia crassipes*) in sub-Saharan Africa (SSA), the IAS are realized and henceforth identified (ICIPE, 2018). IAS are also identified through new uses and applications realized by IPLC through visions, dreams and spiritual means as well as by apprenticeship and mentoring mechanisms across various ethnic backgrounds and during ceremonies and festivals. IPLC also discovers IAS during hunting, fishing and gathering expeditions (Ingold, 2000) and during the emergence of new diseases, pests and aetiologic agents in the ecosystem. Extinctions and the emergence of new species of fauna and flora in the ecosystems result in new uses and applications of ILK in a way that allows IPLC to realize the existence of IAS (Thapa et al., 2014). Nevertheless, beneficial aspects of IAS have been globally recognized (such as cultural artifacts, pesticides, recreation sites, spiritual shrines, weedicides, fuelwood, fodder, timber, food products, cosmetics, pharmaceuticals, windbreaks, shade and beauty) (Shackleton et al., 2019) and IPLC also appreciate, those key benefits originating from the scientific and technological innovations due to research on IAS (Gopal, 2018), henceforth the need for strong and sustainable collaboration and partnership.

4 Indigenous Peoples and Local Communities (IPLC) Approaches to the Impacts of IAS

The impacts of invasive alien species around the world are enormous, treacherous and more often than not multifaceted in a way that Indigenous Peoples and Local Communities could not and cannot ignore the phenomenon and survive sustainably for their livelihoods are directly dependent on the ecosystem that is being degraded by the IAS. Therefore, what are some of the notable responses of IPLC to the impacts of IAS in their respective environments and/or ecological niches in terms of monitoring, evaluation, reporting, prevention, controlling and management of the IAS together with their disastrous effects? Such consideration of collaboration and partnership amongst IPLC, will enable every element in the ecosystem to become accountable and henceforth, responsible. Nevertheless, it is indeed difficult to get relevant literature detailing the responses of IPLC to the impacts of IAS, because they are always ignored in science, law and policy-making process (Grenz, 2020), henceforth, the beginning of our IAS problems in society.

4.1 IAS Utilized as Food

Accepting IAS into native food ethno-practices by IPLC is also one of the concern (Kamelamela, 2019). For example, in Tuawhenua area, forest health and treasured species declined as a result of logging and the presence of IAS such as the deer and the possums. However, the IAS became highly valued for their protein or fur in the absence of other previously harvested species by IPLC (Lyver et al., 2017). In Kenya, Nile perch, *Lates niloticus*, and House sparrow, *Passer domesticus* are utilized by IPLC as food (a good source of proteins) and a source of revenue, while they are serious IAS (Kedera & Kuria, 2005).

4.2 Physical Removal

Like for the case of common water hyacinth (*Eichhornia crassipes* (Mart.) Solms), physical removal approach has proved futile since the plant comes back almost immediately as its seeds spread through animal feaces and can survive 15–20 years (Heuzé et al., 2015), hence ineffective and expensive (Gopal, 2018). For many years, IPLC have been, culturally, uprooting the parasitic purple witchweed (*Striga hermonthica* (Del.) Benth.) from maize plantations without eminent results (Oswald, 2005). However, in the case of Nile perch (*Lates niloticus*), fishing became an economic boost to fishers instead, and reduced the catch of smaller fish species in Lake Victoria (Kedera & Kuria, 2005).

4.3 Adopting Avoidance Behaviour

The IPLC has strategies for avoiding new biological species occurring in the ecosystem by not involving them in their cultural practices and customs, instead associating them with evils in the society. Further to this, some IPLC have set aside land for newly introduced species and IAS so that their original biological diversity is protected (Kamelamela, 2019).

4.4 Changing the Law

In some places, such as Kenya, the colonial masters wrote the country's constitution at independence, including regulations and policies that currently deny the IPLC the right to biointeract with the natural environment they have bonded with for so many years in terms of monitoring, evaluation, reporting, prevention, controlling, and management of the IAS and their effects. To restore glory in their own ethno-knowledge, the law needs to be changed so that it empowers IPLC to be in charge of their own destiny (Wehi et al., 2023). By this strategy, they will be able to differentiate with precision and confidence, the native species and new species that are becoming invasive in their ecosystems.

4.5 Cultural Fires

It is applied by Yellomundee Aboriginal Bushcare in Australia (Barber & Glass, 2015). As per the IPLC, it is "*a cool fire that burns the invasive weed to allow native species a chance to regenerate. Cultural fire removes biomass and allows sunlight to hit the ground where native seedlings fight to compete with exotic weeds. Cool fire also stimulates native seed bank while not burning surrounding native gum trees.*" This approach is meant to improve the livelihoods and well-being of people as well as sustain a healthy approach to natural and cultural resource management.

4.6 Sacred Bathing and Worshipping

Local Rajbanshis people from North Bengal in India perform certain sacred bathing in winter and spring such as *Maghali sinan, Bauni sinan* and also worship the rivers when monsoon begins to ward off invasive pests naturally (Gupta, 2015). While in Africa, to ward off invasive pests, drums, basins, saucepans etc. are hit to produce sound coupled with human noise of refusing the pest.

4.7 Place-Based Weed Management Approach

Working on behalf of the Bunuba people, the Rangers in the Kimberley, Western Australia manages invasive weed species based on the cultural values attached to them and the specific culturally significant sites identified (Aiken et al., 2015).

4.8 Chemical Treatments

The use of ethnobotanicals such as the powder of the Neem tree (*Azadirachta indica*) by some IPLC in Kenya to manage the Larger grain borer (*Prostephanus truncates*) and ash to dust-affected plant species with a view to controlling the Cypress aphid (*Cinara cupressivora* Watson & Voegtlin, 1999). Strategic spraying or stem injecting herbicide is conducted, as is the case in two Indigenous Protected Areas (IPAs) of Wattleridge and Tarriwa Kurrukun in northern New South Wales, Australia (Elone et al., 2015).

4.9 Incorporation of IAS into IPLC's Cultural Ethno-practices

Accepting IAS into material cultural ethnopractices (Kamelamela, 2019), is the case for the Mexican marigold, *Tagetes minuta*, Wild garlic, *Allium vineale, Lantana camara* and *Eucalyptus* spp. in Kenya are used as ethnomedicines by IPLC, while on other hand are serious IAS (Kedera & Kuria, 2005).

4.9.1 Biological Control

Also known as biocontrol, is the utilization of organisms to control pest populations usually from the pest's area of origin (IPPC Secretariat, 2005). The classical science of biological control of invasive species beyond the ethnoknowledge of IPLC, has existed for over 209 years (Seastedt, 2014). The biocontrol was found to be an environmentally friendly tool, a cost-effective and justifiable way of dealing with invasive species and has been used commendably for more than 100 years, since its inception some 209 years ago (CABI, 2023). In its project highlights, CABI (2023) presents very successful stories of the classical science of biological control of invasive species. Although used with some fears, biological control helps to maintain biodiversity and contributes to human wellbeing (CABI, 2023). However, the biological control shall interrupt the normal biointeractions as well as the food-web biorelationships in the ecosystem, thereby degrading the environmental well-being and henceforth, livelihoods (Seastedt, 2014). Nevertheless, the ILPC appreciates this classical science of biological control of invasive alien species in the world.

In Africa, the use of indigenous chickens to control the Larger grain borer (*Prostephanus truncates*) in maize grains during sun drying is a very common phenomenon as well as the use of indigenous chicken (Wanzala et al., 2012) and ox peckers (Wanzala, 2009) to control livestock ticks.

4.9.2 *Economic Utilization of IAS with Special Focus on Common Water Hyacinth (***Eichhornia Crassipes** *(Mart.) Solms [Pontederiaceae])*

Many people in different countries are working to find new uses for water hyacinth. There are so many good uses for just this one plant: feed it to your animals; make paper, good quality organic fertilizer, mats, or compost; burn it for fuel; or build a floating garden of your own, while women in Fiji weave baskets with dried water hyacinth (Casebow, 1967; Farm Radio International, 2000; Wolverton & McDonald, 1976) and also used to make furniture (Rezania et al., 2015). The plant can be used to make biomass briquettes, as well as a bedding material for mushroom cultivation and modified hydroponics (floating agriculture). Together with small amounts of used newspapers and appropriate binders, the plant has also been used to make products such as disposable plates, ready-to-plant biodegradable nursery pots, egg and fruit trays, cartoon models, toys, file boards, multi-purpose boards, special canvas for paintings, etc. (Gopal, 2018). In India, people have blended the coir industry waste fibers popularly known as "baby fibres", along with water hyacinth pulp to prepare the unique canvas for paintings (Gopal, 2018). Water hyacinth can be useful for the bioremediation of water contaminated with heavy metals, as the roots trap and remove from water large amounts of minerals, including heavy metals and radioactive elements (Heuzé et al., 2015; Onyango et al. 2020; Rezania et al., 2015; Wolverton & McDonald, 1976). Upon discovery of their nutritional and medicinal values, a wide range of Invasive alien species, including common water hyacinth are utilized by IPLC to manage human, plant, environmental and livestock health as these species help reduce pressure on heavily harvested indigenous ones (Gakuubi et al., 2016a; Kobisi et al., 2019; Maema et al., 2016; Wagh & Jain, 2018; Wanzala et al., 2012; Wanzala, 2017). This scenario of common water hyacinth also applies to many other IAS, such as the case of Nypa palm (*Nypa fruticans*, Van Wurmb. (Arecaceae)) in Nigeria and in the Gulf of Guinea, where the species is rapidly replacing the native mangrove forests with multiple uses and applications to the Indigenous Peoples and Local Communities (Moudingo et al., 2015).

5 How Some IPLC Use Their ILK to Develop IAS Management Interventions

Just like the internationally recognized IAS interventions involving—legislation, trade regulation, border controls, phytosanitary laws, early detection, eradication, population controls, monitoring and evaluation and restoration of affected ecosystems (Crowley et al., 2017; ICIPE, 2018; Shine, 2005), IPLC has a robust similar system too. The council of elders for a given region monitors and evaluates the entire ecosystem situation and give reports during the IPLC meetings. The council of elders works in harmony with the people's culture and customs (Aiken et al., 2015). For instance, in *Bukusu* community in Kenya and the *Bagisu* in Uganda, for a new species of plant identified in the environment, it is a rule that an elder is notified, who convenes a meeting of the council of elders to confirm the new species of plant, quarantine it and set a date for a ritual ceremony to be performed to determine its destiny. On the set date, a sheep is slaughtered at the site of the new plant and the stomach contents of the sheep together with the aerial parts of a plant called *Vitex doniana* shrub, locally known as *Lufufu* (plural: *Chifufu*) are mixed in water and the elder puts them on the plant and around it while making the oratory oathing statement of the local community. On the 3rd day, they come back to check on the leaves of *Lufufu*, if they are dry, then the plant is a bad one and is uprooted and completely burnt. But if they come back on the 3rd day and find that the leaves have not dried up, then the plant means good for the native ecosystem and it is then given a local name and its uses and applications are associated with those of a similar local plant species in the environment. If it is a new animal species (whether *Esang'i-* the eaten wild animal species or *Esolo*-none eaten wild animal species) that had been seen in the area, the council of elders identifies its footprints and a child is given a concoction of *Kulandula* plant to put in the footprints as the elders curse the animal never to come back again since its effects to the native ecosystem, economy and livelihoods are not known. Surveillance of the native ecosystem is done during hunting and gathering while IPLC encourages reporting of new biological species emerging in the environment. Amongst the community members, the batter trade is strictly conducted with only known biological species of fauna and flora and/or minerals but not foreign ones. This information is held by the council of elders by memory and trust and is passed on from generation to generation by word of mouth, just the way it is done in Kimberley, Northwest Australia by Aboriginal elders (Wanzala et al., 2012; Weir & Duff, 2015).

6 Management Programmes of IAS Applicable by IPLC

Indigenous Peoples and Local Communities (IPLC) have quarantine measures such as restrictions of not bringing certain species of flora and fauna into their environment as a taboo. These species could be either new (not known to that community) and/or prohibited ones and henceforth considered evil. Such newly identified biological species are not used in their cultural practices and customs in any way as a quarantine measure. This follows education, public awareness and capacity building by old men and women in society, targeting the young generation as a way of passing on knowledge from one generation to the next by word of mouth regarding the prohibited species of flora and fauna. Such programs are particularly conducted during festivals and ceremonies like those of the end of the year, Teso and Bukusu/Bagisu Kenya-Uganda transboundary customs and cultural practices (Angujo, 2015;

Barasa, 2012). Monitoring and evaluation through observation and comparing, and contrasting the current and previous environments, can help determine hotspots in any given ecosystem (Adhikari et al., 2015). CABI conducts capacity building and creates awareness of IAS amongst Indigenous Peoples and Local Communities with a view to improving rural poor livelihoods (CABI, 2015).

7 IPLC and ILK Versus Science, Technology and Innovation (STI)

From the very beginning, Indigenous and Local Knowledge (ILK) systems from IPLC have been the foundation and drivers of the evolution and advancement of Science, Technology and Innovation (STI) in society and henceforth, enablers of sustainable development (UNSG's SAB, 2016). The ILK therefore is complementary to STI and integral to knowledge-policy platforms on sustainable development, biodiversity and climate change, as noted by the United Nations Secretary General's (UNSG's) Scientific Advisory Board (SAB). From the foregoing, there is a great need for IPLC—STI partnership involvement to fight this slowly emerging global problem of IAS (Rangi, 2018). Already, there is a realization that ILK is critical to the scientific research and sustainable management of IAS in our local communities (Hill et al., 2020). Many uses and applications of IAS by IPLC are becoming STI-based, for instance like those of *Tagetes minuta* (Wanzala & Wanjala, 2016; Gakuubi et al., 2016b), *Eucalyptus globulus* (Atieno et al., 2019) and common water hyacinth (Gopal, 2018). It therefore follows with logical necessity that the IPLC—STI partnership involvement enhances the development of sustainable home-grown solutions to chronic problems within the frameworks of cultures that value people and their STI within societies (Thaman et al., 2013; UNESCO, 2010).

8 Local Cultural Values Used to Manage Invasive Alien Species (IAS)

Cultural values are critical in the management of Invasive Alien Species (Bonanno, 2016; Ens et al., 2015; Liu et al., 2011) since they enter native ecosystems partly due to their opportunistic utilization of anthropogenically-based disturbances (Kunwar, 2003). Changes in fauna and flora species by the IAS in an ecosystem result in changes in people's cultural values and practices as the ecosystem become transformed (Dickie et al., 2014; Russell et al., 2017) and as the original cultural value of the landscape change (Shackleton et al., 2019). Nevertheless, some IAS are on the contrary integrated into people's cultures and utilized for the benefit of IPLC (Dickie et al., 2014; Moore & Pennington, 2015; Nuñez & Simberloff, 2005) and control of such IAS must therefore consider people's cultural framework (Dickie et al., 2014; Ens et al., 2015; Russell et al., 2017). Some Invasive Alien Species, when grow to maturity, are transformed into worshipping shrines, while others are used as official recreation sites by IPLC (Shackleton et al., 2019) and such IAS become indigenised by succeeding generations in time and space. Traditional regalia for Kingship, eldership and initiation of the youths are ethnic-specific in Africa and traditionally made from specific animal and plant species. In the absence of the original biological species, the alternative IAS have been replaced and utilized. This phenomenon applies to a wide range of African cultural practices, which is strategic in the management of IAS.

9 Conclusions and Future Prospects

The Invasive Alien Species (IAS) are truly problematic worldwide, causing significant socio-economic losses in respective ecosystems. To sustainably mitigate the impacts of IAS, there must be regional and international cooperation, capacity building amongst key stakeholders, strategic research and monitoring mechanisms, development of quarantine measures, enactment and implementation of international and national legislation, prevention/control and containment measures, public education to understand, awareness and mass campaign about the dangers of IAS to biological diversity and related ecosystems. In all these measures, the ethnoknowledge, cooperation, collaboration and partnerships of Indigenous Peoples and Local Communities (IPLC) in monitoring, evaluating, reporting, preventing, controlling and managing IAS is supreme (IPBES, 2022). The threats posed by IAS to biodiversity, ecosystem resilience, policy-making processes and socio-economic stability, are most likely to increase and become complicated in the near future if appropriate and sustainable measures are not put in place to monitor, evaluate, report, prevent, control and manage the alien species together with their disastrous effects. For this to be achieved, there is a great need for affected countries to invest heavily in research on IAS so that concrete evidence-based data and information can be generated for policymakers to help allocate enough resources that can be sufficient to prevent the entry and spread of new IAS and mitigate their impacts.

References

Adhikari, D., Tiwary, R., & Barik, S. K. (2015). Modelling hotspots for invasive alien plants in India. *PLoS ONE, 10*(7), Article e0134665. https://doi.org/10.1371/journal.pone.0134665

Aiken, C., Davey, N., Chungul, K., Bach, T., Rangan, H., & Kull, C., (2015). Managing weeds on Bunuba Country in the Kimberley, Western Australia. In E. Ens, J. Fisher, & O. Costello, (Eds.), *Indigenous people and invasive species: Perceptions, management, challenges and uses*. IUCN Commission on Ecosystem Management Community Report.

Angujo, B. (2015). *Ugandan cultural King arbitrates Kenyan politician.* https://www.standardmedia.co.ke/ureport/article/2000160058/ugandan-cultural-king-arbitrates-kenyan-politician. As retrieved on April 28th, 2015 at 10:51 AM East Africa Time.

Atieno, O. J., Wanzala, W., & Xue, Y. (2019). Production of biomass briquettes from Eichhornia crassipes ((Mart.) Solms) and Eucalyptus globulus (Labill.) leaves. In *The 2019ANSO-MMU-SAJOREC International Conference on "Biodiversity in Drylands and Wetlands: Challenges and Opportunities in the 21st Century"* held between 4th and 6th September, 2019 at Maasai Mara University (MMU), Narok County, Kenya.

Bacher, S., Blackburn, T. M., Essl, F., Genovesi, P., Heikkilä, J., Jeschke, J. M., Jones, G., Keller, R., Kenis, M., Kueffer, C., Martinou, A. F., Nentwig, W., Pergl, J., Pyšek, P., Rabitsch, W., Richardson, D. M., Roy, H. E., Saul, W.-C., Scalera, R., & Kumschick, S. (2018). Socio-economic impact classification of alien taxa (SEICAT). *Methods in Ecology and Evolution, 9*(1), 159–168.

Barasa, M. W. (2012). Celebrating Bukusu tradition at the cultural Festival. https://www.monitor.co.ug/artsculture/Reviews/691232-1306184-124egjiz/index.html. As retrieved on Monday January 16 2012 at 12:27 PM East Africa Time.

Barber, D., & Glass, P. (2015). Yellomundee firesticks. In E. J. Ens, J. Fisher, & O. Costello (Eds.), *Indigenous people and invasive species: Perceptions, management, challenges and uses.* IUCN Commission on Ecosystem Management Community Report.

Barbet-Massin, M., Salles, J.-M., & Courchamp, F. (2020). The economic cost of control of the invasive yellow-legged Asian hornet. *NeoBiota, 55*, 11–25.

Bartz, R., Heink, U., & Kowarik, I. (2010). Proposed definition of environmental damage illustrated by the cases of genetically modified crops and invasive species. *Conservation Biology, 24*, 675–681.

Bartz, R., & Kowarik, I. (2019). Assessing the environmental impacts of invasive alien plants: A review of assessment approaches. *NeoBiota, 43*, 69–99. https://doi.org/10.3897/neobiota.43.30122

Bonanno, G. (2016). Alien species: To remove or not to remove? That is the question. *Environmental Science & Policy, 59*, 67–73.

CABI. (2015). Centre for Agriculture and Biosciences International (CABI). In E. Ens, J. Fisher, & O. Costello, (Eds.). *Indigenous people and invasive species: Perceptions, management, challenges and uses.* IUCN Commission on Ecosystem Management Community Report.

CABI. (2023). Project highlights: Biological control of invasive species. https://www.cabi.org/what-we-do/cabi-centres/biological-control-of-invasive-plants/#:~:text=Classical%20biological%20control%2C%20or%20biocontrol,species%20in%20its%20native%20range

Casebow, A. J. (1967). Letter to the editor on the possibility of converting of water hyacinth to methane. *PANS (C), 13*(3), 243–244.

Crowley, S. L., Hinchliffe, S., & McDonald, R. A. (2017). Invasive species management will benefit from social impact assessment. *Journal of Applied Ecology, 54*, 351–357.

Dickie, I. A., Bennett, B. M., Burrows, L. E., Nuñez, M. A., Peltzer, D. A., Porté, A., Richardson, D. M., Rejmánek, M., Rundel, W. P., Brian, W., & van Wilgen, B. W. (2014). Conflicting values: Ecosystem services and invasive tree management. *Biological Invasions, 16*, 705–719. https://doi.org/10.1007/s10530-013-0609-6

Ellis, E. C., Gauther, N., Goldewijk, K. K., Bird, R. B., Boivin, N., Díaz, S., Fuller, D. Q., Gill, J. L., Kaplan, J. O., Kingston, N., & Locke, H. (2021). People have shaped most of terrestrial nature for at least 12,000 years. *Proceedings of the National Academy of Sciences of the United States of America, 118*, Article e2023483118.

Elone, T., Patterson, T., & McKemey, M. (2015). Weeds, feral animals, fire and research at Wattleridge and Tarriwa Kurrukun Indigenous Protected Areas, Australia. In E. Ens, J. Fisher, & O. Costello (Eds.), *Indigenous people and invasive species: Perceptions, management, challenges and uses.* IUCN Commission on Ecosystem Management Community Report.

Ens, E., Fisher, J., & Costello, O. (Eds.). (2015). *Indigenous people and invasive species: Perceptions, management, challenges and uses.* IUCN Commission on Ecosystem Management Community Report.

Ernst, K. C., Adoka, S. O., Kowuor, D. O. et al. (2006). Malaria hotspot areas in a highland Kenya site are consistent in epidemic and non-epidemic years and are associated with ecological factors. *Malaria Journal, 5*, 78 (2006). https://doi.org/10.1186/1475-2875-5-78

Eschen, R., Beale, T., Bonnin, J. M., et al. (2021). Towards estimating the economic cost of invasive alien species to African crop and livestock production. *CABI Agriculture and Bioscience, 2*, 18. https://doi.org/10.1186/s43170-021-00038-7

Farm Radio International. (2000). The many uses of water hyacinth. From Package 54: Water: Our Precious Resource. http://scripts.farmradio.fm/radio-resource-packs/package-54-water-our-precious-resource/the-many-uses-of-water-hyacinth/

Fèvre, E. M., Bronsvoort, B. M., Hamilton, K. A., & Cleaveland, S. (2006). Animal movements and the spread of infectious diseases. *Trends in Microbiology, 14*(3), 125–31. https://doi.org/10.1016/j.tim.2006.01.004. Epub 2006 Feb 7. PMID: 16460942; PMCID: PMC7119069.

Gakuubi, M. M., Wagacha, J. M., Dossaji, S. F., & Wanzala, W. (2016a). Chemical composition and antibacterial activity of essential oils of *Tagetes minuta* (Asteraceae) against selected plant pathogenic bacteria. *International Journal of Microbiology, 2016*. Article ID 7352509, 9. https://doi.org/10.1155/2016/7352509

Gakuubi, M. M., Wanzala, W., Wagacha, J. M., Dossaji, S. F. (2016b). Bioactive properties of *Tagetes minuta* L. (Asteraceae) essential oils: A review. *American Journal of Essential Oils and Natural Products, 4*(2), 27–36. (ISSN: 2321-9114) http://www.essencejournal.com/vol4issue2.html

Gopal, S. (2018). Researchers innovate to make money out of water hyacinth. https://india.mongabay.com/2018/01/researchers-innovate-to-generate-money-out-of-water-hyacinth/

Grenz, J. B. (2020). *Healing the land by reclaiming an indigenous ecology: A journey exploring the application of the indigenous worldview to invasion biology and ecology.* (PhD Thesis). University of British Columbia.

Gupta, A. D. (2015). Invasive pest attack in North Bengal. In E. Ens, J. Fisher, & O. Costello (Eds.), *Indigenous people and invasive species: Perceptions, management, challenges and uses.* IUCN Commission on Ecosystem Management Community Report.

Heuzé, V., Tran, G., Hassoun, P., Régnier, C., Bastianelli, D., & Lebas, F. (2015). *Water hyacinth (Eichhornia crassipes).* Feedipedia, a programme by INRA, CIRAD, AFZ and FAO. https://www.feedipedia.org/node/160 Last updated on October 13, 2015, 16:25hrs.

Hill, J.E., Copp, G.H., Hardin, S., Lawson, K.M., Lawson, Jr. L.L., Tuckett,Q.M., Vilizzi, L., Watson, C.A. (2020). Comparing apples to oranges and other misrepresentations of the risk screening tools FISK and AS-ISK - a rebuttal of Marcot et al. (2019). *Management of Biological Invasions, 11*, 325–341. https://doi.org/10.3391/mbi.2020.11.2.10

Hufnagel, L. (2004). Forecast and control of epidemics in a globalized world. *Proceedings of the National Academy of Sciences, 101*, 15124–15129.

ICIPE. (2018). *Tackling invasive alien species in Africa. A major forum aimed towards developing a strategy for tackling invasive species in sub-Saharan Africa (SSA)* held between 20 and 21 February 2018 at the International Centre of Insect Physiology Ecology (ICIPE), Duduville Campus, Nairobi, Kenya. http://www.icipe.org/news/tackling-invasive-alien-species-africa. As retrieved on January 30th, 2020 at 23:47 hrs. East Africa Time.

Ingold, T. (2000). Hunting and gathering as ways of perceiving the environment. In T. Ingold (Ed.), *The perception of the environment: Essays on livelihood, dwelling and skill* (pp. 31–54). Routledge.

IPBES. (2019). Global assessment report on biodiversity and ecosystem services of the Intergovernmental Science-Policy Platform on Biodiversity and Ecosystem Services. https://doi.org/10.5281/ZENODO.3831674.

IPBES. (2022). Report of the third indigenous and local knowledge dialogue workshop for the IPBES thematic assessment of invasive alien species and their control. Online,

1–3 February 2022. https://www.ipbes.net/sites/default/files/2023-02/IPBES_IAS_3rdILKDialogue_SPM-SOD_Report_FinalForWeb2.pdf

IPPC Secretariat. (2005). Identification of risks and management of invasive alien species using the IPPC framework. In *Proceedings of the workshop on invasive alien species and the International Plant Protection Convention, Braunschweig, Germany, 22–26 September 2003*. Rome, Italy, FAO. xii + 301 pp.

IUCN. (2020). *Invasive alien species and sustainable development*. The International Union for Conservation of Nature (IUCN). Issues Brief. https://www.iucn.org/resources/issues-briefs/invasive-alien-species-and-sustainable-development. As accessed on February 19th, 2020 at 11:46 PM East Africa Time.

Kamelamela, K. L. (2019). *Contemporary Hawai'i non–timber forest plant gathering practices* (PhD Thesis). University of Mānoa.

Kedera, C., & Kuria, B. (2005). Invasive alien species in Kenya: Status and management. In *Identification of risks and management of invasive alien species using the IPPC framework. Proceedings of the workshop on invasive alien species and the International Plant Protection Convention (IPPC), IPPC Secretariat, Braunschweig, Germany, 22–26 September 2003*. Rome, Italy, FAO. xii + 301 pp.

Kellar, J. A. (1993). The application of risk analysis to international trade in animal products. *Revue Scientifique et Technique, 12*, 1023–1044.

Kimura, M. (1968). Evolutionary rate at the molecular level. *Nature, 217*(5129), 624–626.

Kobisi, K., Seleteng-Kose, L., & Moteetee, A. (2019). Invasive alien plants occurring in Lesotho: Their ethnobotany, potential risks, distribution and origin. *Bothalia, 49*(1), Article a2453. https://doi.org/10.4102/abc.v49i1.2453

Kunwar, R. M. (2003). Invasive alien plants and *Eupatorium*: Biodiversity and livelihood. *Himalayan Journal of Sciences, 1*(2), 129–133.

Lavergne, C., Rameau, J.-C., & Figier, J. (1999). The invasive woody weed *Ligustrum robustum* subsp. walkeri threatens native forests on La Reunion. *Biological Invasions, 1*, 377.

Lawson, L. (2016). Agriculture: Sowing the city. *Nature, 540*, 522–523.

Liu, S., Sheppard, A., Kriticos, D., & Cook, D. (2011). Incorporating uncertainty and social values in managing invasive alien species: A deliberative multi-criteria evaluation approach. *Biological Invasions, 13*, 2323 (2011). https://doi.org/10.1007/s10530-011-0045-4

Lyver, P. O. B., Timoti, P., Jones, C. J., Richardson, S. J., Tahi, B. L., & Greenhalgh, S. (2017). An Indigenous community–based monitoring system for assessing forest health in New Zealand. *Biodiversity and Conservation, 26*, 3183–3212.

Maema, L. P., Potgieter, M., & Mahlo, S. M. (2016). Invasive alien plant species used for the treatment of various diseases in Limpopo Province, South Africa. *African Journal of Traditional, Complementary and Alternative Medicines, 13*(4), 223–231. https://doi.org/10.21010/ajtcam.v13i4.29

Mazaleyrat, A., Le Borgne, H., Lorenzetti, F., & Dupuch, A. (2023). Do disturbance-induced habitat changes drive alien species invasion success and native species decline? A case study based on slug species inhabiting managed deciduous forests. *Biological Invasions, 25*(7), (2343–2359).https://doi.org/10.1007/s10530-023-03046-y

McDonald, A. M., Martin, C. W., Adams, C. R., & Reynolds, L. K. (2023). Competition in a changing world: Invasive aquatic plant is limited by saltwater encroachment. *Ecosphere, 14*(3).https://doi.org/10.1002/ecs2.4394

Moore, T., & Pennington, A. (2015). Feral herbivores in Central Australia. In E. Ens, J. Fisher, & O. Costello, (Eds.), *Indigenous people and invasive species: Perceptions, management, challenges and uses*. IUCN Commission on Ecosystem Management Community Report.

Moudingo, J-H. E., Fon, J. N., Seraphine, E. M., & Elvis, N. M. (2015). Non-native mangrove, *Nypa fruticans* invasion in the Gulf of Guinea. In: E. Ens, J. Fisher, & O. Costello, (Eds.), *Indigenous people and invasive species: Perceptions, management, challenges and uses*. IUCN Commission on Ecosystem Management Community Report.

Mulema, J., Day, R., Nunda, W., et al. (2022). Prioritization of invasive alien species with the potential to threaten agriculture and biodiversity in Kenya through horizon scanning. *Biological Invasions, 24*, 2933–2949. https://doi.org/10.1007/s10530-022-02824-4

Nuñez, M. A., & Simberloff, D. (2005). Invasive species and the cultural keystone species concept. *Ecology and Society, 10*, Article r4.

Onyango, J., Babu, K., Njuguna, S., Wanzala, W., & Yan, X. (2020). Harnessing the potential of common water hyacinth as an industrial raw material for the production of quality biofuel briquettes. *SN Applied Sciences, 2*, 1316. https://doi.org/10.1007/s42452-020-3109-1

Oswald, A. (2005). *Striga* control: Technologies and their dissemination. *Crop Protection, 24*, 333–342.

Polce, C., Cardoso, A. C., Deriu, I., et al. (2023). Invasive alien species of policy concerns show widespread patterns of invasion and potential pressure across European ecosystems. *Science and Reports, 13*, 8124. https://doi.org/10.1038/s41598-023-32993-8

Rangi, D. (2018). *Workshop on invasive species in Africa: A major threat, held in Nairobi, Kenya on March 30th, 2018*. The Director General of Centre for Agriculture and Biosciences International (CABI). Science Africa. https://scienceafrica.co.ke/invasive-species-a-major-threat/ As accessed on January 15th, 2020 at 6:36 PM East Africa Time.

Reo, N. J., & Ogden, L. A. (2018). Anishnaabe Aki: An indigenous perspective on the global threat of invasive species. *Sustainability Sicence, 13*, 1443–1452. https://doi.org/10.1007/s11625-018-0571-4

Rezania, S., Ponraj, M., Din, M. F. M., Songip, S. R., Sairan, F. M., Chelliapan, S. (2015). The diverse applications of water hyacinth with main focus on sustainable energy and production for new era: An overview. *Renewable and Sustainable Energy Reviews, 41*, 943–954. ISSN 1364-0321. https://doi.org/10.1016/j.rser.2014.09.006

Russell, J., Meyer, J., Holmes, N., & Pagad, S. (2017). Invasive alien species on islands: Impacts, distribution, interactions and management. *Environmental Conservation, 44*(4), 359–370. https://doi.org/10.1017/S0376892917000297

Sankaran, K. V., Schwindt, E., Sheppard, A. W., Foxcroft, L. C., Vanderhoeven, S., Egawa, C., Peacock, L., Castillo, M. L., Zenni, R. D., Müllerová, J., González-Martínez, A. I., Bukombe, J. K., Wanzala, W., & Mangwa, D. C. (2023). Management; challenges, opportunities and lessons learned. In H. E. Roy, A. Pauchard, P. Stoett, & T. Renard Truong (Eds.), *Thematic assessment report on invasive alien species and their control of the intergovernmental science-policy platform on biodiversity and ecosystem services*. IPBES secretariat, Bonn, Germany. https://doi.org/10.5281/zenodo.7430733

Seastedt, T. R. (2014). Biological control of invasive plant species: A reassessment for the Anthropocene. *New Phytologist, 205*(2), 490–502. https://doi.org/10.1111/nph.13065

Shackleton, R. T., Shackleton, C. M., & Kull, C. A. (2019). The role of invasive alien species in shaping local livelihoods and human well-being: A review. *Journal of Environmental Management, 229*, 145–157.

Shanks, G. D., Hay, S. I., Omumbo, J. A, Snow, R. W. (2005). Malaria in Kenya's western highlands. *Emerging Infectious Diseases* [PMC3310610]. 2005 Sep. [Wednesday, November 8th, 2023]. https://doi.org/10.3201/eid1109.041131

Shine, C. (2005). Overview of the management of invasive alien species from the environmental perspective. In: *Identification of risks and management of invasive alien species using the IPPC framework. Proceedings of the Workshop on Invasive Alien Species and the International Plant Protection Convention (IPPC), IPPC Secretariat, Braunschweig, Germany, 22–26 September 2003*. Rome, Italy, FAO, xii + 301 pp.

Thaman, R., Lyver, P., Mpande, R., Perez, E., Cariño, J., & Takeuchi, K. (Eds.) (2013). *The contribution of indigenous and local knowledge systems to IPBES: Building synergies with science*. IPBES Expert Meeting Report, UNESCO/UNU. Paris, UNESCO, 49 pp.

Thapa, G. J., Subedi, N., Pandey, M. R., Thapa, S. K., Chapagain, N. R., & Rana A. (Eds.). (2014). *Proceedings of the International Conference on Invasive Alien Species Management*.

Thiermann, A. (2004). Emerging diseases and implications for global trade. *Revue Scientifique et Technique, 23*, 701–707.
Tripathi, S., Bhadouria, R., Srivastava, P., Singh, R., & Batish, D. R. (2023). Plant invasions and global climate change. *Springer*. https://doi.org/10.1007/978-981-99-5910-5
UNESCO. (2010). The power of culture for development. Programme and meeting document. Document code: CLT.2010/WS/14, pp. 16. UNESCO Headquarters, 7 place Fontenoy, 75007 Paris, France. http://unesdoc.unesco.org/images/0018/001893/189382e.pdf
United Nations News Highlights. (2008). *Fighting water weeds in West Africa.* Food and Agriculture Organization of the United Nations. United Nations.
UNSG's SAB. (2016). *Indigenous and local knowledge(s) and science(s) for sustainable development: Policy brief by the Scientific Advisory Board (SAB) of the UN Secretary-General (UNSG). Policy Brief on Science Number: SC/2016/UNSAB/ILK, pp 12.* UNESCO Headquarters, 7 place Fontenoy, 75007 Paris, France.
Wagh, V. V., & Jain, A. K. (2018). Status of ethnobotanical invasive plants in western Madhya Pradesh, India. *South African Journal of Botany, 114*, 171–180.
Watson, G.W. & Voegtlin, S.T. (1999). In: Watson, G.W., Voegtlin, S.T. Murphy & Foottit. Biogeography of the Cinara cupressi complex (*Hemiptera: Aphididae*) on Cupressaceae, with description of a pest species introduced into Africa. *Bulletin of Entomological Research, 89*, 275.
Wanzala, W. (2017). Potential of traditional knowledge of plants in the management of arthropods in livestock industry with focus on (Acari) ticks. *Evidence-Based Complementary and Alternative Medicine, 2017*, Article ID 8647919, 33 pages. https://www.hindawi.com/journals/ecam/2017/8647919/
Wanzala, W. (2009). *Ethnobotanicals for management of the brown ear ticks, Rhipicephalus appendiculatus in western Kenya.* Wageningen University and Research Centre, Wageningen, ISBN: 9085853176 (9789085853176), Printed by Ponsen & Looijen, Wageningen, The Netherlands. pp. 231.
Wanzala, W., Sika, K. F. N., Gule, S., & Hassanali, A. (2004). Attractive and repellent host odours guide ticks to their respective feeding sites. *Chemoecology, 14*, 229–232.
Wanzala, W., Takken, W., Pala, A. O., Mukabana, R. W., & Hassanali, A. (2012). Ethnoknowledge of Bukusu community on livestock tick prevention and control in Bungoma district, western Kenya. *Journal of Ethnopharmacology, 140*(2), 298–324. http://www.sciencedirect.com/science/journal/03788741?oldURL=y
Wanzala, W., & Wanjala W. C. (2016). Tagetes (*Tagetes minuta* L. 1753, Asteraceae) Essential Oil. In Preedy, V. R. (Ed.), *Essential oils in food production, preservation, flavour and safety: Part 2: Essential oils.* King's College London, School of Biomedical and Health Sciences, London, UK. Academic Press. ISBN: 978-0-12-416641-7. https://www.elsevier.com/books/essential-oils-in-food-preservation-flavor-and-safety/preedy/978-0-12-416641-7
Wehi, P. M., Kamelamela, K. L., Whyte, K., Watene, K., & Reo, N. (2023). Contribution of indigenous peoples' understandings and relational frameworks to invasive alien species management. *People and Nature, 5*, 1403–1414. https://doi.org/10.1002/pan3.10508
Wehi, P. M., Wilson, D. J., Stone, C., Ricardo, H., Jones, C. R., Jakob- Hoff, R., & Lyver, P. (2021). Managing for cultural harvest of a valued introduced species (*Rattus exulans*) in New Zealand. *Pacific Conservation Biology, 27*(432–4), 41.
Weir, J., & Duff, N. (2015). Weeds and native title in the kimberley. In E. Ens, J. Fisher, & O. Costello, (Eds.), *Indigenous people and invasive species: Perceptions, management, challenges and uses.* IUCN Commission on Ecosystem Management Community Report.
Wilt, A. (2018). *Workshop on invasive species in Africa: A major threat, held in Nairobi, Kenya on March 30th, 2018.* The Invasive Species Coordinator of Centre for Agriculture and Biosciences International (CABI). Science Africa. https://scienceafrica.co.ke/invasive-species-a-major-threat/ As accessed on January 15th, 2020 at 6:36 PM East Africa Time.
Wolverton, B. C., & McDonald, R. C. (1976). Don't waste waterweeds. *New Scientist, 71*(1013), 318–320.

Part IV
Threat Assessment and Management of Invasive Alien Plants in Urban Ecosystems

Chapter 12
Mapping and Threat Assessment of Invasive Alien Plants in Urban Ecosystem

Pragya Samaridhi, Rahul Kumar, and Shambhu Nath Mishra

Abstract The emergence and dispersion of Invasive Alien Plant Species (IAPS) in the vicinity of metropolitan environments endanger biodiversity, ecological health, and their services. Due to numerous environmental factors and higher propagation pressure, invasive or non-native species tend to be more prevalent in metropolitan regions than the countryside. For the sustainable management of IAPS in urban ecosystems, the importance of integrated trans-disciplinary research is stressed. Geographic Information Systems (GIS) and Remote sensing are tools for evaluating, mapping as well as tracking the IAPS spread and linking to the management of the IAPS habitat. Spatial prioritization techniques such as Species Distribution Modeling (SDM) create predictive models of invasive species distribution based on environmental elements such as climate, topography, and land use change, identifying areas at the highest risk of invasion. Similarly, multi-criteria decision analyses in a GIS (Geographical Information System) establish a strategic, spatial prioritization technique to recognize those places most vulnerable to terrestrial alien species invasions. System for Conservation Planning (SCP) software tool that helps prioritize areas for conservation efforts, including invasive species management. One of the most widely used tools for spatial conservation planning and management, including invasive species control, is Marxan. Similarly, Prioritizr (the R package) can be used to prioritize conservation activities based on their spatial location. Invasive species management initiatives attempt to reduce the impact of invasive species on natural habitats while promoting the health of native plants and wildlife. So, it is crucial to identify their existing dispersion and project future expansion, which is critical for successful management and mitigation strategies.

Keyword Invasive Alien plant species · Multi-criteria decision analyses · Species distribution modeling · Urban ecosystem

---

P. Samaridhi · S. N. Mishra
ICFRE—Institute of Forest Productivity, Ranchi, Indian Council of Forestry Research and Education—Lalgutwa, Ranchi, Jharkhand, India

R. Kumar (✉)
Rajdhani College, University of Delhi, New Delhi, India
e-mail: rahulandc@gmail.com

R. Singh et al. (eds.), *Invasive Alien Plants in Urban Ecosystems*,
https://doi.org/10.1007/978-3-032-00997-5_12

1 Introduction

Invasive alien plant Species (IAPS) refers to plants native to a particular realm or region that have been found by accident or purpose in a region outside their normal distribution. They have occupied or invaded their contemporary habitat, posing an imminent risk to biodiversity (Gaertner et al., 2014; Higgins et al., 1999), ecosystems (Gurevitch & Padilla, 2004), surroundings, and human well-being (Dyer et al., 2017; Kumari & Choudhary, 2016; Thapa & Maharjan, 2014). Global biological invasion poses a hazard to human health, country economies, ecosystem dynamics, the environment, and resource availability (Potgieter et al., 2022). It is widely recognized that IAPS poses a significant threat to forests, threatening their ecosystems (Peerbhay et al., 2016), grazing lands and farming productiveness (Pimentel et al., 2005). In addition, there is evidence suggesting that Invaded Plant Species in a region can reduce the species richness of native plants in that area (Gaertner et al., 2009), change soil characteristics and fire regimes (Pejchar & Mooney, 2009), and homogenise the variety of life or ecological diversity (Joshi et al., 2004; Kimothi & Dasari, 2010; Peerbhay et al., 2016; Tripathi et al., 2023) of surged over the landscapes. Excessive economic losses were revealed as an outcome of alien (non-native) invasion in experimental studies (Ayele, 2007; Karki, 2009). Invasive beings, like invasive organisms, are always threatened with extinction. Invaders are creatures that create a new horizon in which they thrive, expand, and survive at the expense of the ecosystem. The site is the most important ecological effect of the enormous changes in the dispersion of the earth's biota, caused mostly by human mobility and commerce. The outcomes of immigrants are distinctly varied in a world where there are very few regions that are still unaffected by these immigrations. Only a small percentage of survivors of the risks posed by random and persistent forces go on to become naturalised (Lockwood et al., 2013). In turn, some naturalised species do become invasive. The study focuses on how human activities such as transportation, establishment, colonisation, and landscape spread have contributed to the creation of IAPS in new places outside their natural limitations, namely in metropolitan settings (Potgieter et al., 2017). A significant contributor to biodiversity loss, community structure changes, and ecosystem processes are invasive alien plant species (Kumar & Saikia, 2020). These impacts continue to impact human well-being (livelihoods, health, food security, and economy). Invasive species can alter the transport of energy, water, and chemicals through an ecological system, indicating how much native ecosystems are affected by invaders (Bauer, 2012). Here are some of the most popular exotic species in India including *Centaurea maculosa* (Spotted knapweed), *Ageratum conyzoides* (Goat weed), *Parthenium hysterophorus* (Congress), *Lantana camara* (Putus), *Mikania micrantha* (Mile-a-minute), *Argemone mexicana* (Prickly Poppy) and *Eichhornia crassipes* (aquatic plant). The worst, most highly invasive, and most challenging species of these are *Lantana camara*, *Parthenium hysterophorus*, and *Ageratum conyzoides* (Kohli et al., 2004). Table 1 shows a few major IAPS, their origins, and their impact. The two forces that led to the emergence of invasion ecology were

(a) the formulation of the scientific basis for invasion biology and (b) the urgency of the issue of invasive species (Reichard & White, 2003).

In the past few decades, there has been plenty of focus on recognizing, preventing, and managing invasive species because of the potential and proven loss to the environment and economy (Pyšek & Richardson, 2010). When a species transforms from an immigrant to an invader, it usually goes through an incubation or lag phase, after there is a period of rapid rise of plant number, which persists till the species approaches the boundaries of the fresh area, and its rate of population expansion falls (Mack, 2000). All invasive organisms have a few general characteristics: Accommodate a broad spectrum of surroundings, including the state of the soil, water quality, temperatures, and moisture; induce a high germination success rate in a large number of seedlings; grow quickly; outlive native species in the invaded area by virtue of vitality, the plant matter, height, and lifespan (Khaeim et al., 2022).

In the 1980s, SCOPE (Scientific Committee on Problems of the Environment) evaluated approximately two thousand invading alien woody plant species from twenty-four restricted natural reserve zones (Foxcroft et al., 2013, 2017). In addition, the Global Invasive Species Programme (GISP) revealed that four hundred eighty-seven protected areas could have been overtaken by alien plants, threatening vernacular forest biodiversity (Foxcroft et al., 2017). In the 2010 Conference of Parties (COP10) of the Convention on Biological Diversity it was stated that "Human activity (habitat loss, overexploitation, and pollution) is the primary cause of species extinction, but invasive species are the second biggest threat." Some examples include the noxious plant witch weed, which threatens significant agronomic yields in the southern United States, millions of hectares of spotted knapweed and leafy spurge on western rangelands, and the recent influx of large salvinia, which creates a serious environmental and fiscal risk to the lower Colorado River system. Another example demonstrates how the number and diversity of soil-litter insects and their associations to the physio-chemical characteristics of the soil under various land uses in Rwanda explore the connections between the characteristics of the soil under native and exotic varieties of trees as well as in tree plantations. Studies conducted by the team found that native tree species forests are critical to maintaining improved soil conditions and preventing soil-litter arthropod infestations (Gentili et al., 2021).

According to Hassan and Nawchoo (2020), more than five thousand alien plants have been introduced to US natural habitats, displacing many. Among these nuisance weeds is *Lythrum salicaria* (European purple loosestrife), which first appeared as a plant for ornamental purposes in the beginning of the nineteenth century (Shaw & Hatcher, 2017). Lindgren (2013) claims that it has invaded and expanded over 115,000 hectares a year, changing the basic structure of the majority of wetlands. A single explanation cannot explain the invasion of IAPS; nonetheless, various theories have been put forward to describe the invasion of IAPS in new environments. Examples include enemy release (ERH), novel weapon (NWH), and empty niche (EN) (Rai & Singh, 2020). Exotic species that turn out to be invasive are the primary direct drivers of biodiversity loss around the world (Jaureguiberry et al., 2022). Managing IAPS seems as a major task in the field of biodiversity management and conservation. The ecological price is typically the irreversible demise of indigenous

Table 1 List of some major IAPS and their impact

Sl.No.	Species	Common name, family and habitat	Native	Threats	References
1	*Ageratina adenophora*	Crofton weed, Asteraceae, Herb	Mexico and Central America	Invasive in farmland and bushland. It is toxic, so livestock usually avoid it. Reducing biodiversity in invaded areas	Yadav et al. (2024), LA OE (2023), Sun et al. (2018)
2	*Ageratum conyzoides*	Goat weed, Asteraceae, Herb	Tropical America	Allelopathic, highly invasive threat to the croplands of the Himalayan region	Negi et al (2023), Gupta et al. (2021), Yadav et al. (2016)
3	*Argemone Mexicana*	Mexican Prickly Poppy, Papaveraceae, Herb	Tropical Central and South America	Harm native flora through allelopathy, rapid growth, and spread disrupt ecosystems	Ooka and Owens (2018), Yadav et al. (2016), Namkeleja et al. (2014)
4	*Centaurea maculosa*	Spotted knapweed, Asteraceae, Herb	Western central Europe and western Asia	States produce phytotoxin (-) -catechin from its roots, a significant increase in ammonia-oxidizing bacteria and soil nitrate	Callaway and Ridenour (2004)
5	*Chromolaena odorata*	(Siam weed, Asteraceae, Shrub	Central and South America	Novel weapons/ Allelochemicals are responsible for its invasion and native vegetation decline	Kato-Noguchi and Kato (2023), Rai and Singh (2020)
6	*Eichhornia crassipes*	Water hyacinth, Pontederiaceae, Aquatic herb	Tropical South America	Serious aquatic weed, allelopathic in nature, causes hindrance in navigation and reduces water quality and algal growth	Nihal et al (2021), Yadav et al. (2016)
7	*Eupatorium adenophorum*	(Crofton weed), Asteraceae, Shrub	Mexico	The allelopathic effect causes serious threats to native flora	Rai and Singh (2020), Rawat et al. (2019), Yadav et al. (2016)

(continued)

Table 1 (continued)

Sl.No.	Species	Common name, family and habitat	Native	Threats	References
8	*Fallopio japonica*	Japanese knotweed, Polygonaceae, Herb	East Asia	Causes substantial economic and environmental damage and alters food webs through secondary metabolites	Cucu et al. (2021), Rai and Singh (2020)
9	*Lantana camara*	Putus, Verbeneaceae, Shrub	Tropical America	Strongly allelopathic, a serious threat to medicinal plants, and responsible for forest fires	Barik et al. (2020), Negi et al. (2019), Yadav et al. (2016)
10	*Mikania micrantha*	(Mile-a-minute, Asteraceae, Herb	Sub-tropical zone of America	Known for its allelopathic potential, highly invaded forest areas	Poudel et al. (2019), Wu et al. (2009)
11	*Opuntia stricta*	Erect prickly pear Cactaceae, Shrubs	Mexico, Caribbean, South America	Negatively affecting environment, health and socio-economy/livelihood	Rai and Singh (2020), Shackleton et al. (2017)
12	*Parthenium hysterophorus*	Congress or Carrot Grass Asteraceae, Herb	Tropical America	Aggressive coloniser, highly allelopathic, allergic to animals and human beings, a threat to crops and other native flora	Yadav et al. (2016), Gnanavel and Natarajan (2013)

species and habitats. Additionally, it causes significant monetary damage in terms of lower harvest and livestock productivity, decreased local plant and animal life, rising production expenses, and so on. Biodiversity has grown to be among the most popular themes for consideration at the local, countrywide, and worldwide levels. Managing invasive species requires a combination of prevention, control, and eradication techniques. Preventing the entry and propagation of noxious plants is frequently the most successful and economical technique. Detecting invasive species early and intervening quickly can help prevent them from establishing and spreading. Management measures must be tailored to specific invasive species and local ecological conditions to achieve the best outcomes. Powell et al. (2023) examined collaborative efforts to control invasive woody plants like Ligustrum lucidum and Acacia melanoxylon in Argentina. The study highlighted that incorporating various stakeholder viewpoints and tailoring management strategies to fit local socio-economic and ecological conditions have incredibly improved the success of invasive species management. Study conducted by Caughlin et al. (2023), the plant was more common in places with a lots of bare ground and lower property values. This implies that ecological conditions and socioeconomic variables are important in the spread of invasive species, emphasizing the necessity of management approaches that take social and environmental settings into account.

Remote sensing (RS) has evolved as a potential method for measuring and observing the dynamics of IAPS, giving information fast, efficiently, constantly, reliably, and reproducibly over broad ranges. High spatial resolution photos offer a user-friendly and straightforward RS approach to visually determine the geographical spreading of IAPS (Huang & Asner, 2009). By gathering pertinent period images and good time series, very high spatial resolution (VHR) imagery from the present and the past provides crucial information on how the terrain has changed over period, aiding in identification of the target species (Brook & Bowman, 2006; Laliberte & Ripplc, 2004). This technique has potential to support the conventional IAPS field monitoring, which can be expensive as well as technically challenging. Worldwide, there are very few ecosystems that are devoid of invaded species, and as the percentage of ecosystems, habitats, and biomes increases, introduced species are taking over. Plant species can be identified and distinguished within the same functional groupings thanks to modern technologies like "Light detection and ranging" (LIDAR) and hyperspectral imaging spectroscopy. Since RS can provide synoptic views over wide geographic areas, it has long been preferred as a means for IAS mapping, particularly for flora. When paired with developments in algorithms for image processing, these innovations have made RS precise and reliable. Compared to field surveys, which are sometimes restricted to a minor region also sometimes located in challenging-to-access areas, this offers a benefit.

The main agenda of the current chapter was to map and assess invasive alien plants in urban ecosystems, which are essential for protecting biodiversity, maintaining ecosystem services, preventing economic losses, ensuring public health and safety, and supporting sustainable urban planning and management.

2 Factors Responsible for the Establishment of IAPS in Urban Areas

The invasion of plant species into new settings is largely triggered by human actions that disturb the natural biological balance. The benefit of this method is that it is not restricted to a small region and is not located in difficult-to-access areas, as is the case with field surveys. Figure 1 shows the prime factors of the invasion of plant species. These introduced species often exhibit traits such as fast growth, high propagative capacity, and their ability to endure a variety of environmental factors, which gives them an advantage over indigenous species.

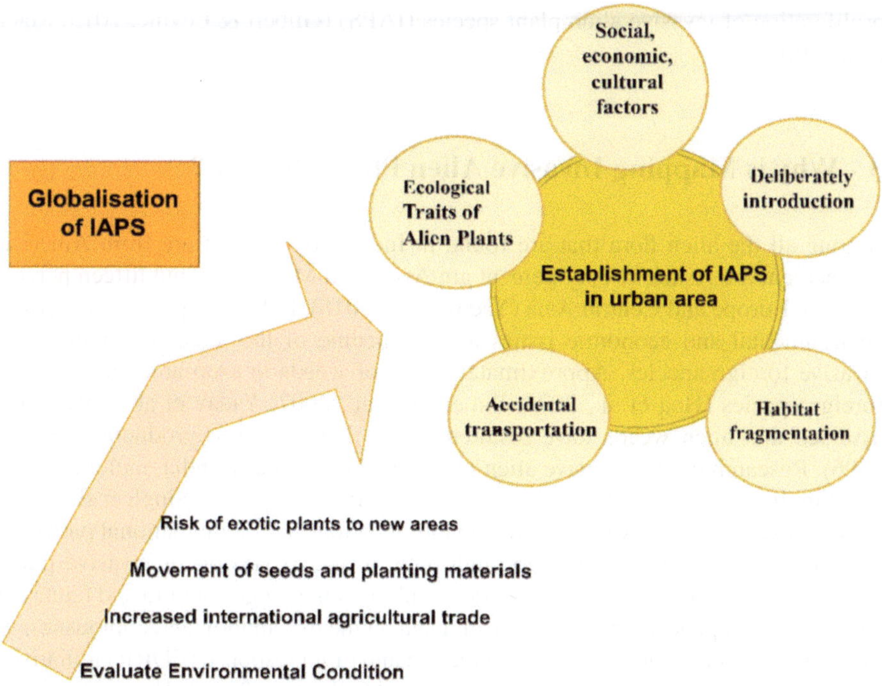

Fig. 1 Factors responsible for the introduction of Invasive Alien Plant Species (IAPS). *Source* Authors

Accidental introductions through trade, transit, or botched bio-control programs can show a major part in plant invasions. Changes in land use, habitat fragmentation, and climate change all contribute to the spread of invasive species. Once introduced, these alien species have the probability to disrupt native ecosystems, diminish biodiversity, and alter soil chemistry, providing significant challenges to conservation and ecosystem management. Understanding and controlling the principal causes of plant invasion is critical for maintaining the integrity and functionality of the natural environments. According to Williamson's "tens rule," ten per cent of imported species become invasive, ten per cent of introduced species establish themselves, and ten per cent of established species become a nuisance (Jarić et al., 2019). Only 0.5–0.7% of the world's shrub and tree species are truly invasive outside their indigenous range, despite the fact that woody plant invasions are on the rise globally (Richardson & Rejmanek, 2011). Migration, trade, and human-mediated transportation contribute to spreading an ever-widening variety of IAPS via hitherto unbreakable environmental barriers like mountain ranges, difficult temperature zones, and fresh and marine aquatic habitats (Rai, 2015; Vaz et al., 2017). As a result of modern-day intensive farming for food security, retailers and wholesalers often do not know how to distinguish between native and exotic species, which has significantly accelerated the proliferation of invasive alien plant species (IAPS) (Gilbert & Levine, 2013; Mack et al., 2000).

3 Why is Mapping Invasive Alien Plants Needed?

Among all the alien flora that are found in India, fifty percent are from America, Ten percent are Asian, twenty percent are Asian and Malaysian, and fifteen percent are from Europe and Central Asia (Yadav et al., 2016). India is experiencing serious environmental and economic issues as an outcome of the massive expansion of invasive foreign species. Approximately 42% of weeds in crop areas in India are foreign species (Haq et al., 2021; Pimentel et al., 2007; Yadav et al., 2016). It is assessed that alien weeds have resulted a 30% harm in crop production (Singh, 1996). Research on the invasive alien plants in Uttar Pradesh, India, includes forty-four families, one hundred nine genera, and one fifty-two species (Singh et al., 2010). Based on vegetation surveys, several smaller locations, including national parks and some counties, have gathered thorough data on the expansion of invasive plants (Foxcroft et al., 2009). Insufficient understanding of the ranges of flora and fauna can significantly impede attempts at control. Data on the distribution and copiousness of invasive plants can help guide early detection and rapid response (EDRR) initiatives, which are designed to treat initial infestations (Abella et al., 2016). These data can be utilized at the regional and landscape levels to build invasion risk models (Bradley & Marvin, 2011; Cabra-Rivas et al., 2016). Maps of regional distribution are helpful in a lot of ways. They are an essential source of information for informing the public about the severity of the issue caused by invasive species. They offer critical financial support for programs focused on managing and preventing invasive plant species. Maps can help with administration efforts across county and state lines by highlighting invasion hot regions for priority monitoring. Distribution maps are also necessary for evaluating regional invasion risk and guiding management priority (Booy et al., 2017).

4 Remote Sensing and GIS Approach for Mapping and Threat Assessment of Invasive Species

Remote Sensing (RS) and Geographic Information Systems (GIS) are very prominent, feasible, as well as cost-effective tools in forest mapping, quantitative assessment, vegetation classification, and natural resource management (Sonti, 2015). Different multispectral imageries with GIS maps are considered add-on tools for the protection and management of the earth's resources (Freeman & Fox, 1994). Similarly, different temporal satellite images can be used to study forest cover change detection, forest clearing and forest fragmentation surveying (Dash et al., 2018) as it is useful for the representation of data in the form of graphs, maps, and other GIS-based statistical modeling functionalities (Abousaeidi et al., 2016). Scientific and effective forest management information could be operationalized, utilising the GIS technique on a small to large scale (Mani & Varghese, 2018). The GIS technique is effective in building forest cover maps and creating a geodatabase for spatial analysis, including monitoring plant species, forest fires, invasion of exotic species and forest management (Hasmadi et al., 2010).

The application of RS in urban regions to identify IAPS presents distinct obstacles. Because the majority of the terrain is covered by human-made features, locating green areas, mapping, and identifying particular species is challenging (Alonzo et al., 2014). These issues are manageable with adequate spatial resolution. The best strategy to date is a mix of hyperspectral and Lidar, which provides spectral, structural, and height data (Bolch et al., 2020).

Increasingly taking over native plants, invasive species put some directly at risk and indirectly affect others as ecological conditions change. Invasive species cause harm to almost half of the species listed under the Endangered Species Act and play a major role in contributing to global change. Because of their rapid expansion, the threat to biodiversity, and harm to ecosystems, invasive species are currently of concern to ecologists, biological conservationists, and managers of natural resources (Estévez et al., 2015; Lockwood et al., 2013; Vitousek et al., 1996). Over the past decade, rapid technology improvements have made the employing of remote sensing for the identification and management of invasive alien plants is becoming increasingly important.

Managers utilise an early detection and rapid response (EDRR) method to avoid invasions by detecting and eradicating new infestations before they can develop a lasting foothold in a new environment (Huang & Asner, 2009; Paz-Kagan et al., 2019; Royimani et al., 2019). Well-established benefits of remote sensing technology include affordability, a synoptic view, multispectral imaging data, and multitemporal surveillance (Dahy et al., 2021). When researching biological invasions, SDMs, are becoming more and more common. For example, prediction techniques that employ statistical models or theoretically designed response curves to identify relationships between observable events and environmental conditions are used to evaluate the impact of SDMs (Guisan & Thuiller, 2005; Hui, 2023; Srivastava et al., 2019). SDM are often classified into correlative and 'process-based' or 'mechanistic models. They are also known as statistical or correlative, habitat, or ecological niche models (Dormann et al., 2012). Correlative Species distribution models (SDM) create maps of a species' relative environmental appropriateness or probability of occurrence using species occurrence data, while process-based, or mechanistic, niche models fit their models using physiological thresholds and functional traits of the species (Kearney & Porter, 2009). The correlative SDM uses existing species' occurrence data collected from surveys, online databases, or museums/herbaria records for fitting (Bloom et al., 2018). Practically, the majority of mechanistic models necessitate comprehensive experimental information on the functional features of species, which presents difficulties for the particular species being studied (Shabani et al., 2016).

In this study, a modelling approach with the help of remote sensing was used to predict the spread and niche suitability of IAPS. The process of modeling may take place in a GIS or use a GIS tool and computer program.

4.1 Species Distribution Model (SDM)

Machine learning technologies can considerably speed up the processing and investigation of massive repositories of unprocessed information (Nath & Kumar, 2024; Nguyen et al., 2019; Zhou et al., 2017), which can accelerate ecological analysis and broaden its geographic extent. Spatial data, such as aerial photos, national forestry inventories, and cartographic layers have been important assets for statistical mapping since the advent of modelling techniques (Di Biase et al., 2018; Fleischer et al., 2017). It is important to develop spatial decision support systems that incorporate a multitude of forest attributes and land suitability datasets from various locations around the globe (Johnson et al., 2014). There are three types of statistical modelling techniques: empirical, correlative, and mechanistic. Correlative SDM entail gathering species occurrence information, linking these occurrences to environmental parameters, and producing maps that forecast past, current, or future geographical distributions of species (Pecchi et al., 2019). BIOCLIM was developed in 1996; these 19 bioclimatic variables are still broadly used in various simulation-based modeling studies. Historically, SDM was developed from Hutchinson's ecological niche idea in the 1950s, which was improved by Booth (1988).

There is a rising need for SDM to investigate many problems in conservation biology as well as evolution (Guisan et al., 2013), to evaluate biogeographical theories (Graham et al., 2004), anticipate invasions of species and development (Peterson, 2003), and predict novel incidences of rare species (Williams et al., 2009). The effect of alterations to land use, climate, as well as other environmental factors on the distribution of species has also been evaluated using distribution models (Thuiller et al., 2005), furthermore to aid in reserve establishment and conservation strategy (Seoane et al., 2011). An ecological understanding of the species has taken a backseat in most modelling efforts, which have focused on finding functional connections between the species' locations and the environment around them (Austin, 2007).

4.1.1 Model and Its Components

SDMs are restricted because distributional data solely illuminates the realized niche (Jiménez, 2021); however, they have been frequently used in conservation, invading ecology, and global change biology (Franklin, 2010; Peterson, 2003). The increased availability of distributions of species and fine-scale meteorological information permissible to the development of a substantial number of SDMs for a variety of species (Kearney & Porter, 2009), employing an inclusive range of ecological factors (Braunisch et al., 2013; Elith et al., 2006).

It has two main components.

(A) Entropy: It is calibrated to find the distribution that is the most spread out throughout the study area or the closest to uniform across the entire study area.
(B) Constraints: The rules governing the expected distribution. These guidelines are based on the values of environmental factors (known as characteristics) in areas where the particular species has been seen.

Environmental factors are frequently selected based on historical expected performance, a known link with the species of interest, or using a variable screening method (Bradie & Leung, 2017). Variables included in SDMs essentially are environmentally appropriate to guarantee models can provide ecological knowledge and are transferable across areas (Elith & Leathwick, 2009; Elith et al., 2011).

In the present research, we emphasize on research that utilize the Maximum Entropy Algorithm (MaxEnt), as it is an efficient, widely used technique that employs publicly available presence-only datasets (Elith et al., 2011). MaxEnt gives reliable rankings for variable relevance throughout validation data sets (Braunisch et al., 2013) and takes environmental variables into account independently when developing models (Phillips et al., 2006). Furthermore, varied contributions have been documented in a typical manner in numerous peer-reviewed research since MaxEnt provides an integrated function that examines the impact of each environmental factor on model gain (Phillips et al., 2006). Zhu et al. (2018) investigated the probable distributions and fluctuation of three *Stipa tianschanica* varieties (var. *tianschanica*, var. *gobica*, and var. *klemenzii*) as a result of climate change in the dry steppe of the Mongolian Plateau along with Central Asia using the Maxent approach.

Chaudhary et al. (2021) conducted research on the current distribution of *Ageratina adenophora* and *Lantana camara* in the Kailash sacred region of India. The researchers used MaxEnt modelling to investigate potential future climates for the two species. Both the training data set and the test data set under the AUC curve for present and future scientific research on *Ageratina adenophora* were 0.81 and 0.82, respectively. Also, the AUC values for *Lantana camara* for the present year and the future years for the test and training data set were 0.93 and 0.93 present, and 0.92 and 0.93, respectively.

The MaxEnt Programme (Phillips et al., 2006) is especially prominent in species distribution/ecological niche modelling, with more than a thousand applications reported since 2006. MaxEnt modeling decisions and their biological explanations are lacking with the prominent exceptions (Elith et al., 2011). Because MaxEnt is used to handle more complicated problems (rather than simple exploratory studies), it is crucial to ensure that modelling options are biologically justified by particular theories, the research aims, and species-specific factors, and they represent the desired priori presumptions (Araujo & Peterson, 2012). Initially, MaxEnt was used to assess the density of the presence throughout the landscape (Phillips et al., 2006). When the entire population dimensions were understood, such models forecast the rate of occurrence in an area, which can be expressed as the expected number of individuals within that cell (Fithian & Hastie, 2012). In addition, since the population count is often unknown, only relative comparisons between these rates are

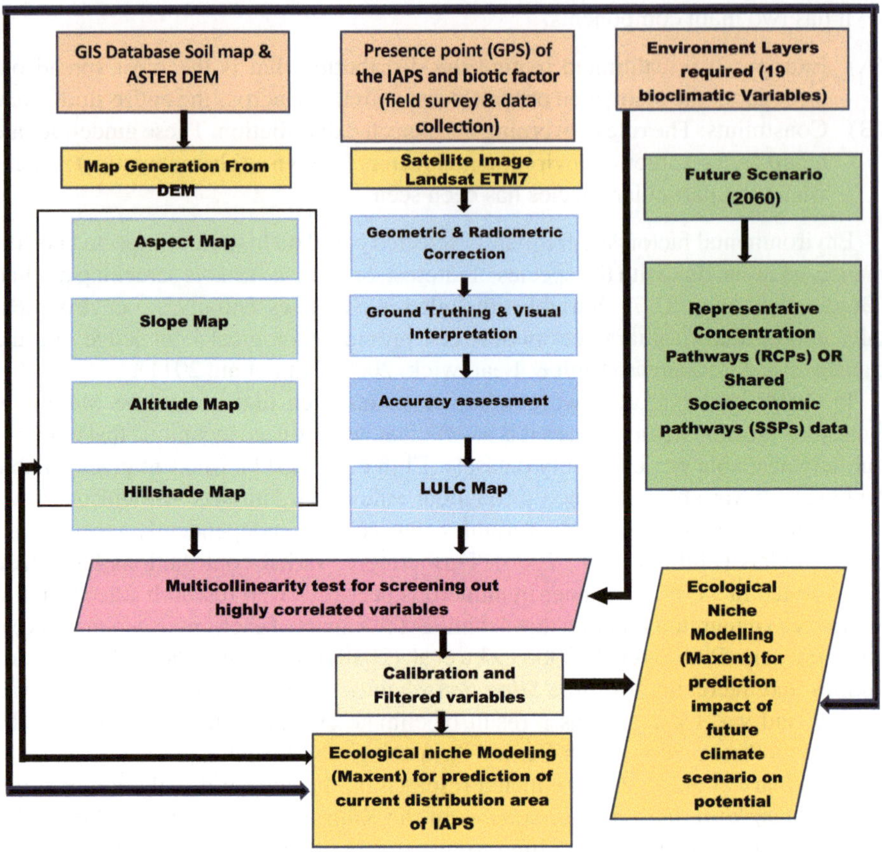

Fig. 2 Predictions of the distribution of Invasive alien plant Species (IAPS) using MaxEnt. *Sources* Authors

significant, leading to an approximate occurrence rate (Fithian & Hastie, 2012). MaxEnt can forecast the likelihood of presence solely by utilizing a transformation of the Receiver Operating Characteristic (ROC), known as logistic output (Phillips & Dudík, 2008), which utilizes robust presumptions that have been criticised (Royle et al., 2012). The working principle of MaxEnt is shown in Fig. 2.

4.2 Invasive or Non-native Species Are More Prevalent in Urban Regions Than in Rural Ones

Disturbances brought on by human actions, like urbanization have the power to alter ecosystems drastically. Urbanization increases pollutants and replaces open spaces with constructed areas and impermeable surfaces. Because of this, biodiversity is decreased, and a very simplified community is created in which generalist tolerant taxa thrive and specialized sensitive species decline (Faeth et al., 2011). Cities serve as important entry and establishment points for IAPS. Urban ecosystems frequently have higher populations of invasive species than the surrounding landscapes. According to Padayachee et al. (2017), the majority of metropolitan invasive species are purposely introduced and then dispersed to neighboring regions. The study also revealed that the richness of alien species in urban zones is probably continue to upsurge universally, particularly if introduced species have already acclimated to urban surroundings. Town regions are also hubs for human mobility and the transportation of commodities, which is able to facilitate a swift, long-distance spreading of local species through human activity. The act of deliberately introducing multiple species into new environments can result in the selection of highly adaptable characteristics that can withstand the harsh circumstances encountered during transportation. Better plasticity can aid in adaptation since it is linked to increased invasion risk. During transit, assortment burdens for more plastic species may favor those that are more likely to become invasive. With the possibility that human-mediated spreading and evolution may mutually increase their spread and its foundation, species that humans move are, therefore, probably more susceptible to urban evolution (Buczkowski, 2010; van Kleunen et al., 2018).

Global urban ecosystems share many of the same environmental factors (Grimm et al., 2008). Because of this, species that have adapted to live in an urban environment in one place may serve as an origin for invasive populations that proliferate in other places due to its urban evolution. For instance, *Lepidium virginianucum*, a plant native to the United States, showed more genetic similarity between distant urban populations than geographically closer rural populations. Due to their capacity to migrate quickly across urban regions and their ability to adapt to new urban environments, urbanized invaders can propagate geographically more quickly (Yakub & Tiffin, 2017).

As per Grimm et al. (2008) and Lahr et al. (2018), urban ecosystems have the ability to predict future conditions and serve as intermediaries for global changes. Urban areas tend to mimic the expected circumstances of global change since they have a higher concentration of carbon dioxide (CO_2) and are typically warmer than non-urban locations (Lahr et al., 2018). The ability to examine how native and invasive species have evolved to acclimatize to the conditions anticipated via future global variation is thus unmatched in urban ecosystems, and this has enormous consequences for the distribution of species in the future.

4.3 Multi-criteria Decision Analysis (MCDA) for Habitat Suitability of IAPS

Multi-Criteria Decision Making (MCDM), or Multi-Criteria Decision Analysis (MCDA), is considered to be among the most precise ways of decision-making. It seeks to discover the optimal choice by taking into account more than one factor in the selection procedure. This method evaluates many qualitative and quantitative criteria that must be met in order to discover the optimal answer. For instance, expenditure or cost, and manufacturing quality are two of the most commonly used parameters in various decision-making challenges. Furthermore, in certain situations, expert groups assign varying weights to the criteria based on how significant each criterion is in that particular scenario.

In the conservation policy field, IAPS are becoming a key topic due to their depletion of biodiversity (Caffrey et al., 2014; Dechoum et al., 2018). Despite the fact that invasions occur in highly human-influenced contexts; we need to learn more about how mankind is influencing invasion processes (McNeely, 2001; Vitousek et al., 1997). In most management-oriented studies to date, invasion processes have been explored in terms of their impact on humans (Binimelis et al., 2007). For example, economic studies (Dehnen-Schmutz et al., 2004; Shogren & Tschirhart, 2005) and the use of economic mechanisms like input/output models (Julia et al., 2007) have brought attention to the economic costs linked with the spread of exotic species. Furthermore, risk analysis approaches (Holt, 2006; Shameli-Sendi et al., 2016) have been used to estimate threats in support of policy decisions on alien species management (Monterroso et al., 2011). MCDA mechanisms have been investigated as useful ways for assisting IAPS management choices that integrate the perceptions of different groups of people in the surroundings of competing interests and beliefs. Risk assessment involving MCDA principles has also been beneficial when assessing management choices in cases of unwanted plant proliferation, including risks from exotic plant-related pests (Cook & Proctor, 2007; Maguire, 2004). All these types of research investigations encourage the use of MCDA as a useful framework for making decisions in invasion activities because of the complexity of the invaders' species and perceptions of risk (Monterroso et al., 2011).

The first step for MCDA is to define criteria, which is crucial to identify and determine their potential for invasion and impact. By doing so, we can take proactive measures to prevent their spread and protect our natural habitats. These criteria may include ecological variables (e.g., habitat suitability, spread rate), economic factors (e.g., control measure cost), and social factors (e.g., influence on human health).

As part of the next step, data will be collected from a variety of sources in order to evaluate each of the criteria, such as field surveys, remote sensing, and expert knowledge. Ideally, it should ensure data quality and completeness. Standardization of collected data to a common scale (e.g., 0–1) is done to facilitate comparison. Normalizing data within each criterion is to mitigate the effect of different measurement units. Now, each criterion is given a weighting based on how important it is to the assessment process and how important each criterion is to the others. It uses stakeholder consultations or analytical methods like the Analytic Hierarchy Process (AHP) to determine weights. Each invasive plant species assesses its threat based on the aggregate scores. Then, identifying high-threat species is based on their overall scores. Geographic Information Systems (GIS) are used to represent the results of the assessment spatially, and maps are generated to show the distribution and threat levels of invasive plant species based on MCDA outputs. The mapped results are verified by comparing them to known invasive plant distributions. Sensitivity analysis is conducted to assess the robustness of the results to changes in criteria weights or data inputs. How variations in criteria weights affect the final threat assessment is evaluated. Mapped outputs and threat assessments support the decision-making processes related to introduced species management. The overall methodology for MCDA is shown in Fig. 3.

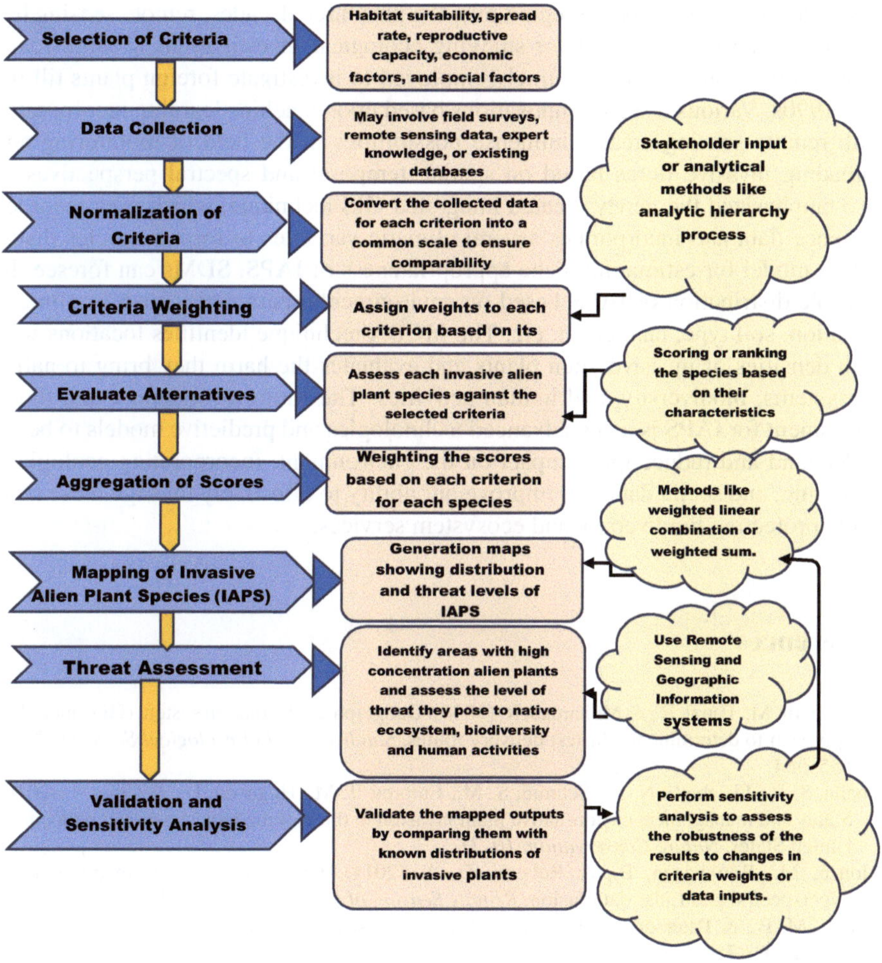

Fig. 3 Methodology for MCDA for habitat suitability assessment of IAPS. *Source* Author

5 Conclusions and Future Prospects

Rapid expansion of urban areas has increased the threat of invasive alien plant species, which modify soil quality, promote land degradation, and disrupt important ecological services by changing species composition. Accidentally carried foreign flora with rapid growth have significant consequences for rural livelihoods in the initial phases of their invasion. Both adverse effects along with the proliferation of invasive plants are estimated to be reduced with time as humans respond to circumstances that jeopardize their monetary well-being and strive to adjust to variations in the dynamics of the ecosystem. Biological invaders can impact ecosystems under a broad range of bioclimatic conditions. Thus, it is every so often vital to thoroughly check the growth of species over a widespread region over the past three decades; remote sensing has grown into an essential tool for studying ecological investigations at large-scale, although the issue was not routinely employed to investigate foreign plants till the mid-1990s. Various research innovations based upon machine learning incorporated with remote sensing creates immense possibilities in the field of monitoring and spreading invasive floras based on spatial, temporal and spectral perspectives. It also emphasized the newly created integrated GIS technique, which uses recorded presence data and incorporates several climatic variables to form a species distribution model for estimating niche appropriateness of IAPS. SDMs can foresee the possible distribution of IAPS based on environmental parameters such as climate, elevation, soil type, land cover, etc. The MCDA technique identifies locations with high densities of invasive alien plants and evaluates the harm they bring to native ecosystems, biodiversity, and human activities. The future of mapping and threat assessment for IAPS is using advanced technologies and predictive models to better understand and reduce their impact on the environment. Incorporating ecological, economic, and social data will improve our ability to effectively manage these risks while protecting biodiversity and ecosystem services.

References

Abousaeidi, M., Fauzi, R., & Muhamad, R. (2016). Geographic information system (GIS) modeling approach to determine the fastest delivery routes. *Saudi Journal of Biological Sciences, 23*(5), 555–564.

Abella, S. R., Fisichelli, N. A., Schmid, S. M., Embrey, T. M., Hughson, D., & Cipra, J. (2016). Status and management of non-native plant invasion in three of the largest national parks in the United States. *Nature Conservation, 10*, 71.

Alonzo, M., Bookhagen, B., & Roberts, D. A. (2014). Urban tree species mapping using hyperspectral and Lidar data fusion. *Remote Sensing of Environment, 148*, 70–83.

Araújo, M. B., & Peterson, A. T. (2012). Uses and misuses of bioclimatic envelope modeling. *Ecology, 93*(7), 1527–1539.

Austin, M. (2007). Species distribution models and ecological theory: A critical assessment and some possible new approaches. *Ecological Modelling, 200*(1–2), 1–9.

Ayele, S. (2007). *The impact of Parthenium (Parthenium hysterophorus L.) on the range ecosystem dynamics of the Jijiga rangeland, Ethiopia.* Department of Animal Sciences, School of Graduate Studies, Haramaya University (p. 134).

Barik, S. S., Sahoo, R. P., & Barik, S. S. (2020). *Lantana camara* L.: An emerging threat to native flora and livestock: A review. *Journal of Pharmacognosy and Phytochemistry, 9*(5), 2363–2366.

Bauer, J. T. (2012). Invasive species: "Back-seat drivers" of ecosystem change. *Biological Invasions, 14*, 1295–304.

Binimelis, R., Born, W., Monterroso, I., & Rodríguez-Labajos, B. (2007). Socio-economic impact and assessment of biological invasions. *Ecological Studies, 193*(13), 331.

Bloom, T. D., Flower, A., & DeChaine, E. G. (2018). Why georeferencing matters: Introducing a practical protocol to prepare species occurrence records for spatial analysis. *Ecology and Evolution, 8*(1), 765–777.

Bolch, E. A., Santos, M. J., Ade, C., Khanna, S., Basinger, N. T., Reader, M. O., & Hestir, E. L. (2020). Remote detection of invasive alien species. *Remote Sensing of Plant Biodiversity, 2020*, 267–307.

Booth, W. C. (1988). *The company we keep: An ethics of fiction.* University of California Press.

Booy, O., Mill, A. C., Roy, H. E., Hiley, A., Moore, N., Robertson, P., Baker, S., Brazier, M., Bue, M., Bullock, R., & Campbell, S. (2017). Risk management to prioritise the eradication of new and emerging invasive non-native species. *Biological Invasions, 19*, 2401–2417.

Bradie, J., & Leung, B. (2017). A quantitative synthesis of the importance of variables used in MaxEnt species distribution models. *Journal of Biogeography, 44*(6), 1344–1361.

Bradley, B. A., & Marvin, D. C. (2011). Using expert knowledge to satisfy data needs: Mapping invasive plant distributions in the western United States. *Western North American Naturalist, 71*(3), 302–315.

Braunisch, V., Coppes, J., Arlettaz, R., Suchant, R., Schmid, H., & Bollmann, K. (2013). Selecting from correlated climate variables: A major source of uncertainty for predicting species distributions under climate change. *Ecography, 36*(9), 971–983.

Brook, B. W., & Bowman, D. M. (2006). Postcards from the past: charting the landscape-scale conversion of tropical Australian savanna to closed forest during the 20th century. *Landscape Ecology, 21*, 1253–1266.

Buczkowski, G. (2010). Extreme life history plasticity and the evolution of invasive characteristics in a native ant. *Biological Invasions, 12*, 3343–3349.

Cabra-Rivas, I., Saldaña, A., Castro-Díez, P., & Gallien, L. (2016). A multi-scale approach to identify invasion drivers and invaders' future dynamics. *Biological Invasions, 18*, 411–426.

Caffrey, J. M., Baars, J. R., Barbour, J. H., Boets, P., Boon, P., Davenport, K., Dick, J. T., Early, J., Edsman, L., Gallagher, C., & Gross, J. (2014). Tackling invasive alien species in Europe: The top 20 issues. *Management of Biological Invasions, 5*(1), 1.

Callaway, R. M., & Ridenour, W. M. (2004). Novel weapons: invasive success and the evolution of increased competitive ability. *Frontiers in Ecology and the Environment, 2*(8), 436–443.

Caughlin, T. T., Clark, M., Jochems, L. W., Kolarik, N., Zaiats, A., Hall, C., et al. (2023). Socio-ecological interactions promote outbreaks of a harmful invasive plant in an urban landscape. *Ecological Solutions and Evidence, 4*(2), Article e12247.

Chaudhary, A., Sarkar, M. S., Adhikari, B. S., & Rawat, G. S. (2021). *Ageratina adenophora* and *Lantana camara* in Kailash Sacred Landscape, India: Current distribution and future climatic scenarios through modeling. *PLoS ONE, 16*(5), Article e0239690.

Cook, D., & Proctor, W. (2007). Assessing the threat of exotic plant pests. *Ecological Economics, 63*(2–3), 594–604.

Cucu, A. A., Baci, G. M., Dezsi, Ş, Nap, M. E., Beteg, F. I., Bonta, V., Bobiş, O., Caprio, E., & Dezmirean, D. S. (2021). New approaches on Japanese knotweed (*Fallopia japonica*) bioactive compounds and their potential of pharmacological and beekeeping activities: Challenges and future directions. *Plants, 10*(12), 2621.

Dahy, B., Issa, S., & Saleous, N. (2021). A review of land change modelling techniques using remote sensing and GIS. In *42nd Asian Conference on Remote Sensing, ACRS 2021*. Asian Association on Remote Sensing.

Dash, C. J., Adhikary, P. P., Madhu, M., Mukhopadhyay, S., Singh, S. K., & Mishra, P. K. (2018). Assessment of spatial changes in forest cover and deforestation rate in Eastern Ghats Highlands of Odisha, India. *Journal of Environmental Biology, 39*(2), 196–203.

Dechoum, M. D., Sampaio, A. B., Ziller, S. R., & Zenni, R. D. (2018). Invasive species and the global strategy for plant conservation: How close has Brazil come to achieving Target 10? *Rodriguésia, 69*, 1567–1576.

Dehnen-Schmutz, K., Perrings, C., & Williamson, M. (2004). Controlling Rhododendron ponticum in the British Isles: An economic analysis. *Journal of Environmental Management, 70*(4), 323–332.

Di Biase, R. M., Fattorini, L., & Marchi, M. (2018). Statistical inferential techniques for approaching forest mapping. A review of methods. *Annals of Silvicultural Research, 42*(2), 46–58.

Dormann, C. F., Schymanski, S. J., Cabral, J., Chuine, I., Graham, C., Hartig, F., Kearney, M., Morin, X., Römermann, C., Schröder, B., & Singer, A. (2012). Correlation and process in species distribution models: Bridging a dichotomy. *Journal of Biogeography, 39*(12), 2119–2131.

Dyer, E. E., Cassey, P., Redding, D. W., Collen, B., Franks, V., Gaston, K. J., Jones, K. E., Kark, S., Orme, C. D., & Blackburn, T. M. (2017). The global distribution and drivers of alien bird species richness. *PLoS Biology, 15*(1), Article e2000942.

Elith, J., & Leathwick, J. R. (2009). Species distribution models: Ecological explanation and prediction across space and time. *Annual Review of Ecology, Evolution, and Systematics, 40*, 677–697.

Elith, J., Phillips, S. J., Hastie, T., Dudík, M., Chee, Y. E., & Yates, C. J. (2011). A statistical explanation of MaxEnt for ecologists. *Diversity and Distributions, 17*(1), 43–57.

Elith, J. H., Graham, C. P., Anderson, R., Dudík, M., Ferrier, S., Guisan, A., & J. Hijmans R, Huettmann F, Leathwick J, Lehmann A, Li J. (2006). Novel methods improve prediction of species' distributions from occurrence data. *Ecography, 29*(2), 129–151.

Estévez, R. A., Anderson, C. B., Pizarro, J. C., & Burgman, M. A. (2015). Clarifying values, risk perceptions, and attitudes to resolve or avoid social conflicts in invasive species management. *Conservation Biology, 29*(1), 19–30.

Faeth, S. H., Bang, C., & Saari, S. (2011). Urban biodiversity: Patterns and mechanisms. *Annals of the New York Academy of Sciences, 1223*(1), 69–81.

Fithian, W., & Hastie, T. (2012). *Statistical models for presence-only data: Finite-sample equivalence and addressing observer bias*. Stanford University.

Fleischer, V., Gröger, A., Koirala, N., Droby, A., Muthuraman, M., Kolber, P., Reuter, E., Meuth, S. G., Zipp, F., & Groppa, S. (2017). Increased structural white and grey matter network connectivity compensates for functional decline in early multiple sclerosis. *Multiple Sclerosis Journal, 23*(3), 432–441.

Foxcroft, L. C., Pyšek, P., Richardson, D. M., Genovesi, P., & MacFadyen, S. (2017). Plant invasion science in protected areas: progress and priorities. *Biological Invasions, 19*, 1353–1378.

Foxcroft, L. C., Richardson, D. M., Pyšek, P., & Genovesi, P. (2013). Plant invasions in protected areas: Outlining the issues and creating the links. In *Plant Invasions in Protected Areas: Patterns, Problems and Challenges* (pp. 3–18).

Foxcroft, L. C., Richardson, D. M., Rouget, M., & MacFadyen, S. (2009). Patterns of alien plant distribution at multiple spatial scales in a large national park: implications for ecology, management, and monitoring. *Diversity and Distributions, 15*(3), 367–378.

Franklin, J. (2010). *Mapping species distributions: Spatial inference and prediction*. Cambridge University Press.

Freeman, P. H., & Fox, R. (1994). Satellite mapping of tropical forest cover and deforestation: A review with recommendations for USAID. Environment and Natural Resources Information Center, DATEX, Arlington, VA.

Gaertner, J. P., Forstner, M. R., O'Donnell, L., & Hahn, D. (2009). Detection of *Batrachochytrium dendrobatidis* in endemic salamander species from Central Texas. *EcoHealth, 6*, 20–26.

Gaertner, M., Biggs, R., Te Beest, M., Hui, C., Molofsky, J., & Richardson, D. M. (2014). Invasive plants as drivers of regime shifts: Identifying high-priority invaders that alter feedback relationships. *Diversity and Distributions, 20*(7), 733–744.

Gentili, R., Schaffner, U., Martinoli, A., & Citterio, S. (2021). Invasive alien species and biodiversity: Impacts and management. *Biodiversity, 22*(1–2), 1–3.

Gilbert, B., & Levine, J. M. (2013). Plant invasions and extinction debts. *Proceedings of the National Academy of Sciences, 110*(5), 1744–1749.

Gnanavel, I., & Natarajan, S. K. (2013). *Parthenium hysterophorus* L.: A major threat to natural and agro eco-systems in India. *International Journal of Agriculture, Environment and Biotechnology, 6*(2):261–269

Graham, A. J., Atkinson, P. M., & Danson, F. M. (2004). Spatial analysis for epidemiology. *Acta Tropica, 91*(3), 219–225.

Grimm, N. B., Faeth, S. H., Golubiewski, N. E., Redman, C. L., Wu, J., Bai, X., & Briggs, J. M. (2008). Global change and the ecology of cities. *Science, 319*(5864), 756–760.

Guisan, A., & Thuiller, W. (2005). Predicting species distribution: Offering more than simple habitat models. *Ecology Letters, 8*(9), 993–1009.

Guisan, A., Tingley, R., Baumgartner, J. B., Naujokaitis-Lewis, I., Sutcliffe, P. R., Tulloch, A. I., Regan, T. J., Brotons, L., McDonald-Madden, E., Mantyka-Pringle, C., & Martin, T. G. (2013). Predicting species distributions for conservation decisions. *Ecology Letters, 16*(12), 1424–1435.

Gupta, S., Kumar, N., Verma, D., Bachheti, A., Arya, A. K., Joshi, K. K., & Bachheti, R. K. (2021). Impacts of invasive alien plant species on biodiversity in the regions of Western Himalayas, India: An overview. *Biological Diversity: Current Status and Conservation Policies, 1.*

Gurevitch, J., & Padilla, D. K. (2004). Are invasive species a major cause of extinctions. *Trends in Ecology & Evolution, 19*(9), 470–474.

Haq, S. M., Hamid, M., Lone, F. A., & Singh, B. (2021). Himalayan hotspot with Alien Weeds: A case study of biological spectrum, phenology, and diversity of weedy plants of high-altitude mountains in District Kupwara of J&K Himalaya, India. *Proceedings of the National Academy of Sciences, India Section B: Biological Sciences, 91*, 139–152.

Hasmadi, I. M., Pakhriazad, H. Z., & Mohamad, F. S. (2010). Geographic information system-allocation model for forest path: A case study in Ayer Hitam forest reserve, Malaysia. *American Journal of Applied Sciences, 7*(3), 376.

Hassan, A., & Nawchoo, I. A. (2020). Impact of invasive plants in aquatic ecosystems. *Bioremediation and Biotechnology: Sustainable Approaches to Pollution Degradation, 2020*, 55–73.

Higgins, S. I., Richardson, D. M., Cowling, R. M., & Trinder-Smith, T. H. (1999). Predicting the landscape-scale distribution of alien plants and their threat to plant diversity. *Conservation Biology, 13*(2), 303–313.

Holt, J. (2006). Score averaging for alien species risk assessment: A probabilistic alternative. *Journal of Environmental Management, 81*(1), 58–62.

Huang, C. Y., & Asner, G. P. (2009). Applications of remote sensing to alien invasive plant studies. *Sensors, 9*(6), 4869–4889.

Hui, C. (2023). The dos and don'ts for predicting invasion dynamics with species distribution models. *Biological Invasions, 25*(4), 947–953.

Jarić, I., Heger, T., Monzon, F. C., Jeschke, J. M., Kowarik, I., McConkey, K. R., Pyšek, P., Sagouis, A., Essl, F. (2019). Crypticity in biological invasions. *Trends in Ecology & Evolution, 34*(4), 291–302.

Jaureguiberry, P., Titeux, N., Wiemers, M., Bowler, D. E., Coscieme, L., Golden, A. S., Guerra, C. A., Jacob, U., Takahashi, Y., Settele, J., & Díaz, S. (2022). The direct drivers of recent global anthropogenic biodiversity loss. *Science Advances, 8*(45), IBM 9982.

Jiménez, L. J. (2021). *Estimation of the Fundamental Niche of a Species Using Presence Data, Physiological Tolerances, and Accessible Areas, and the Evaluation of Niche and Species Distribution Models* (Doctoral dissertation, University of Kansas).

Johnson, D. W., Johnson, R. T., & Smith, K. A. (2014). Cooperative learning: Improving university instruction by basing practice on validated theory. *Journal on Excellence in University Teaching, 25*(4), 1–26.

Joshi, C., De Leeuw, J., & van Duren, I. C. (2004). Remote sensing and GIS applications for mapping and spatial modelling of invasive species. In *Proceedings of ISPRS* (Vol. 35, p. B7).

Julia, R., Holland, D. W., & Guenthner, J. (2007). Assessing the economic impact of invasive species: The case of yellow starthistle (*Centaurea solsitialis* L.) in the rangelands of Idaho, USA. *Journal of Environmental Management, 85*(4):876–82.

Karki, D. (2009). Ecological and socio-economic impacts of *Parthenium hysterophorus* L. invasion in two urban areas in Nepal.

Kato-Noguchi, H., & Kato, M. (2023). Evolution of the secondary metabolites in invasive plant species *Chromolaena odorata* for the defense and allelopathic functions. *Plants, 12*(3), 521.

Kearney, M., & Porter, W. (2009). Mechanistic niche modelling: Combining physiological and spatial data to predict species' ranges. *Ecology Letters, 12*(4), 334–350.

Khaeim, H., Kende, Z., Jolánkai, M., Kovács, G. P., Gyuricza, C., & Tarnawa, Á. (2022). Impact of temperature and water on seed germination and seedling growth of maize (*Zea mays* L.). *Agronomy, 12*(2), 397.

Kimothi, M. M., & Dasari, A. (2010). Methodology to map the spread of an invasive plant (*Lantana camara* L.) in forest ecosystems using Indian remote sensing satellite data. *International Journal of Remote Sensing, 31*(12), 3273–3289.

Kohli, R. K., Dogra, K. S., Batish, D. R., & Singh, H. P. (2004). Impact of invasive plants on the structure and composition of the natural vegetation of northwestern Indian Himalayas1. *Weed Technology* 1296–300.

Kumar, R., & Saikia, P. (2020). Wild edible plants of Jharkhand and their utilitarian perspectives. *Indian Journal of Traditional Knowledge, 19*(2), 237–250.

Kumari, P., & Choudhary, A. K. (2016). Exotic species invasion threats to forests: A case study from the Betla national park, Palamu, Jharkhand, India. *Tropical Plant Research, 3*(3), 592–599.

Lahr, E. C., Dunn, R. R., & Frank, S. D. (2018). Getting ahead of the curve: Cities as surrogates for global change. *Proceedings of the Royal Society B: Biological Sciences, 285*(1882), 20180643.

Laliberte, A. S., & Ripple, W. J. (2004). Range contractions of North American carnivores and ungulates. *BioScience, 54*(2), 123–138.

Lindgren, C. J. (2013). *Addressing the risks of invasive plants through spatial predictive modelling.* University of Manitoba (Canada).

Lockwood, J. L., Hoopes, M. F., & Marchetti, M. P. (2013). *Invasion ecology*. Wiley.

Mack, R. N. (2000). Assessing the extent, status, and dynamism of plant invasions: current and emerging approaches. *Invasive Species in a Changing World*, 141–168.

Mack, R. N., Simberloff, D., Mark Lonsdale, W., Evans, H., Clout, M., & Bazzaz, F. A. (2000). Biotic invasions: Causes, epidemiology, global consequences, and control. *Ecological Applications, 10*(3), 689–710.

Maguire, L. A. (2004). What can decision analysis do for invasive species management? *Risk Analysis: An International Journal, 24*(4), 859–868.

Mani, J. K., & Varghese, A. O. (2018). Remote sensing and GIS in agriculture and forest resource monitoring. *Geospatial Technologies in Land Resources Mapping, Monitoring and Management, 2018*, 377–400.

McNeely, J. A. (2001). *Global strategy on invasive alien species*. IUCN.

Monterroso, I., Binimelis, R., & Rodríguez-Labajos, B. (2011). New methods for the analysis of invasion processes: Multi-criteria evaluation of the invasion of *Hydrilla verticillata* in Guatemala. *Journal of Environmental Management, 92*(3), 494–507.

Namkeleja, H. S., Tarimo, M. T., & Ndakidemi, P. A. (2014). Allelopathic effects of *Argemone mexicana* to growth of native plant species. *American Journal of Plant Sciences.*

Nath, S. N., & Kumar, R. (2024). Factors Responsible for Sal (*Shorea robusta*) Mortality in Jharkhand State, Bishen Singh Mahendra Pal Singh, Dehradun.

Negi, B., Khatri, K., Bargali, S. S., & Bargali, K. (2023). Invasive *Ageratina adenophora* (Asteraceae) in Agroecosystems of Kumaun Himalaya, India: A threat to plant diversity and sustainable crop yield. *Sustainability, 15*(14), 10748.

Negi, G. C., Sharma, S., Vishvakarma, S. C., Samant, S. S., Maikhuri, R. K., Prasad, R. C., & Palni, L. M. (2019). Ecology and use of Lantana camara in India. *The Botanical Review, 85*(2), 109–130.

Nguyen, G., Dlugolinsky, S., Bobák, M., Tran, V., López García, Á., Heredia, I., Malík, P., & Hluchý, L. (2019). Machine learning and deep learning frameworks and libraries for large-scale data mining: A survey. *Artificial Intelligence Review, 52*, 77–124.

Nihal, D., Prabhakaran, M. P., Abhirami, N., & Naseem, N. M. (2021). Role of hydrological parameters and environmental disturbances in steering bioinvasion and biohomogenization of *Eichhornia crassipes* in Cochin Backwaters-A mini review. *International Journal of Ecology and Environmental Sciences, 47*(4), 345–354.

OE, L. E. (2023). Pest risk analysis for *Ageratina adenophora*. EPPO Technical Document No. 1090, Paris.

Ooka, J. K., & Owens, D. K. (2018). Allelopathy in tropical and subtropical species. *Phytochemistry Reviews, 17*, 1225–1237.

Padayachee, A., Day, L., Howell, K., & Gidley, M. J. (2017). Complexity and health functionality of plant cell wall fibers from fruits and vegetables. *Critical Reviews in Food Science and Nutrition, 57*(1), 59–81.

Paz-Kagan, T., Silver, M., Panov, N., & Karnieli, A. (2019). Multispectral approach for identifying invasive plant species based on flowering phenology characteristics. *Remote Sensing, 11*(8), 953.

Pecchi, M., Marchi, M., Burton, V., Giannetti, F., Moriondo, M., Bernetti, I., Bindi, M., & Chirici, G. (2019). Species distribution modelling to support forest management. A literature review. *Ecological Modelling*, 108817.

Peerbhay, K., Mutanga, O., Lottering, R., & Ismail, R. (2016). Mapping *Solanum mauritianum* plant invasions using WorldView-2 imagery and unsupervised random forests. *Remote Sensing of Environment, 182*, 39–48.

Pejchar, L., & Mooney, H. A. (2009). Invasive species, ecosystem services and human well-being. *Trends in Ecology & Evolution, 24*(9), 497–504.

Peterson, A. T. (2003). Predicting the geography of species' invasions via ecological niche modeling. *The Quarterly Review of Biology, 78*(4), 419–433.

Phillips, S. J., Anderson, R. P., & Schapire, R. E. (2006). Maximum entropy modeling of species geographic distributions. *Ecological Modelling, 190*(3–4), 231–259.

Phillips, S. J., & Dudík, M. (2008). Modeling of species distributions with Maxent: New extensions and a comprehensive evaluation. *Ecography, 31*(2), 161–175.

Pimentel, D., Pimentel, M., & Wilson, A. (2007). Plant, animal, and microbe invasive species in the United States and world. *Biological Invasions,* 315–330.

Pimentel, D., Zuniga, R., & Morrison, D. (2005). Update on the environmental and economic costs associated with alien-invasive species in the United States. *Ecological Economics, 52*(3), 273–288.

Potgieter, L. J., Aronson, M. F., Brandt, A. J., Cook, C. N., Gaertner, M., Mandrak, N. E., Richardson, D. M., Shrestha, N., & Cadotte, M. W. (2022). Prioritization and thresholds for managing biological invasions in urban ecosystems. *Urban Ecosystems, 25*(1), 253–271.

Poudel, M., Adhikari, P., & Thapa, K. (2019). Biology and control methods of the alien invasive weed *Mikania micrantha*: A review. *Environmental Contaminants Reviews (ECR), 2*(2), 06–12.

Potgieter, L. J., Gaertner, M., Kueffer, C., Larson, B. M., Livingstone, S. W., O'Farrell, P. J., & Richardson, D. M. (2017). Alien plants as mediators of ecosystem services and disservices in urban systems: A global review. *Biological Invasions, 19*, 3571–3588.

Powell, P. A., García-Díaz, P., Fernández Cánepa, G., Grau, A., Herrera, L., Nuñez, C., et al. (2023). Insights from experiences comanaging woody invasive alien plants in Argentina. *Ecological Solutions and Evidence, 4*(4), Article e12272.

Pyšek, P., & Richardson, D. M. (2010). Invasive species, environmental change and management, and health. *Annual Review of Environment and Resources, 35*, 25–55.

Rai, P. K. (2015). What makes the plant invasion possible? Paradigm of invasion mechanisms, theories and attributes. *Environmental Skeptics and Critics, 4*(2), 36.

Rai, P. K., & Singh, J. S. (2020). Invasive alien plant species: Their impact on environment, ecosystem services and human health. *Ecological Indicators, 111*, Article 106020.

Rawat, L. S., Maikhuri, R. K., Bahuguna, Y. M., Maletha, A., Phondani, P. C., Jha, N. K., & Pharswan, D. S. (2019). Interference of Eupatorium adenophorum (Spr.) and its allelopathic effect on growth and yield attributes of traditional food crops in Indian Himalayan Region. *Ecological Research, 5*, 587–599.

Reichard, S. H., & White, P. S. (2003). Invasion biology: An emerging field of study. *Annals of the Missouri Botanical Garden, 1*, 64–66.

Richardson, D. M., & Rejmánek, M. (2011). Trees and shrubs as invasive alien species–A global review. *Diversity and Distributions, 17*(5), 788–809.

Royimani, L., Mutanga, O., Odindi, J., Dube, T., & Matongera, T. N. (2019). Advancements in satellite remote sensing for mapping and monitoring of alien invasive plant species (AIPs). *Physics and Chemistry of the Earth, Parts a/b/c, 112*, 237–245.

Royle, J. A., Chandler, R. B., Yackulic, C., & Nichols, J. D. (2012). Likelihood analysis of species occurrence probability from presence-only data for modelling species distributions. *Methods in Ecology and Evolution, 3*(3), 545–554.

Seoane, J., Carrascal, L. M., & Palomino, D. (2011). Assessing the ecological basis of conservation priority lists for bird species in an island scenario. *Journal for Nature Conservation, 19*(2), 103–115.

Shabani, F., Kumar, L., & Ahmadi, M. (2016). A comparison of absolute performance of different correlative and mechanistic species distribution models in an independent area. *Ecology and Evolution, 6*(16), 5973–5986.

Shackleton, C. M., Hurley, P. T., Dahlberg, A. C., Emery, M. R., & Nagendra, H. (2017). Urban foraging: A ubiquitous human practice overlooked by urban planners, policy, and research. *Sustainability, 9*(10),1884.

Shameli-Sendi, A., Aghababaei-Barzegar, R., & Cheriet, M. (2016). Taxonomy of information security risk assessment (ISRA). *Computers & Security, 57*, 14–30.

Shaw, R. H., & Hatcher, P. E. (2017). Weed biological control. Weed Research: Expanding Horizons 215–243.

Shogren, J. F., & Tschirhart, J. (2005). Integrating ecology and economics to address bioinvasions. *Ecological Economics, 52*(3), 267–271.

Singh, S. P. (1996). Biological control. In: Paroda, R. S., & Chadha, K. L. (Eds.), *50 years of crop science research in India* (pp. 88–116). Indian Council of Agricultural Research, New Delhi.

Singh, K. P., Shukla, A. N., & Singh, J. S. (2010). State-level inventory of invasive alien plants, their source regions, and use potential. *Current Science, 10*, 107–114.

Sonti, S. H. (2015). Application of geographic information system (GIS) in forest management. *Journal of Geography & Natural Disasters, 5*(3), 1000145.

Srivastava, V., Lafond, V., & Griess, V. C. (2019). Species distribution models (SDM): Applications, benefits and challenges in invasive species management. *CABI Reviews, 24*(2019), 1–3.

Sun, W., Zeng, C., Liu, S., Fu, J., Hu, L., Shi, Z., Yue, D., Ren, Z., Zhong, Z., Zuo, Z., & Cao, S. (2018). *Ageratina adenophora* induces mice hepatotoxicity via ROS-NLRP3-mediated pyroptosis. *Scientific Reports, 8*(1), 16032.

Thapa, N., & Maharjan, M. (2014). Invasive alien species: Threats and challenges for biodiversity conservation-A case study of Annapurna Conservation Area, Nepal. *Invasive Alien Species Management, 25*, 18–22.

Thuiller, W., Lavorel, S., Araújo, M. B., Sykes, M. T., & Prentice, I. C. (2005). Climate change threats to plant diversity in Europe. *Proceedings of the National Academy of Sciences, 102*(23), 8245–8250.

Tripathi, S., Bhadouria, R., Srivastava, P., Singh, R., Batish, D. R. (Eds.). (2023). *Plant invasions and global climate change*. Springer Nature. https://link.springer.com/book/10.1007/978-981-99-5910-5. Accessed Apr 28, 2024.

Van Kleunen, M., Essl, F., Pergl, J., Brundu, G., Carboni, M., Dullinger, S., Early, R., González-Moreno, P., Groom, Q. J., Hulme, P. E., & Kueffer, C. (2018). The changing role of ornamental horticulture in alien plant invasions. *Biological Reviews, 93*(3), 1421–1437.

Vaz, A. S., Kueffer, C., Kull, C. A., Richardson, D. M., Schindler, S., Muñoz-Pajares, A. J., Vicente, J. R., Martins, J., Hui, C., Kühn, I., & Honrado, J. P. (2017). The progress of interdisciplinarity in invasion science. *Ambio, 46*, 428–442.

Vitousek, P. M., d'Antonio, C. M., Loope, L. L., & Westbrooks, R. (1996). Biological invasions as global environmental change. *American Scientist, 84*(5), 468–478.

Vitousek, P. M., Mooney, H. A., Lubchenco, J., & Melillo, J. M. (1997). Human domination of Earth's ecosystems. *Science, 277*(5325), 494–499.

Williams, S. E., Williams, Y. M., VanDerWal, J., Isaac, J. L., Shoo, L. P., & Johnson, C. N. (2009). Ecological specialization and population size in a biodiversity hotspot: How rare species avoid extinction. *Proceedings of the National Academy of Sciences, 106*, 19737–19741.

Wu, A. P., Yu, H., Gao, S. Q., Huang, Z. Y., He, W. M., Miao, S. L., & Dong, M. (2009). Differential belowground allelopathic effects of leaf and root of *Mikania micrantha*. *Trees, 23*, 11–17.

Yadav, S. P., Mehata, D. K., Pokhrel, S., Ghimire, N. P., Gyawali, P., Katel, S., & Timsina, U. (2024). A comprehensive study of banmara (invasive alien plant species): Understanding the invasive potential and ecological consequences for biodiversity conservation, and management strategies. *Journal of Agriculture and Food Research., 8*, Article 101031.

Yadav, V., Singh, N. B., Singh, H., Singh, A., & Hussain, I. (2016). Allelopathic invasion of alien plant species in India and their management strategies: A review. *Tropical Plant Research, 3*(1), 87–101.

Yakub, M., & Tiffin, P. (2017). Living in the city: Urban environments shape the evolution of a native annual plant. *Global Change Biology, 23*(5), 2082–2089.

Zhou, L., Pan, S., Wang, J., & Vasilakos, A. V. (2017). Machine learning on big data: Opportunities and challenges. *Neurocomputing, 237*, 350–361.

Zhu, Y., Wei, W., Li, H., Wang, B., Yang, X., & Liu, Y. (2018). Modelling the potential distribution and shifts of three varieties of *Stipa tianschanica* in the eastern Eurasian Steppe under multiple climate change scenarios. *Global Ecology and Conservation, 16*, Article e00501.

Chapter 13
Incorporating Phenological Traits for Mapping of Invasive Plant Species in Urban Ecosystems

Kariya Ishita Bhaveshkumar and Laxmi Kant Sharma

Abstract The urban ecosystem consists of many complex socio-economic processes, which lead to the removal of existing vegetation and biological introductions in the ecosystem. These introductions can lead to the decline of the native population due to intra/inter-specific interactions. Remote sensing is a fundamental tool that is prominent in mapping these invasions in the ecosystem. This chapter summarizes how incorporating phenology in mapping invasive plant species has been used to characterize ecosystem functions. The potential of the phenological aspect in urban ecosystems is under-discussed which has been a highlight in this chapter. An extensive review of the literature addresses how phenological trait mapping has been used in urban mapping in the past and how it can be extremely important for comprehending plant physiological processes. The evidence of using remote sensing in this process is emerging to study changes in the ecosystem, vegetation structure and plant indicators.

Keywords Biological invasion · Phenology · Discrepancies · Urban ecology · Satellite remote sensing

1 Remote Sensing for Invasion Studies

Biological Invasion is the disintegration of biogeographic barriers and the fuzziness of the regional distinctiveness of Earth's biota. It is a serious concern for it effects of human-caused environmental change. Invaders modify the composition and functionality of the ecosystem in which they establish themselves and alter the homes of existing native species (Asner & Vitousek, 2005). Invasive plant species have been a serious threat to biodiversity which can lead to extinction, endangerment and the alternation of ecosystem processes (Tripathi et al., 2023). Areas which are at the

K. I. Bhaveshkumar · L. K. Sharma (✉)
Remote Sensing and GIS Lab, Department of Environmental Sciences, Central University of Rajasthan, Kishangarh, Rajasthan, India
e-mail: laxmikant_evs@curaj.ac.in

R. Singh et al. (eds.), *Invasive Alien Plants in Urban Ecosystems*,
https://doi.org/10.1007/978-3-032-00997-5_13

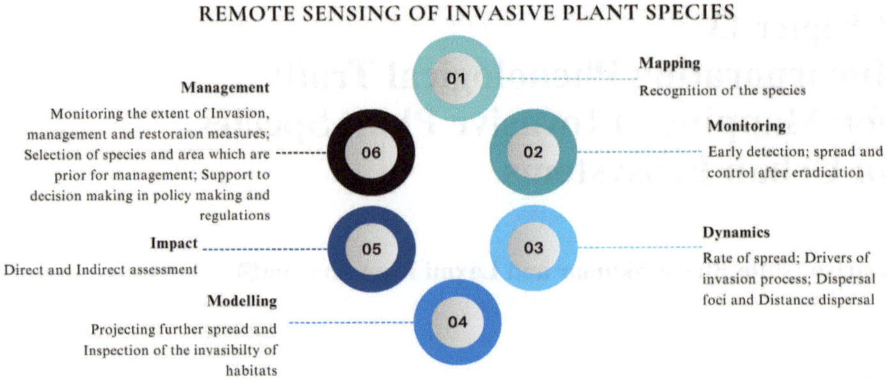

Fig. 1 Systematic diagrammatic representation indicating the iterative role of remote sensing in mapping plant species invasion

risk of potential biological invasion can be identified and effectively managed by geospatial technology (Truong et al., 2017).

In order to evaluate the impacts on ecosystem functional properties, attributes, and services, remote sensing aids in understanding the drivers, processes, and effects of plant invasion (Vaz et al., 2018). Spatiotemporal data can provide a perspective on mapping the invasion dynamics of invasive plant species. Remote sensing applications range from mapping and tracking invasion; providing data on the environmental and landscape features for studies on invasion processes (including drivers, mechanism and impacts); and upscaling invasion science to effectively manage invasion at various temporal and spatial scales (Muellerova et al., 2023). Figure 1 showed the systematic diagram representation indicating iterative role of remote sensing in mapping plant species invasion.

The chapter discuss the issues faced in monitoring plant invasion in different ecosystem, its mapping in varied landscape, past research regarding considering phenology as functional trait for mapping invasion and lastly summarizing with integrated approach defining the need and requirement for incorporating phenology in mapping and monitoring invasive plant species in urban ecosystem.

2 Issues Faced in Invasive Plant Species Monitoring

Scale is one of the primary factors taken into account in invasion ecology. The size of the study area, the sensor's resolution, and the spatial scale all affect the field observations. The significance of spectral characteristics, image resolution, and the superiority of lower-scaled aerial photographs demonstrates emphatically how crucial it is to aim aerial photo or image captures with the phenological stage in mind for precise mapping of invasive species (Klosterman et al., 2017). While species that are canopy dominant can be classified because of its high spectral and spatial resolution, but its precision is still a challenging task. Numerous investigations into the spectral characteristics of invasive species have been conducted, primarily using field spectrographs or low-altitude aerial photography. But very little information has actually reached the distant observer. Additional elements that contribute to the confusion and difficulty in differentiating between different species, including soil contribution, humidity, shadows, and noise in the atmosphere. Finding the best wavelengths for discrimination, however, is not practically possible when there are a lot of invasive species present. Moreover, the difficulty of distinguishing the distinct elements that contribute to the mixed spectrum also increases with the number of invasive species present in each pixel. If there is significant species variability in spectral signatures, these issues will worsen (Bhaveshkumar et al., 2024). Therefore, it seems unlikely that direct mapping will be able to accurately identify every individual in a large-scale monitoring program involving direct remotely sensed monitoring of multiple invasive species (Joshi et al., 2004).

In all areas of conservation, the task of prioritizing species and areas for management is growing more difficult due to the numerous interrelated issues and outside influences. One major complicating factor is that the majority of the most invasive species are undesirable, but in some areas of the landscape, these species are important commercially. Solutions may involve traditional, mechanical or chemical methods of controlling plants; alternatively, biological control may be employed, employing agents that lower seed production and, consequently, the risk of invasive species spreading (Moran et al., 2005). Pragmatic approaches to invasions are gaining popularity and support. One such approach is the concept of novel ecosystems, which suggests managing some ecosystems by taking sustainability and ecosystem services into account rather than strictly applying native-alien or invasive-non-invasive criteria (Hobbs et al., 2011). Controlling the invaders may be desirable and feasible in many other areas where there has not been as much landscape transformation as in others, where the presence of non-native species clearly threatens native biota and ecosystems' services (Richardson et al., 2013).

3 Mapping Biological Invasion in Varied Land Cover Types

Two aspects of global change that can influence each other are biological invasions and changes in land use. Invaders have the power to drastically change how the land is used, which could have an impact on the functioning of the ecosystem. Furthermore, it is evident that human-dominated landscapes with embedded natural ecosystems experience higher rates of invasion than do expansive, uninterrupted wilderness areas (Leyva et al., 2006; McKinney, 2006). The fact states that, in addition to the kind of land use, the surrounding landscape elements also impact the frequency of regional plant invasions. Theoretical research indicates that factors functioning at the landscape levels are closely linked to the spread of invasive species and a number of empirical studies have supported this concept (Ibáñez et al., 2009; Minor et al., 2009). Furthermore, ecologists are investigating how landscape dynamics affect invasion, showing that large-scale, repeated introductions combined with disturbances to the land surface have been crucial in the spread of certain invasive species (DeGasperis & Motzkin, 2007; Vilà & Ibáñez, 2011).Urban areas exhibit high non-native species richness, particularly for plants, because non-native species have lower complementarity than native species (Wilsey et al., 2009). Thus, the functioning of invasive species' biodiversity ecosystems can contribute to our knowledge of how they disturb the urban ecology (Schwarz et al., 2017; Weiskopf et al., 2024).

Invasive plant species' can be identified and mapped, understanding the changes in its spectral signatures through remote sensing (Höft et al., 2007). Distinct spectral reflectance in different electromagnetic spectrum areas indicates differences in the biophysical (such as leaf structure, texture, canopy, native and nonnative orientation) and biochemical (such as water content, chlorophyll) features of species. Spectral signatures studied with a phenological aspect can help researchers to disseminate alien invasive species from indigenous plants that invade the landscape (Royimani et al., 2019). Santos and Ustin (2018) successfully detected and mapped fennel in grassland ecosystems due to its phenological differences.

For the purpose of mapping invasive life-forms using remotely sensed data, phenological traits are mostly used to detect shrubs, herbs and trees (Table 1; Fig. 2). For tracking invasion, Very High Resolution (VHR) and Medium spatial resolution data are most commonly employed, although coarse spatial resolution is less frequently used. VHR data has the advantage that it provides detailed information regarding how invasive plant species are spatially explicit. Optical sensors are most commonly used to map phenological traits, but the use of hyperspectral sensors is increased recently. The narrow spectral band configuration in hyperspectral sensors can measure the subtle changes that can discriminate slightest changes in the phenology of species (Niphadkar & Nagendra, 2016). 4-band aerial photography obtained during the flowering season was utilized to map and identify the Callery pear (*Pyrus calleryana*) with its distinct phenological traits. To classify these species high resolution imagery was integrated with CNN-model for pixel-based identification (Ardohain et al., 2024).

Table 1 Compiling different aspects of the publications that have been studied in order to map phenological trait groups

Attribute	Trait group	Papers
Lifeform	Shrub	21
	Herb	9
	Tree	18
Spatial resolution	Very high resolution (VHR)	23
	Medium	10
Type of RS data	Hyperspectral	12
	Multispectral	33
	Active (Lidar + Radar)	7

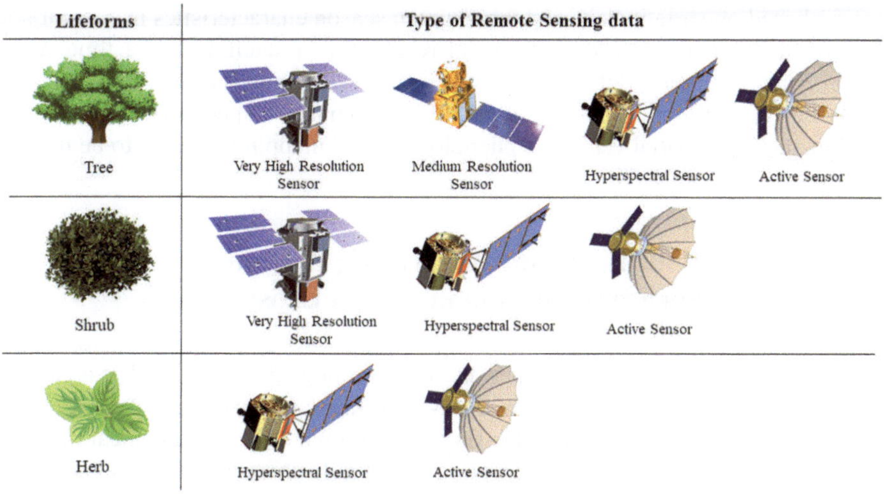

Fig. 2 Picturesque representation of various habitat or life-form types and indicative sensors which are mostly used for mapping of invasive plant species

4 Phenological Trait-Based Research for Invasion

The way in which plants adjust to changes in their biotic interactions and physical surroundings is reflected in their functional traits. Therefore, to predict the effects of climate and environmental stressors that affect distributions of species, biodiversity and ecosystem process including phenological traits into understanding mechanics of plant ecosystem functioning can be improved (Fajardo & Siefert, 2016; Kumar et al., 2024).

The timing of recurring events like leaf budding, flowering, and flushing is known as phenology. Additionally, it is crucial to the growth of invasive plant species that are impacted by climate change. The phenology of plants can be tied to many important characteristics which stands a role in determining when the plants are at risk of herbivory and when plant species are in complete stage to be pollinators (Wolkovich & Cleland, 2010).

If an invasive species differs from native species in its seasonal or interannual growth pattern, it can be identified from native species using phenology. Remote detection of invasive plants has distinct phenology, having an advantage over native plants, which makes identification easier (Willis et al., 2010) (Table 2). When an invasive plant species of evergreen nature invades a deciduous landscape, then these species show different interannual signals such as higher inter-annual variability and are phenologically distinct as compared to the native species, it stays green longer or has earlier green up than native species (Müllerová et al., 2013). Understory invasive plants in the forest canopy can be discriminated if they are phenologically distinct characteristics from overstory species. For instance, Becker et al. (2012) identified buckthorn, an understory species, using time series analysis of Landsat images based on variations in prolonged green season characteristics in comparison to the forest canopy. The coarse spatial resolution of satellite image time series influences phenological detection as it can be utilised at later stages of invasion, so it can be detected. When an invasive plant differs from native species functionally and exhibits minimal spatial variation, phenology-based mapping is ought to be used.

5 Mapping Plant Invasion: Integrating Bottom Up and Top-Down Information in Urban Ecosystem

The urban ecosystem is highly influenced by anthropogenic activities. The areas like agriculture and forests are replaced by built-up surfaces, which impacts the ecosystem for the people living in the cities (De Barros Ruas et al., 2022). Biological invasion in this landscape has a nature of adaptation and establishment with the availability of light, soil and water nutrients. Because of ecological factors like propagule pressure and environmental factors like light availability owing to canopy structure, elevation, soil, climate, etc., invasive alien species will thrive, grow, and spread (Daehler, 2003; Lockwood et al., 2005;). Phenology is the essential biodiversity variable that can explain changes in vegetation responses due to changes in climatic and environmental variables (Cleland et al., 2007; Wolf et al., 2017; Wolkovich et al., 2013). Phenological effects emerge in landscape gradients of floristic and phenological similarity may additionally react to soil, geological, and topographic factors (Afreen et al., 2023; Madonsela et al., 2017). Tracking and incorporating phenology in your study across regional and global scales may prove important evidence, which can track changes in the dominance, diversity and composition of plant communities (Dronova & Taddeo, 2022; Geissler et al., 2023). Studies using remote sensing have concentrated on mapping the later stages of plant invasion relative to the earlier stages. The identification of established invasive populations of plants is largely dependent on the discrimination of spectral signatures used to detect plant invasion from native species (Große-Stoltenberg et al., 2016; Hill et al., 2016; Kumar et al., 2024; Taboada et al., 2017). The chapter explains that incorporating phenological traits can improve the mapping of invasive plant species in urban ecosystems.

As a result, mapping invasions cannot be done solely by using data from remote sensing; ground observations, and a thorough ecological knowledge of the invasive plant species present in a particular landscape are also essential. Monitoring these structural responses of invasive alien plants is crucial under future management scenarios, as it will facilitate their identification and characterization in response to variability in their physical and biotic environments. Therefore, it is suggested that phenological traits be included in identification in order to comprehend the mechanism of monitoring and mapping invasion using an integrated bottom-up and top-down, bottom-up approach.

Table 2 Review table on remote sensing techniques, sensor and image characteristics and methods for detecting invasive plant species of different life forms based on phenological trait group

Traits		Sensor type	Spatial resolution	Remote sensing method	Similarity/dissimilarity used for discrimination	References
Phenology	Before the flowering period	Hyperspectral imagery (Aerial photos)	2 m	Pixel based classification using spectral angle mapper (SAM) algorithm	Spectral reflectance curves for species noticed differences in the flowering period which made them separable from native species	Lass et al. (2005)
	Seasonal progression of leaf greenness	Multispectral imagery (Landsat 5 & 7)	30 m	Vegetation indices, Field measurements, Tasseled Cap Greenness transformation	Time series analysis shows the prolonging greenness of species from native species	Becker et al. (2012)
	Presence of yellow flowering period	Multispectral imagery (WorldView-2 images), hyperspectral imagery (Airborne AisaFENIX)	2 m, 1 m	Pixel-based classification	Yellow and red edge bands in imagery that improve detection by separating spectral signatures of flowering period in species	Paz-Kagan et al. (2019)
	Progression of leaf senescence	Multispectral Imagery (Sentinel-2 & Pléiades 1A)	10 m, 2 m	Segmentation, Vegetation Indices, Feature Selection and Classification	Date of acquisition has played role map the dry conditions of the study site to map selected species	Ng et al. (2017)

(continued)

Table 2 (continued)

Traits		Sensor type	Spatial resolution	Remote sensing method	Similarity/dissimilarity used for discrimination	References
	Seasonal change in leaf progression	Multispectral imagery (Landsat 8 & Sentinel-2)	30 m, 10 m	Field data measurement, Feature selection and classification	Spectral discrimination based on the inflorescence period during spring and summer. Optimum band analysis identified the most influential bands for seasonal variation	Rajah et al. (2018)
	Seasonal progression of leaf senescence	Multispectral imagery (Landsat-8)	30 m	Model optimization with machine and deep learning algorithms	NDVI plays a huge role in determining the dry conditions of the site, thus creating a spectrally indistinct signal from native species	Shiferaw et al. (2019)
	Start and end of the peak flowering period	Aerial imagery (UAV)	1 m	Field sampling and ML classification	Temporal variations in phenology made invasive plant species discriminate from native species	De Sá et al. (2018)
	Seasonal phenology occurrence in dormant phase, early green-up, and senescence onset stages	Multispectral imagery (Landsat surface reflectance collection)	30 m	Vegetation and landscape indices, Field data sampling and classification	Differences in intra-annual phenological characteristics in addition to indices enhanced the mapping of invasive plant species	Singh et al. (2018)

(continued)

Table 2 (continued)

Traits		Sensor type	Spatial resolution	Remote sensing method	Similarity/dissimilarity used for discrimination	References
	Seasonal change in leaf progression	Aerial photographs, multispectral imagery (Sentinel-2), microwave imagery (Sentinel-1)	0.5 m, 10 m, 20 m	Field data collection, Vegetation indices, Image fusion and classification	Spectrally derived vegetation indices show a significant difference in white inflorescence phenological traits for invasive species	Rajah et al. (2019)
	Leaf on and off season at peak of senescence and abscission	Multispectral imagery (IKONOS), Airborne LiDAR data	1 m, 5 m	Field data collection, Vegetation indices and classification	Spectral differences can unravel vertical structure on the basis of the leaf-on and off period for heterotypic forest stands	Singh et al. (2015)
	Dry and wet season progression	Multispectral imagery (Landsat 8), elevation data (ASTER GDEM)	30 m, 30 m	Field data collection, Feature selection, Pixel and object-based classification	Spectral differences observed in mapping of invasive species for dry, wet and combined data, noticing phenological variations of the site	Ng et al. (2016)
	Start and end of the peak flowering period	UAV photographs, Multispectral Imagery (Pléiades 1B)	50 cm, 2 m	Image pre-processing for CHM generation, classification	Differences in phenology can help select appropriate period for generation of accurate results	Martin et al. (2018)

6 Conclusions and Future Prospects

The effectual management of invasive plant species is difficult as detecting these species is a task using traditional remote sensing methods. Classification of these species from native species depends on the availability of a high-resolution dataset, climatic and environmental conditions of that place. This chapter draws attention towards the potential of incorporating phenological traits in the mapping of invasive plant species. Additionally, some remote sensing non-parametric algorithms were put forward to exploring how these techniques aid in mapping and management of invasive species. A lot of functional traits have been developed and implemented for mapping invasion, the integration of plant phenological traits and remote sensing methods seems to be more effective in accordance to study site conditions and has potential to progress and expand if incorporated in urban ecosystem. Therefore, phenology can enhance the ability to map, predict and monitor the process of invasion which provides important information to the practitioners, land-managers and policy-makers. Ecologists need to work with remote sensing experts to identify the phenological conditions of study site and selecting the suitable remote sensing method in the area of biological invasion.

Acknowledgements The author wishes to acknowledge the support from the Department of Science and Technology (DST) for providing funds under the INSPIRE fellowship [**IF220128**]. The author express gratitude for Department of Environmental Science, Central University of Rajasthan for its infrastructure and resources. The review of literature and statistical analysis has been carried out on the software platform, i.e., Microsoft Word and Excel are used to prepare the diagrams and content.

References

Afreen, T., Kumari, S., Bhadouria, R., Devi, R. S., Singh, S., & Tripathi, S. (2023). Plant invasion and soil processes: A mechanistic understanding. In *Plant invasions and global climate change* (pp. 227–246). https://doi.org/10.1007/978-981-99-5910-5_10

Ardohain, C., Wingren, C., Thapa, B., & Fei, S. (2024). Invasive species identification from high-resolution 4-band multispectral imagery. *Biological Invasions, 26*(11), 3603–3619. https://doi.org/10.1007/s10530-024-03397-0

Asner, G. P., & Vitousek, P. M. (2005). Remote analysis of biological invasion and biogeochemical change. *Proceedings of the National Academy of Sciences of the United States of America, 102*(12), 4383–4386.

Becker, R. H., Zmijewski, K. A., & Crail, T. D. (2012). Seeing the forest for the invasives: Mapping buckthorn in the Oak Openings. *Biological Invasions, 15*(2), 315–326. https://doi.org/10.1007/s10530-012-0288-8

Bhaveshkumar, K. I., Sharma, L. K., & Verma, R. K. (2024). Applicability of phenological indices for mapping of understory invasive species using machine learning algorithms. *Biological Invasions, 26*(9), 2901–2921. https://doi.org/10.1007/s10530-024-03361-y

Cleland, E. E., Chuine, I., Menzel, A., Mooney, H. A., & Schwartz, M. D. (2007). Shifting plant phenology in response to global change. *Trends in Ecology & Evolution, 22*(7), 357–365. https://doi.org/10.1016/j.tree.2007.04.003

Daehler, C. C. (2003). Performance comparisons of Co-Occurring native and alien invasive plants: Implications for conservation and restoration. *Annual Review of Ecology, Evolution, and Systematics, 34*(1), 183–211. https://doi.org/10.1146/annurev.ecolsys.34.011802.132403

De Barros Ruas, R., Costa, L. M. S., & Bered, F. (2022). Urbanization driving changes in plant species and communities—A global view. *Global Ecology and Conservation, 38*, Article e02243. https://doi.org/10.1016/j.gecco.2022.e02243

De Sá, N. C., Castro, P. M., Carvalho, S., Marchante, E., López-Núñez, F. A., & Marchante, H. (2018). Mapping the flowering of an invasive plant using unmanned aerial vehicles: Is there potential for biocontrol monitoring? *Frontiers in Plant Science.* https://doi.org/10.3389/fpls.2018.00293

DeGasperis, B. G., & Motzkin, G. (2007). Windows of opportunity: historical and ecological controls on *Berberis thunbergi iinvasions. Ecology, 88*(12), 3115–3125. https://doi.org/10.1890/06-2014.1

Dronova, I., & Taddeo, S. (2022). Remote sensing of phenology: Towards the comprehensive indicators of plant community dynamics from species to regional scales. *Journal of Ecology, 110*(7), 1460–1484. https://doi.org/10.1111/1365-2745.13897

Fajardo, A., & Siefert, A. (2016). Phenological variation of leaf functional traits within species. *Oecologia, 180*(4), 951–959. https://doi.org/10.1007/s00442-016-3545-1

Geissler, C., Davidson, A., & Niesenbaum, R. A. (2023). The influence of climate warming on flowering phenology in relation to historical annual and seasonal temperatures and plant functional traits. *PeerJ, 11*, Article e15188. https://doi.org/10.7717/peerj.15188

Große-Stoltenberg, A., Hellmann, C., Werner, C., Oldeland, J., & Thiele, J. (2016). Evaluation of continuous VNIR-SWIR spectra versus narrowband hyperspectral indices to discriminate the invasive *Acacia longifolia* within a Mediterranean Dune ecosystem. *Remote Sensing, 8*(4), 334. https://doi.org/10.3390/rs8040334

Hill, D., Tarasoff, C. S., Whitworth, G. E., Baron, J., Bradshaw, J. L., & Church, J. S. (2016). Utility of unmanned aerial vehicles for mapping invasive plant species: a case study on yellow flag iris (*Iris pseudacorus* L.). *International Journal of Remote Sensing, 38*(8–10), 2083–2105. https://doi.org/10.1080/01431161.2016.1264030

Hobbs, R. J., Hallett, L. M., Ehrlich, P. R., & Mooney, H. A. (2011). Intervention ecology: Applying ecological science in the twenty-first century. *Bioscience (Print), 61*(6), 442–450. https://doi.org/10.1525/bio.2011.61.6.6

Höft, R., Strand, H., & Strittholt, J. R. (2007). Sourcebook on remote sensing and biodiversity indicators. Convention on Biological Diversity Technical Series 32. *Montreal: Secretariat of the Convention on Biological Diversity. Technical Series, 2007.* https://www.biodiversitylibrary.org/item/119137

Ibáñez, I., Silander, J. A., Allen, J. M., Treanor, S. A., & Wilson, A. (2009). Identifying hotspots for plant invasions and forecasting focal points of further spread. *Journal of Applied Ecology, 46*(6), 1219–1228. https://doi.org/10.1111/j.1365-2664.2009.01736.x

Joshi, C., De Leeuw, J., & Van Duren, I. (2004). Remote sensing and GIS applications for mapping and spatial modelling of invasive species. In *Proceedings of ISPRS* (Vol. 35, No. B7), 669–677. https://www.cartesia.org/geodoc/isprs2004/comm7/papers/132.pdf

Klosterman, S., Melaas, E., Wang, J. A., Martinez, A., Frederick, S., O'Keefe, J., Orwig, D. A., Wang, Z., Sun, Q., Schaaf, C., Friedl, M., & Richardson, A. D. (2017). Fine-scale perspectives on landscape phenology from unmanned aerial vehicle (UAV) photography. *Agricultural and Forest Meteorology, 248*, 397–407. https://doi.org/10.1016/j.agrformet.2017.10.015

Kumar, A., Singh, S., Kumar, D., Singh, R. K., Gupta, A. K., Premkumar, K., Chand, H. B., & Kewat, A. K. (2024). Investigating the phenology and interactions of competitive plant species co-occurring with invasive *Lantana camara* in Indian Himalayan Region. *Scientific Reports, 14*(1). https://doi.org/10.1038/s41598-023-50287-x

Lass, L. W., Prather, T. S., Glenn, N. F., Weber, K. T., Mundt, J. T., & Pettingill, J. (2005). A review of remote sensing of invasive weeds and example of the early detection of spotted knapweed

(*Centaurea maculosa*) and babysbreath (*Gypsophila paniculata*) with a hyperspectral sensor. *Weed Science, 53*(2), 242–251. https://doi.org/10.1614/WS-04-044R2

Leyva, C., Espejel, I., Escofet, A., & Bullock, S. H. (2006). Coastal landscape fragmentation by tourism development: Impacts and conservation alternatives. *Natural Areas Journal, 26*(2), 117–125. https://doi.org/10.3375/0885-8608(2006)26

Lockwood, J. L., Cassey, P., & Blackburn, T. M. (2005). The role of propagule pressure in explaining species invasions. *Trends in Ecology & Evolution, 20*(5), 223–228. https://doi.org/10.1016/j.tree.2005.02.004

Madonsela, S., Cho, M. A., Mathieu, R., Mutanga, O., Ramoelo, A., Kaszta, Ż, Van De Kerchove, R., & Wolff, É. (2017). Multi-phenology WorldView-2 imagery improves remote sensing of savannah tree species. *International Journal of Applied Earth Observation and Geoinformation, 58*, 65–73. https://doi.org/10.1016/j.jag.2017.01.018

Martin, F., Müllerová, J., Borgniet, L., Dommanget, F., Breton, V., & Évette, A. (2018). Using single- and multi-date UAV and satellite imagery to accurately monitor invasive knotweed species. *Remote Sensing, 10*(10), 1662. https://doi.org/10.3390/rs10101662

McKinney, M. L. (2006). Urbanization as a major cause of biotic homogenization. *Biological Conservation, 127*(3), 247–260. https://doi.org/10.1016/j.biocon.2005.09.005

Minor, E. S., Tessel, S. M., Engelhardt, K. a. M., & Lookingbill, T. R. (2009). The role of landscape connectivity in assembling exotic plant communities: a network analysis. *Ecology, 90*(7), 1802–1809.https://doi.org/10.1890/08-1015.1

Moran, V. C., Hoffmann, J. H., & Zimmermann, H. G. (2005). Biological control of invasive alien plants in South Africa: Necessity, circumspection, and success. *Frontiers in Ecology and the Environment (Print), 3*(2), 71–77. https://doi.org/10.1890/1540-9295(2005)003

Müllerová, J., Pergl, J., & Pyšek, P. (2013). Remote sensing as a tool for monitoring plant invasions: Testing the effects of data resolution and image classification approach on the detection of a model plant species *Heracleum mantegazzianum* (giant hogweed). *International Journal of Applied Earth Observation and Geoinformation, 25*, 55–65. https://doi.org/10.1016/j.jag.2013.03.004

Muellerova, J., Brundu, G., Grosse-Stoltenberg, A., Kattenborn, T., & Richardson, D. M. (2023). Pattern to process, research to practice: Remote sensing of plant invasions. *Biol Invasions, 25*(12), 3651–3676.

Ng, W., Meroni, M., Immitzer, M., Böck, S., Leonardi, U., Rembold, F., Gadain, H., & Atzberger, C. (2016). Mapping Prosopis spp. with Landsat 8 data in arid environments: Evaluating effectiveness of different methods and temporal imagery selection for Hargeisa, Somaliland. *International Journal of Applied Earth Observation and Geoinformation, 53*, 76–89. https://doi.org/10.1016/j.jag.2016.07.019

Ng, W., Rima, P., Einzmann, K., Immitzer, M., Atzberger, C., & Eckert, S. (2017). Assessing the potential of Sentinel-2 and Pléiades data for the detection of *Prosopis* and *Vachellia* spp. in Kenya. *Remote Sensing, 9*(1), 74. https://doi.org/10.3390/rs9010074

Niphadkar, M., & Nagendra, H. (2016). Remote sensing of invasive plants: Incorporating functional traits into the picture. *International Journal of Remote Sensing, 37*(13), 3074–3085. https://doi.org/10.1080/01431161.2016.1193795

Novel ecosystems: intervening in the new ecological world order. (2013). *Choice Reviews Online, 51*(02), 51–0871. https://doi.org/10.5860/choice.51-0871

Paz-Kagan, T., Silver, M., Panov, N., & Karnieli, A. (2019). Multispectral approach for identifying invasive plant species based on flowering phenology characteristics. *Remote Sensing, 11*(8), 953. https://doi.org/10.3390/rs11080953

Rajah, P., Odindi, J., & Mutanga, O. (2018). Evaluating the potential of freely available multispectral remotely sensed imagery in mapping American bramble (*Rubus cuneifolius*). *South African Geographical Journal, 100*(3), 291–307. https://doi.org/10.1080/03736245.2018.1461683

Rajah, P., Odindi, J., Mutanga, O., & Kiala, Z. (2019). The utility of Sentinel-2 vegetation indices (VIs) and Sentinel-1 synthetic aperture radar (SAR) for invasive alien species detection and mapping. *Nature Conservation, 35*, 41–61. https://doi.org/10.3897/natureconservation.35.29588

Richardson, D. M., Hui, C., Núñez, M. A., & Pauchard, A. (2013). Tree invasions: Patterns, processes, challenges and opportunities. *Biological Invasions, 16*(3), 473–481. https://doi.org/10.1007/s10530-013-0606-9

Royimani, L., Mutanga, O., Odindi, J., Dube, T., & Matongera, T. N. (2019). Advancements in satellite remote sensing for mapping and monitoring of alien invasive plant species (AIPs). *Physics and Chemistry of the Earth/Physics and Chemistry of the Earth. Parts a/B/C, 112*, 237–245. https://doi.org/10.1016/j.pce.2018.12.004

Santos, M. J., & Ustin, S. L. (2018). Spectral identification of native and non-native plant species for biodiversity assessments. In *Proceedings of ASD and IEEEGRS; Art, Science and Applications of Reflectance Spectroscopy Symposium*. https://doi.org/10.1109/igarss.2018.8517397

Schwarz, N., Moretti, M., Bugalho, M. N., Davies, Z. G., Haase, D., Hack, J., Hof, A., Melero, Y., Pett, T. J., & Knapp, S. (2017). Understanding biodiversity-ecosystem service relationships in urban areas: A comprehensive literature review. *Ecosystem Services, 27*, 161–171. https://doi.org/10.1016/j.ecoser.2017.08.014

Shiferaw, H., Bewket, W., & Eckert, S. (2019). Performances of machine learning algorithms for mapping fractional cover of an invasive plant species in a dryland ecosystem. *Ecology and Evolution, 9*(5), 2562–2574. https://doi.org/10.1002/ece3.4919

Singh, K., Chen, Y., Smart, L. S., Gray, J. V., & Meentemeyer, R. K. (2018). Intra-annual phenology for detecting understory plant invasion in urban forests. *ISPRS Journal of Photogrammetry and Remote Sensing, 142*, 151–161. https://doi.org/10.1016/j.isprsjprs.2018.05.023

Singh, K., Davis, A. J., & Meentemeyer, R. K. (2015). Detecting understory plant invasion in urban forests using LiDAR. *International Journal of Applied Earth Observation and Geoinformation, 38*, 267–279. https://doi.org/10.1016/j.jag.2015.01.012

Taboada, M. F. Á., Paredes, C., & Julián-Pelaz, J. (2017). Mapping of the invasive species *Hakea sericea* using unmanned aerial vehicle (UAV) and worldview-2 imagery and an object-oriented approach. *Remote Sensing, 9*(9), 913. https://doi.org/10.3390/rs9090913

Tripathi, S., Bhadouria, Srivastava, Singh, & Batish (Eds.). (2023). *Plant invasions and global climate change*. Springer Nature.

Truong, T., St J Hardy, G. E., & Andrew, M. E. (2017). Contemporary remotely sensed data products refine invasive plants risk mapping in data poor regions. *Frontiers in Plant Science, 8*.https://doi.org/10.3389/fpls.2017.00770

Vaz, A. S., Alcaraz-Segura, D., Campos, J. C., Vicente, J. R., & Honrado, J. P. (2018). Managing plant invasions through the lens of remote sensing: A review of progress and the way forward. *Science of the Total Environment, 642*, 1328–1339. https://doi.org/10.1016/j.scitotenv.2018.06.134

Vilà, M., & Ibáñez, I. (2011). Plant invasions in the landscape. *Landscape Ecology, 26*(4), 461–472. https://doi.org/10.1007/s10980-011-9585-3

Weiskopf, S. R., Lerman, S. B., Isbell, F., & Morelli, T. L. (2024). Biodiversity promotes urban ecosystem functioning. *Ecography, 2024*(9). https://doi.org/10.1111/ecog.07366

Willis, C. G., Ruhfel, B. R., Primack, R. B., Miller-Rushing, A. J., Losos, J. B., & Davis, C. C. (2010). Favorable climate change response explains non-native species' success in Thoreau's woods. *PLoS ONE, 5*(1), Article e8878. https://doi.org/10.1371/journal.pone.0008878

Wilsey, B. J., Teaschner, T. B., Daneshgar, P. P., Isbell, F. I., & Polley, H. W. (2009). Biodiversity maintenance mechanisms differ between native and novel exotic-dominated communities. *Ecology Letters, 12*(5), 432–442. https://doi.org/10.1111/j.1461-0248.2009.01298.x

Wolf, A. A., Zavaleta, E. S., & Selmants, P. C. (2017). Flowering phenology shifts in response to biodiversity loss. *Proceedings of the National Academy of Sciences of the United States of America, 114*(13), 3463–3468. https://doi.org/10.1073/pnas.1608357114

Wolkovich, E. M., & Cleland, E. E. (2010). The phenology of plant invasions: A community ecology perspective. *Frontiers in Ecology and the Environment (Print), 9*(5), 287–294. https://doi.org/10.1890/100033

Wolkovich, E. M., Cook, B. I., & Davies, T. J. (2013). Progress towards an interdisciplinary science of plant phenology: Building predictions across space, time and species diversity. *New Phytologist, 201*(4), 1156–1162. https://doi.org/10.1111/nph.12599

Chapter 14
Management Approaches and Policy Regulation for Invasive Alien Plants in Urban Ecosystems

Shaopeng Zhao, Omar Aziz, and Muhammad Riaz

Abstract Invasive alien plant species (IAPS) are non-native plants introduced to new environments where they can cause significant harm to ecosystems, economies, and human well-being. These species are recognized as major drivers of biodiversity loss and can alter ecosystem services and socio-economic conditions. Urban ecosystems provide particular difficulties in the management of IAPS because of their diverse habitats, increased connectedness, and disruptions. To control IAPS in urban ecosystems, this chapter discusses developing technology, management strategies, and regulatory issues. IAPS has the potential to seriously disrupt ecological processes, lower biodiversity, hinder agriculture, harm infrastructure, and result in large financial losses. Proactive steps, early identification, and prompt action, as well as integrated approaches such as biological, chemical, mechanical, and physical control measures adapted to the species and habitat are necessary for effective management. Policies and laws offer frameworks for IAPS prevention and control. Public education initiatives, adaptable, context-specific tactics, and an inclusive decision-making process are necessary for success. Moreover, this chapter discusses different methods to control alien species e.g., chemical methods, physical methods, biological methods, integrated GIS techniques, and public awareness. It can be easier to establish and enforce policies when they are transparent, decentralized, and span several sectors. In conclusion, urban ecosystems may reduce ecological and economic risks while maintaining ecosystem services through integrated, adaptive IAPS management that incorporates stakeholder input, suitable control systems, supporting policies, and public participation.

S. Zhao · M. Riaz (✉)
Guangdong Engineering and Technology Center for Environmental Pollution Prevention and Control in Agricultural Producing Areas, College of Resources and Environment, Zhongkai University of Agriculture and Engineering, Guangzhou, China
e-mail: Riaz1480@hotmail.com

S. Zhao
e-mail: zhaoshaopeng@zhku.edu.cn

O. Aziz
Soil & Water Testing Laboratory, Lahore, Pakistan

R. Singh et al. (eds.), *Invasive Alien Plants in Urban Ecosystems*,
https://doi.org/10.1007/978-3-032-00997-5_14

Keywords Alien plants · Integrated management · Invasive species · GIS · Urban ecosystems

1 Introduction

The spread of alien plant species beyond their usual ranges due to human actions is a significant ecological concern. Human activities, both intentional and inadvertent, facilitate the crossing of biogeographic barriers by these plants, leading to their naturalization and potential invasiveness in a new region (Blackburn et al., 2011; Essl et al., 2018). These plants are known as Invasive Alien Species (IAS) because they disperse quickly across large distances (Blackburn et al., 2011; Occhipinti-Ambrogi & Galil, 2004). A plant is considered an invasive alien plant species (IAPS) if its effective vegetative or seed-based reproduction mechanisms allow it to maintain self-replacing populations. The "*Intergovernmental Science-Policy on Biodiversity and Ecosystem Services*" estimates that biological invaders pose a hazard to around one-fifth of the planet's surface. Furthermore, high-income countries also face IAS (Seebens et al., 2018).

High levels of urbanization are related to denser networks of trade, communication, and transportation (by land, air, and sea). Other than invasion, risks to native biodiversity include anthropogenic global changes (e.g., climate change), environmental pollution, and habitat degradation. It is interesting to note that invasion ecologists disagree on whether IAPS should be ranked even lower or if it is the first or second the most serious issue (only 27.3% believe this) (Young & Larson, 2011). These ratings may differ by location when it comes to risks to or extinctions of biodiversity. In this regard, most invasion ecologists concur that human disturbances are making the global IAPS issue worse (Young & Larson, 2011). Notably, the IAPS/alien invaders are the reason for biodiversity loss in the "*UN's Global Assessment Report on Biodiversity and Ecosystem Services*" (IPBES) (Watson et al., 2019). IAPS has been a bigger problem for all taxa and regions in recent years (Seebens et al., 2017, 2018). Additionally, land area (17%) and biodiversity hotspots (16%) are extremely susceptible to invasion (Early et al., 2016). Some alien species have been introduced through different methods besides anthropogenic IAPS invasions. For instance, it is thought that ocean currents brought *Limnocharis flava* to Kerala (Abhilash et al., 2008). Similar to this, waves and ocean currents have carried coconuts to many parts of the planet, including islands (Harries & Clement, 2014). As a result, ecological IAPS research is progressively growing. IAPS can transcend many biogeographical barriers to anthropogenic influences. It is interesting to note that depending on where they are on the naturalization invasion continuum, some species may or may not be invasive. The local flora in the overrun area (those that have organically grown there and without human interference) can be severely impacted by IAPS.

1.1 Influences of IAPS

When an invading species introduces a new feature or function to an ecosystem, such as the ability to fix nitrogen or distinct fuel qualities (Brooks et al., 2004), the impact of the species is usually highest. Many people have benefited socially and economically from globalization, but it has also brought forth new difficulties, a critical part is IAS. There has never been such a large-scale biological invasion as seen today (Reaser et al., 2007). Invasive plant species lower development returns by blocking industrial pipelines, threatening hydroelectric projects, and obstructing irrigation canals. Thus, constraints are placed on economic growth, sustainable development, food security, and poverty alleviation. This, in turn, intensifies social discontent and economic suffering (Abdulahi et al., 2017; Kefelegn, 2015). Combinations of features that influence both intruder abundance and ecosystem processes can result in impacts. Table 1 summarizes the types and severity of ecological, economic, and social impacts associated with various invasive alien plant species (IAPS). Controlling biological invasions may be aided by micro-scale social factors, such as the attitudes and social interactions of landowners that support their willingness to control invading species. However, in contrast to our ecological comprehension of invasion dynamics, little is understood about how micro-level social factors affect the effectiveness of controlling invading species.

Ecologists are becoming more concerned about invasive plants as they may pose a global hazard to ecosystems (Vilà et al., 2011). The spread of invasive plant species is a growing concern in the context of human-induced climate change. These species threaten biodiversity, disrupt ecosystems, and can have significant socio-economic impacts. Climate change, with its associated temperature fluctuations and extreme weather events, exacerbates these issues by altering the habitats and distribution of both native and invasive species. These occurrences are likely to create favorable conditions for the rapid proliferation of these invasive species. Research has indicated that invasive plants have detrimental effects on soil nitrogen cycles. Wolf et al. (2004) revealed that low soil N levels in grasslands are linked to invasive species in "*Rocky Mountain National Park*" (*Melilotus officinalis, Melilotus alba*). The plant lengthy growing season, low nitrogen availability, aggressive growth habit, and/or low nitrogen usage efficiency may be the primary causes of the low nitrogen availability identified in the invaded patches (Evans et al., 2001; Wolf et al., 2004). Some invasive species alter nitrogen cycles, which may lead to imbalances in native plant communities (Wolf et al., 2004). The makeup of soil microbial communities in invaded environments can be changed by invasive plants (Sanon et al., 2012). According to Vogelsang and Bever (2009), by upsetting the mutualistic interactions, invasive vegetation may alter soil ecosystems. Studies by Gaertner et al. (2009), Hejda et al. (2009), Powell et al. (2011), and Liao et al. (2008) indicate that invasive vegetation affects a range of consequences on human welfare and the environmental services. They also increase ecosystem productivity.

Numerous IAPS that were implemented for human well-being have a history of wreaking destruction on the environment and the economy (Souza et al., 2018).

Table 1 Nature and extent of impacts caused by invasive alien plant species (IAPS)

IAPS	Nature of Impact	Effects	References
Centaurea maculosa, Centaurea stoebe (spotted knapweed)	Economic	Its roots produce phytotoxin. It produces ammonia-oxidizing bacteria and soil nitrate in significant amounts	McLeod et al. (2016)
Mikania micrantha (mile-a-minute weed)	Environmental/ economic	In the last decade, IAPS spread tremendously, especially because of global warming	Guo et al. (2018)
Solidago gigantean	Environmental	The impact of an IAPS on ecosystems in Europe	Stefanowicz et al. (2018)
Opuntia stricta	Health	Irritations to the eyes and skin	Shackleton et al. (2017a), (2017b)
Rosa rugosa	Health	Wounds or injuries brought on by thick, sharp thorns	Mazza et al. (2014)
Ageratina adenophora	Land use change	deadliest IAPS in China and Northeast India	Fu et al. (2018), Kong et al. (2017), Yu et al. (2014)
Prosopis juliflora, Senna didymobotrya, and *Tecoma stans*	Health	Attract the malarial parasites and act as a host to contribute to the disease's propagation	Müller-Schärer et al. (2018), Shiferaw et al. (2019), Stone et al. (2018)
Arundo donax	Environmental	Disrupts ecosystem function by affecting native species and changing fire patterns	Plaza et al. (2018)
Fallopia japonica (Japanese knotweed)	Economical/ environmental	Modifies food chains utilizing secondary metabolites	Abgrall et al. (2018), Rouifed et al. (2012)
Ailanthus altissima	Health	Myocarditis	Pyšek and Richardson (2010)
Acacia, Acer	Health	Risks to human health	Belmonte and Vilà (2004), Mazza et al. (2014)
Phragmites australis and *Typha* spp.	Health	Health risks to people	Mackay et al. (2016)

Notably, IAS burdens food production, human health, and other important goods, in addition to ecological services necessary for human welfare (Shackleton et al., 2019; Simberloff et al., 2013; Vilà et al., 2011). They also endanger the natural biodiversity of a place globally (Pyšek et al., 2012; Watson et al., 2019). The economic effects of invasions have not received as much attention as they might, despite IAS being an ecological and economic concerns. For instance, it was found that many experiments did not statistically assess the financial effects of managing invasive plants (Kettenring & Adams, 2011). Furthermore, research was conducted over a relatively short period and at relatively small regional scales (Kettenring & Adams, 2011). Nevertheless, determining losses and controlling expenses at the federal level is crucial for the order of importance for IAS management initiatives. Due to its alteration of water quality, *Acacia mangium*, is known for ill consequences on the economy, indigenous people, and the environment (Souza et al., 2018). "*Global Climate Models and Ecological Niche Models*" (CLIMEX) predict that water hyacinth will move into the European and Mediterranean regions under the climate change regime. This illustrates the enormous financial consequences (Kriticos & Brunel, 2016). In the USA, invasions due to *Tamarix ramosissima* have caused massive water losses (1.4–3.0 B cm^3, valued at US $26.3–67.8 M), leaving many human requirements unmet (Pejchar & Mooney, 2009). Groundwater is severely utilized by Eucalyptus species in California and *Melaleuca quinquenervia* in Florida due to their deep tap roots (Schmitz, 1997). The aquatic macrophyte *Myriophyllum spicatum*, sometimes known as water milfoil, has been linked to a 1% annual loss in recreational value in Lake Tahoe, Sierra Nevada (Eiswerth et al., 2005; Pejchar & Mooney, 2009). Pathogenic IAPS like *Euphorbia esula,* caused an annual economic damage of around $200 M (Andersen et al., 2004). According to Pimentel et al. (2005), the financial damage brought on by pathogenic intruders was between $120 to $138 B in the USA. The global estimate was even higher, coming in at $1.5 T annually. A total of 283 plant and fauna invaders caused an economic loss of $14.45 B in China (Xu et al., 2006). Using the participatory rural appraisal (PRA) technique, it was determined that *Opuntia stricta,* an IAPS at high risk in Africa, would result in an annual economic loss of $500–$1000 per household (Shackleton et al., 2017a, 2017b). Furthermore, it was estimated that foreign invaders would harm agricultural harvests in African nations, resulting in an annual economic loss of US $1 billion (Sileshi et al., 2019). According to Nghiem et al., disease-spreading foreign invaders caused an economic loss of $1.85 B to the human health sector alone in Southeast Asia. The alien invasive species caused economic damage of $33.5 B to the agriculture and health sectors combined. Thus, the agricultural sector had a greater degree of economic damage from invaders than the human health sector (around 90% of monetary loss). To control IAPS in urban ecosystems, this chapter discusses developing technology, management strategies, and regulatory issues,

2 Management of Invasive Alien Plants

Preventive measures should be prioritized on land that has been recognized as comparatively free of invasive weeds to maximize costs and benefits. The simplest approach to lessen the negative effects of IAPs now and in the future, as well as the management expenses related to them, appears to be to stop their introduction and further spread. IAP outlines (Brunel, 2009; Hussner et al., 2014), are the aquarium trade, which includes mail order and online shipping (Maki & Galatowitsch, 2004). However, other less controlled pathways are also engaged (Cohen et al., 2007). In addition to lowering the possibility of secondary releases, international and national regulations can both aid in preventing the introduction of novel IAPs. Programs for "*Early Warning and Rapid Response*" (EWRR) and "*Early Detection and Rapid Response*" (EDRR) are essential for the efficient administration and successful removal of plant invasions since they can drastically reduce their harmful effects (Genovesi et al., 2010).

Key measures under many national initiatives include mapping and reporting new infestations and monitoring sensitive locations (Adriaens et al., 2015; Brunel, 2009). Unfortunately, it is difficult to identify species that limit early detection methods. It is more appropriate to prevent, detect, and eradicate problems early on. Using a clean crop seeds, cleaning up agricultural tools, and controlling weeds on roadways, levees, and bunds are a few weed prevention techniques. According to a Montana State University study, a car travelling through an infestation of *Centaurea biebersteinii* DC, collected over 2000 seeds, 90% of which dispersed within ten miles (Reinhart & Rinella, 2011; Sheley & Krueger-Mangold, 2003). The study also highlights that limiting soil disturbance by cattle, automobiles, machinery, and wildlife is essential to reducing the growth of noxious weeds. *R. raphanistrum* and *L. rigidum* mature. Most of the seeds enter the harvester because they are kept on the plant. While over up to 80% of *R. raphanistrum* and 95% of *L. rigidum* seeds were captured in grains (Walsh & Newman, 2007). Consequently, alternative techniques to weed seeds are seed direct chaff bailing, chaff carts, and cleaning. Preventive measures involve limiting the time and space in which weed propagules (seeds and vegetative propagules) can disperse and survive, as well as establishing conditions that hinder weed emergence inside the crop.

A comprehensive research program is needed, encompassing everything from fundamental investigations of the biological and physical mechanisms that influence the efficiency of preventive interventions to field efficacy trials of preventive systems. Participating in this way will guarantee that the management of preventive invasive species will work well with other components of a farming system, establish credibility, encourage adoption, and—above all—harvest the abundance of knowledge and the expertise that producers and experts have to offer. For ecosystems, biodiversity, and human activity to be minimized, it is essential to control invasive alien species. Various control strategies, as shown in Table 2, are employed. Here are some common control strategies for invasive alien species.

2.1 Chemical Control

The program to manage invasive species in main crops is chemical control (Ahmadi et al., 2016). Achieving good control requires careful consideration of the herbicide to use, as well as the right dosage and time of application (Chauhan et al., 2012; Kumar & Ladha, 2011). Herbicides can be categorized based on the chemical family,

Table 2 Various strategies for managing invasive plant species

Target plant species	Technique	Control agent	Effect	References
Azolla filiculoides	Chemical	Glyphosate	Within seven days, 2·88 kg a.i. ha^{-1} of mats were killed in treated plots	Kathiresan and Deivasigamani (2015)
Alternanthera philoxeroides	Mechanical	Hand weeding	Intensive labor	de Winton et al. (2013)
Salvinia molesta	Chemical	Imazapyr	A.I. ha^{-1} of 560 g decrcased biomass	Emerine et al. (2010)
Lagarosiphon major	Mechanical	Using substrate and benthic barriers to shade the bottom	Eliminated, but regeneration from native charophytes with rather fine stems happened	Hofstra and Clayton (2012), Schooler (2008)
Hygrophila polysperma	Chemical	Flumioxazin	90% less biomass was produced at 158 g a.i. L^{-1}	Haller and Gettys (2013)
Alternanthera philoxeroides	Chemical	Metsulfuron-methyl	The biomass was 100% reduced by 0.06 g a.i. L^{-1}	Clements et al. (2014)
Crassula helmsii	Chemical	Glyphosate	50% decrease in coverage	Bridge (2005)
Lagarosiphon major	Mechanical	Hand weeding	Eradicate a species	Clements et al. (2014), de Winton et al. (2013)
Surface-reaching submerged species	Mechanical	Mechanical harvesting and excavation	– Not species-specific	Clayton (1996)
Myriophyllum spicatum	Biological	*Euhrychiopsis lecontei*	Additionally consumes other *Myriophyllum* species	Newman and Inglis (2009), Newman et al. (2001)
Ludwigia grandiflora	Chemical	Imazamox	a.i. ha^{-1} at 560 g decreased biomass by 80%	Emerine et al. (2010)
Submerged species	Mechanical	Water-level drawdown	Typically, significant biomass decreases	de Winton et al. (2013)
Myriophyllum heterophyllum	Chemical	Bispyribac–sodium	No effect	Glomski and Netherland (2008)
All species	Mechanical	Nutrient reduction	Less algae in the water, which would improve its clarity	de Winton et al. (2013)

formulation, site of absorption, mode of action, timing of administration, and selectivity (Teasdale & Zimdahl, 2018). To minimize the usage of herbicides, it is essential to apply them in accordance with the economic threshold level of a particular weed (Mehmood et al., 2018). Herbicide-assisted invasive plant control is gaining popularity worldwide, mostly due to its ability to manage invasive species with flexibility, less labor-intensive application, and the ability to control plants that are challenging to control. However, overusing a single herbicide carelessly and persistently over an extended period can lead to resistant biotypes emerging, changes in the weed flora, and adverse consequences on the environment and the next crop. It is possible to apply herbicides prior to or following crop planting. Application must be applied as near to crop as feasible to reduce weed appearance before crop development.

Growers only employ glyphosate herbicide for weed control due to its efficacy and the economic viability of glyphosate-tolerant crop technology. For instance, in zero-till Roundup Ready corn-soybean in the United States, *Conyza canadensis* (L.) developed resistance against glyphosate (Mueller et al., 2003). Tolerant crop cultivars to herbicides should not be viewed as a weed management strategy as a result. To guarantee that this crucial weed control instrument continues to be lucrative and environmentally sound for an extended period, an integrated management approach should be implemented.

Many conventionally bred (CHT) and genetically modified herbicide-tolerant (GMHT) have grown in recent years due to their low cost, flexibility, ease of use, and selective weed management (Lamichhane et al., 2017). Herbaceous field crops that are genetically modified (GMHT) including cotton, soybean, corn, sugarbeet, oilseed rape, canola, alfalfa, and rice. However, the selection of herbicide-resistant (HR) invasive plants has been accelerated by the usage of CHT and GMHT crops and has expanded significantly over the past ten years.

Not only are gardens and coastal regions traditional users of salt as a "natural" and "total (i.e., non-selective)" pesticide, but salt treatment may also be a viable strategy to control the populations of *M. aquaticum* and *L. grandiflora*, according to experiments on their susceptibility to it (Thouvenot et al., 2012). In order to combat invasive macrophytes, brackish water is utilized to reduce their populations (Dandelot et al., 2004), and is employed to control *E. crassipes* (Hussner et al., 2017) which can limit the use of chemicals. The safe and successful use of herbicides requires a relatively high level of training, and the use of chemical control is frequently restricted by regulation.

2.2 Mechanical Control

Mechanical management harms invasive plants through three methods: (1) cutting, (2) uprooting, and (3) burial. In the case of early seedlings, mechanical control may result in wounds to any of the invasive plant tissues, depleting plant reserves and causing the plants to wither and desiccate (Chicouene, 2007). The performance of these methods varies greatly depending on the habitat (Lambert et al., 2010) and the target species. Mechanical management approaches typically generate large numbers of plant fragments. Some of the commonly used mechanical control methods are described below.

2.2.1 Inter-Row Cultivators

By flicking small-seeded invasive species out of the earth, the minimum tillage rotary hoe controls their spread (Leblanc et al., 2011). Little weeds with weak anchoring are suppressed by the harrow by being uprooted and buried in shallow earth (Armengot et al., 2013). Weeders (finger and torsion) most popular low-tech tools which are American inventions that have been made simpler by several businesses (Van der Weide et al., 2005).

2.2.2 Hand Weeding

The most effective way to manage a specific area is by manually removing unwanted plants, using hand hoes, push hoes, or by hand-picking them. Recently, the implementation of non-chemical weed control techniques is widely used, such as environmentally conscious methods or safety measures. When it comes to annual and biennial weeds, hand weeding or hoeing is a very safe and efficient method. As a final defense, onions and carrots are often utilized to control pests in crops. Perennial weeds in grassland are still manually removed using tools (Bond et al., 2007) for removing deep-rooted weeds. There are hoes with strong blades for removing huge invasive plants and hoes with plough-shaped blades for moving dirt and hiding little weeds. Examples include the Dutch hoe and the Swiss oscillating hoe. When working on vegetable crops, the operator of these tools stands or kneels and uses the instruments to chop and shift dirt.

However, human labor is becoming more and more expensive due to the fast industrialization and urbanization of developing nations. Leinonen and Närkki (2004) have noted that hand labor is frequently necessary in (organically maintained) horticulture areas. Hard-to-mechanize jobs include hand weeding, hand planting, and hand harvesting of cucumbers and strawberries. They have concluded that the primary elements that frequently restrict the production of organic vegetables on farms is manual weed control. Another significant limitation is the low capacity (ha hour^{-1}). The kind and severity of the issue at different phases, the cost–benefit ratio, the length and pattern of crop growth, and the labor pool are taken into consideration when determining the number of weedings.

2.2.3 Stale Seed Bed Method

Before planting crops, the germination zone, or the shallow soil layers from which weeds can grow, is cleared of non-dormant invasive species, resulting in a stale seedbed. Creating a stale seedbed before planting effectively controls weed's density throughout the early crop season for both narrow- and broad-leaved species. Compared to crops cultivated in no-till cropping systems, crops in shallow-tilled seedbeds produced higher yields and fewer weeds (Burnside et al., 1980). The crop is under less pressure from invading species as a result of the germination zone diminished seed bank. Exposure to light during the seedbed preparation process is a crucial factor in improved seed germination. Many species' seeds are vulnerable to light exposure at specific stages of their life cycles (Andersson et al., 1997). A well-prepared seedbed is essential for three reasons: (1) faster canopy closure; (2) improved crop establishment (sowing and/or transplanting); and (3) efficient mechanical control. It is also possible to make stale seedbeds with a lot of shallow tillage tools. To create stable seedbeds, a range of organic-compliant methods can be employed with good results. Improved use of stale seedbeds would result from more research on the possibility of measuring soil moisture levels, developing emergence models (degree-days) to schedule activities to coincide with the maximal invasive plant emergence, and double pre-irrigation to suppress multiple weed flushes. The primary benefit of using stale seedbeds is that the crop arises in surroundings devoid of IAPS, giving it a competitive edge over seedlings that emerge later.

2.2.4 Tillage

Since the beginning of agriculture, tillage has been utilized to eradicate undesirable species. Weed emergence and the soil seed bank are significantly impacted by tillage. Dormancy, duration, size, and the capacity of seedlings to emerge are critical characteristics of invasive plant seeds (Batlla & Benech-Arnold, 2007). Tillage alters a soil profile's vertical seed delivery, and this distribution influences soil atmosphere nearby seeds, which in turn affects invasive species' ability to germinate (Chauhan & Johnson, 2009).

Most IAPS are found near the soil surface in undisturbed environments like MT "minimum tillage" and ZT "zero tillage" while CT "conventional tillage" systems, which involve plowing, distribute seeds evenly (Chauhan & Johnson, 2009). By tilling soil, and creating optimal conditions for seed growth, certain types of weed seeds may be stimulated to sprout (Chauhan & Johnson, 2009; Chauhan et al., 2012). Seeds present on the surface are susceptible to being eaten or drying out. Invasive plants with small, hardly dormant seeds and short lifespans may spread quickly if they are not properly managed. Comparing reduced tillage to conventional tillage, more weeds emerged because 40–80% of seeds were at 0–5 cm of soil (Bàrberi et al., 2001).

Researchers are focusing on the management of cultural practices, i.e., tillage, as a crucial part of integrated management programs. By changing the physicochemical properties of the seedbed, tillage techniques are known to affect invasive species and crops (Arif et al., 2007). Herbicide adsorption, transport, persistence, and efficacy can all be impacted by tillage-induced changes in soil parameters such as pH, temperature, soil moisture, and organic matter (Blevins et al., 1983). In the non-chemical management of IAPS, tillage, and crop variety are essential strategies that have a significant impact on crop performance. Tillage impact on IAPS may vary according to the species and tillage intensity.

2.3 Physical Control

Physical control involves using various manual or mechanical methods to manage the spread of vegetation. These methods aim to physically remove or disrupt the growth of invasive plants, limiting their impact on native ecosystems.

2.3.1 Soil Solarization

To manage invasive plants and soil-borne diseases, soil solarization is applied before planting and was initially introduced in Israel in 1976 (Katan, 2015). It has been successfully demonstrated that soil solarization is a safe and effective way to prevent invasive species (Abouziena et al., 2015). Using this technique, solar radiation is used to warm damp soil that is protected by transparent film (polyethylene, 0.05 mm thick) or mulch. A high temperature is required, and mulch should be applied on the surface for four to twelve weeks. Although the primary uses of this system are the control of nematodes and fungal diseases, it also effectively manages invasive plants (Gill et al., 2009).

According to Chase et al. (1999), solarization makes it possible to keep soil temperatures over 40 °C, which kills invasive plant seeds and weed seedlings. Certain species can break their dormancy and produce a fresh flush of weed seedlings if the lethal temperature is not attained. This phenomenon may happen at the top of the soil (Vidotto et al., 2013). Solarization is commonly used in greenhouses for growing vegetables and minor crops like colewort, pepper, cucumber, fennel, and cauliflower. However, studies have demonstrated its effectiveness in both greenhouse and field environments (Campiglia et al., 1998; Candido et al., 2011; Lombardo et al., 2012).

According to Candido et al. (2011), in both greenhouses and the field, the average marketable yield of lettuce was consistently observed to be considerably greater in solarized soil. According to Schreiner et al. (2001), solarization can be the potentially effective non-pesticide way to lower the number of invasive species and soil-borne pests. Adequate moisture, warm air, strong sunlight, and sufficient time are all required for it to be effective. After removing the plastic, the effects of solarization on emergence were briefly observed. Less than 15% of an untreated check had developing annuals over the first two months following excision. Soil solarization is an environmentally friendly technique that decreases the reliance on chemical pesticides, making it a popular option for farmers and gardeners seeking natural methods to manage invasive plants.

2.3.2 Mulching

Throughout the main crop growth period, live mulches are planted as cover crops, which can be either annual or perennial. Live mulches have been proven by various studies (Qi et al., 2011) to have positive effects on soil health, reducing the need for intensive ploughing and preventing nitrate leaching and soil erosion. Some ways that living mulches can hinder the growth of invasive alien plants include allelopathy, competition for resources underground and light, and obstructing seed germination through shading.

The best organic mulches for weed management in orchards seem to be wood-chip mulches (Granatstein & Mullinix, 2008). Wood-chip mulch can reduce irrigation water use by more than 20% and suppress weeds for a duration of one to three years (Granatstein & Sánchez, 2009). Materials that are not plastic are likewise fascinating. When thoroughly combined with soil, paper mulches effectively eradicate weeds (Granatstein & Sánchez, 2009), from common beans, and tomatoes (Radics and Szné Bognár, 2002). Because plastic is inexpensive and effective over the long term, it can reduce the costs associated with managing invasive species in fruit orchards (Hammermeister, 2016). For the best results and adequate control of invasive plants and insects in small orchards, 15 mm thick cellulose sheets are used (Benoit et al., 2006). When the competing vegetation is more active, it must be applied to a minimum area of 1 m^2 to be effective. Nonetheless, mulching has become very popular because it can produce favorable environmental conditions, particularly microclimatic variables (temperature, soil moisture), around plants that are mulched. These advantages, which are unavailable from other methods, support plant establishment and growth.

2.3.3 Crop Rotation

In monoculture, invasive plant species often become resistant to herbicides and reduce yields through early seed shattering, crop mimicry, and other adaptations to management techniques. IAPS is prevented from adapting and surviving in the crop rotations by a variety of control techniques (Anderson, 2007). Globally, crop production decreases due to IAPS, averaging 40%, and the cost of management inputs is a significant financial burden for farmers (Atwood & Paisley-Jones, 2017; Oerke, 2006). Increasing crop diversity may be crucial for mitigating the threat of herbicide resistance (Beckie & Harker, 2017; Teasdale & Zimdahl, 2018). Compared to simpler rotations, diverse rotations are more successful at suppressing IAPS. According to Abbas et al. (2016), diversified rotation is essential for sustainable weed management because it makes weed circumstances unstable and lowers the amount of weed seed bank that weeds can access. A recent analysis by Zhao et al. (2020) included 45 research studies revealed that crop rotation increased crop yields by 20%. In addition, 54 studies found that the rotation of crops decreased weed density by 49% (Weisberger et al., 2019).

Crop rotation reduces IAPS pressure and increases crop yield (Bilalis et al., 2003) by an allelopathic crop. Allelochemicals are released in the soil by allelopathic crops in rotation. Growing wheat after sorghum can decrease weed infestation due to the release of allelochemicals. To reduce the negative impact on crops, rotating with allelopathic crops, and carefully timing tillage can be effective (Cheema et al., 2004). Effective crop rotation, which involves the use of allelopathic crops can help prevent autotoxicity and reduce chemical methods.

Crop rotations are well acknowledged to break the life cycles of IAPS and aid in their control. Given that IAPS seeds frequently have a long survival rate, it is also realistic to anticipate that several years should be required in most cases to achieve significant decreases in the seed bank and emergence. Researchers combine various crop rotations with other cultural and herbicide strategies to increase the ideal rotation period to six years, leading to high crop yield and long-term weed control with minimal pesticide use (Adeux et al., 2019). Crop rotations' ability to control weed populations is greatly influenced by the design of the rotations, especially the crop sequence (Anderson, 2005; Westerman et al., 2005). This is so because the seed bank and the pace of population increase of IAPS are determined by the crop phase (Leon & Wright, 2018; Leon et al., 2015). Therefore, crop rotation can be used to control weeds in agricultural systems.

2.4 Biological Control

It is believed that using biological management to control invasive plants is a reasonably safe and successful strategy (McFadyen, 1998). Biological control involves bringing in, establishing, and colonizing foreign natural enemies to keep the exotic pest inhabitants at economically insignificant densities (McFadyen, 1998) to create self-sustaining populations, where natural enemies are released either all at once or over a short period. Conservation, which involves changing the environment to assist (mainly native) natural enemies, is rarely used. In general, biocontrol agents can inhibit the development of IAPS; however, for more focused control, specialized agents such as plant diseases and arthropods (insects, mites), are more suitable and often utilized. Since not all IAPS are good candidates for biocontrol, it is advised to prioritize issues by combining specialist knowledge with stakeholder participation. Additionally, it is asserted that biological control is an affordable strategy for regionally managing invasive weeds. This is especially true when weeds become chemically resistant.

The idea behind biological weed control is that non-native plants become problematic when they don't have to deal with the natural checks and balances that keep them in check in their original habitats. The propensity of plant species to spread may also be influenced by other variables. Natural enemies can also indirectly control IAPS by killing the plant, damaging important parts of it, or decreasing its ability to reproduce. They can also control IAPS indirectly by increasing the plant's vulnerability to pathogenic or saprophytic organism attacks or by stressing it to the point where it loses out to other valuable plants in the marketplace.

Biological management is a cost–benefit analysis, weighing the costs, benefits, and ecological effects (Culliney, 2005; Louda & Stiling, 2004). This is one example of low effectiveness (Hulme, 2009; Hulme et al., 2009). Over 500 biological control agents were used to control 200 different types of weeds. The release of at least one biological control agent has occurred in over 90 countries and 42 biological control agents have been purposefully introduced by several Asian nations against 19 invasive weed species; 11 of these species (or 58% of the total) are thought to be under some degree of control (Schwarzländer et al., 2018).

There are numerous examples of in-between biocontrol scenarios, such as complete failure (e.g., India with *Lantana camara* L.) and complete control (*Azolla filiculoides* Lam., South Africa) (Hinz et al., 2019; Schwarzländer et al., 2018). The aforementioned cases and others were included in the more recent global analysis (Van Driesche & Heinz, 2016) on agriculture and wildlands. *Ageratina adenophora* causes significant ecological harm as well as financial losses (Baral et al., 2017; Ding et al., 2007). In certain places, (*Passalora ageratinae*) leaf spot fungus and (*Procecidochares utilis*) Gall fly have been successful in inhibiting the growth of this weed. *A. adenophora* may be competitively suppressed by a few perennial native grasses (Poudel et al., 2019).

There is still little knowledge about the conditions of invasive plants at restoration sites worldwide and how they are controlled (Kettenring & Adams, 2011). Effective control strategies for invasive species with low environmental impact and operating costs need to be devised, and current approaches need to be improved. Gaining more insight into the fundamental causes of the spatial or temporal variability in biocontrol success is an important subject that requires additional research. The following steps are taken to increase the predictability of IAPS biocontrol success: (1) select target weeds that have a higher chance of being successfully controlled (Paynter et al., 2012), (2) identify the target's most sensitive life stage (Erickson et al., 2020), and (3) identify the biocontrol agent or agents that have the best chance of slowing the targeted IAPS's population growth (Davis et al., 2006).

2.5 Integrated Pest Management

Various weed control methods are available, but none can guarantee long-lasting effectiveness. Hence, to expand the range of weed control and enhance sustainable crop production, it is advisable to mix numerous weed management methods. Using herbicides at the right time and incorporating proper farming techniques, such as fertilization, water management, and crop residue retention, can greatly enhance their effectiveness. By implementing a high seeding proportion, top-quality cultivars, and a weed elimination schedule in Canada, researchers found a significant 40% increase in crop yield compared to using weaker cultivars, the lowest seeding rate, and a delayed weed removal schedule (Harker et al., 2003). Developing effective IWM "Integrated Weed Management" methods needs the biology and ecology of invasive plants (Radosevich & Ghersa, 1992). Using ecological and traditional crop-weed competition theories, Smith et al. (2010) put forth the resource pool diversity hypothesis, contending that crop competitiveness with weeds is increased in a more varied cropping system with a more diverse soil resource pool.

Monitoring, prevention, and control should be combined in a systematic approach, not necessarily eradicating invasive plants, but controlling them below agronomically, environmentally, and economically acceptable thresholds. There have been numerous definitions of IWM, and is seen as a medium-to-long-term approach to pest control that combines biological, mechanical, physical, and chemical methods. Based on this principle, none of these methods alone, except for the chemical ones, can adequately control a situation. They should be implemented and integrated in a multidimensional manner. Additionally, rainfall regime, solar radiation, temperature, wind intensity, plant farming methods, soil texture, exposure, socioeconomic constraints, and farmer expectations determine how indirect and direct control methods are integrated. To remain flexible, IWM must change in socio-economic factors and environment as they develop over time. Research is necessary since integrating control approaches that are highly dissimilar from one another is undoubtedly a difficult undertaking, especially in the context of long-term incentives, policies, and experiments. Observing that to succeed in stable and profitable cropping systems, Radosevich and Ghersa (1992) noted that economics and sociology also need to be influenced by researchers.

2.6 Role of Emerging Technologies

To effectively manage and control IAPS, excellent observation and data-gathering procedures are necessary (Groom et al., 2017). In current surveys, plant data are collected through time-consuming and resource-intensive methods, and the results are slow to appear (Adriaens et al., 2015), preventing alien plants (Groom et al., 2017). As alien species colonize new sites rapidly, monitoring should require methodologies capable of keeping up with this process. A useful contribution to reducing the detrimental impacts that these species have on ecosystems would be the easy, economical, and efficient monitoring of IAPS. To manage and intervene against invasive species, it is important to identify their geographic locations. Detecting and identifying invasive plants usually entail thorough field surveys. Remote sensing is one possible detection technique. As a general rule, direct remote sensing measures textural, spectral, and phenological differences (Bradley, 2014). When used with hyperspectral pictures, spectral differentiation primarily highlights invasive vegetation (Asner et al., 2008; Barbosa et al., 2016; Underwood et al., 2003).

Textural distinction compares the spatial forms of invasive plants and the background terrestrial plants based on plant size or aggregation. This usually uses high-resolution imagery (Lishawa et al., 2013; Pearlstine et al., 2005). According to various studies, Bradley et al. (2018) and Somers and Asner (2013), phenological distinction distinguishes between invasive and native plants' distinct growth patterns which include the time and rate of greening date in the middle of the season, and other factors. Repeat imaging is necessary for phenological differentiation to obtain sufficient time data to identify spectral variations. The main hurdles in creating accurate classifications in these studies involve selecting the appropriate radiometric resolution to differentiate between the target weeds and the surrounding vegetation, and choosing the proper spectral bands.

The risk of invasion can be predicted through spatial modeling. Deforestation, urbanization, and vegetation types have been predicted using spatially explicit models. A geographic information system can be used to predict invasion patterns such as topography, land use, and geomorphology (Store & Kangas, 2001). Land use and landforms can guide conservation and management strategies. NDVI (normalized difference vegetation index), a measure of plant health, has been used to create predictive models for species occurrence (Kerr & Ostrovsky, 2003). Precisely determining the time frame in which invasive species spread is crucial for utilizing remote sensing data to guess species movement. The successful remote identification of an invasive plant, such as cheatgrass, was achieved through the use of Landsat data (Peterson, 2005). It is becoming increasingly common to detect invasive plants using remote sensing (Niphadkar, 2016; Niphadkar & Nagendra, 2016). It is especially possible to monitor areas with other unmanned aerial vehicles (UAVs) and drones (Dvořák et al., 2015). UAVs cannot monitor alien species found along roads due to regulations that prohibit fuel limits and aircraft travel above roadways. IAPS can be monitored by car surveying along roads (Abella et al., 2009; Barone et al., 2021; Kraaij & Baard, 2019; McAvoy et al., 2012; McDougall et al., 2018). Dyrmann et al. (2021) revealed that deep learning algorithms were used to automatically register the IAPS along highways. Despite its potential, this approach needs a significant amount of processing power, specialized tools, and operator skills to be successful. There are potential limitations to their use on a broader scale due to these aspects. As a consequence of "citizen science" initiatives, smartphones have become useful data collection tools for scientists (Miller-Rushing et al., 2012). By using ICT, the ability of lay volunteers to collect data and communicate with scientists has increased, e.g., via mobile applications or web interfaces (Chandler et al., 2017; Goodchild, 2007; Miller-Rushing et al., 2012). Smartphones have advanced capabilities for precise positioning. The SPM method's ability to be reproduced, accessed, and completed quickly allows it to be easily implemented, widely used, and adjusted to various geographic and environmental situations (Bradter et al., 2018; Chandler et al., 2017; Klemann-Junior et al., 2017).

Numerous modeling methods are used to forecast the probable spread of the plant species (Guisan & Thuiller, 2005; Segurado & Araujo, 2004). Peterson and Cohoon (1999) developed the GARP modeling system, a useful method known as the "*Genetic Algorithm for Rule-Set Production*". This method has been extensively assessed for its predictive capabilities in areas, e.g., conservation and biogeography (Ortega-Huerta & Peterson, 2004), and global climate change (Parra-Olea et al., 2005). Scientists are now studying the potential of drones for mapping invasive species, as they are affordable and have advanced imaging capabilities. Drones can improve the mapping of invasive species. As canopies obscure traditional remote sensing techniques, mapping and identifying invasive plants in the understory can be challenging. The proposed method is an innovative and valuable tool for tracking highly invasive plants in the context of biological invasions.

2.7 Policy Regulations

Management of invasive species adopts numerous policy measures (Butchart et al., 2010; Perrings et al., 2010; Vilà et al., 2011). For management to be successful, all relevant stakeholders must accept and support control and eradication efforts, as well as governing policies. When dealing with invasions, it is important to do a stakeholder analysis because many invasive species are brought in on purpose or by accident due to commercial reasons. Invading species also bring about increased social expenses (Pimentel et al., 2005; Xu et al., 2006). The study revealed that people have a heightened understanding of invasive species. Andreu et al. (2009) discovered that natural resource managers prioritize controlling species that are widespread in their respective regions and are believed to have significant effects.

Within the limits of conservation areas, invasive plant management, or IPM, is funded and implemented by numerous organizations. Due to their high degree of mobility and ability to spread via a variety of social and physical barriers, invasive plants pose a transboundary problem that has hampered the effectiveness of conservation efforts. Therefore, tenure-blind collaborative action, in which land managers collaborate to achieve their shared objective of controlling invasive plants, is essential to successful IPM (Graham, 2013; Meinzen-Dick et al., 2004). Thus far, previous studies on rural social issues have shown that to successfully manage invasive species through collective action, long-term organizational and individual commitments, supportive institutional arrangements, and strong working relationships are necessary (Graham, 2013; Graham & Rogers, 2017; McAllister et al., 2017). Only then can efforts be coordinated and collaborated upon (McAllister et al., 2015). To enable collaboration at the right scales, institutional policies, and networks must meet social-ecological conditions, offer stability, and provide sufficient resources (Epanchin-Niell & Wilen, 2015; McAllister et al., 2015).

The best action for controlling invasive, when considering all stakeholders collectively, is to promote social awareness and education about invasive plants (Dehnen-Schmutz et al., 2007; Dehnen-Schmutz, 2011) which is consistent with earlier research (Vanderhoeven et al., 2011), which calls for an approach that is both the most doable and has the fewest direct effects on various stakeholders. It is crucial to eliminate IAPS to avoid potential costs of damage, along with implementing preventative measures and establishing early warning systems. To lessen the negative impact of eradication on nearby communities, it is important to implement aggressive eradication efforts that gradually decrease customer reliance and influence their preferences, along with prohibiting the sale of IAPS. Removing the IAPS requires considering the species distribution, abundance, management costs, and effectiveness, as well as the perspectives of both public and private entities towards eradication methods (Booy et al., 2017). Biodiversity and ecosystem services are maintained through efficient decision-making, which can be aided by cost–benefit analysis (Courtois et al., 2018; Shackleton et al., 2007). In the future, this species could be integrated into policy decisions by utilizing eradication techniques. People can choose to consider the benefits, such as increased income for their household, and drawbacks, like the negative impacts on the environment and expensive maintenance, which come with a particular species (Shackleton et al., 2017a, 2017b). A decentralized framework that is transparent and dynamic is required to establish a "*National Invasive Species Strategy and Action Plan*" (NISSAP) which aims to foster a cohesive and varied community by swiftly and efficiently upholding rules, guidelines, and legislation. To accomplish this, we must spread awareness, combat ongoing invasions, involve citizens in addressing IAPS problems, and enhance our response capabilities through research and knowledge exchange. Establishing a cooperative alliance among various groups to create a uniform method for assessing risk and implementing regulations, ensuring adherence to laws, raising public awareness, and improving preparedness for emergencies by exchanging research and knowledge.

Every metropolitan region has a different collection of problems that call for management strategies appropriate to that place. These methods must take into account the opinions of stakeholders, the social repercussions of management decisions, and the requirement for sophisticated, flexible management frameworks (Gaertner et al., 2016, 2017). When a species' potential to have major negative effects (EDS) is evaluated against its actual or perceived advantages (ES), conflicts of interest frequently result. Conflicts about their use and management are typically the result of IAPs that offer ES and EDS (Dickie et al., 2014).

3 Conclusions

In summary, non-native plants are a threat to global infrastructure, economies, ecosystems, and agriculture. Urban areas pose unique challenges due to their high interconnectedness, frequent disruptions, and diverse ecosystems that facilitate the introduction and proliferation of invasive species. Once invasions have taken place, certain physical, chemical, mechanical, and biological control techniques are employed to restrict abundance and spread. Administration at suitable scales is made possible by coordination between authorities. Real-time monitoring and precise targeting are enhanced by developments in AI and remote sensing. Programs that are sustainably supported and have higher compliance rates are the results of including all stakeholders via inclusive decision-making and education. For the majority of extensive urban infestations, eradication is not practical; nevertheless, long-term suppression, containment, and gradual removal can lessen the negative effects. The harm that invasive plants due to structures, decreased ecological services, and declining biodiversity frequently outweighs the expense of control. By combining these components, invasive species' negative effects on ecosystems and human health may be reduced while establishing a resilient and sustainable urban environment.

Funding and Acknowledgments This project is financially supported by the National Natural Science Foundation of China (32402675).

References

Abbas, T., et al. (2016). Mulching with allelopathic crops to manage herbicide resistant littleseed canarygrass. *Herbologia, 16*.

Abdulahi, M. M., et al. (2017). Prosopis juliflora L: Distribution, impacts and available control methods in Ethiopia. *Tropical and Subtropical Agroecosystems, 20*, 75–89.

Abella, S. R., et al. (2009). Assessing an exotic plant surveying program in the Mojave Desert, Clark County, Nevada, USA. *Environmental Monitoring and Assessment., 151*, 221–230.

Abgrall, C., et al. (2018). Invasion by *Fallopia japonica* alters soil food webs through secondary metabolites. *Soil Biology and Biochemistry, 127*, 100–109.

Abhilash, P., et al. (2008). Eco-distribution mapping of invasive weed *Limnocharis flava* (L.) Buchenau using geographical information system: Implications for containment and integrated weed management for ecosystem conservation. *Taiwania, 53*, 30–41.

Abouziena, H., et al. (2015). Comparison of potato yield, quality and weed control obtained with different plastic mulch colours. *Middle East Journal of Applied Science, 5*, 374–382.

Adeux, G., et al. (2019). Diversified grain-based cropping systems provide long-term weed control while limiting herbicide use and yield losses. *Agronomy for Sustainable Development, 39*, 1–13.

Adriaens, T., et al. (2015). Trying to engage the crowd in recording invasive alien species in Europe: Experiences from two smartphone applications in northwest Europe. *Management of Biological Invasions, 6*, 215.

Ahmadi, A. R., et al. (2016). Efficacy of five herbicides for weed control in rain-fed lentil (*Lens culinaris* Medik.). *Weed Technology, 30*, 448–455.

Andersen, M. C., et al. (2004). Risk assessment for invasive species. *Risk Analysis: An International Journal., 24*, 787–793.

Anderson, R. L. (2005). A multi-tactic approach to manage weed population dynamics in crop rotations. *Agronomy Journal, 97*, 1579–1583.

Anderson, R. L. (2007). Managing weeds with a dualistic approach of prevention and control. A review. *Agronomy for Sustainable Development, 27*, 13–18.

Andersson, L., et al. (1997). Germination response of weed seeds to light, light of short duration and darkness after stratification in soil. *Swedish Journal of Agricultural Research, 27*, 113–120.

Andreu, J., et al. (2009). An assessment of stakeholder perceptions and management of noxious alien plants in Spain. *Environmental Management, 43*, 1244–1255.

Arif, M., et al. (2007). Effect of tillage and zinc application methods on weeds and yield of maize. *Pakistan Journal of Botany, 39*, 1583–1591.

Armengot, L., et al. (2013). Weed harrowing in organically grown cereal crops avoids yield losses without reducing weed diversity. *Agronomy for Sustainable Development, 33*, 405–411.

Asner, G. P., et al. (2008). Invasive species detection in Hawaiian rainforests using airborne imaging spectroscopy and LiDAR. *Remote Sensing of Environment, 112*, 1942–1955.

Atwood, D., & Paisley-Jones, C. (2017). Pesticides industry sales and usage: 2008–2012 market estimates. *US Environmental Protection Agency, Washington, DC, 20460*, 2017–2101.

Baral, S., et al. (2017). Invasion of alien plant species and their impact on different ecosystems of Panchase Area, Nepal. *Banko Janakari, 27*, 31–42.

Bàrberi, P., et al. (2001). Weed density and composition in winter wheat as influenced by tillage systems. Conservation agriculture, a worldwide challenge. *First World Congress on conservation agriculture*, Madrid, Spain, 1–5 October, 2001. Volume 2: offered contributions (pp. 451–455). XUL, 2001.

Barbosa, J. M., et al. (2016). Determining subcanopy *Psidium cattleianum* invasion in Hawaiian forests using imaging spectroscopy. *Remote Sensing, 8*, 33.

Barone, G., et al. (2021). Comparison of different methods to assess the distribution of alien plants along the road network and use of Google Street View panoramas interpretation in Sicily (Italy) as a case study. *Biodiversity Data Journal, 9*.

Batlla, D., & Benech-Arnold, R. L. (2007). Predicting changes in dormancy level in weed seed soil banks: Implications for weed management. *Crop Protection, 26*, 189–197.

Beckie, H. J., & Harker, K. N. (2017). Our top 10 herbicide-resistant weed management practices. *Pest Management Science, 73*, 1045–1052.

Belmonte, J., & Vilà, M. (2004). Atmospheric invasion of non-native pollen in the Mediterranean region. *American Journal of Botany, 91*, 1243–1250.

Benoit, D. L., et al. (2006). Management of weeds, apple sawfly (*Hoplocampa testudinea* Klug) and plum curculio (*Conotrachelus nenuphar* Herbst) with cellulose sheeting. *Crop Protection, 25*, 331–337.

Bilalis, D., et al. (2003). Effect of different levels of wheat straw soil surface coverage on weed flora in *Vicia faba* crops. *Journal of Agronomy and Crop Science, 189*, 233–241.

Blackburn, T. M., et al. (2011). A proposed unified framework for biological invasions. *Trends in Ecology & Evolution, 26*, 333–339.

Blevins, R., et al. (1983). Changes in soil properties after 10 years continuous non-tilled and conventionally tilled corn. *Soil and Tillage Research, 3*, 135–146.

Bond, W., et al. (2007). A review of mechanical weed control. Coventry, UK: HDRA: The Organic Organization, 1–23.

Booy, O., et al. (2017). Risk management to prioritise the eradication of new and emerging invasive non-native species. *Biological Invasions, 19*, 2401–2417.

Bradley, B. A. (2014). Remote detection of invasive plants: A review of spectral, textural and phenological approaches. *Biological Invasions, 16*, 1411–1425.

Bradley, B. A., et al. (2018). Cheatgrass (*Bromus tectorum*) distribution in the intermountain Western United States and its relationship to fire frequency, seasonality, and ignitions. *Biological Invasions, 20*, 1493–1506.

Bradter, U., et al. (2018). Can opportunistically collected Citizen Science data fill a data gap for habitat suitability models of less common species? *Methods in Ecology and Evolution., 9*, 1667–1678.

Bridge, T. (2005). Controlling New Zealand pygmyweed Crassula helmsii using hot foam, herbicide and by burying at Old Moor RSPB Reserve, South Yorkshire, England. *Conservation Evidence, 2*, 33–34.

Brooks, M. L., et al. (2004). Effects of invasive alien plants on fire regimes. *BioScience, 54*, 677–688.

Brunel, S. (2009). Pathway analysis: Aquatic plants imported in 10 EPPO countries. *EPPO Bulletin, 39*, 201–213.

Burnside, O., et al. (1980). Control of weeds in an oat (*Avena sativa*)—soybean (*Glycine max*) ecofarming rotation. *Weed Science, 28*, 46–50.

Butchart, S. H., et al. (2010). Global biodiversity: Indicators of recent declines. *Science, 328*, 1164–1168.

Campiglia, E., et al. (1998). Soil solarization in the Mediterranean environment: Effect on weed control and yield of cos lettuce (*Lactuca sativa* L. var. longifolia Lam.). *Italus-Hortus, 5*, 36–42.

Candido, V., et al. (2011). Weed control and yield response of soil solarization with different plastic films in lettuce. *Scientia Horticulturae, 130*, 491–497.

Chandler, M., et al. (2017). Contribution of citizen science towards international biodiversity monitoring. *Biological Conservation, 213*, 280–294.

Chase, C. A., et al. (1999). Heat-retentive films for increasing soil temperatures during solarization in a humid, cloudy environment. *HortScience, 34*, 1085–1089.

Chauhan, B. S., & Johnson, D. E. (2009). Influence of tillage systems on weed seedling emergence pattern in rainfed rice. *Soil and Tillage Research, 106*, 15–21.

Chauhan, B. S., et al. (2012). Ecology and management of weeds under conservation agriculture: A review. *Crop Protection, 38*, 57–65.

Cheema, Z., et al. (2004). Weed control in maize (*Zea mays* L.) through sorghum allelopathy. *Journal of Sustainable Agriculture, 23*, 73–86.

Chicouene, D. (2007). Mechanical destruction of weeds. A review. *Agronomy for Sustainable Development, 27*, 19–27.

Clayton, J. S. (1996). Aquatic weeds and their control in New Zealand lakes. *Lake and Reservoir Management, 12*, 477–486.

Clements, D., et al. (2014). Management of aquatic alligator weed (*Alternanthera philoxeroides*) in an early stage of invasion. *Management of Biological Invasions, 5*, 327.

Cohen, J., et al. (2007). Thousands introduced annually: The aquarium pathway for non-indigenous plants to the St Lawrence Seaway. *Frontiers in Ecology and the Environment, 5*, 528–532.

Courtois, P., et al. (2018). A cost–benefit approach for prioritizing invasive species. *Ecological Economics, 146*, 607–620.

Culliney, T. W. (2005). Benefits of classical biological control for managing invasive plants. *Critical Reviews in Plant Sciences, 24*, 131–150.

Dandelot, S., et al. (2004). Analyse de l'efficacité d'un traitement par le sel pour lutter contre la prolifération de la Jussie Ludwigia peploides Kunth (Raven) en Camargue. *Parc Naturel Régional De Camargue/université Aix-Marseille, 3*, 25p.

Davis, A. S., et al. (2006). Demographic models inform selection of biocontrol agents for garlic mustard (*Alliaria petiolata*). *Ecological Applications, 16*, 2399–2410.

de Winton, M., et al. (2013). *Review of best management practices for aquatic vegetation control in stormwater ponds, wetlands, and lakes*. Prepared by NIWA and the University of Waikato for Auckland Council. Auckland Council technical report, TR2013/026.

Dehnen-Schmutz, K. (2011). Determining non-invasiveness in ornamental plants to build green lists. *Journal of Applied Ecology, 48*, 1374–1380.

Dehnen-Schmutz, K., et al. (2007). A century of the ornamental plant trade and its impact on invasion success. *Diversity and Distributions, 13*, 527–534.

Dickie, I. A., et al. (2014). Conflicting values: Ecosystem services and invasive tree management. *Biological Invasions, 16*, 705–719.

Ding, H., et al. (2007). Impacts of invasion of Eupatorium adenophorum on vegetation diversity. *Journal of Ecology and Rural Environment, 23*, 29–75.

Dvořák, P., et al. (2015). Unmanned aerial vehicles for alien plant species detection and monitoring. *The International Archives of the Photogrammetry, Remote Sensing and Spatial Information Sciences, 40*, 83–90.

Dyrmann, M., et al. (2021). Camera assisted roadside monitoring for invasive alien plant species using deep learning. *Sensors, 21*, 6126.

Early, R., et al. (2016). Global threats from invasive alien species in the twenty-first century and national response capacities. *Nature Communications, 7*, 12485.

Eiswerth, M. E., et al. (2005). Input–output modeling, outdoor recreation, and the economic impacts of weeds. *Weed Science, 53*, 130–137.

Emerine, S. E., et al. (2010). Greenhouse response of six aquatic invasive weeds to imazamox. *Journal of Aquatic Plant Management (JAPM), 48*, 105.

Epanchin-Niell, R. S., & Wilen, J. E. (2015). Individual and cooperative management of invasive species in human-mediated landscapes. *American Journal of Agricultural Economics, 97*, 180–198.

Erickson, K. D., et al. (2020). Seedling maturation drives spatial variability in demographic dynamics of an invader with multiple introductions: Insights from an LTRE analysis. *Biological Invasions, 22*, 2185–2203.

Essl, F., et al. (2018). Which taxa are alien? Criteria, applications, and uncertainties. *BioScience, 68*, 496–509.

Evans, R. D., et al. (2001). Exotic plant invasion alters nitrogen dynamics in an arid grassland. *Ecological Applications, 11*, 1301–1310.

Fu, D., et al. (2018). Effects of the invasive herb *Ageratina adenophora* on understory plant communities and tree seedling growth in *Pinus yunnanensis* forests in Yunnan, China. *Journal of Forest Research, 23*, 112–119.

Gaertner, M., et al. (2009). Impacts of alien plant invasions on species richness in Mediterranean-type ecosystems: A meta-analysis. *Progress in Physical Geography, 33*, 319–338.

Gaertner, M., et al. (2016). Managing invasive species in cities: A framework from Cape Town, South Africa. *Landscape and Urban Planning, 151*, 1–9.

Gaertner, M., et al. (2017). *Non-native species in urban environments: patterns, processes, impacts and challenges* (Vol. 19, pp. 3461–3469). Springer.

Genovesi, P., et al. (2010). *Towards an early warning and information system for invasive alien species (IAS) threatening biodiversity in Europe* (p. 47). European Environment Agency.

Gill, H. K., et al. (2009). Comparative performance of different plastic films for soil solarization and weed suppression. *HortTechnology, 19*, 769–774.

Glomski, L. M., & Netherland, M. D. (2008). Efficacy of fluridone, penoxsulam, and bispyribac-sodium on variable-leaf milfoil. *Journal of Aquatic Plant Management, 46*, 193–196.

Goodchild, M. F. (2007). Citizens as sensors: The world of volunteered geography. *GeoJournal, 69*, 211–221.

Graham, S. (2013). Three cooperative pathways to solving a collective weed management problem. *Australasian Journal of Environmental Management, 20*, 116–129.

Graham, S., & Rogers, S. (2017). How local landholder groups collectively manage weeds in south-eastern Australia. *Environmental Management, 60*, 396–408.

Granatstein, D., & Mullinix, K. (2008). Mulching options for northwest organic and conventional orchards. *HortScience, 43*, 45–50.

Granatstein, D., & Sánchez, E. (2009). Research knowledge and needs for orchard floor management in organic tree fruit systems. *International Journal of Fruit Science, 9*, 257–281.

Groom, Q. J., et al. (2017). Seven recommendations to make your invasive alien species data more useful. *Frontiers in Applied Mathematics and Statistics, 3*, 13.

Guisan, A., & Thuiller, W. (2005). Predicting species distribution: Offering more than simple habitat models. *Ecology Letters, 8*, 993–1009.

Guo, W., et al. (2018). Comparative transcriptome analysis of the invasive weed *Mikania micrantha* with its native congeners provides insights into genetic basis underlying successful invasion. *BMC Genomics, 19*, 1–17.

Haller, W. T., & Gettys, L. (2013). Efficacy of experimental and recently registered herbicides on hygrophila. *Journal of Aquatic Plant Management, 51*, 84–87.

Hammermeister, A. M. (2016). Organic weed management in perennial fruits. *Scientia Horticulturae, 208*, 28–42.

Harker, K., et al. (2003). Seeding rate, herbicide timing and competitive hybrids contribute to integrated weed management in canola (*Brassica napus*). *Canadian Journal of Plant Science, 83*, 433–440.

Harries, H. C., & Clement, C. R. (2014). Long-distance dispersal of the coconut palm by migration within the coral atoll ecosystem. *Annals of Botany, 113*, 565–570.

Hejda, M., et al. (2009). Impact of invasive plants on the species richness, diversity and composition of invaded communities. *Journal of Ecology, 97*, 393–403.

Hinz, H. L., et al. (2019). How safe is weed biological control? A global review of direct nontarget attack. *The Quarterly Review of Biology, 94*, 1–27.

Hofstra, D. E., & Clayton, J. S. (2012). Assessment of benthic barrier products for submerged aquatic weed control. *Journal of Aquatic Plant Management, 50*, 101–105.

Hulme, P. E. (2009). Trade, transport and trouble: Managing invasive species pathways in an era of globalization. *Journal of Applied Ecology, 46*, 10–18.

Hulme, P. E., et al. (2009). A pan-European inventory of alien species: rationale, implementation and implications for managing biological invasions. *Handbook of Alien Species in Europe* (pp. 1–14).
Hussner, A., et al. (2014). From first reports to successful control: A plea for improved management of alien aquatic plant species in Germany. *Hydrobiologia, 737*, 321–331.
Hussner, A., et al. (2017). Management and control methods of invasive alien freshwater aquatic plants: A review. *Aquatic Botany., 136*, 112–137.
Katan, J. (2015). Soil solarization: The idea, the research and its development. *Phytoparasitica, 43*, 1–4.
Kathiresan, R., & Deivasigamani, S. (2015). Herbicidal control of water hyacinth (*Eichhornia crassipes*) and its impact on aquatic environment. *Indian Journal of Agronomy, 60*, 606–609.
Kefelegn, H. (2015). Invasive alien weed species impacts on biodiversity and socio-economic aspect in Ethiopia: A review. *International Journal of Science and Research, 4*, 2179–2185.
Kerr, J. T., & Ostrovsky, M. (2003). From space to species: Ecological applications for remote sensing. *Trends in Ecology & Evolution, 18*, 299–305.
Kettenring, K. M., & Adams, C. R. (2011). Lessons learned from invasive plant control experiments: A systematic review and meta-analysis. *Journal of Applied Ecology, 48*, 970–979.
Klemann-Junior, L., et al. (2017). Traditional scientific data vs. uncoordinated citizen science effort: A review of the current status and comparison of data on avifauna in Southern Brazil. *PloS one. 12*, e0188819.
Kong, Y., et al. (2017). Effect of Ageratina adenophora invasion on the composition and diversity of soil microbiome. *The Journal of General and Applied Microbiology, 63*, 114–121.
Kraaij, T., & Baard, J. A. (2019). Use of a rapid roadside survey to detect potentially invasive plant species along the Garden Route, South Africa. *Koedoe: African Protected Area Conservation and Science, 61*, 1–10.
Kriticos, D. J., & Brunel, S. (2016). Assessing and managing the current and future pest risk from water hyacinth, (*Eichhornia crassipes*), an invasive aquatic plant threatening the environment and water security. *PLoS ONE, 11*, Article e0120054.
Kumar, V., & Ladha, J. K. (2011). Direct seeding of rice: Recent developments and future research needs. *Advances in Agronomy, 111*, 297–413.
Lambert, E., et al. (2010). Relationships between the biomass production of invasive Ludwigia species and physical properties of habitats in France. *Hydrobiologia, 656*, 173–186.
Lamichhane, J. R., et al. (2017). Integrated weed management systems with herbicide-tolerant crops in the European Union: Lessons learnt from home and abroad. *Critical Reviews in Biotechnology, 37*, 459–475.
Leblanc, M., et al. (2011). Mechanical weed control in cereal crops in Eastern Canada. Physical weed control: Progress and challenges. *Topics in Canadian Weed Science, 6*, 35–42.
Leinonen, P., Närkki, V. (2004). Lay-down working cart improves efficacy of hand weeding. European weed research society. In *Proceedings of the 6th EWRS workshop on physical and cultural weed control*, Lillehammer, Norway, 8–10 March, 2004. European Weed Research Society.
Leon, R. G., & Wright, D. L. (2018). Recurrent changes of weed seed bank density and diversity in crop—livestock systems. *Agronomy Journal, 110*, 1068–1078.
Leon, R. G., et al. (2015). Weed seed banks are more dynamic in a sod-based, than in a conventional, peanut–cotton rotation. *Weed Science, 63*, 877–887.
Liao, C., et al. (2008). Altered ecosystem carbon and nitrogen cycles by plant invasion: A meta-analysis. *New Phytologist, 177*, 706–714.
Lishawa, S. C., et al. (2013). Reconstructing plant invasions using historical aerial imagery and pollen core analysis: Typha in the Laurentian Great Lakes. *Diversity and Distributions, 19*, 14–28.
Lombardo, S., et al. (2012). The effect of soil solarization and fumigation on pests and yields in greenhouse tomatoes. *Crop Protection, 37*, 59–64.
Louda, S. M., & Stiling, P. (2004). The double-edged sword of biological control in conservation and restoration. *Conservation Biology, 18*, 50–53.

Mackay, A. J., et al. (2016). Cascade of ecological consequences for West Nile virus transmission when aquatic macrophytes invade stormwater habitats. *Ecological Applications, 26*, 219–232.

Maki, K., & Galatowitsch, S. (2004). Movement of invasive aquatic plants into Minnesota (USA) through horticultural trade. *Biological Conservation, 118*, 389–396.

Mazza, G., et al. (2014). Biological invaders are threats to human health: An overview. *Ethology Ecology & Evolution, 26*, 112–129.

McAllister, R., et al. (2017). Balancing collaboration with coordination: contesting eradication in the Australian plant pest and disease biosecurity system. *International Journal of the Commons, 11*.

McAllister, R. R., et al. (2015). From local to central: a network analysis of who manages plant pest and disease outbreaks across scales. *Ecology and Society, 20*.

McAvoy, T. J., et al. (2012). Road survey of the invasive tree-of-heaven (*Ailanthus altissima*) in Virginia. *Invasive Plant Science and Management, 5*, 506–512.

McDougall, K. L., et al. (2018). Running off the road: Roadside non-native plants invading mountain vegetation. *Biological Invasions, 20*, 3461–3473.

McFadyen, R. E. C. (1998). Biological control of weeds. *Annual Review of Entomology, 43*, 369–393.

McLeod, M. L., et al. (2016). Exotic invasive plants increase productivity, abundance of ammonia-oxidizing bacteria and nitrogen availability in intermountain grasslands. *Journal of Ecology, 104*, 994–1002.

Mehmood, A., et al. (2018). Estimation of economic threshold level of alligator weed (*Alternanthera philoxeroides* (Mart.) Griseb.) to tackle grain quality and yield losses in rice. *Archives of Agronomy and Soil Science, 64*, 208–218.

Meinzen-Dick, R., et al. (2004). *Methods for studying collective action in rural development* (Vol. 82, pp. 197–214). Elsevier.

Miller-Rushing, A., et al. (2012). The history of public participation in ecological research. *Frontiers in Ecology and the Environment, 10*, 285–290.

Mueller, T. C., et al. (2003). Shikimate accumulates in both glyphosate-sensitive and glyphosate-resistant horseweed (*Conyza canadensis* L. Cronq.). *Journal of Agricultural and Food Chemistry, 51*, 680–684.

Müller-Schärer, H., et al. (2018). Cross-fertilizing weed science and plant invasion science to improve efficient management: A European challenge. *Basic and Applied Ecology, 33*, 1–13.

Newman, R. M., & Inglis, W. (2009). Distribution and abundance of the milfoil weevil, *Euhrychiopsis lecontei*, in Lake Minnetonka and relation to milfoil harvesting. *Journal of Aquatic Plant Management, 47*, 21–25.

Newman, R. M., et al. (2001). Overwinter habitat and the relationship of overwinter to in-lake densities of the milfoil weevil, *Euhrychiopsis lecontei*, a Eurasian watermilfoil biological control agent. *Journal of Aquatic Plant Management, 39*, 63–67.

Niphadkar, M. (2016). *Mapping invasive species Lantana camara in a high diversity tropical ecosystem of Western Ghats, India*. Manipal University, Ashoka Trust for Research in Ecology and the Environment….

Niphadkar, M., & Nagendra, H. (2016). Remote sensing of invasive plants: Incorporating functional traits into the picture. *International Journal of Remote Sensing, 37*, 3074–3085.

Occhipinti-Ambrogi, A., & Galil, B. S. (2004). A uniform terminology on bioinvasions: A chimera or an operative tool? *Marine Pollution Bulletin, 49*, 688–694.

Oerke, E.-C. (2006). Crop losses to pests. *The Journal of Agricultural Science, 144*, 31–43.

Ortega-Huerta, M. A., & Peterson, A. (2004). Modelling spatial patterns of biodiversity for conservation prioritization in north-eastern Mexico. *Diversity and Distributions, 10*, 39–54.

Parra-Olea, G., et al. (2005). Forecasting climate change effects on Salamander distribution in the highlands of central Mexico 1. *Biotropica: The Journal of Biology and Conservation. 37*, 202–208.

Paynter, Q., et al. (2012). Plant traits predict the success of weed biocontrol. *Journal of Applied Ecology, 49*, 1140–1148.

Pearlstine, L., et al. (2005). Textural discrimination of an invasive plant, *Schinus terebinthifolius*, from low altitude aerial digital imagery. *Photogrammetric Engineering & Remote Sensing, 71*, 289–298.

Pejchar, L., & Mooney, H. A. (2009). Invasive species, ecosystem services and human well-being. *Trends in Ecology & Evolution, 24*, 497–504.

Perrings, C., et al. (2010). Ecosystem services for 2020. *Science, 330*, 323–324.

Peterson, A. T., & Cohoon, K. P. (1999). Sensitivity of distributional prediction algorithms to geographic data completeness. *Ecological Modelling, 117*, 159–164.

Peterson, E. (2005). Estimating cover of an invasive grass (*Bromus tectorum*) using Tobit regression and phenology derived from two dates of Landsat ETM+ data. *International Journal of Remote Sensing, 26*, 2491–2507.

Pimentel, D., et al. (2005). Update on the environmental and economic costs associated with alien-invasive species in the United States. *Ecological Economics, 52*, 273–288.

Plaza, P. I., et al. (2018). Rubbish dumps as invasive plant epicentres. *Biological Invasions, 20*, 2277–2283.

Poudel, A. S., et al. (2019). Biology and management of the invasive weed *Ageratina adenophora* (Asteraceae): Current state of knowledge and future research needs. *Weed Research, 59*, 79–92.

Powell, K. I., et al. (2011). A synthesis of plant invasion effects on biodiversity across spatial scales. *American Journal of Botany, 98*, 539–548.

Pyšek, P., et al. (2012). A global assessment of invasive plant impacts on resident species, communities and ecosystems: The interaction of impact measures, invading species' traits and environment. *Global Change Biology, 18*, 1725–1737.

Pyšek, P., & Richardson, D. M. (2010). Invasive species, environmental change and management, and health. *Annual Review of Environment and Resources, 35*, 25–55.

Qi, Z., et al. (2011). Nitrate-nitrogen losses through subsurface drainage under various agricultural land covers. *Journal of Environmental Quality, 40*, 1578–1585.

Radics, L., & Szné Bognár, E. (2002). Comparison of different mulching methods for weed control in organic green bean and tomato. In *XXVI International Horticultural Congress: Sustainability of Horticultural Systems in the 21st Century 638* (pp. 189–196).

Radosevich, S. R., & Ghersa, C. M. (1992). Weeds, crops, and herbicides: A modern-day "neckriddle." *Weed Technology, 6*, 788–795.

Reaser, J. K., et al. (2007). Ecological and socioeconomic impacts of invasive alien species in island ecosystems. *Environmental Conservation, 34*, 98–111.

Reinhart, K. O., & Rinella, M. (2011). Comparing susceptibility of eastern and western US grasslands to competition and allelopathy from spotted knapweed [*Centaurea stoebe* L. subsp. micranthos (Gugler) Hayek]. *Plant Ecology, 212*, 821–828.

Rouifed, S., et al. (2012). Invasive knotweeds are highly tolerant to salt stress. *Environmental Management, 50*, 1027–1034.

Sanon, S., et al. (2012). Quantifying ecosystem service trade-offs: The case of an urban floodplain in Vienna, Austria. *Journal of Environmental Management., 111*, 159–172.

Schmitz, D. C. (1997). The ecological impact of nonindigenous plants. *Stranger in Paradise: Impact and Management of Nonindigenous Species in Florida*. 39–61.

Schooler, S. S. (2008). Shade as a management tool for the invasive submerged macrophyte, *Cabomba caroliniana. Journal of Aquatic Plant Management, 46*, 168–171.

Schreiner, P. R., et al. (2001). Soil solarization reduces arbuscular mycorrhizal fungi as a consequence of weed suppression. *Mycorrhiza, 11*, 273–277.

Schwarzländer, M., et al. (2018). Biological control of weeds: An analysis of introductions, rates of establishment and estimates of success, worldwide. *BioControl, 63*, 319–331.

Seebens, H., et al. (2018). Global rise in emerging alien species results from increased accessibility of new source pools. *Proceedings of the National Academy of Sciences, 115*, E2264–E2273.

Seebens, H., et al. (2017). No saturation in the accumulation of alien species worldwide. *Nature Communications, 8*, 14435.

Segurado, P., & Araujo, M. B. (2004). An evaluation of methods for modelling species distributions. *Journal of Biogeography, 31*, 1555–1568.
Shackleton, C. M., et al. (2007). Assessing the effects of invasive alien species on rural livelihoods: Case examples and a framework from South Africa. *Human Ecology, 35*, 113–127.
Shackleton, R. T., et al. (2019). The role of invasive alien species in shaping local livelihoods and human well-being: A review. *Journal of Environmental Management, 229*, 145–157.
Shackleton, R. T., et al. (2017). Distribution of the invasive alien weed, *Lantana camara*, and its ecological and livelihood impacts in eastern Africa. *African Journal of Range & Forage Science, 34*, 1–11.
Shackleton, R. T., et al. (2017). Distribution and socio-ecological impacts of the invasive alien cactus *Opuntia stricta* in eastern Africa. *Biological Invasions, 19*, 2427–2441.
Sheley, R. L., & Krueger-Mangold, J. (2003). Principles for restoring invasive plant-infested rangeland. *Weed Science, 51*, 260–265.
Shiferaw, H., et al. (2019). Modelling the current fractional cover of an invasive alien plant and drivers of its invasion in a dryland ecosystem. *Scientific Reports, 9*, 1576.
Sileshi, G. W., et al. (2019). The threat of alien invasive insect and mite species to food security in Africa and the need for a continent-wide response. *Food Security, 11*, 763–775.
Simberloff, D., et al. (2013). Impacts of biological invasions: What's what and the way forward. *Trends in Ecology & Evolution, 28*, 58–66.
Smith, R. G., et al. (2010). A new hypothesis for the functional role of diversity in mediating resource pools and weed–crop competition in agroecosystems. *Weed Research, 50*, 37–48.
Somers, B., & Asner, G. P. (2013). Multi-temporal hyperspectral mixture analysis and feature selection for invasive species mapping in rainforests. *Remote Sensing of Environment, 136*, 14–27.
Souza, A. O., et al. (2018). Local ecological knowledge concerning the invasion of Amerindian lands in the northern Brazilian Amazon by *Acacia mangium* (Willd.). *Journal of Ethnobiology and Ethnomedicine, 14*, 1–14.
Stefanowicz, A. M., et al. (2018). Differential influence of four invasive plant species on soil physicochemical properties in a pot experiment. *Journal of Soils and Sediments, 18*, 1409–1423.
Stone, C. M., et al. (2018). Would the control of invasive alien plants reduce malaria transmission? A review. *Parasites & Vectors, 11*, 1–18.
Store, R., & Kangas, J. (2001). Integrating spatial multi-criteria evaluation and expert knowledge for GIS-based habitat suitability modelling. *Landscape and Urban Planning, 55*, 79–93.
Teasdale, J. R., & Zimdahl, R. (2018). The use of rotations and cover crops to manage weeds. *Integrated Weed Management for Sustainable Agriculture* (pp. 227–260). Burleigh Dodds Science Publishing.
Thouvenot, L., et al. (2012). Responses of two invasive macrophyte species to salt. *Hydrobiologia, 686*, 213–223.
Underwood, E., et al. (2003). Mapping nonnative plants using hyperspectral imagery. *Remote Sensing of Environment, 86*, 150–161.
Van der Weide, R., et al. (2005). Innovation in mechanical weed control in row crops. *13th EWRS Symposium*, Bari, Vol. 174.
Van Driesche, R. G., & Heinz, K. M. (2016). Economics of biological control for species invading wildlands. *Integrating Biological Control into Conservation Practice* 294–313.
Vanderhoeven, S., et al. (2011). Perception and understanding of invasive alien species issues by nature conservation and horticulture professionals in Belgium. *Environmental Management, 47*, 425–442.
Vidotto, F., et al. (2013). Effect of short-duration high temperatures on weed seed germination. *Annals of Applied Biology, 163*, 454–465.
Vilà, M., et al. (2011). Ecological impacts of invasive alien plants: A meta-analysis of their effects on species, communities and ecosystems. *Ecology Letters, 14*, 702–708.
Vogelsang, K. M., & Bever, J. D. (2009). Mycorrhizal densities decline in association with nonnative plants and contribute to plant invasion. *Ecology, 90*, 399–407.

Walsh, M., & Newman, P. (2007). Burning narrow windrows for weed seed destruction. *Field Crops Research, 104*, 24–30.

Watson, R., et al., 2019. Summary for policymakers of the global assessment report on biodiversity and ecosystem services of the Intergovernmental Science-Policy Platform on Biodiversity and Ecosystem Services. IPBES Secretariat: Bonn, Germany. 22–47.

Weisberger, D., et al. (2019). Does diversifying crop rotations suppress weeds? A meta-analysis. *Plos One, 14*, Article e0219847.

Westerman, P. R., et al. (2005). Are many little hammers effective? Velvetleaf (*Abutilon theophrasti*) population dynamics in two-and four-year crop rotation systems. *Weed Science, 53*, 382–392.

Wolf, J., et al. (2004). Soil characteristics of Rocky Mountain National Park grasslands invaded by *Melilotus officinalis* and *M. alba. Journal of Biogeography, 31*, 415–424.

Xu, H., et al. (2006). The distribution and economic losses of alien species invasion to China. *Biological Invasions, 8*, 1495–1500.

Young, A. M., & Larson, B. M. (2011). Clarifying debates in invasion biology: A survey of invasion biologists. *Environmental Research, 111*, 893–898.

Yu, F.-K., et al. (2014). Impacts of *Ageratina adenophora* invasion on soil physical–chemical properties of Eucalyptus plantation and implications for constructing agro-forest ecosystem. *Ecological Engineering, 64*, 130–135.

Zhao, J., et al. (2020). Does crop rotation yield more in China? *A Meta-Analysis. Field Crops Research, 245*, Article 107659.

Chapter 15
Management Approaches for the Restoration of Invaded Urban Plant Communities

Leeladarshini Sujeeun

Abstract The rapid growth of urban human populations has accelerated the introduction of exotic plants in cities globally. Urban landscapes create habitats with distinct environmental conditions that promote the establishment and spread of non-native plant species, some of which can become highly invasive in urban ecosystems and threaten native plant communities. Invasive plants often displace native species from urban plant communities and reduce the quality of ecological, economic, and social benefits that urban ecosystems provide. Conventional methods (e.g., mechanical, chemical, and biological control) to mitigate the spread of invasive species are very labor-intensive and expensive. In addition, management approaches that only target the removal of invasive plants are unlikely to succeed in the long term because invasive species can easily reclaim managed sites. Restoration strategies should therefore also aim to decrease the competitive advantage of invasive species by planting strong native competitors, increasing the functional diversity of native plants, suppressing the invasive seed bank, and adding soil amendments to enhance the recovery of native plant communities in invaded urban ecosystems and increase their resilience against re-invasion by exotic species. This chapter gives an overview of the environmental factors and characteristics of non-native plants that facilitate their invasion in urban environments and discusses active management strategies for the restoration of invaded urban ecosystems.

Keywords Allelopathy · Anthropogenic disturbances · Exotic species introductions · Native species recovery

L. Sujeeun (✉)
University of Toronto, Institute of Forestry and Conservation, Toronto, ON, Canada
e-mail: lutchmee.sujeeun@mail.utoronto.ca

R. Singh et al. (eds.), *Invasive Alien Plants in Urban Ecosystems*,
https://doi.org/10.1007/978-3-032-00997-5_15

1 Introduction

Anthropogenic introductions of exotic plants have increased the establishment of invasive species that threaten native ecosystems globally (Kühn et al., 2017; Riera et al., 2024; Ruas et al., 2022). Exotic plant invasions and their detrimental impacts are becoming more rampant as a result of an increase in global trade, climate change, and anthropogenic disturbances (Henderson et al., 2006; Skultety & Matthews, 2017; Tripathi et al., 2023). Local natural communities have become more susceptible to invasions due to urbanization (McKinney, 2006). Exotic species are closely associated with urban areas, which are often the first point of entry and hotspots for the establishment of invasive plants (Padayachee et al., 2017; Ruas et al., 2022; Skultety & Matthews, 2017). Several exotic plant species have been introduced intentionally into urban areas for horticultural purposes and to provide, enhance, or restore certain ecosystem services (Barton et al., 2004; Gavier-Pizarro et al., 2010; Potgieter et al., 2019); however, some of these species can become invasive (Blackburn et al., 2011) and have harmful effects on native species (Gurevitch & Padilla, 2004; Pyšek et al., 2012). The accidental introductions of exotic invasive plants, e.g., through international shipments (Hulme, 2009), can further exacerbate the negative effects on native species (Lehan et al., 2013). Urban environments create conditions that promote the establishment and proliferation of invasive plants in urban areas and the surrounding natural ecosystems, such as high levels of disturbance (Chytrý et al., 2008; Davis et al., 2000; Johnston-Bates et al., 2025), habitat heterogeneity (Deutschewitz et al., 2003; Pyšek, 1995), higher temperatures as a result of the 'urban heat island' effect (Lososová et al., 2012; McKinney, 2006), and facilitation of the introduction and transportation of propagules of invasive species (Lockwood et al., 2005; Malavasi et al., 2014).

The ecological impacts of invasive plants in urban areas are particularly severe because the local native communities already exist in a fragmented state due to anthropogenic activities (Malavasi et al., 2014; McKinney, 2002) and may thus have little resilience against aggressive non-native invaders. Invasive plants can decrease species diversity in local communities (Gaertner et al., 2009; Hejda et al., 2009; Powell et al., 2011; Vilà et al., 2006) and negatively impact the ecosystem services that native communities provide (Charles & Dukes, 2007). Exotic invaders can also lead to the homogenization of biotic communities (Kühn & Klotz, 2006; McKinney, 2006; Trentanovi et al., 2013) by replacing locally unique native species with common exotic species (Godefroid, 2001; Shochat et al., 2010; Wania et al., 2006). In addition to their negative ecological effects in urban ecosystems, invasive plants can negatively impact the economy by increasing crop losses and decreasing forage yields and land values (DiTomaso, 2000) and due to the substantial costs of removing invasive plants through herbicide use and other control methods (Pimentel et al., 2000). The presence of exotic invasive species in urban areas can also reduce the quality of life by contributing to the loss of undisturbed natural spaces and unique cultural landscapes (Henderson et al., 2006; Reichard, 2010).

Management strategies for the control of invasive plants generally include chemical, biological, and mechanical treatments (Annighöfer et al., 2012; Seastedt, 2015; Weidlich et al., 2020). However, these passive restoration strategies that only target the removal of invasive species are often ineffective and costly and can sometimes have unintended negative impacts on the environment and native communities (Smith et al., 2006; Stricker et al., 2015). For example, mechanical control methods can increase the proliferation of invasive plants through accidental seed dispersal if not handled carefully when the species is removed from a site (Francis & Chadwick, 2015). The use of herbicides can potentially be harmful to humans and the environment (Cronk & Fuller, 1995). In addition, herbicides and biocontrol agents can have harmful impacts on non-target species; these risks are particularly severe when the invasive species occur alongside closely related native species (Cronk & Fuller, 1995; Webb et al., 2001) and when native species are locally rare in the fragmented plant communities in urban areas. Biocontrol agents may also be less effective in urban areas due to their heterogeneous nature which can reduce their dispersal between different invasive plant populations (Francis & Chadwick, 2015).

In some cases, mechanical control, herbicides, and biocontrol agents can promote the production of inhibitory compounds in allelopathic invasive plants (Siemens et al., 2002; Thelen et al., 2005), which comprise approximately half of the invasive plant species globally (Kalisz et al., 2021); the allelopathic compounds that are released in response to the control treatments can thus suppress native species in the managed areas. Another challenging aspect of invasive plant management is that invasive plants can re-invade a site after removal if the conditions promoting its establishment are not addressed. The removal of invasive plants can even create disturbed conditions that lead to re-invasion of the site (Iannone III & Galatowitsch, 2008; Kettenring & Adams, 2011; Singh & Byun, 2023). Major changes to the plant communities through invasive plant removal can increase the availability of resources such as light, nutrients, water, and space to invaders that occur in the regional species pool (D'Antonio & Meyerson, 2002; Zedler & Kercher, 2004). Although passive restoration, i.e., the natural recolonization of a site by native plant species after invasive plant removal, is achievable (Bierzychudek, 2020), its success can be highly variable, often failing to restore the pre-invasion assemblage of native plant species in the ecosystem (Kettenring & Adams, 2011; Ogden & Rejmánek, 2005; Rohal et al., 2019). Consequently, management strategies that include active restoration measures are more likely to ensure the success of invasive plant control programs in the long term.

Active restoration strategies that involve additional intervention to promote the re-establishment of native species after the removal of invasive plants can facilitate the recovery of native plant communities and decrease the likelihood of re-invasions (Hulme, 2006; Mostert et al., 2018; Schuster et al., 2018; Singh & Byun, 2023). Active restoration measures after the removal of invasive plants include revegetation with native species (Hall et al., 2021; Schuster et al., 2018), increasing species and functional diversity in native communities (Byun et al., 2023; Funk et al., 2008; Oakley & Knox, 2013), and soil remediation using carbon amendments such as the additions of activated carbon and biochar (Perry et al., 2004; Sujeeun & Thomas,

2022). The application of soil amendments like biochar is also a safer method of controlling the proliferation of invasive plants in urban areas because it does not pose the same risks as herbicides and biocontrol agents. In addition, biochar has the potential to mitigate the allelopathic effects of many invasive plants on native species through the sorption of their allelochemicals (Mahall & Callaway, 1992; Rogovska et al., 2012; Sujeeun & Thomas, 2017).

This chapter addresses the environmental factors and characteristics of invasive plant species that facilitate their invasion in urban areas and active restoration strategies to enhance the recovery of native plant communities and decrease re-invasion after the removal of invasive species. We specifically address evidence that demonstrates that habitat fragmentation, climatic changes, and changes to soil conditions caused by urbanization facilitate the invasion of exotic plants in urban environments. In addition, we discuss invasive plant traits that contribute to their dominance outside their native range, such as prolific propagation, high phenotypic plasticity, and allelopathy. Finally, we examine the potential of native plant revegetation, trait-based restoration, suppression of invasive seed banks, and soil amendments to increase the recovery of native plant communities in urban areas and increase their resilience against re-invasion by exotic species.

2 Environmental Factors that Facilitate Exotic Plant Invasion in Urban Areas

2.1 Habitat Fragmentation

Urbanization has resulted in the fragmentation of native plant communities and an overall reduction in the abundance and diversity of native plant species (Bozkurt & Basaraner, 2024; Malavasi et al., 2014; McKinney, 2002; Threlfall et al., 2016). Fragmentation and edge effects can greatly facilitate the transition of native communities to ones that are dominated by non-native invasive plants (Cilliers et al., 2008; Forman, 1995; Mavimbela et al., 2018). In forest ecosystems, large gaps in the canopy can promote the spread of shade-intolerant invasive plants while reducing the establishment of late-successional native plants (Mavimbela et al., 2018). In addition, larger forest patches and forest areas with higher connectivity are less susceptible to invasion because the ratio of edge area to patch size is lower (Hansen & Clevenger, 2005). Fragmentation can also reduce the ability of native plants to re-colonize a site after being displaced by invasive species as a result of the limited availability of a regional species pool; the likelihood of re-colonization by a native species decreases substantially as the distance from the nearest seed source increases (Komuro & Koike, 2005; Yao et al., 1999). Therefore, plant invasions facilitated by fragmentation can potentially result in the local extinctions of native species.

2.2 Climatic Changes

Urban environments can drastically change the local climate, often favoring non-native invasive plants over native species (Ruas et al., 2022; Von der Lippe et al., 2005). Urban areas are typically associated with warmer temperatures, higher temperature minima, fewer frost days, as well as lower frost intensity, which increases the length of the growing season and causes a shift in phenological phases (Sukopp & Wurzel, 2003; Von der Lippe et al., 2005). The warmer microclimate of urban regions has enhanced the spread of some highly invasive species, including *Ailanthus altissima* (Kowarik & Böcker, 1984), *Robinia pseudoacacia* (Sukopp & Wurzel, 2003), *Acer negundo* (Von der Lippe et al., 2005), and *Fallopia japonica* (Turbelin & Catford, 2021). In addition, the changes in temperature have been found to prolong and advance the timing of phenological events like flowering, fruit ripening, and leaf fall in urban areas (Davis et al., 2016; Jochner & Menzel, 2015; Su et al., 2019), which can promote exotic plant invasion (Wolkovich et al., 2013; Zettlemoyer et al., 2019).

2.3 Changes in Soil Properties

Urban soils may experience increased inputs of certain nutrients (particularly N) through agricultural run-off (Rickey & Anderson, 2004) or via atmospheric deposition (Brooks, 2003), and higher levels of pollutants (Pickett et al., 2011), which can increase the spread of invasive species (Grella et al., 2018; Lopez et al., 2018). Several fast-growing and highly competitive invasive species are disproportionately promoted by high soil N availability (Afreen et al., 2023; Perry et al., 2010; Yu et al., 2021) and can outcompete native plants in urban soils with elevated N levels (King & Hovick, 2020). Increased soil N availability in urban areas has been shown to promote the dominance of invasive species and reduce native plant diversity (Bidwell et al., 2006; Brooks, 2003). In addition to enhancing the growth of invasive plants, large N inputs can dramatically increase the seed production and germination of invaders, further increasing their ability to outcompete and displace native species (Bidwell et al., 2006; Vincent & Roberts, 1977). In some instances, invasive plants are even able to establish and proliferate rapidly in contaminated soils, whereas native species that are negatively affected by soil contaminants are excluded from these sites (Ariori et al., 2017; Prabakaran et al., 2019; Sun et al., 2023; Yang et al., 2007). Some invasive species have greater salt tolerance than native species and are able to outcompete the native vegetation under high salinity conditions (Tang et al., 2014). Increased salinity in wetlands due to road salt runoff from urban areas can increase the invasion of exotic species such as *Phalaris arundinacea* (Prasser & Zedler, 2010) and *Phragmites australis* (Richburg et al., 2001). Therefore, soil contaminants from urban sources can provide a competitive advantage to invasive species.

3 Characteristics of Invasive Plant Species

3.1 *Prolific Propagation*

The higher reproductive capacity of non-native invasive species compared to native plants can greatly increase the ability of exotic species to establish and invade new habitats (Mason et al., 2008; Rouget & Richardson, 2003; Witkowski & Wilson, 2001). Large seed banks of invasive species are strongly associated with a reduction in the performance and diversity of native plant communities and can enhance their susceptibility to reinvasions (Ibáñez et al., 2021; Nsikani et al., 2018; Wang et al., 2022). Invasive species that escape from their natural enemies when they are introduced to a new environment are able to reallocate their resources from defence to growth and increase their seed production and/or seed size (Crawley et al., 1996). Invasive species also have a greater seed mass in their invasive habitats compared to their native habitats (Daws et al., 2007). Furthermore, invasive species have a higher seed mass and produce more seeds than co-occurring native species of the same family (Daws et al., 2007; Ferreras & Galetto, 2010). Greater seed production in invasive species can substantially increase the success of long-distance dispersal because higher levels of seed production by the invasive plant increase the likelihood of its successful establishment as the distance from the parent plant increases (Clark et al., 2001; Mason et al., 2008).

Higher seed production capacity of invaders compared to native species can promote their proliferation rate and even reduce the future regeneration of native species. For example, in an urban restoration forest in Brazil, propagules of non-native invasive species in adjacent invaded areas dominated the seed bank and seedling emergence in the area undergoing restoration (Londe et al., 2017). Seed banks in urban soils have also been found to contain a higher density of seeds of invasive species in more densely populated areas, likely due to larger and more frequent disturbances (Figueroa et al., 2020). High propagule pressure (Daws et al., 2007; Ferreras & Galetto, 2010) and seed persistence (Passos et al., 2017), which is enhanced by soil disturbance (Fenner & Thompson, 2005; Presotto et al., 2020), can thus promote exotic plant invasion in urban environments and reduce the success of restoration if active measures are not taken to suppress the seed density of invasive species.

3.2 *Phenotypic Plasticity and Adaptation Strategies*

Higher phenotypic plasticity in invasive plants can increase the likelihood of invasion in new environments (Annapurna & Singh, 2003; Daehler, 2003; Rejmánek & Richardson, 1996; Williams et al., 1995). Invasive species generally show greater plasticity in functional traits, such as size, fitness, and growth rate, compared to native and non-invasive species (Van Kleunen et al., 2010). Invasive species can quickly adapt to disturbed environments and successfully invade new habitats due to their higher phenotypic plasticity (Guo et al., 2024; Kobayashi et al., 1999). For example, the invasive species *Calotropis procera* shows high plasticity in both morphological and reproductive traits in response to changes in soil edaphic factors in urban environments. The ability of *C. procera* to increase self-shading and accumulate Na in its leaves can allow it to tolerate drought and salinity stresses, respectively (Farahat et al., 2015). The exotic invasive species *Alternanthera philoxeroides* has higher phenotypic plasticity than its co-occurring native congener *Alternanthera sessilis*, allowing it to grow and reproduce under unfavorable conditions and form dense stands that suppress the germination and early growth of native plants (Pan et al., 2006). In its native range (Germany), *Lythrum salicaria* flowers earlier and increases flower production, whereas in its introduced habitat (North America), it undergoes a longer growing period before flowering in order to increase its height and clonal reproduction, which can contribute to its invasion in its non-native range (Chun et al., 2007). Similarly, invasive populations of *Spartina alterniflora* had higher plasticity in growth and morphological traits at different soil N levels compared to native populations of the species, suggesting that its adaptation to different environments contributed to its successful invasion (Qing et al., 2011). In addition, invasive species have higher phenotypic plasticity than native species when disturbances occur (Davidson et al., 2011), which allows the invasive plants to tolerate stressful environmental conditions and make use of available resources effectively to promote their dominance in invaded plant communities (El-Barougy et al., 2021; Granata et al., 2020; Ruprecht et al., 2014).

3.3 Allelopathy

Several invasive species have been shown to release allelochemicals in order to suppress native plants in their introduced habitat (Hagan et al., 2013; Hu & Zhang, 2012; Sujeeun & Thomas, 2022). Allelopathy provides a major competitive advantage to invasive species over neighboring native species, which is thought to increase their invasiveness in their non-native range (Callaway & Hierro, 2006; Lodhi, 1978; van Kleunen et al., 2015). Several invasive species that are widespread in urban areas are allelopathic, including *Lantana camara* (Kato-Noguchi & Kurniadie, 2021) and *Mikania micrantha* (Wu et al., 2009) in the tropics, *Acer platanoides* (Sujeeun & Thomas, 2023a), *Rhamnus cathartica* (Warren et al., 2017), and *Alliaria petiolata* (Cipollini, 2016) in North America, and *Fallopia japonica* in Europe and North America (Šoln et al., 2022). In an urban forest invaded by *R. cathartica*, native species richness and abundance were significantly suppressed by the allelochemicals in the leaves and roots of this invasive species (Warren et al., 2017). Similarly, soils amended with the leaf litter of *A. platanoides* from urban areas severely inhibited the germination and growth of seedlings of co-occurring native tree species (Sujeeun & Thomas, 2023a). Urban soils from sites invaded by *Paspalum dilatatum* had higher concentrations of phenolic acids and flavonoids than uninvaded sites, which was associated with a decrease in germination and growth of native species (Hassan & Mohamed, 2020). The allelopathic effects of invasive species can also be enhanced under elevated soil N levels (Sun et al., 2022; Wang et al., 2016, 2017). For example, the invasive perennial *Alternanthera philoxeroides* had stronger allelopathic effects on the co-occurring native species *Ludwigia peploides* in a eutrophicated wetland due to higher nutrient availability (Xiao et al., 2019). Therefore, N enrichment in urban soils as a result of anthropogenic activities can potentially increase the inhibitory effects of allelopathic invasive species.

4 Active Restoration of Urban Invaded Plant Communities

Passive management strategies to restore native plant communities by relying on natural re-establishment of native species are common but require certain conditions to be effective, such as low rates of disturbance after invasive plant removal, high resilience of native communities, and sufficient native propagule pressure for colonization of the restoration site (Holl & Aide, 2011; Schuster et al., 2018). However, in many cases, passive approaches to restoration fail to re-establish native plant communities after invasive plant removal and can even promote re-invasions by other non-native species as a result of the increase in available resources (Bauer & Reynolds, 2016; Holmes, 2001; Kettenring & Adams, 2011), thus requiring active management strategies to ensure the success of restoration projects in the long term (Table 1) (Hulme, 2006; Mostert et al., 2018; Singh & Byun, 2023). Potential approaches to increase the re-establishment of native species and reduce re-invasion after removal include revegetation with native species (Schuster et al., 2018), reducing the seed banks of invasive species (Richardson & Kluge, 2008), planting native species with high competitive ability and diverse functional traits (Funk et al., 2008), additions of soil amendments to reduce nutrient imbalances (e.g., high N levels) that favor invaders (Perry et al., 2004), and biochar additions to mitigate the allelopathic effects of invasive plants on native species (Sujeeun & Thomas, 2023b).

4.1 Revegetation with Native Species

After invasive plants are removed from a site, native species can be planted to re-introduce species that were excluded from the community after invasion and to increase the number of competitors, which can decrease the population size, recruitment, and propagation of both the targeted invasive species and other potential non-native species (Bakker et al., 2003; Schuster et al., 2018) that can establish before native species are able to re-colonize the site (Iannone III & Galatowitsch, 2008; Kettenring & Adams, 2011; Singh & Byun, 2023). In urban areas, where anthropogenic activities have created highly fragmented native plant communities, passive recovery is less likely to be successful due to a lack of seed sources of native species

Table 1 Summary of implementation methods and mechanisms for reducing invasion under different active restoration strategies

Restoration approach	Implementation method	Mechanism for reducing invasion	References
Revegetation with native plants	Addition of seeds or seedlings of native species after removing invasive plants	Increasing the number of competitors and reducing the resources available to invasive plants	Bucharova and Krahulec (2020)
Trait-based restoration	Addition of native species with high competitive ability and tolerance to urban conditions	Increasing the competitive ability of native species in order to resist invasion	Funk et al. (2008)
Reduction in invasive seed bank	Application of herbicides, tillage, or fire	Decreasing the spread of invasive species by suppressing their regeneration	Singh and Byun (2023)
Mitigation of soil disturbance	Application of soil amendments (e.g., activated carbon and biochar)	Decreasing the competitive advantage of invasive species by reducing the availability of soil N, salts, and heavy metals	Perry et al. (2004), Zhang et al. (2013), Brändli et al. (2008), Thomas et al. (2013)
Mitigation of allelopathy	Application of biochar	Decreasing the competitive advantage of invasive species by reducing the availability of allelochemicals	Sujeeun and Thomas (2023b)

nearby (Komuro & Koike, 2005; Yao et al., 1999); therefore, native species must be planted after removing invasive plants to facilitate the recovery of native communities. Active sowing has been shown to promote the establishment of native shrub cover and diversity, whereas passive restoration resulted in limited recovery of native plant communities due to the lower density of native species that were established only through natural processes (Hall et al., 2021).

Revegetation with native species has great potential to reduce the growth and spread of invasive plants by increasing competition against the invaders (Bhattacharjee et al., 2009; Li et al., 2015). For instance, the addition of seeds of native species after herbicide treatment on the invasive species *Rumex alpinus* suppressed its regeneration and growth, which can impede its invasion (Bucharova & Krahulec, 2020). However, in some cases, active sowing may not have any beneficial impacts on native species richness and diversity because the planted species are stronger competitors than naturally regenerating native species; therefore, revegetation approaches must include a diverse seed mixture to obtain an assemblage of native species similar to the pre-invasion community (Boxriker et al., 2022; Bucharova & Krahulec, 2020).

4.2 Trait-Based Approach to Restoration

A main challenge of restoring invaded ecosystems is to establish native plant communities that can resist invasion by exotic species (Funk et al., 2008). Potential invaders are less likely to become dominant in a new site if species with similar traits are present in the community or available niches are already occupied by native species (Funk et al., 2008; Gilbert & Lechowicz, 2005; Macarthur & Levins, 1967). For example, the invasion of a European grassland by non-native legumes was lower in native legume communities compared to communities containing native grasses and non-leguminous forbs (Turnbull et al., 2005). Stands of *Acer saccharum* that are able to create heavy shade were more resistant to the invasion of the moderately shade-tolerant *Rhamnus cathartica* (McCay & McCay, 2009) than stands with lower amounts of shade (Kurylo et al., 2007). Similarly, in several other plant communities, the invasion of non-native species was reduced in the presence of native species with similar functional traits (Dukes, 2001; Emery, 2007; Fargione et al., 2003). However, in an urban restoration area, native species that were functionally dissimilar to invasive species had greater survival and growth than native species that were functionally similar to the invader (Hallett et al., 2017). Native species with traits that are dissimilar to those of invasive species may be able to occupy different niches than invaders and thus experience reduced competition (Laughlin, 2014). Native species that have similar functional traits, as invasive species are more likely to be successful if they can strongly compete for resources against invaders, and they may also be able to competitively exclude invaders or reduce their spread (Funk et al., 2008; Laughlin, 2014). On the other hand, native species that are weaker competitors than invaders may be more likely to succeed if they are functionally dissimilar to invasive species. Post-removal restoration must also select for traits associated with high tolerance to the climatic conditions (e.g., warmer temperatures) of urban areas, such as heat and drought tolerance (Teixeira et al., 2022).

4.3 Reducing Seed Density of Invasive Species

The management of invasive plants often only targets the removal of biomass, leaving behind a large persistent seed bank that allows the invasive species to regenerate rapidly and re-invade the restoration site after removal (Bucharova & Krahulec, 2020; Clements & Bierzychudek, 2017; Daws et al., 2007; Passos et al., 2017). Due to the higher persistence of seeds of invasive species compared to that of native species, particularly in disturbed conditions such as urban areas (Fenner & Thompson, 2005; Presotto et al., 2020), soil seed banks often comprise a large proportion of invasive species (Drake, 1998; Gioria et al., 2012). For example, in an urban forest in Brazil, planting native species was not sufficient to enhance the recovery of native plant communities because the seed bank was dominated by exotic invasive plants (Londe et al., 2017). Therefore, active restoration must aim to decrease the abundance of invasive seeds to suppress seedling emergence that can compete with regenerating native plants for resources (Singh & Byun, 2023).

The seed density of invasive species can be reduced by removing the invasive plants before the seed setting by tilling the soil (Marushia & Allen, 2011) and through the application of herbicides (Bucharova & Krahulec, 2020). However, some of these methods may have unintended negative effects on native plants, such as an increased likelihood of invasion due to disturbance caused by tillage (Singh & Byun, 2023) or suppression of native seeds due to herbicide use (Hazelton et al., 2018; Maclean et al., 2018; Wotton et al., 2020). The detrimental effects of herbicides can be managed by using activated carbon, e.g., as a seed coating to protect native seeds from herbicides (Brown et al., 2023; Clenet et al., 2019; Terry et al., 2021). Fire can also be used to destroy invasive seeds in communities dominated by fire-adapted species (Holmes, 1989; Pieterse & Cairns, 1988), although this approach may not be practical in urban environments that are in close proximity to human populations. Given that most methods to suppress invasive seeds are also likely to harm the seeds of native plants, additional measures must be taken to restore the native seed bank (Singh & Byun, 2023). Depletion of the invasive seed bank should also be supplemented with revegetation to increase the recovery of native plant communities, particularly when conditions for germination are not favorable for the regeneration of native species (Middleton et al., 2010).

4.4 Soil Amendments for Urban Soil Restoration

Soil amendments can be an effective tool to reduce the competitive advantage of invasive plants over native species in urban plant communities by restoring soil N levels and reducing the availability of soil contaminants (Hulme, 2006; Perry et al., 2004; Shen et al., 2015). Plant invasions in urban environments are often promoted by increased soil N availability (Bidwell et al., 2006; King & Hovick, 2020) and can thus be reduced by lowering N availability. Soil C addition in the form of sawdust, wood chips, activated carbon, sucrose, or biochar (Nguyen et al., 2017; Perry et al., 2004) can decrease N availability by increasing N immobilization (Averett et al., 2004) and denitrification by microbes (Ullah & Faulkner, 2006), which decrease the growth of invasive plants and promote native species due to reduced competition from the invaders (Blumenthal et al., 2003; Eschen et al., 2007; Perry et al., 2010; Prober et al., 2005). Soil amendments like compost, activated carbon, and biochar can also be used to decrease the availability of contaminants in urban areas, including Cu, Zn, and Pb (Brändli et al., 2008; Kargar et al., 2015; Shen et al., 2015; Zhang et al., 2013). In addition, biochar has been found to substantially mitigate the negative impact of salt stress on plant growth by reducing Na uptake (Ali et al., 2017; Amini et al., 2016; Thomas et al., 2013) and has great potential to alleviate the salinity of urban soils, which can reduce the competitive advantage of salt-tolerant invaders over native species.

4.5 *Biochar Addition for Mitigation of Allelopathy in Invasive Plants*

Biochar is increasingly being used for soil restoration given its potential to immobilize inorganic compounds, e.g., heavy metals and salts, but it also has a high capacity to decrease the availability of allelochemicals released by invasive plants due to its sorptive properties (Gundale & DeLuca, 2006; Hall et al., 2014; Sujeeun & Thomas, 2017; Thomas et al., 2013). Similarly, activated carbon has the potential to reduce the allelopathic effects of invasive plants because it has a high affinity for organic compounds (Hale et al., 2011); however, biochar is a more practical restoration tool due to its low cost and similarity to naturally occurring charcoal (Bruckman & Pumpanen, 2019; Qi et al., 2017). Biochar contains micropores that can enhance its capacity to sorb a wide variety of organic compounds that are chemically similar to common allelochemicals (Zhao et al., 2020; Zhou et al., 2022). In addition, biochar can increase the microbial breakdown of allelochemicals into compounds that are non-inhibitory (Yang et al., 2016; Zhao et al., 2020). Biochar can also alleviate the inhibitory effects of highly invasive tropical and temperate species on the growth of native plants (Sujeeun & Thomas, 2022, 2023a). Biochar has been shown to mitigate the allelopathic effects of the leaf litter of *Acer platanoides*, a common invasive species in urban plant communities in North America; biochar had large positive responses on native species by reducing the availability of allelochemicals released by *A. platanoides* (Sujeeun & Thomas, 2023a). In urban plant communities dominated by allelopathic invasive species, biochar addition can be an effective tool to reduce the spread of invaders by reducing their competitive advantage over native plants.

5 Conclusion and Future Perspectives

Invasive plants are becoming a growing threat to native plant communities in urban environments as a result of the rapid expansion of human populations in urban areas. Urbanization has facilitated plant invasions due to increased habitat fragmentation, soil disturbance, and climatic changes. Invasive species also have several characteristics that allow them to exploit the modified habitat, including large persistent seed banks, high phenotypic plasticity, and allelopathy. Passive recovery of native plant communities is unlikely to succeed in urban environments. Instead, active restoration strategies are necessary to decrease the spread of invaders and enhance the resilience of native communities against invasion. After removing invasive plants from a site, native species can be planted to increase the number of competitors and reduce the resources that are available to invaders. The addition of native species is particularly important where human disturbance has created highly fragmented landscapes that prevent the natural colonization of native plants from other populations. The resilience of native plant communities can be further enhanced by using species with traits that promote their competitive ability and tolerance to urban conditions. The spread of invasive plants can potentially be mitigated by reducing their seed abundance through herbicide application, tillage, and fire. Restoration of native plant communities must include measures to decrease the seed bank of invaders whenever possible to decrease competition from seedlings of invasive species. The use of soil amendments is another potential approach to decrease soil N availability and reduce soil contaminant concentrations, which can reduce the competitive advantage of invasive plants that are favored over native species under elevated N levels and high concentrations of salts and heavy metals. For example, biochar can be used to alleviate the negative effects of several allelopathic invasive plants that occur in urban areas globally, thus reducing their spread and promoting native species. Overall, this chapter has demonstrated that active management strategies are highly effective at promoting native plant species and reducing the establishment of invasive plants in urban ecosystems. The success of active restoration approaches will depend on the environmental factors facilitating invasion, composition of the native plant community, and identity of the invaders. Further research is necessary to determine which native species traits, soil amendments, and method(s) of reducing invasive soil seed banks are most likely to succeed in different native plant communities.

References

Afreen, T., Kumari, S., Bhadouria, R., Devi, R. S., Singh, S., & Tripathi, S. (2023). Plant invasion and soil processes: A mechanistic understanding. In S. Tripathi, R. Bhadouria, P. Srivastava, R. Singh & D. R. Batish (Eds.), *Plant invasions and global climate change* (pp. 227–246). Springer Nature Singapore.

Ali, S., Rizwan, M., Qayyum, M. F., Ok, Y. S., Ibrahim, M., Riaz, M., Arif, M. S., Hafeez, F., Al-Wabel, M. I., & Shahzad, A. N. (2017). Biochar soil amendment on alleviation of drought and salt stress in plants: A critical review. *Environmental Science and Pollution Research, 24*(14), 12700–12712. https://doi.org/10.1007/s11356-017-8904-x

Amini, S., Ghadiri, H., Chen, C., & Marschner, P. (2016). Salt-affected soils, reclamation, carbon dynamics, and biochar: A review. *Journal of Soils and Sediments, 16*(3), 939–953. https://doi.org/10.1007/s11368-015-1293-1

Annapurna, C., & Singh, J. S. (2003). Phenotypic plasticity and plant invasiveness: Case study of congress grass. *Current Science, 85*(2), 197–201.

Annighöfer, P., Schall, P., Kawaletz, H., Mölder, I., Terwei, A., Zerbe, S., & Ammer, C. (2012). Vegetative growth response of black cherry (*Prunus serotina*) to different mechanical control methods in a biosphere reserve. *Canadian Journal of Forest Research, 42*(12), 2037–2051. https://doi.org/10.1139/cjfr-2012-0257

Ariori, C., Aiello-Lammens, M. E., Silander, J. A. Jr. (2017). Plant invasion along an urban-to-rural gradient in northeast Connecticut. *Journal of Urban Ecology, 3*(1):jux008. https://doi.org/10.1093/jue/jux008

Averett, J. M., Klips, R. A., Nave, L. E., Frey, S. D., & Curtis, P. S. (2004). Effects of soil carbon amendment on nitrogen availability and plant growth in an experimental tallgrass prairie restoration. *Restoration Ecology, 12*(4), 568–574. https://doi.org/10.1111/j.1061-2971.2004.00284.x

Bakker, J. D., Wilson, S. D., Christian, J. M., Li, X., Ambrose, L. G., & Waddington, J. (2003). Contingency of grassland restoration on year, site, and competition from introduced grasses.

Ecological Applications, 13(1), 137–153. https://doi.org/10.1890/1051-0761(2003)013[0137:COGROY]2.0.CO;2

Barton, A. M., Brewster, L. B., Cox, A. N., & Prentiss, N. K. (2004). Non-indigenous woody invasive plants in a rural New England town. *Biological Invasions, 6*(2), 205–211. https://doi.org/10.1023/B:BINV.0000022139.03833.e5

Bauer, J. T., & Reynolds, H. L. (2016). Restoring native understory to a woodland invaded by *Euonymus fortunei*: Multiple factors affect success. *Restoration Ecology, 24*(1), 45–52. https://doi.org/10.1111/rec.12285

Bhattacharjee, J., Taylor, J. P., Smith, L. M., & Haukos, D. A. (2009). Seedling competition between native cottonwood and exotic saltcedar: Implications for restoration. *Biological Invasions, 11*(8), 1777–1787. https://doi.org/10.1007/s10530-008-9357-4

Bidwell, S., Attiwill, P. M., & Adams, M. A. (2006). Nitrogen availability and weed invasion in a remnant native woodland in urban Melbourne. *Austral Ecology, 31*(2), 262–270. https://doi.org/10.1111/j.1442-9993.2006.01575.x

Bierzychudek, P. (2020). Passive recovery of an urban forest in the Pacific Northwest after removal of invasive plants. *Urban Ecosyst, 23*(5), 1023–1038. https://doi.org/10.1007/s11252-020-009 66-8

Blackburn, T. M., Pyšek, P., Bacher, S., Carlton, J. T., Duncan, R. P., Jarošík, V., Wilson, J. R. U., & Richardson, D. M. (2011). A proposed unified framework for biological invasions. *Trends in Ecology & Evolution, 26*(7), 333–339. https://doi.org/10.1016/j.tree.2011.03.023

Blumenthal, D. M., Jordan, N. R., & Russelle, M. P. (2003). Soil carbon addition controls weeds and facilitates prairie restoration. *Ecological Applications, 13*(3), 605–615. https://doi.org/10.1890/1051-0761(2003)013[0605:SCACWA]2.0.CO;2

Boxriker, M., Privett, S. D. J., Richardson, D. M., & Gaertner, M. (2022). Active restoration in South African fynbos—A long-term perspective from the Agulhas Plain. *Transactions of the Royal Society of South Africa, 77*(2), 133–143. https://doi.org/10.1080/0035919X.2022.2087116

Bozkurt, S. G., & Basaraner, M. (2024). Spatio-temporal investigation of urbanization and its impact on habitat fragmentation in natural ecosystems of Istanbul using Shannon's entropy and landscape metrics in GIS. *Environment, Development and Sustainability, 26*(10), 26879–26907.

Brändli, R. C., Hartnik, T., Henriksen, T., & Cornelissen, G. (2008). Sorption of native polyaromatic hydrocarbons (PAH) to black carbon and amended activated carbon in soil. *Chemosphere, 73*(11), 1805–1810. https://doi.org/10.1016/j.chemosphere.2008.08.034

Brooks, M. L. (2003). Effects of increased soil nitrogen on the dominance of alien annual plants in the Mojave Desert. *Journal of Applied Ecology, 40*(2), 344–353. https://doi.org/10.1046/j.1365-2664.2003.00789.x

Brown, V. S., Ritchie, A. L., Stevens, J. C., Hanks, T. D., Hobbs, R. J., & Erickson, T. E. (2023). Seed positioning in extruded pellets: Does it matter? *Restoration Ecology, 31*(1), Article e13784. https://doi.org/10.1111/rec.13784

Bruckman, V. J., & Pumpanen, J. (2019). Chapter 17—Biochar use in global forests: Opportunities and challenges. In M. Busse, C. P. Giardina, D. M. Morris, & D. S. Page-Dumroese (Eds.), *Developments in soil science* (pp. 427–453). Elsevier.

Bucharova, A., & Krahulec, F. (2020). Native seed addition as an effective tool for post-invasion restoration. *Basic and Applied Ecology, 42*, 54–61. https://doi.org/10.1016/j.baae.2020.01.002

Byun, C., Kettenring, K. M., Tarsa, E. E., & de Blois, S. (2023). Applying ecological principles to maximize resistance to invasion in restored plant communities. *Ecological Engineering, 190*, Article 106926. https://doi.org/10.1016/j.ecoleng.2023.106926

Callaway, R. M., & Hierro, J. L. (2006). Resistance and susceptibility of plant communities to invasion: Revisiting Rabotnov's ideas about community homestasis. In N. Pedrol & L. González (Eds.), *Reigosa MJ* (pp. 395–414). Kluwer Academic Publishers.

Charles, H., & Dukes, J. S. (2007). Impacts of invasive species on ecosystem services. In W. Nentwig (Ed.), *Biological invasions* (pp. 217–237). Springer.

Chun, Y. J., Collyer, M. L., Moloney, K. A., & Nason, J. D. (2007). Phenotypic plasticity of native vs. invasive purple loosestrife: A two-state multivariate approach. *Ecology, 88*(6):1499–1512. https://doi.org/10.1890/06-0856

Chytrý, M., Jarošík, V., Pyšek, P., Hájek, O., Knollová, I., Tichý, L., & Danihelka, J. (2008). Separating habitat invasibility by alien plants from the actual level of invasion. *Ecology, 89*(6), 1541–1553. https://doi.org/10.1890/07-0682.1

Cilliers, S. S., Williams, N. S. G., & Barnard, F. J. (2008). Patterns of exotic plant invasions in fragmented urban and rural grasslands across continents. *Landscape Ecology, 23*(10), 1243–1256. https://doi.org/10.1007/s10980-008-9295-7

Cipollini, D. (2016). A review of garlic mustard (*Alliaria petiolata*, Brassicaceae) as an allelopathic plant. *tbot, 143*(4):339–348. https://doi.org/10.3159/TORREY-D-15-00059

Clark, J. S., Lewis, M., & Horvath, L. (2001). Invasion by extremes: Population spread with variation in dispersal and reproduction. *The American Naturalist, 157*(5), 537–554. https://doi.org/10.1086/319934

Clements, H., & Bierzychudek, P. (2017). Can the persistent seed bank contribute to the passive restoration of urban forest fragments after invasive species removal? *Ecological Restoration, 35*(2), 156–166. https://doi.org/10.3368/er.35.2.156

Clenet, D. R., Davies, K. W., Johnson, D. D., & Kerby, J. D. (2019). Native seeds incorporated into activated carbon pods applied concurrently with indaziflam: A new strategy for restoring annual-invaded communities? *Restoration Ecology, 27*(4), 738–744. https://doi.org/10.1111/rec.12927

Crawley, M. J., Harvey, P. H., & Purvis, A. (1996). Comparative ecology of the native and alien floras of the British Isles. *Philosophical Transactions of the Royal Society B, 351*(1345), 1251–1259.

Cronk, Q. C. B., & Fuller, J. L. (1995). *Plant invaders: The threat to natural ecosystems*. Routledge.

Daehler, C. C. (2003). Performance comparisons of co-occurring native and alien invasive plants: Implications for conservation and restoration. *Annual Review of Ecology Evolution and Systematics, 34*, 183–211.

D'Antonio, C., & Meyerson, L. A. (2002). Exotic plant species as problems and solutions in ecological restoration: A synthesis. *Restoration Ecology, 10*(4), 703–713. https://doi.org/10.1046/j.1526-100X.2002.01051.x

Davidson, A. M., Jennions, M., & Nicotra, A. B. (2011). Do invasive species show higher phenotypic plasticity than native species and if so, is it adaptive? A meta-analysis. *Ecology Letters, 14*(4), 419–431. https://doi.org/10.1111/j.1461-0248.2011.01596.x

Davis, A., Major, R. E., & Taylor, C. E. (2016). Do trees flower longer in the city? A comparison of flowering eucalyptus trees in streets, remnants and continuous forest and their association with nectarivorous birds. *Urban Ecosystem, 19*(2), 735–747. https://doi.org/10.1007/s11252-015-0515-z

Davis, M. A., Grime, J. P., & Thompson, K. (2000). Fluctuating resources in plant communities: A general theory of invasibility. *Journal of Ecology, 88*(3), 528–534. https://doi.org/10.1046/j.1365-2745.2000.00473.x

Daws, M. I., Hall, J., Flynn, S., & Pritchard, H. W. (2007). Do invasive species have bigger seeds? Evidence from intra- and inter-specific comparisons. *South+A1443 African Journal of Botany, 73*(1), 138–143. https://doi.org/10.1016/j.sajb.2006.09.003

Deutschewitz, K., Lausch, A., Kühn, I., & Klotz, S. (2003). Native and alien plant species richness in relation to spatial heterogeneity on a regional scale in Germany. *Global Ecology and Biogeography, 12*(4), 299–311. https://doi.org/10.1046/j.1466-822X.2003.00025.x

DiTomaso, J. M. (2000). Invasive weeds in rangelands: Species, impacts, and management. *Weed Science, 48*(2), 255–265. https://doi.org/10.1614/0043-1745(2000)048[0255:IWIRSI]2.0.CO;2

Drake, D. R. (1998). Relationships among the seed rain, seed bank and vegetation of a Hawaiian forest. *Journal of Vegetation Science, 9*(1), 103–112. https://doi.org/10.2307/3237228

Dukes, J. S. (2001). Biodiversity and invasibility in grassland microcosms. *Oecologia, 126*(4), 563–568. https://doi.org/10.1007/s004420000549

El-Barougy, R. F., Dakhil, M. A., Abdelaal, M., El-Keblawy, A., & Bersier, L.-F. (2021). Trait-environment relationships reveal the success of alien plants invasiveness in an urbanized landscape. *Plants, 10*(8), 1519. https://doi.org/10.3390/plants10081519

Emery, S. M. (2007). Limiting similarity between invaders and dominant species in herbaceous plant communities? *Journal of Ecology, 95*(5), 1027–1035. https://doi.org/10.1111/j.1365-2745.2007.01274.x

Eschen, R., Mortimer, S. R., Lawson, C. S., Edwards, A. R., Brook, A. J., Igual, J. M., Hedlund, K., & Schaffner, U. (2007). Carbon addition alters vegetation composition on ex-arable fields. *Journal of Applied Ecology, 44*(1), 95–104. https://doi.org/10.1111/j.1365-2664.2006.01240.x

Farahat, E., Galal, T., El-Midany, M., & Hassan, L. (2015). Effect of urban habitat heterogeneity on functional traits plasticity of the invasive species *Calotropis procera* (Aiton) W.T. Aiton. *Rendconti Fis Acc Lincei, 26*(2), 193–201. https://doi.org/10.1007/s12210-015-0408-3

Fargione, J., Brown, C. S., & Tilman, D. (2003). Community assembly and invasion: An experimental test of neutral versus niche processes. *Proceedings of the National Academy of Sciences, 100*(15), 8916–8920. https://doi.org/10.1073/pnas.1033107100

Fenner, M., & Thompson, K. (2005). *The ecology of seeds*. Cambridge University Press.

Ferreras, A. E., & Galetto, L. (2010). From seed production to seedling establishment: Important steps in an invasive process. *Acta Oecologica, 36*(2), 211–218. https://doi.org/10.1016/j.actao.2009.12.005

Figueroa, J. A., Saldías, G., Teillier, S., Carrera, E., & Castro, S. A. (2020). Seed banks in urban vacant lots of a Latin American megacity are easily germinable and strongly dominated by exotic flora. *Urban Ecosystem, 23*(5), 945–955. https://doi.org/10.1007/s11252-020-00986-4

Forman, R. T. T. (1995). *Land mosaics: The ecology of landscapes and regions*. Cambridge University Press.

Francis, R. A., & Chadwick, M. A. (2015). Urban invasions: Non-native and invasive species in cities. *Geography, 100*(3), 144–151.

Funk, J. L., Cleland, E. E., Suding, K. N., & Zavaleta, E. S. (2008). Restoration through reassembly: Plant traits and invasion resistance. *Trends in Ecology & Evolution, 23*(12), 695–703. https://doi.org/10.1016/j.tree.2008.07.013

Gaertner, M., Den Breeyen, A., Hui, C., & Richardson, D. M. (2009). Impacts of alien plant invasions on species richness in Mediterranean-type ecosystems: A meta-analysis. *Progress in Physical Geography, 33*(3), 319–338. https://doi.org/10.1177/0309133309341607

Gavier-Pizarro, G. I., Radeloff, V. C., Stewart, S. I., Huebner, C. D., & Keuler, N. S. (2010). Housing is positively associated with invasive exotic plant species richness in New England, USA. *Ecological Applications, 20*(7), 1913–1925. https://doi.org/10.1890/09-2168.1

Gilbert, B., & Lechowicz, M. J. (2005). Invasibility and abiotic gradients: The positive correlation between native and exotic plant diversity. *Ecology, 86*(7), 1848–1855. https://doi.org/10.1890/04-09997

Gioria, M., Pyšek, P., & Moravcová, L. (2012). Soil seed banks in plant invasions: Promoting species invasiveness and long-term impact on plant community dynamics. *Preslia, 84*(2), 327–350.

Godefroid, S. (2001). Temporal analysis of the Brussels flora as indicator for changing environmental quality. *Landscape and Urban Planning, 52*(4), 203–224. https://doi.org/10.1016/S0169-2046(00)00117-1

Granata, M. U., Bracco, F., & Catoni, R. (2020). Phenotypic plasticity of two invasive alien plant species inside a deciduous forest in a strict nature reserve in Italy. *Journal of Sustainable Forestry, 39*(4), 346–364. https://doi.org/10.1080/10549811.2019.1670678

Grella, C., Renshaw, A., & Wright, I. A. (2018). Invasive weeds in urban riparian zones: The influence of catchment imperviousness and soil chemistry across an urbanization gradient. *Urban Ecosystem, 21*(3), 505–517. https://doi.org/10.1007/s11252-018-0736-z

Gundale, M. J., & DeLuca, T. H. (2006). Temperature and source material influence ecological attributes of ponderosa pine and Douglas-fir charcoal. *Forest Ecology and Management, 231*(1–3), 86–93. https://doi.org/10.1016/j.foreco.2006.05.004

Guo, Y.-J., Wang, S.-J., Xu, Y.-H., Liu, Y.-Y., Yan, X.-L., & Wang, Y.-J. (2024). Maternal effects of human-mediated disturbances on the growth and reproductive traits of different ecotypes of an annual invasive plant *Erigeron annuus*. *Environmental and Experimental Botany, 218*, Article 105588. https://doi.org/10.1016/j.envexpbot.2023.105588

Gurevitch, J., & Padilla, D. K. (2004). Are invasive species a major cause of extinctions? *Trends in Ecology & Evolution, 19*(9), 470–474. https://doi.org/10.1016/j.tree.2004.07.005

Hagan, D. L., Jose, S., & Lin, C.-H. (2013). Allelopathic exudates of cogongrass (*Imperata cylindrica*): Implications for the performance of native pine savanna plant species in the Southeastern US. *Journal of Chemical Ecology, 39*(2), 312–322. https://doi.org/10.1007/s10886-013-0241-z

Hale, S. E., Hanley, K., Lehmann, J., Zimmerman Andrew, R., & Cornelissen, G. (2011). Effects of chemical, biological, and physical aging as well as soil addition on the sorption of pyrene to activated carbon and biochar. *Environmental Science and Technology, 45*(24), 10445–10453.https://doi.org/10.1021/es202970x

Hall, K. E., Calderon, M. J., Spokas, K. A., Cox, L., Koskinen, W. C., Novak, J., & Cantrell, K. (2014). Phenolic acid sorption to biochars from mixtures of feedstock materials. *Water, Air, and Soil Pollution, 225*(7), 2031. https://doi.org/10.1007/s11270-014-2031-9

Hall, S. A., Holmes, P. M., Gaertner, M., & Esler, K. J. (2021). Active seed sowing can overcome constraints to passive restoration of a critically endangered vegetation type. *South+A1443 African Journal of Botany, 138*, 249–261. https://doi.org/10.1016/j.sajb.2020.12.020

Hallett, L. M., Chapple, D. E., Bickart, N., Cherbowsky, A., Fernandez, L., Ho, C. H., Alexander, M., Schwab, K., & Suding, K. N. (2017). Trait complementarity enhances native plant restoration in an invaded urban landscape. *Ecological Restoration, 35*(2), 148–155. https://doi.org/10.3368/er.35.2.148

Hansen, M. J., & Clevenger, A. P. (2005). The influence of disturbance and habitat on the presence of non-native plant species along transport corridors. *Biological Conservation, 125*(2), 249–259. https://doi.org/10.1016/j.biocon.2005.03.024

Hassan, M. O., & Mohamed, H. Y. (2020). Allelopathic interference of the exotic naturalized *Paspalum dilatatum* Poir. threatens diversity of native plants in urban gardens. *Flora, 266*, 151593. https://doi.org/10.1016/j.flora.2020.151593

Hazelton, E. L. G., Downard, R., Kettenring, K. M., McCormick, M. K., & Whigham, D. F. (2018). Spatial and temporal variation in Brackish Wetland seedbanks: Implications for wetland restoration following phragmites control. *Estuaries and Coasts: Journal of the Estuarine Research Federation, 41*(1), 68–84. https://doi.org/10.1007/s12237-017-0289-z

Hejda, M., Pyšek, P., & Jarošík, V. (2009). Impact of invasive plants on the species richness, diversity and composition of invaded communities. *Journal of Ecology, 97*(3), 393–403. https://doi.org/10.1111/j.1365-2745.2009.01480.x

Henderson, S., Dawson, T. P., & Whittaker, R. J. (2006). Progress in invasive plants research. *Progress in Physical Geography, 30*(1), 25–46. https://doi.org/10.1191/0309133306pp468ra

Holl, K. D., & Aide, T. M. (2011). When and where to actively restore ecosystems? *Forest Ecology Management, 261*(10), 1558–1563. https://doi.org/10.1016/j.foreco.2010.07.004

Holmes, P. M. (1989). Effects of different clearing treatments on the seed-bank dynamics of an invasive Australian shrub, *Acacia cyclops*, in the Southwestern Cape, South Africa. *Forest Ecology Management, 28*(1), 33–46. https://doi.org/10.1016/0378-1127(89)90072-8

Holmes, P. M. (2001). Shrubland restoration following woody alien invasion and mining: Effects of topsoil depth, seed source, and fertilizer addition. *Restoration Ecology, 9*(1), 71–84. https://doi.org/10.1046/j.1526-100x.2001.009001071.x

Hu, G., & Zhang, Z. (2012). Allelopathic effects of *Chromolaena odorata* on native and non-native invasive herbs. *Journal of Food, Agriculture and Environment, 11*(1), 878–882.

Hulme, P. E. (2009). Trade, transport and trouble: Managing invasive species pathways in an era of globalization. *Journal of Applied Ecology, 46*(1), 10–18. https://doi.org/10.1111/j.1365-2664.2008.01600.x

Hulme, P. E. (2006). Beyond control: Wider implications for the management of biological invasions. *Journal of Applied Ecology, 43*(5), 835–847. https://doi.org/10.1111/j.1365-2664.2006.01227.x

Iannone, B. V., III., & Galatowitsch, S. M. (2008). Altering light and soil N to limit *Phalaris arundinacea* reinvasion in sedge meadow restorations. *Restoration Ecology, 16*(4), 689–701. https://doi.org/10.1111/j.1526-100X.2008.00481.x

Ibáñez, I., Liu, G., Petri, L., Schaffer-Morrison, S., & Schueller, S. (2021). Assessing vulnerability and resistance to plant invasions: A native community perspective. *Invasive Plant Science and Management, 14*(2), 64–74. https://doi.org/10.1017/inp.2021.15

Jochner, S., & Menzel, A. (2015). Urban phenological studies—Past, present, future. *Environmental Pollution, 203*, 250–261. https://doi.org/10.1016/j.envpol.2015.01.003

Johnston-Bates, J., Capon, S. J., & Castley, J. G. (2025). Propagule bank dynamics in urban subtropical forest remnants: Edge effects, fragmentation and plant invasion. *Urban for Urban Green, 103*, Article 128563.

Kalisz, S., Kivlin, S. N., & Bialic-Murphy, L. (2021). Allelopathy is pervasive in invasive plants. *Biological Invasions, 23*(2), 367–371. https://doi.org/10.1007/s10530-020-02383-6

Kargar, M., Clark, O. G., Hendershot, W. H., Jutras, P., & Prasher, S. O. (2015). Immobilization of trace metals in contaminated urban soil amended with compost and biochar. *Water, Air, and Soil Pollution, 226*(6), 191. https://doi.org/10.1007/s11270-015-2450-2

Kato-Noguchi, H., & Kurniadie, D. (2021). Allelopathy of *Lantana camara* as an invasive plant. *Plants, 10*(5), 1028. https://doi.org/10.3390/plants10051028

Kettenring, K. M., & Adams, C. R. (2011). Lessons learned from invasive plant control experiments: A systematic review and meta-analysis. *Journal of Applied Ecology, 48*(4), 970–979. https://doi.org/10.1111/j.1365-2664.2011.01979.x

King, C. M., & Hovick, S. M. (2020). Wetland plant community variation across replicate urban to rural gradients: Non-native species as both drivers and passengers in systems impacted by anthropogenic land-use. *Urban Ecosystems, 23*(6), 1209–1226. https://doi.org/10.1007/s11252-020-01012-3

Kobayashi, T., Ikeda, H., & Hon, Y. (1999). Growth analysis and reproductive allocation of Japanese forbs and grasses in relation to organ toughness under trampling. *Plant Biology, 1*(4), 445–452. https://doi.org/10.1111/j.1438-8677.1999.tb00727.x

Komuro, T., & Koike, F. (2005). Colonization by woody plants in fragmented habitats of a suburban landscape. *Ecological Applications, 15*(2), 662–673. https://doi.org/10.1890/03-5232

Kowarik, I., & Böcker, R. (1984). Zur verbreitung, vergesellschaftung und einbürgerung des götterbaumes (*Ailanthus altissima* [Mill.] Swingle) in Mitteleuropea. *Tuexenia, 4*, 9–29.

Kühn, I., & Klotz, S. (2006). Urbanization and homogenization–Comparing the floras of urban and rural areas in Germany. *Biological Conservation, 127*(3), 292–300. https://doi.org/10.1016/j.biocon.2005.06.033

Kühn, I., Wolf, J., & Schneider, A. (2017). Is there an urban effect in alien plant invasions? *Biological Invasions, 19*(12), 3505–3513. https://doi.org/10.1007/s10530-017-1591-1

Kurylo, J. S., Knight, K. S., Stewart, J. R., & Endress, A. G. (2007). *Rhamnus cathartica*: Native and naturalized distribution and habitat preferences. *The Journal of the Torrey Botanical Society, 134*(3), 420–430. https://doi.org/10.3159/1095-5674(2007)134[420:RCNAND]2.0.CO;2

Laughlin, D. C. (2014). Applying trait-based models to achieve functional targets for theory-driven ecological restoration. *Ecology Letters, 17*(7), 771–784. https://doi.org/10.1111/ele.12288

Lehan, N. E., Murphy, J. R., Thorburn, L. P., & Bradley, B. A. (2013). Accidental introductions are an important source of invasive plants in the continental United States. *American Journal of Botany, 100*(7), 1287–1293. https://doi.org/10.3732/ajb.1300061

Li, W., Luo, J., Tian, X., Soon Chow, W., Sun, Z., Zhang, T., Peng, S., & Peng, C. (2015). A new strategy for controlling invasive weeds: Selecting valuable native plants to defeat them. *Science and Reports, 5*(1), 11004. https://doi.org/10.1038/srep11004

Lockwood, J. L., Cassey, P., & Blackburn, T. (2005). The role of propagule pressure in explaining species invasions. *Trends in Ecology & Evolution, 20*(5), 223–228. https://doi.org/10.1016/j.tree.2005.02.004

Lodhi, M. A. K. (1978). Allelopathic effects of decaying litter of dominant trees and their associated soil in a lowland forest community. *American Journal of Botany, 65*(3), 340–344. https://doi.org/10.2307/2442275

Londe, V., de Sousa, H. C., & Kozovits, A. R. (2017). Exotic and invasive species compromise the seed bank and seed rain dynamics in forests undergoing restoration at urban regions. *Journal of Forest Research, 28*(5), 1019–1026. https://doi.org/10.1007/s11676-017-0370-2

Lopez, B. E., Urban, D., & White, P. S. (2018). Testing the effects of four urbanization filters on forest plant taxonomic, functional, and phylogenetic diversity. *Ecological Applications, 28*(8), 2197–2205. https://doi.org/10.1002/eap.1812

Lososová, Z., Chytrý, M., Tichý, L., Danihelka, J., Fajmon, K., Hájek, O., Kintrová, K., Láníková, D., Otýpková, Z., & Řehořek, V. (2012). Biotic homogenization of Central European urban floras depends on residence time of alien species and habitat types. *Biological Conservation, 145*(1), 179–184. https://doi.org/10.1016/j.biocon.2011.11.003

Macarthur, R., & Levins, R. (1967). The limiting similarity, convergence, and divergence of coexisting species. *The American Naturalist, 101*(921), 377–385.

Maclean, J. E., Mitchell, R. J., Burslem, D. F. R. P., Genney, D., Hall, J., & Pakeman, R. J. (2018). Invasion by *Rhododendron ponticum* depletes the native seed bank with long-term impacts after its removal. *Biological Invasions, 20*(2), 375–384. https://doi.org/10.1007/s10530-017-1538-6

Mahall, B. E., & Callaway, R. M. (1992). Root communication mechanisms and intracommunity distributions of two Mojave desert shrubs. *Ecology, 73*(6), 2145–2151. https://doi.org/10.2307/1941462

Malavasi, M., Carboni, M., Cutini, M., Carranza, M. L., & Acosta, A. T. R. (2014). Landscape fragmentation, land-use legacy and propagule pressure promote plant invasion on coastal dunes: A patch-based approach. *Landscape Ecology, 29*(9), 1541–1550. https://doi.org/10.1007/s10980-014-0074-3

Marushia, R. G., & Allen, E. B. (2011). Control of exotic annual grasses to restore native forbs in abandoned agricultural land. *Restoration Ecology, 19*(1), 45–54. https://doi.org/10.1111/j.1526-100X.2009.00540.x

Mason, R. A. B., Cooke, J., Moles, A. T., & Leishman, M. R. (2008). Reproductive output of invasive versus native plants. *Global Ecology and Biogeography, 17*(5), 633–640. https://doi.org/10.1111/j.1466-8238.2008.00402.x

Mavimbela, L. Z., Sieben, E. J. J., & Procheş, Ş. (2018). Invasive alien plant species, fragmentation and scale effects on urban forest community composition in Durban, South Africa. *New Zealand Journal of Forest Science, 48*(1), 19. https://doi.org/10.1186/s40490-018-0124-8

McCay, T. S., & McCay, D. H. (2009). Processes regulating the invasion of European buckthorn (*Rhamnus cathartica*) in three habitats of the northeastern United States. *Biological Invasions, 11*(8), 1835–1844. https://doi.org/10.1007/s10530-008-9362-7

McKinney, M. L. (2006). Urbanization as a major cause of biotic homogenization. *Biological Conservation, 127*(3), 247–260. https://doi.org/10.1016/j.biocon.2005.09.005

McKinney, M. L. (2002). Urbanization, biodiversity, and conservation: The impacts of urbanization on native species are poorly studied, but educating a highly urbanized human population about these impacts can greatly improve species conservation in all ecosystems. *BioScience, 52*(10), 883–890. https://doi.org/10.1641/0006-3568(2002)052[0883:UBAC]2.0.CO;2

Middleton, E. L., Bever, J. D., & Schultz, P. A. (2010). The effect of restoration methods on the quality of the restoration and resistance to invasion by exotics. *Restoration Ecology, 18*(2), 181–187. https://doi.org/10.1111/j.1526-100X.2008.00501.x

Mostert, E., Gaertner, M., Holmes, P. M., O'Farrell, P. J., & Richardson, D. M. (2018). A multi-criterion approach for prioritizing areas in urban ecosystems for active restoration following invasive plant control. *Environmental Management, 62*(6), 1150–1167. https://doi.org/10.1007/s00267-018-1103-9

Nguyen, T. T. N., Xu, C.-Y., Tahmasbian, I., Che, R., Xu, Z., Zhou, X., Wallace, H. M., & Bai, S. H. (2017). Effects of biochar on soil available inorganic nitrogen: A review and meta-analysis. *Geoderma, 288*, 79–96. https://doi.org/10.1016/j.geoderma.2016.11.004

Nsikani, M. M., van Wilgen, B. W., & Gaertner, M. (2018). Barriers to ecosystem restoration presented by soil legacy effects of invasive alien N2-fixing woody species: Implications for ecological restoration. *Restoration Ecology, 26*(2), 235–244. https://doi.org/10.1111/rec.12669

Oakley, C. A., & Knox, J. S. (2013). Plant species richness increases resistance to invasion by non-resident plant species during grassland restoration. *Applied Vegetation Science, 16*(1), 21–28. https://doi.org/10.1111/j.1654-109X.2012.01202.x

Ogden, J. A. E., & Rejmánek, M. (2005). Recovery of native plant communities after the control of a dominant invasive plant species, *Foeniculum vulgare*: Implications for management. *Biological Conservation, 125*(4), 427–439. https://doi.org/10.1016/j.biocon.2005.03.025

Padayachee, A. L., Irlich, U. M., Faulkner, K. T., Gaertner, M., Procheş, Ş, Wilson, J. R. U., & Rouget, M. (2017). How do invasive species travel to and through urban environments? *Biological Invasions, 19*(12), 3557–3570. https://doi.org/10.1007/s10530-017-1596-9

Pan, X., Geng, Y., Zhang, W., Li, B., & Chen, J. (2006). The influence of abiotic stress and phenotypic plasticity on the distribution of invasive *Alternanthera philoxeroides* along a riparian zone. *Acta Oecologica, 30*(3), 333–341. https://doi.org/10.1016/j.actao.2006.03.003

Passos, I., Marchante, H., Pinho, R., & Marchante, E. (2017). What we don't seed: The role of long-lived seed banks as hidden legacies of invasive plants. *Plant Ecology, 218*(11), 1313–1324. https://doi.org/10.1007/s11258-017-0770-6

Perry, L. G., Blumenthal, D. M., Monaco, T. A., Paschke, M. W., & Redente, E. F. (2010). Immobilizing nitrogen to control plant invasion. *Oecologia, 163*(1), 13–24. https://doi.org/10.1007/s00442-010-1580-x

Perry, L. G., Galatowitsch, S. M., & Rosen, C. J. (2004). Competitive control of invasive vegetation: A native wetland sedge suppresses *Phalaris arundinacea* in carbon-enriched soil. *Journal of Applied Ecology, 41*(1), 151–162. https://doi.org/10.1111/j.1365-2664.2004.00871.x

Pickett, S. T. A., Cadenasso, M. L., Grove, J. M., Boone, C. G., Groffman, P. M., Irwin, E., Kaushal, S. S., Marshall, V., McGrath, B. P., Nilon, C. H., Pouyat, R. V., Szlavecz, K., Troy, A., & Warren, P. (2011). Urban ecological systems: Scientific foundations and a decade of progress. *Journal of Environmental Management, 92*(3), 331–362. https://doi.org/10.1016/j.jenvman.2010.08.022

Pieterse, P. J., & Cairns, A. L. P. (1988). The population dynamics of the weed *Acacia longifolia* (Fabacae) in the absence and presence of fire. *Southern African Forestry Journal, 145*(1), 25–27. https://doi.org/10.1080/00382167.1988.9630330

Pimentel, D., Lach, L., Zuniga, R., & Morrison, D. (2000). Environmental and economic costs of nonindigenous species in the United States. *BioScience, 50*(1), 53–65. https://doi.org/10.1641/0006-3568(2000)050[0053:EAECON]2.3.CO;2

Potgieter, L. J., Gaertner, M., O'Farrell, P. J., & Richardson, D. M. (2019). Perceptions of impact: Invasive alien plants in the urban environment. *Journal of Environmental Management, 229*, 76–87. https://doi.org/10.1016/j.jenvman.2018.05.080

Powell, K. I., Chase, J. M., & Knight, T. M. (2011). A synthesis of plant invasion effects on biodiversity across spatial scales. *American Journal of Botany, 98*(3), 539–548. https://doi.org/10.3732/ajb.1000402

Prabakaran, K., Li, J., Anandkumar, A., Leng, Z., Zou, C. B., & Du, D. (2019). Managing environmental contamination through phytoremediation by invasive plants: A review. *Ecological Engineering, 138*, 28–37. https://doi.org/10.1016/j.ecoleng.2019.07.002

Prasser, N., & Zedler, J. B. (2010). Salt tolerance of invasive *Phalaris arundinacea* exceeds that of native *Carex stricta* (Wisconsin). *Ecological Restoration, 28*(3), 238–240.

Presotto, A., Hernández, F., Casquero, M., Vercellino, R., Pandolfo, C., Poverene, M., & Cantamutto, M. (2020). Seed bank dynamics of an invasive alien species, *Helianthus Annuus* l. *J Plant Ecol, 13*(3), 313–322. https://doi.org/10.1093/jpe/rtaa016

Prober, S. M., Thiele, K. R., Lunt, I. D., & Koen, T. B. (2005). Restoring ecological function in temperate grassy woodlands: Manipulating soil nutrients, exotic annuals and native perennial grasses through carbon supplements and spring burns. *Journal of Applied Ecology, 42*(6), 1073–1085. https://doi.org/10.1111/j.1365-2664.2005.01095.x

Pyšek, P. (1995). *Approaches to studying spontaneous settlement flora and vegetation in central Europe: A review* (pp. 23–29). Urban Ecology As the Basis of Urban Planning. SPB Academic Publishing.

Pyšek, P., Jarošík, V., Hulme, P. E., Pergl, J., Hejda, M., Schaffner, U., & Vilà, M. (2012). A global assessment of invasive plant impacts on resident species, communities and ecosystems: The interaction of impact measures, invading species' traits and environment. *Global Change Biology, 18*(5), 1725–1737. https://doi.org/10.1111/j.1365-2486.2011.02636.x

Qi, F., Kuppusamy, S., Naidu, R., Bolan, N. S., Ok, Y. S., Lamb, D., Li, Y., Yu, L., Semple, K. T., & Wang, H. (2017). Pyrogenic carbon and its role in contaminant immobilization in soils. *Critical Reviews in Environment Science and Technology, 47*(10), 795–876. https://doi.org/10.1080/10643389.2017.1328918

Qing, H., Yao, Y., Xiao, Y., Hu, F., Sun, Y., Zhou, C., & An, S. (2011). Invasive and native tall forms of *Spartina alterniflora* respond differently to nitrogen availability. *Acta Oecol, 37*(1), 23–30. https://doi.org/10.1016/j.actao.2010.11.002

Reichard, S. H. (2010). Inside out: Invasive plants and urban environments. In J. Aitkenhead-Peterson & A. Volder (Eds.), *Urban ecosystem ecology* (pp. 241–251). Wiley.

Rejmánek, M., & Richardson, D. M. (1996). What attributes make some plant species more invasive? *Ecology, 77*(6), 1655–1661. https://doi.org/10.2307/2265768

Richardson, D. M., & Kluge, R. L. (2008). Seed banks of invasive Australian Acacia species in South Africa: Role in invasiveness and options for management. *Perspectives in Plant Ecology, Evolution and Systematics, 10*(3), 161–177. https://doi.org/10.1016/j.ppees.2008.03.001

Richburg, J. A., Patterson, W. A., & Lowenstein, F. (2001). Effects of road salt and Phragmites australis invasion on the vegetation of a Western Massachusetts calcareous lake-basin fen. *Wetlands, 21*(2), 247–255. https://doi.org/10.1672/0277-5212(2001)021[0247:EORSAP]2.0.CO;2

Rickey, M. A., & Anderson, R. C. (2004). Effects of nitrogen addition on the invasive grass *Phragmites australis* and a native competitor *Spartina pectinata. Journal of Applied Ecology, 41*(5), 888–896. https://doi.org/10.1111/j.0021-8901.2004.00948.x

Riera, M., Pino, J., Sáez, L., Aymerich, P., & Melero, Y. (2024). Effect of introduction pathways on the invasion success of non-native plants along environmental gradients. *Biological Invasions, 26*(5), 1561–1580.

Rogovska, N., Laird, D., Cruse, R. M., Trabue, S., & Heaton, E. (2012). Germination tests for assessing biochar quality. *Journal of Environmental Quality, 41*(4), 1014–1022. https://doi.org/10.2134/jeq2011.0103

Rohal, C. B., Cranney, C., & Kettenring, K. M. (2019). Abiotic and landscape factors constrain restoration outcomes across spatial scales of a widespread invasive plant. *Frontiers in Plant Science, 10*, 481.

Rouget, M., & Richardson, D. M. (2003). Inferring process from pattern in plant invasions: A semimechanistic model incorporating propagule pressure and environmental factors. *The American Naturalist, 162*(6), 713–724. https://doi.org/10.1086/379204

Ruas, R. D. B., Costa, L. M. S., & Bered, F. (2022). Urbanization driving changes in plant species and communities—A global view. *Global Ecology and Conservation, 38*, Article e02243. https://doi.org/10.1016/j.gecco.2022.e02243

Ruprecht, E., Fenesi, A., & Nijs, I. (2014). Are plasticity in functional traits and constancy in performance traits linked with invasiveness? An experimental test comparing invasive and naturalized plant species. *Biological Invasions, 16*(7), 1359–1372. https://doi.org/10.1007/s10530-013-0574-0

Schuster, M. J., Wragg, P. D., & Reich, P. B. (2018). Using revegetation to suppress invasive plants in grasslands and forests. *Journal of Applied Ecology, 55*(5), 2362–2373. https://doi.org/10.1111/1365-2664.13195

Seastedt, T. R. (2015). Biological control of invasive plant species: A reassessment for the Anthropocene. *New Phytologist, 205*(2), 490–502. https://doi.org/10.1111/nph.13065

Shen, Z., Jin, F., Wang, F., McMillan, O., & Al-Tabbaa, A. (2015). Sorption of lead by Salisbury biochar produced from British broadleaf hardwood. *Bioresource Technology, 193*, 553–556. https://doi.org/10.1016/j.biortech.2015.06.111

Shochat, E., Lerman, S. B., Anderies, J. M., Warren, P. S., Faeth, S. H., & Nilon, C. H. (2010). Invasion, competition, and biodiversity loss in urban ecosystems. *BioScience, 60*(3), 199–208. https://doi.org/10.1525/bio.2010.60.3.6

Siemens, D. H., Garner, S. H., Mitchell-Olds, T., & Callaway, R. M. (2002). Cost of defense in the context of plant competition: *Brassica Rapa* may grow and defend. *Ecology, 83*(2), 505–517. https://doi.org/10.1890/0012-9658(2002)083[0505:CODITC]2.0.CO;2

Singh, K., & Byun, C. (2023). Ecological restoration after management of invasive alien plants. *Ecological Engineering, 197*, Article 107122. https://doi.org/10.1016/j.ecoleng.2023.107122

Skultety, D., & Matthews, J. W. (2017). Urbanization and roads drive non-native plant invasion in the Chicago Metropolitan region. *Biological Invasions, 19*(9), 2553–2566. https://doi.org/10.1007/s10530-017-1464-7

Smith, R. G., Maxwell, B. D., Menalled, F. D., & Rew, L. J. (2006). Lessons from agriculture may improve the management of invasive plants in wildland systems. *Frontiers in Ecology and the Environment, 4*(8), 428–434. https://doi.org/10.1890/1540-9295(2006)4[428:LFAMIT]2.0.CO;2

Šoln, K., Žnidaršič, N., & Dolenc Koce, J. (2022). Root growth inhibition and ultrastructural changes in radish root tips after treatment with aqueous extracts of *Fallopia japonica* and *F.* ×*bohemica* rhizomes. *Protoplasma, 259*(2), 343–355. https://doi.org/10.1007/s00709-021-01668-4

Stricker, K. B., Hagan, D., & Flory, S. L. (2015). Improving methods to evaluate the impacts of plant invasions: lessons from 40 years of research. *AoB PLANTS, 7*, plv028. https://doi.org/10.1093/aobpla/plv028

Su, Y., Wang, X., Wang, X., Cui, B., & Sun, X. (2019). Leaf and male cone phenophases of Chinese pine (*Pinus tabulaeformis* Carr.) along a rural-urban gradient in Beijing, China. *Urban for Urban Green, 42*, 61–71. https://doi.org/10.1016/j.ufug.2019.05.009

Sujeeun, L., & Thomas, S. C. (2022). Biochar rescues native trees in the biodiversity hotspot of Mauritius. *Forests, 13*(2), 277. https://doi.org/10.3390/f13020277

Sujeeun, L., & Thomas, S. C. (2017). Potential of biochar to mitigate allelopathic effects in tropical island invasive plants: Evidence from seed germination trials. *Tropical Conservation Science, 10*, 194008291769726. https://doi.org/10.1177/1940082917697264

Sujeeun, L., & Thomas, S. C. (2023). Biochar mitigates allelopathic effects in temperate trees. *Ecological Applications, 33*(4), Article e2832. https://doi.org/10.1002/eap.2832

Sujeeun, L., & Thomas, S. C. (2023b). Biochar: A tool for combatting both invasive species and climate change. In S. Tripathi, R. Bhadouria, P. Srivastava, R. Singh & D. R. Batish (Eds.), *Plant invasions and global climate change* (pp. 367–393). Springer Nature Singapore.

Sukopp, H., & Wurzel, A. (2003). The effects of climate change on the vegetation of central european cities. *Urban Habitats, 1*, 1.

Sun, J., He, Y., Wu, N., Javed, Q., Xiang, Y., & Du, D. (2022). Nitrogen deposition elevated the allelopathic effects of three Compositae invasive species on indigenous *Lactuca sativa*. *Polish Journal of Environmental Studies, 31*. https://doi.org/10.15244/pjoes/146940

Sun, J., Khattak, W. A., Abbas, A., Nawaz, M., Hameed, R., Javed, Q., Bo, Y., Khan, K. A., & Du, D. (2023). Ecological adaptability of invasive weeds under environmental pollutants: A review. *Environmental and Experimental Botany, 215*, Article 105492. https://doi.org/10.1016/j.envexpbot.2023.105492

Tang, L., Gao, Y., Li, B., Wang, Q., Wang, C.-H., & Zhao, B. (2014). *Spartina alterniflora* with high tolerance to salt stress changes vegetation pattern by outcompeting native species. *Ecosphere, 5*(9):art116. https://doi.org/10.1890/ES14-00166.1

Teixeira, C. P., Fernandes, C. O., & Ahern, J. (2022). Adaptive planting design and management framework for urban climate change adaptation and mitigation. *Urban for Urban Green, 70*, Article 127548. https://doi.org/10.1016/j.ufug.2022.127548

Terry, T. J., Madsen, M. D., Gill, R. A., Anderson, V. J., & St. Clair, S. B. (2021) Selective herbicide control: Using furrows and carbon seed coatings to establish a native bunchgrass while reducing cheatgrass cover. *Restoration Ecology, 29*(5), e13351. https://doi.org/10.1111/rec.13351

Thelen, G. C., Vivanco, J. M., Newingham, B., Good, W., Bais, H. P., Landres, P., Caesar, A., & Callaway, R. M. (2005). Insect herbivory stimulates allelopathic exudation by an invasive plant and the suppression of natives. *Ecology Letters, 8*(2), 209–217. https://doi.org/10.1111/j.1461-0248.2004.00713.x

Thomas, S. C., Frye, S., Gale, N., Garmon, M., Launchbury, R., Machado, N., Melamed, S., Murray, J., Petroff, A., & Winsborough, C. (2013). Biochar mitigates negative effects of salt additions on two herbaceous plant species. *Journal of Environmental Management, 129*, 62–68. https://doi.org/10.1016/j.jenvman.2013.05.057

Threlfall, C. G., Ossola, A., Hahs, A. K., Williams, N. S. G., Wilson, L., & Livesley, S. J. (2016). Variation in vegetation structure and composition across urban green space types. *Frontiers in Ecology and Evolution, 4*, 66.

Trentanovi, G., von der Lippe, M., Sitzia, T., Ziechmann, U., Kowarik, I., & Cierjacks, A. (2013). Biotic homogenization at the community scale: Disentangling the roles of urbanization and plant invasion. *Diversity and Distributions, 19*(7), 738–748. https://doi.org/10.1111/ddi.12028

Tripathi, S., Bhadouria, R., Srivastava, P., Singh, R., & Batish, D. R. (2023). *Plant invasions and global climate change*. Springer Nature Singapore.

Turbelin, A., & Catford, J. A. (2021). Chapter 25—Invasive plants and climate change. In T. M. Letcher (Ed.), *Climate change (Third Edition)* (pp. 515–539). Elsevier.

Turnbull, L. A., Rahm, S., Baudois, O., Eichenberger-Glinz, S., Wacker, L., & Schmid, B. (2005). Experimental invasion by legumes reveals non-random assembly rules in grassland communities. *Journal of Ecology, 93*(6), 1062–1070. https://doi.org/10.1111/j.1365-2745.2005.01051.x

Ullah, S., & Faulkner, S. P. (2006). Use of cotton gin trash to enhance denitrification in restored forested wetlands. *Forest Ecology Management, 237*(1), 557–563. https://doi.org/10.1016/j.foreco.2006.09.075

van Kleunen, M., Dawson, W., & Maurel, N. (2015). Characteristics of successful alien plants. *Molecular Ecology, 24*(9), 1954–1968. https://doi.org/10.1111/mec.13013

Van Kleunen, M., Weber, E., & Fischer, M. (2010). A meta-analysis of trait differences between invasive and non-invasive plant species. *Ecology Letters, 13*(2), 235–245. https://doi.org/10.1111/j.1461-0248.2009.01418.x

Vilà, M., Tessier, M., Suehs, C. M., Brundu, G., Carta, L., Galanidis, A., Lambdon, P., Manca, M., Médail, F., Moragues, E., Traveset, A., Troumbis, A. Y., & Hulme, P. E. (2006). Local and regional assessments of the impacts of plant invaders on vegetation structure and soil properties of Mediterranean islands. *Journal of Biogeography, 33*(5), 853–861. https://doi.org/10.1111/j.1365-2699.2005.01430.x

Vincent, E. M., & Roberts, E. H. (1977). The interaction of light, nitrate and alternating temperature in promoting the germination of dormant seeds of common weed species. *Seed Science and Technology, 5*(2), 659–670.

Von der Lippe, M., Säumel, I., & Kowarik, I. (2005). Cities as drivers for biological invasions - The role of urban climate and traffic. *Die Erde, 136*, 123–143.

Wang, C., Liu, J., & Zhou, J. (2017). N deposition affects allelopathic potential of *Amaranthus retroflexus* with different distribution regions. *Anais Da Academia Brasileira De Ciências, 89*, 919–926. https://doi.org/10.1590/0001-3765201720160513

Wang, H., Liu, T., Dong, H., Zhao, W., Liu, X., Wang, R., & Xu, W. (2022). Changes in the composition of the soil seed bank of grassland after giant ragweed (*Ambrosia trifida* L.) invasion. *Journal of Environmental and Management, 317*, 115468. https://doi.org/10.1016/j.jenvman.2022.115468

Wang, R. L., Chen, S., Qin, Z., Feng, Z. L., Xia, W. N., Su, Y. J., Zhang, J. E. (2016). Effects of simulated nitrogen deposition on the seeding growth, photosynthetic characteristics and allelopathic potential of invasive plant *Ipomoea cairica* (L.) Sweet. *Allelopathy Journal, 38*(2), 183–192.

Wania, A., Kühn, I., & Klotz, S. (2006). Plant richness patterns in agricultural and urban landscapes in Central Germany—Spatial gradients of species richness. *Landscape and Urban Planning, 75*(1), 97–110. https://doi.org/10.1016/j.landurbplan.2004.12.006

Warren, R. J., Labatore, A., & Candeias, M. (2017). Allelopathic invasive tree (*Rhamnus cathartica*) alters native plant communities. *Plant Ecology, 218*(10), 1233–1241. https://doi.org/10.1007/s11258-017-0766-2

Webb, S. L., Pendergast, T. H., & Dwyer, M. E. (2001). Response of native and exotic maple seedling banks to removal of the exotic, invasive Norway Maple (*Acer platanoides*). *The Journal of the Torrey Botanical Society, 128*(2), 141–149. https://doi.org/10.2307/3088736

Weidlich, E. W. A., Flórido, F. G., Sorrini, T. B., & Brancalion, P. H. S. (2020). Controlling invasive plant species in ecological restoration: A global review. *Journal of Applied Ecology, 57*(9), 1806–1817. https://doi.org/10.1111/1365-2664.13656

Williams, D. G., Mack, R. N., & Black, R. A. (1995). Ecophysiology of Introduced *Pennisetum Setaceum* on Hawaii: The Role of Phenotypic Plasticity. *Ecology, 76*(5), 1569–1580. https://doi.org/10.2307/1938158

Witkowski, E. T. F., & Wilson, M. (2001). Changes in density, biomass, seed production and soil seed banks of the non-native invasive plant, *Chromolaena odorata*, along a 15 year chronosequence. *Plant Ecology, 152*(1), 13–27.

Wolkovich, E. M., Davies, T. J., Schaefer, H., Cleland, E. E., Cook, B. I., Travers, S. E., Willis, C. G., & Davis, C. C. (2013). Temperature-dependent shifts in phenology contribute to the success of exotic species with climate change. *American Journal of Botany, 100*(7), 1407–1421. https://doi.org/10.3732/ajb.1200478

Wotton, D. M., Grove, P. B., & Kelly, D. (2020). Does habitat manipulation enhance native woody seedling recruitment in a dryland river floodplain? *New Zealand Journal of Ecology, 44*(2), 1–9.

Wu, A.-P., Yu, H., Gao, S.-Q., Huang, Z.-Y., He, W.-M., Miao, S.-L., & Dong, M. (2009). Differential belowground allelopathic effects of leaf and root of *Mikania micrantha*. *Trees, 23*(1), 11–17. https://doi.org/10.1007/s00468-008-0249-0

Xiao, T., Yu, H., Song, Y.-B., Jiang, Y.-P., Zeng, B., & Dong, M. (2019). Nutrient enhancement of allelopathic effects of exotic invasive on native plant species. *PLoS ONE, 14*(1), Article e0206165. https://doi.org/10.1371/journal.pone.0206165

Yang, F., Zhao, L., Gao, B., Xu, X., & Cao, X. (2016). The interfacial behavior between biochar and soil minerals and its effect on biochar stability. *Environmental Science and Technology, 50*(5), 2264–2271. https://doi.org/10.1021/acs.est.5b03656

Yang, R.-Y., Tang, J.-J., Yang, Y.-S., & Chen, X. (2007). Invasive and non-invasive plants differ in response to soil heavy metal lead contamination. *Botanical Studies, 48*(4), 453–458.

Yao, J., Holt, R. D., Rich, P. M., & Marshall, W. S. (1999). Woody plant colonization in an experimentally fragmented landscape. *Ecography, 22*(6), 715–728. https://doi.org/10.1111/j.1600-0587.1999.tb00521.x

Yu, H., Le Roux, J. J., Jiang, Z., Sun, F., Peng, C., & Li, W. (2021). Soil nitrogen dynamics and competition during plant invasion: Insights from *Mikania micrantha* invasions in China. *New Phytologist, 229*(6), 3440–3452. https://doi.org/10.1111/nph.17125

Zedler, J. B., & Kercher, S. (2004). Causes and consequences of invasive plants in wetlands: Opportunities, opportunists, and outcomes. *Critical Reviews in Plant Sciences, 23*(5), 431–452. https://doi.org/10.1080/07352680490514673

Zettlemoyer, M. A., Schultheis, E. H., & Lau, J. A. (2019). Phenology in a warming world: Differences between native and non-native plant species. *Ecology Letters, 22*(8), 1253–1263. https://doi.org/10.1111/ele.13290

Zhang, X., Wang, H., He, L., Lu, K., Sarmah, A., Li, J., Bolan, N. S., Pei, J., & Huang, H. (2013). Using biochar for remediation of soils contaminated with heavy metals and organic pollutants.

Environmental Science and Pollution Research, 20(12), 8472–8483. https://doi.org/10.1007/s11356-013-1659-0

Zhao, L., Xiao, D., Liu, Y., Xu, H., Nan, H., Li, D., Kan, Y., & Cao, X. (2020). Biochar as simultaneous shelter, adsorbent, pH buffer, and substrate of *Pseudomonas citronellolis* to promote biodegradation of high concentrations of phenol in wastewater. *Water Research, 172*, Article 115494. https://doi.org/10.1016/j.watres.2020.115494

Zhou, X., Shi, L., Moghaddam, T. B., Chen, M., Wu, S., & Yuan, X. (2022). Adsorption mechanism of polycyclic aromatic hydrocarbons using wood waste-derived biochar. *Journal of Hazardous Materials, 425*, Article 128003. https://doi.org/10.1016/j.jhazmat.2021.128003

MIX
Papier aus verantwortungsvollen Quellen
Paper from responsible sources
FSC® C105338

If you have any concerns about our products,
you can contact us on
ProductSafety@springernature.com

In case Publisher is established outside the EU,
the EU authorized representative is:
Springer Nature Customer Service Center GmbH
Europaplatz 3, 69115 Heidelberg, Germany

Printed by Libri Plureos GmbH
in Hamburg, Germany